T0189155

Lecture Notes in Computer Science 13344

Editors
Ying Tan
Peking University
Beijing, China

Ben Niu
Shenzhen University
Shenzhen, China

Yuhui Shi
Southern University of Science and Technology
Shenzhen, China

ISSN 0302-9743 ISSN 1611-3349 (electronic)
Lecture Notes in Computer Science
ISBN 978-3-031-09676-1 ISBN 978-3-031-09677-8 (eBook)
https://doi.org/10.1007/978-3-031-09677-8

This Springer imprint is published by the registered company Springer Nature Switzerland AG
The registered company address is: Gewerbestrasse 11, 6330 Cham, Switzerland

Preface

This book and its companion volumes, LNCS vols. 13344 and 13345, constitute the proceedings of The Thirteenth International International Conference on Swarm Intelligence (ICSI 2022) held during July 15–19, 2022 in Xi'an, China, both onsite and online.

The theme of ICSI 2022 was "Serving Life with Swarm Intelligence." ICSI 2022 provided an excellent opportunity for academics and practitioners to present and discuss the latest scientific results and methods, innovative ideas, and advantages in theories, technologies, and applications in swarm intelligence. The technical program covered a number of aspects of swarm intelligence and its related areas. ICSI 2022 was the thirteenth international gathering for academics and researchers working on aspects of swarm intelligence, following successful events in Qingdao (ICSI 2021), Serbia (ICSI 2020) virtually, ChiangMai (ICSI 2019), Shanghai (ICSI 2018), Fukuoka (ICSI 2017), Bali (ICSI 2016), Beijing (ICSI-CCI 2015), Hefei (ICSI 2014), Harbin (ICSI 2013), Shenzhen (ICSI 2012), Chongqing (ICSI 2011), and Beijing (ICSI 2010), which provided a high-level academic forum for participants to disseminate their new research findings and discuss emerging areas of research. The conference also created a stimulating environment for participants to interact and exchange information on future challenges and opportunities in the field of swarm intelligence research.

Due to the continuous global COVID-19 pandemic, ICSI 2022 provided both online and offline presentations. On one hand, ICSI 2022 was normally held in Xi'an, China. On the other hand, the ICSI 2022 technical team enabled the authors of accepted papers who were restricted from traveling overseas to present their work through an interactive online platform or video replay. The presentations by accepted authors were available to all registered attendees onsite and online.

The host city of ICSI 2022, Xi'an in China, is the capital of Shaanxi Province. A sub-provincial city on the Guanzhong Plain in Northwest China, it is one of the oldest cities in China, the oldest prefecture capital, and one of the Chinese Four Great Ancient Capitals, having held the position under several of the most important dynasties in Chinese history, including Western Zhou, Qin, Western Han, Sui, Northern Zhou, and Tang. The city is the starting point of the Silk Road and home to the UNESCO World Heritage site of the Terracotta Army of Emperor Qin Shi Huang.

The ICSI 2022 received a total of 171 submissions and invited submissions from about 368 authors in 15 countries and regions (Brazil, China, the Czech Republic, Germany, India, Italy, Japan, Mexico, Portugal, Russia, South Africa, Taiwan (China), Thailand, the UK, and the USA) across five continents (Asia, Europe, North America, South America, and Africa). Each submission was reviewed by at least 2 reviewers, and on average 2.6 reviewers. Based on rigorous reviews by the Program Committee members and reviewers, 85 high-quality papers were selected for publication in this proceedings volume with an acceptance rate of 49.7%. The papers are organized into 13 cohesive sections covering major topics of swarm intelligence research and its development and applications along with a competition session entitled "Competition on Single Objective Bounded Optimization Problems (ICSI-OC 2022)."

On behalf of the Organizing Committee of ICSI 2022, we would like to express our sincere thanks to the International Association of Swarm and Evolutionary Intelligence (IASEI), which is the premier international scholarly society devoted to advancing the theories, algorithms, real-world applications, and developments of swarm intelligence and evolutionary intelligence (iasei.org). We would also like to thank Peking University, Xi'an Jiaotong University, Shaanxi Normal University, Xi'dan University, Xi'an University of Posts & Telecommunications, and the Southern University of Science and Technology for their co-sponsorships, the Computational Intelligence Laboratory of Peking University and IEEE Beijing Chapter for their technical co-sponsorships, and Nanjing Kanbo iHealth Academy for its technical and financial co-sponsorship, as well as our supporters: the International Neural Network Society, the World Federation on SoftComputing, MDPI's journal 'Entropy', the Beijing Xinghui Hi-Tech Co., and Springer.

We would also like to thank the members of the Advisory Committee for their guidance, the members of the international Program Committee and additional reviewers for reviewing the papers, and the members of the Publication Committee for checking the accepted papers in a short period of time. We are particularly grateful to Springer for publishing the proceedings in the prestigious series of Lecture Notes in Computer Science. Moreover, we wish to express our heartfelt appreciation to the plenary speakers, session chairs, and student helpers. In addition, there are many more colleagues, associates, friends, and supporters who helped us in immeasurable ways; we express our sincere gratitude to them all. Last but not the least, we would like to thank all the speakers, authors, and participants for their great contributions that made ICSI 2022 successful and all the hard work worthwhile.

May 2022

Ying Tan
Yuhui Shi
Ben Niu

Organization

Honorary Co-chairs

Zongben Xu	Xi'an Jiaotong University, China
Russell C. Eberhart	IUPUI, USA

General Chair

Ying Tan	Peking University, China

Program Committee Chair

Yuhui Shi	Southern University of Science and Technology, China

Advisory Committee Chairs

Xingui He	Peking University, China
Gary G. Yen	Oklahoma State University, USA

Technical Committee Co-chairs

Haibo He	University of Rhode Island, USA
Kay Chen Tan	City University of Hong Kong, China
Nikola Kasabov	Auckland University of Technology, New Zealand
Ponnuthurai Nagaratnam Suganthan	Nanyang Technological University, Singapore
Xiaodong Li	RMIT University, Australia
Hideyuki Takagi	Kyushu University, Japan
M. Middendorf	University of Leipzig, Germany
Yaochu Jin	University of Surrey, UK
Qirong Tang	Tongji University, China
Milan Tuba	Singidunum University, Serbia

Plenary Session Co-chairs

Andreas Engelbrecht	University of Pretoria, South Africa
Chaoming Luo	University of Mississippi, USA

Invited Session Co-chairs

Andres Iglesias	University of Cantabria, Spain
Haibin Duan	Beihang University, China

Special Sessions Co-chairs

Ben Niu	Shenzhen University, China
Yan Pei	University of Aizu, Japan
Shaoqiu Zheng	China Electronics Technology Group Corporation, China

Tutorial Co-chairs

Junqi Zhang	Tongji University, China
Gaige Wang	Ocean University of China, China

Publications Co-chairs

Swagatam Das	Indian Statistical Institute, India
Radu-Emil Precup	Politehnica University of Timisoara, Romania
Pengfei Guo	Xi'an Jiaotong University Press, China

Publicity Co-chairs

Yew-Soon Ong	Nanyang Technological University, Singapore
Carlos Coello	CINVESTAV-IPN, Mexico
Mengjie Zhang	Victoria University of Wellington, New Zealand
Dongbin Zhao	Institute of Automation, CAS, China
Rossi Kamal	GERIOT, Bangladesh

Finance and Registration Chairs

Andreas Janecek	University of Vienna, Austria
Suicheng Gu	Google Corporation, USA

Local Arrangement Chairs

Liangjun Ke	Xi'an Jiaotong University, China
Shi Cheng	Shanxi Normal University, China
Yongzhi Zhe	Xi'an University of Posts and Telecommunications, China

Conference Secretariat

Yifan Liu	Peking University, China

Program Committee

Abdelmalek Amine	Tahar Moulay University of Saida, Algeria
Sabri Arik	Istanbul University, Turkey
Helio Barbosa	Laboratório Nacional de Computação Científica, Spain
Carmelo J. A. Bastos Filho	University of Pernambuco, Brazil
Heder Bernardino	Universidade Federal de Juiz de Fora, Brazil
Sandeep Bhongade	G.S. Institute of Technology, India
Sujin Bureerat	Khon Kaen University, Thailand
Angelo Cangelosi	University of Manchester, UK
Mu-Song Chen	Da-Yeh University, Taiwan, China
Walter Chen	National Taipei University of Technology, Taiwan, China
Long Cheng	Institute of Automation, CAS, China
Shi Cheng	Shaanxi Normal University, China
Prithviraj Dasgupta	U. S. Naval Research Laboratory, USA
Khaldoon Dhou	Texas A&M University–Central Texas, USA
Haibin Duan	Beijing University of Aeronautics and Astronautics, China
Wei Fang	Jiangnan University, China
Liang Feng	Chongqing University, China
Philippe Fournier-Viger	Shenzhen University, China
Hongyuan Gao	Harbin Engineering University, China
Shangce Gao	University of Toyama, Japan
Zhigao Guo	Queen Mary University of London, UK
Guosheng Hao	Jiangsu Normal University, China
Mo Hongwei	Harbin Engineering University, China
Changan Jiang	Osaka Institute of Technology, Japan
Mingyan Jiang	Shandong University, China
Qiaoyong Jiang	Xi'an University of Technology, China
Colin Johnson	University of Nottingham, UK
Yasushi Kambayashi	Nippon Institute of Technology, Japan
Liangjun Ke	Xi'an Jiaotong University, China
Waqas Haider Khan	University of Gujrat, Pakistan
Vivek Kumar	Università degli Studi di Cagliari, Italy
Germano Lambert-Torres	PS Solutions, USA
Xiujuan Lei	Shaanxi Normal University, China
Bin Li	University of Science and Technology of China, China

Jing Liang	Zhengzhou University, China
Fernando B. De Lima Neto	University of Pernambuco, Brazil
Peng Lin	Capital University of Economics and Business, China
Jia Liu	University of Surrey, UK
Ju Liu	Shandong University, China
Qunfeng Liu	Dongguan University of Technology, China
Wenlian Lu	Fudan University, China
Chaomin Luo	Mississippi State University, USA
Dingsheng Luo	Peking University, China
Wenjian Luo	Harbin Institute of Technology, China
Lianbo Ma	Northeastern University, China
Chengying Mao	Jiangxi University of Finance and Economics, China
Yi Mei	Victoria University of Wellington, New Zealand
Bernd Meyer	Monash University, Australia
Carsten Mueller	Baden-Wuerttemberg Cooperative State University, Mosbach, Germany
Sreeja N. K.	PSG College of Technology, India
Qingjian Ni	Southeast University, China
Ben Niu	Shenzhen University, China
Lie Meng Pang	Southern University of Science and Technology, China
Bijaya Ketan Panigrahi	IIT Delhi, India
Endre Pap	Singidunum University, Serbia
Om Prakash Patel	Mahindra University, India
Mario Pavone	University of Catania, Spain
Yan Pei	University of Aizu, Japan
Danilo Pelusi	University of Teramo, Italy
Radu-Emil Precup	Politehnica University of Timisoara, Romania
Quande Qin	Guangzhou University, China
Robert Reynolds	Wayne State University, USA
Yuji Sato	Hosei University, Japan
Carlos Segura	Centro de Investigación en Matemáticas, A.C. (CIMAT), Mexico
Ke Shang	Southern University of Science and Technology, China
Zhongzhi Shi	Institute of Computing Technology, CAS, China
Joao Soares	Polytechnic Institute of Porto, Portugal
Wei Song	North China University of Technology, China
Yifei Sun	Shaanxi Normal University, China
Ying Tan	Peking University, China

Qirong Tang	Tongji University, China
Eva Tuba	University of Belgrade, Serbia
Mladen Veinović	Singidunum University, Serbia
Kaifang Wan	Northwestern Polytechnical University, China
Gai-Ge Wang	Ocean University of China, China
Guoyin Wang	Chongqing University of Posts and Telecommunications, China
Lei Wang	Tongji University, China
Liang Wang	Northwestern Polytechnical University, China
Yuping Wang	Xidian University, China
Ka-Chun Wong	City University of Hong Kong, Hong Kong SAR, China
Man Leung Wong	Lingnan University, Hong Kong SAR, China
Ning Xiong	Mälardalen University, Sweden
Benlian Xu	Changshu Institute of Technology, China
Rui Xu	Hohai University, China
Yu Xue	Nanjing University of Information Science & Technology, China
Xuesong Yan	China University of Geosciences, China
Yingjie Yang	De Montfort University, UK
Guo Yi-Nan	China University of Mining and Technology, China
Peng-Yeng Yin	National Chi Nan University, Taiwan, China
Jun Yu	Niigata University, Japan
Ling Yu	Jinan University, China
Zhi-Hui Zhan	South China University of Technology, China
Fangfang Zhang	Victoria University of Wellington, New Zealand
Jie Zhang	Newcastle University, UK
Junqi Zhang	Tongji University, China
Xiangyin Zhang	Beijing University of Technology, China
Xingyi Zhang	Anhui University, China
Zili Zhang	Deakin University, Australia
Xinchao Zhao	Beijing University of Posts and Telecommunications, China
Shaoqiu Zheng	Peking University, China
Yujun Zheng	Zhejiang University of Technology, China
Miodrag Zivkovic	Singidunum University, Serbia

Additional Reviewers

Contents – Part I

Swarm Intelligence and Nature-Inspired Computing

Swarm-Based Computing Algorithms for Optimization

Particle Swarm Optimization

Ant Colony Optimization

Brain Storm Optimization Algorithm

Swarm Intelligence Approach-Based Applications

Multi-objective Optimization

Swarm Intelligence and Nature-Inspired Computing

Information Utilization Ratio in Heuristic Optimization Algorithms

Junzhi Li and Ying Tan(✉)

Key Laboratory of Machine Perception (MOE), School of Artificial Intelligence, Peking University, Beijing 100871, China
{ljz,ytan}@pku.edu.cn

Abstract. Heuristic algorithms are able to optimize objective functions efficiently because they use intelligently the information about the objective functions. Thus, information utilization is critical to the performance of heuristics. However, the concept of information utilization has remained vague and abstract because there is no reliable metric to reflect the extent to which the information about the objective function is utilized by heuristic algorithms. In this paper, the metric of information utilization ratio (IUR) is defined, which is the ratio of the utilized information quantity over the acquired information quantity in the search process. The IUR proves to be well-defined. Several examples of typical heuristic algorithms are given to demonstrate the procedure of calculating the IUR. Empirical evidences on the correlation between the IUR and the performance of a heuristic are also provided. The IUR can be an index of how sophisticated an algorithm is designed and guide the invention of new heuristics and the improvement of existing ones.

Keywords: Optimization algorithm · Information utilization · Swarm intelligence

1 Introduction

In the field of computer science, many heuristic algorithms have been developed to solve complex non-convex optimization problems. Although optimal solutions are not guaranteed to be found, heuristics can often find acceptable solutions at affordable cost. The key to designing a heuristic algorithm is to use heuristic information about the objective function. Many algorithms [9,33,34] are claimed to be reasonably designed because they use heuristic information intelligently. Even more algorithmic improvement works [17,25,38] are claimed to be significant because they use more heuristic information or use heuristic information more thoroughly than the original algorithms.

Empirically, heuristic information is used more thoroughly in more advanced algorithms. Suppose there are two search algorithms A and B for one dimensional optimization. Algorithm A compares the evaluation values of the solutions x_1

Y. Tan et al. (Eds.): ICSI 2022, LNCS 13344, pp. 3–22, 2022.
https://doi.org/10.1007/978-3-031-09677-8_1

and x_2 to decide which direction (left or right) is more promising, while algorithm B uses their evaluation values to calculate both the direction and the step size for the next search. If the underlying distribution of objective functions is already known, then algorithm B is able to search faster than algorithm A if they are both reasonably designed because more information is utilized by algorithm B. It has been a common sense in the field of heuristic search that the extent of information utilization in a heuristic algorithm is crucial to its performance.

However, so far there is no reliable metric to reflect the extent of information utilization because unlike direct performance analyses [18,21], this issue seems abstract. Especially, it is very difficult to measure how much information is used by an optimization algorithm.

In this paper, based on some basic concepts in the information theory, a formal definition of the information utilization ratio (IUR) is proposed, which is defined as the ratio of the utilized information quantity over the acquired information quantity in the search process. It is shown theoretically that IUR is well-defined. Examples of typical heuristic algorithms are also given to demonstrate the procedure of calculating IURs.

Theoretically, IUR itself is a useful index of how sophisticated an algorithm is designed, but we still expect it to be practically serviceable, that is, we need to study the correlation between IUR and performance. However, the correlation between IUR and performance of heuristics is not so straightforward as some may expect. The performance of an optimization algorithm depends not only on the extent of information utilization but also on the manner of information utilization. Still, for algorithms that utilize information in similar manners, the influence of the IUR is often crucial, as is illustrated in the experiments.

After all, the metric of IUR helps researchers construct a clear (but not deterministic) relationship between the design and the performance of an optimization algorithm, which makes it possible that researchers can to some extent predict the performance of an algorithm even before running it. Thus, the IUR can be a useful index for guiding the design and the improvement of heuristic optimization algorithms.

2 Information Utilization Ratio

Definition 1 (Information Entropy). *The information entropy of a discrete random variable X with possible values x_i and probability density $p(x_i)$ is defined as follows.*

$$H(X) = -\sum_i p(x_i) \log p(x_i). \tag{1}$$

Definition 2 (Conditional Entropy). *The conditional entropy of two discrete random variables X and Y with possible values x_i and y_j respectively and joint probability density $p(x_i, y_j)$ is defined as follows.*

$$H(X|Y) = -\sum_{i,j} p(x_i, y_j) \log \frac{p(x_i, y_j)}{p(y_j)}. \tag{2}$$

Some elementary properties of information entropy and conditional entropy are frequently used in this paper, which however cannot be present here due to the limitation of space. We refer readers who are unfamiliar with the information theory to the original paper [32] or other tutorials.

The following lemma defines a useful function for calculating the IURs of various algorithms.

Lemma 1. *If $\eta_1, \eta_2, \ldots, \eta_{g+1} \in \mathbb{R}$ are independent identically distributed random variables,*

$$H(I(\min(\eta_1, \eta_2, \ldots, \eta_g) < \eta_{g+1}))$$
$$= -\frac{g}{g+1}\log\frac{g}{g+1} - \frac{1}{g+1}\log\frac{1}{g+1} \triangleq \pi(g). \quad (3)$$

where $I(x < y) = \begin{cases} 1 & \text{if } x < y \\ 0 & \text{otherwise} \end{cases}$ is the indicator function.

$\pi(g) \in (0, 1]$ is a monotonic decreasing function of g.

Definition 3 (Objective Function). *The objective function is a mapping $f : \mathcal{X} \mapsto \mathcal{Y}$, where $\mathcal{Y}$ is a totally ordered set.*

$\mathcal{X}$ is called the search space. The target of an optimization algorithm is to find a solution $x \in \mathcal{X}$ with the best evaluation value $f(x) \in \mathcal{Y}$.

Definition 4 (Optimization Algorithm). *An optimization algorithm $\mathscr{A}$ is defined as follows.*

Algorithm 1. Optimization Algorithm $\mathscr{A}$

1: $i \leftarrow 0$.
2: $D_0 \leftarrow \emptyset$.
3: **repeat**
4: $\quad i \leftarrow i + 1$.
5: $\quad$ Sample $X_i \in 2^{\mathcal{X}}$ with distribution $\mathscr{A}_i(D_{i-1})$.
6: $\quad$ Evaluate $f(X_i) = \{f(x) | x \in X_i\}$.
7: $\quad D_i \leftarrow D_{i-1} \cup \bigcup_{x \in X_i} \{x, f(x)\}$.
8: **until** $i = g$.

In each iteration, $\mathscr{A}_i$ is a mapping from $2^{\mathcal{X} \times \mathcal{Y}}$ to the set of all distributions over $2^{\mathcal{X}}$. $\mathscr{A}_1(D_0)$ is a pre-fixed distribution for sampling solutions in the first iteration. g is the maximal iteration number.

In each iteration, the input of the algorithm D_{i-1} is the historical information, which is a subset of $\mathcal{X} \times \mathcal{Y}$, and the output $\mathscr{A}_i(D_{i-1})$ is a distribution over $2^{\mathcal{X}}$, with which the solutions to be evaluated next are drawn. Note that the output $\mathscr{A}_i(D_{i-1})$ is deterministic given D_{i-1}.

By randomizing the evaluation step (consider $y = f(x)$ as a random variable), we are able to investigate how much acquired information is used in an optimization algorithm. That is, to what extent the action of the algorithm will change when the acquired information changes. Review the example in the introduction. It is clear that the algorithm A only uses the information of "which one is better", while the information of evaluation values are fully utilized by the algorithm B. But how to express such an observation? Any change in y_1 or y_2 would cause the algorithm B to search a different location, while only when $I(y_1 > y_2)$ changes would the action of the algorithm A change. So, the quantity of utilized information can be expressed by the information entropy of an algorithm's action. The entropy of the action of the algorithm A is one bit, while the entropy of the action of the algorithm B is equal to the entropy of the evaluation values. Assume Z is the "action" of the algorithm, X is the positions of the solutions, Y is the evaluation values, (they are all random variables), then we can roughly think the information utilization ratio is $H(Z|X)/H(Y|X)$. However, optimization algorithms are iterative processes, so the formal definition is more complicated.

Definition 5 (Information Utilization Ratio). *If $\mathscr{A}$ is an optimization algorithm, the information utilization ratio of $\mathscr{A}$ is defined as follows.*

$$IUR_{\mathscr{A}}(g) = \frac{\sum_{i=1}^{g} H(Z_i|\overline{X}_{i-1}, \overline{Z}_{i-1})}{\sum_{i=1}^{g} H(Y_i|\overline{X}_i, \overline{Y}_{i-1})}. \tag{4}$$

where g is the maximal iteration number, $X = \{X_1, X_2, \dots, X_g\}$ is the set of all sets of evaluated solutions, $Y = \{f(X_1), f(X_2), \dots, f(X_g)\}$ is the set of all sets of evaluation values, $Z = \{\mathscr{A}_1(D_0), \mathscr{A}_2(D_1), \dots, \mathscr{A}_g(D_{g-1})\}$ is the output distributions in all iterations of algorithm $\mathscr{A}$, $\overline{X}_i \triangleq \{X_1, \dots, X_i\}, \overline{Y}_i \triangleq \{Y_1, \dots, Y_i\}, \overline{Z}_i \triangleq \{Z_1, \dots, Z_i\}$, $\overline{X}_0 = \overline{Y}_0 = \overline{Z}_0 = \emptyset$.

Figure 1 shows the relationship among these random variables. Generally, X_i is acquired by sampling with the distribution Z_i, Y_i is acquired by evaluating X_i, and Z_i is determined by the algorithm according to the historical information $\overline{X}_{i-1}$ and $\overline{Y}_{i-1}$.

For deterministic algorithms (i.e., $H(X_i|Z_i) = 0$), the numerator degenerates to $H(Z)$. If function evaluations are independent, the denominator degenerates to $\sum_{i=1}^{g} H(Y_i|X_i)$.

The following theorem guarantees that IUR is well defined.

Theorem 1. *If $0 < \sum_{i=1}^{g} H(Y_i|\overline{X}_i, \overline{Y}_{i-1}) < \infty$, then $0 \le IUR_{\mathscr{A}}(g) \le 1$.*

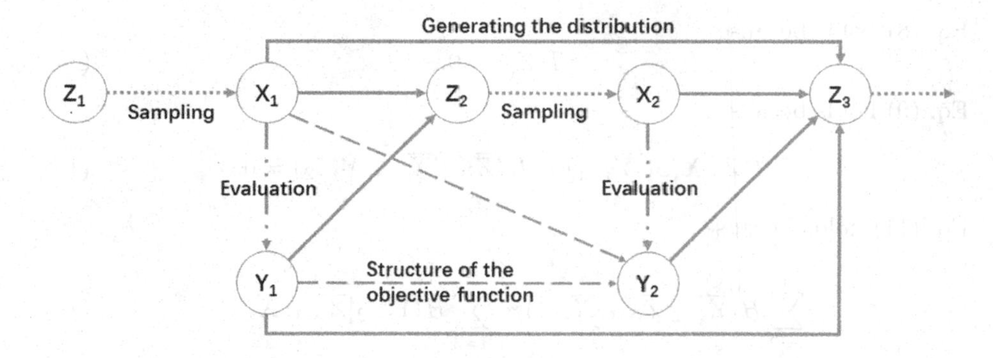

Fig. 1. Graphic model

Proof.

$$H(X,Z) - \sum_{i=1}^{g} H(X_i|Z_i) \tag{5}$$

$$= \sum_{i=1}^{g} H(X_i, Z_i|\overline{X}_{i-1}, \overline{Z}_{i-1}) - \sum_{i=1}^{g} H(X_i|\overline{X}_{i-1}, \overline{Z}_i) \tag{6}$$

$$= \sum_{i=1}^{g} H(Z_i|\overline{X}_{i-1}, \overline{Z}_{i-1}) \tag{7}$$

$$= \sum_{i=2}^{g} H(\overline{Z}_i|\overline{X}_{i-1}) - \sum_{i=2}^{g} H(\overline{Z}_{i-1}|\overline{X}_{i-1}) \tag{8}$$

$$= \sum_{i=2}^{g} H(\overline{Z}_i|\overline{X}_{i-1}) - \sum_{i=2}^{g} H(\overline{Z}_i|\overline{X}_{i-1}, \overline{Y}_{i-1}) - \sum_{i=2}^{g} H(\overline{Z}_{i-1}|\overline{X}_{i-1})$$

$$+ \sum_{i=2}^{g} H(\overline{Z}_{i-1}|\overline{X}_{i-1}, \overline{Y}_{i-2}) \tag{9}$$

$$= \sum_{i=2}^{g} -H(\overline{Y}_{i-1}|\overline{Z}_i, \overline{X}_{i-1}) + \sum_{i=2}^{g} H(\overline{Y}_{i-1}|\overline{X}_{i-1}) + \sum_{i=2}^{g} H(\overline{Y}_{i-2}|\overline{Z}_{i-1}, \overline{X}_{i-1})$$

$$- \sum_{i=2}^{g} H(\overline{Y}_{i-2}|\overline{X}_{i-1}) \tag{10}$$

$$= \sum_{i=2}^{g} -H(\overline{Y}_{i-1}|\overline{Z}_i, \overline{X}_{i-1}) + \sum_{i=2}^{g} H(\overline{Y}_{i-2}|\overline{Z}_{i-1}, \overline{X}_{i-2}) + \sum_{i=2}^{g} H(Y_{i-1}|\overline{X}_{i-1}, \overline{Y}_{i-2}) \tag{11}$$

$$= -H(\overline{Y}_{g-1}|Z, \overline{X}_{g-1}) + \sum_{i=1}^{g} H(Y_i|\overline{X}_i, \overline{Y}_{i-1}) - H(Y_g|\overline{X}_g, \overline{Y}_{g-1}) \tag{12}$$

$$\leq \sum_{i=1}^{g} H(Y_i|\overline{X}_i, \overline{Y}_{i-1}). \tag{13}$$

Eq. (8) holds because

$$H(Z_1) = 0. \tag{14}$$

Eq. (9) holds because

$$H(\overline{Z}_i|\overline{X}_{i-1}, \overline{Y}_{i-1}) = H(\overline{Z}_{i-1}|\overline{X}_{i-1}, \overline{Y}_{i-2}) = 0. \tag{15}$$

Eq. (11) holds because

$$\sum_{i=2}^{g} H(\overline{Y}_{i-2}|\overline{Z}_{i-1}, \overline{X}_{i-1}) = \sum_{i=2}^{g} H(\overline{Y}_{i-2}|\overline{Z}_{i-1}, \overline{X}_{i-2}). \tag{16}$$

Eq. (12) is by dislocation subtraction.

The denominator in the definition $\sum_{i=1}^{g} H(Y_i|\overline{X}_i, \overline{Y}_{i-1})$ represents the information quantity that is acquired in the search process. If function evaluations are independent, then $H(Y_i|\overline{X}_i, \overline{Y}_{i-1}) = H(Y_i|X_i, \overline{X}_{i-1}, \overline{Y}_{i-1}) = H(Y_i|X_i)$. While the numerator is more obscure. Actually it represents the quantity of the information about the objective function which is utilized by the algorithm (or in other words, the minimal information quantity that is needed to run the algorithm). Firstly, $\sum_{i=1}^{g} H(Z_i|\overline{X}_{i-1}, \overline{Z}_{i-1}) = \sum_{i=1}^{g} H(Z_i|\overline{X}_{i-1}, \overline{Z}_{i-1}) - \sum_{i=1}^{g} H(Z_i|\overline{X}_{i-1}, \overline{Z}_{i-1}, \overline{Y}_{i-1})$ is similar to the concept of information gain in classification problems [30], which indicates the contribution of the information of Y to the algorithm. Secondly, the uncertainty of X and Z only lies in two aspects: the random sampling step and the lack of the information from Y. Thus $H(X, Z) - \sum_{i=1}^{g} H(X_i|Z_i)$ can be regarded as the objective function's information that is utilized by the algorithm. And in fact, it is equal to the numerator. Thirdly, the numerator equals the denominator minus $H(Y_g|\overline{X}_g, \overline{Y}_{g-1}) + H(\overline{Y}_{g-1}|Z, \overline{X}_{g-1})$ which can be seen as the wasted information of Y, because 1) the evaluation values in the last iteration Y_g cannot be utilized and 2) the information of previous evaluation values $\overline{Y}_{g-1}$ is fully utilized only if $H(\overline{Y}_{g-1}|Z, \overline{X}_{g-1}) = 0$, i.e., $\overline{Y}_{g-1}$ can be reconstructed with Z given $\overline{X}_{g-1}$.

3 IURs of Heuristic Optimization Algorithms

In order to calculate the IURs of algorithms, we further assume $f(x) \in \mathcal{Y}$ is identically and independently distributed (i.i.d). In most cases, it is unwise to calculate the IUR by definition. To calculate the denominator is quite straightforward under the above assumption, which equals the number of evaluations times $H(f(x))$. For example, if there are 100 cities in a travelling salesman problem [24] and $f(x)$ obey uniform distribution, then $|\mathcal{Y}| = 100!, H(f(x)) = \log 100!$. While on the other hand, to directly calculate the numerator is difficult and unnecessary. In each iteration, the output $\mathscr{A}_i(D_{i-1})$ is a certain distribution, which is usually determined by some parameters in the algorithm. In fact we can certainly find (or construct) the set of intermediate parameters M_i such that 1) there is a bijection from M_i to Z_i given $\overline{X}_{i-1}$ and 2) M_i is determined only by $\overline{Y}_{i-1}$ (otherwise $H(Z_i|\overline{X}_{i-1}, \overline{Y}_{i-1}) > 0$), then

$$\sum_{i=2}^{g} H(Z_i|\overline{X}_{i-1}, \overline{Z}_{i-1}) = \sum_{i=2}^{g} H(M_i|\overline{M}_{i-1}) = H(M). \tag{17}$$

We only have to know the information quantity that is required to determine these intermediate parameters.

In the following, we investigate the IURs of several heuristics to show the procedure of calculating the IUR. Although these algorithms are designed for continuous (domain) optimization, the IURs of any kind of (discrete, combinatorial, dynamic, multi-objective) optimization algorithms can be calculated in the same way as long as there are a domain and a codomain. Without loss of generality, the following algorithms are all minimization algorithms, that is, they all intend to find the solution with the minimal evaluation value in the search space.

3.1 Random Search Algorithms

Monte Carlo. The Monte Carlo (MC) method is often considered as a baseline for optimization algorithms. It is not a heuristic algorithm and usually fails to find acceptable solutions. If the maximal evaluation number is m, MC just uniformly randomly sample m solutions from $\mathcal{X}$.

MC does not utilize any information about the objective function because Z is fixed.

Proposition 1.

$$IUR_{MC} = 0. \tag{18}$$

Luus-Jaakola. Luus-Jaakola (LJ) [27] is a heuristic algorithm based on MC. In each iteration, the algorithm generates a new individual y with the uniform distribution within a hypercube whose center is the position of the current individual x. If $f(y) < f(x)$, x is replaced by y; otherwise, the radius of the hypercube is multiplied by a parameter $\gamma < 1$.

The output of LJ in each iteration is the uniform distribution within the hypercube, which is determined by the position x and the radius. They are both controlled by the comparison result, i.e., $I(f(y) < f(x))$. $f(y)$ is i.i.d, but $f(x)$ is the best in the history. Thus, $H(M_i|\overline{M}_{i-1}) = H(I(f(y) < f(x))|\overline{M}_{i-1}) = \pi(i-1)$.

Proposition 2.

$$IUR_{LJ}(g) = \frac{\sum_{i=1}^{g-1} \pi(i)}{gH(f(x))}. \tag{19}$$

3.2 Evolution Strategies

(μ, λ)-Evolution Strategy. (μ, λ)-evolution strategy (ES) [3] is an important heuristic algorithm in the family of evolution strategies. In each generation, λ new offspring are generated from μ parents by crossover and mutation with normal distribution, and then the parents of a new generation are selected from these λ offspring. As a self-adaptive algorithm, the step size of the mutation is itself mutated along with the position of an individual.

The distribution for generating new offspring is determined by the μ parents, namely the indexes of the best μ of the λ individuals. Each set of μ candidates has the same probability to be the best. $H(M_i|\overline{M}_{i-1}) = H(M_i) = \log\binom{\lambda}{\mu}$, where $\binom{\lambda}{\mu} = \frac{\lambda!}{\mu!(\lambda-\mu)!}$.

Proposition 3.

$$IUR_{(\mu,\lambda)\text{-}ES}(g) = \frac{(g-1)\log\binom{\lambda}{\mu}}{g\lambda H(f(x))}. \tag{20}$$

Covariance Matrix Adaptation Evolution Strategy. In order to more adaptively control the mutation parameters in (μ, λ)-ES, a covariance matrix adaptation evolution strategy (CMA-ES) was proposed [15]. CMA-ES is a very complicated estimation of distribution algorithm [23], which adopts several different mechanisms to adapt the mean, the covariance matrix and the step size of the mutation operation. It is very efficient on benchmark functions especially when restart mechanisms are adopted. CMA-ES cannot be introduced here in detail. We refer interested readers to an elementary tutorial: [15].

Given $\overline{X}_{i-1}$, the mean, the covariance matrix and the step size of the distribution is determined by the indexes and the rankings of the best μ individuals in each iteration in history. $H(M_i|\overline{M}_{i-1}) = \log\frac{\lambda!}{(\lambda-\mu)!}$.

Proposition 4.

$$IUR_{CMA\text{-}ES}(g) = \frac{(g-1)\log\frac{\lambda!}{(\lambda-\mu)!}}{g\lambda H(f(x))}. \tag{21}$$

Compared with (μ, λ)-ES, it is obvious that $IUR_{CMA\text{-}ES} \geq IUR_{(\mu,\lambda)\text{-}ES}$, because not only the indexes of the μ best individuals, but also their rankings are used in CMA-ES (to calculate their weights, for example). By utilizing the information of the solutions more thoroughly, CMA-ES is able to obtain more accurate knowledge of the objective function and search more efficiently.

The IURs of Particle Swarm algorithms [5,11] and Differential Evolution algorithms [34,38] are also investigated, shown in the appendix. If readers are interested in the IURs of other algorithms, we encourage you to conduct an investigation on your own which can be usually done with limited effort.

4 IUR Versus Performance

The IUR is an intrinsic property of a heuristic algorithm, but the performance is not. Besides the algorithm itself, the performance of a heuristic also depends on the termination criterion, the way to measure the performance, and most importantly the distribution of the objective functions. A well-known fact about performance is that no algorithm outperforms another when there is no prior distribution [37], which is quite counter-experience. The objective functions in the real world usually subject to a certain underlying distribution. Although it is usually very difficult to precisely describe this distribution, we know that it has a much smaller information entropy than the uniform distribution and hence there is a free lunch [2,14,35]. In this case, the objective function (and resultantly its optimal point) can be identified with limited information quantity (the entropy of the distribution).

Reconsider the setting of the no free lunch (NFL) theorem from the perspective of information utilization. Assume $|\mathcal{X}| = m$ and $|\mathcal{Y}| = n$. Under the setting of NFL (no prior distribution), the total uncertainty of the objective function is $\log n^m = m \log n$. In each evaluation, the information acquired is $\log n$. Therefore, no algorithm is able to certainly find the optimal point of the objective function within less than m times of evaluation even if all acquired information is thoroughly utilized. In this case, enumeration is the best algorithm [13]. On the contrary, if we already know the objective function is a sphere function, which is determined only by its center, then the required information quantity is $\log m$, and the least required number of evaluations is (more than) $\log m / \log n = \log_n m$. Suppose the dimensionality of the search space is d, then n is $O(m^{\frac{1}{d}})$, $\log_n m$ is $O(d)$, which is usually acceptable. If information is fully utilized ($IUR \approx 1$), the exact number is $d + 1$ [2]. While for algorithms with smaller IURs, more evaluations are needed. For example, if the IUR of another algorithm is half of the best algorithm (with half of the acquired information wasted), then at least about $2d + 2$ evaluations are needed.

How much information is utilized by the algorithm per each evaluation determines the lower bound of the required evaluation number to locate the optimal point. **In this sense, IUR determines the upper bound of an algorithm's performance.** That is, algorithms with larger IURs have greater potential. However, the actual performance also depends on the manner of information utilization and how it accords with the underlying distribution of the objective function. For instance, one can easily design an algorithm with the same IUR as CMA-ES but does not work.

In the following, we will give empirical evidences on the correlation between IUR and performance. The preconditions of the experiments include: 1) the algorithms we investigate here are reasonably designed to optimize the objective functions from the underlying distribution; 2) the benchmark suite is large and comprehensive enough to represent the underlying distribution. The following conclusions may not hold for algorithms that are not reasonably designed or for a narrow or special range of objective functions. In other words, if the manner of information utilization does not accord with the underlying distribution of objective functions, utilizing more information is not necessarily advantageous.

The theoretical correctness of IUR does not rely on these experimental results, but these examples may help readers understand how and to what extent IUR influences performance.

4.1 Different Parameters of the Same Algorithm

Sometimes for a certain optimization algorithm the IUR is influenced by only a few parameters. For these algorithms, we may adapt these parameters to show the correlation between the tendency of IUR and the tendency of performance.

(μ, λ)-ES. Intuitively, using $\mu = \lambda$ is not a sensible option for (μ, λ)-ES (commonly used μ/λ values are in the range from 1/7 to 1/2 [4]) because it makes the selection operation invalid. Now we have a clearer explanation: the IUR of (μ, λ)-ES is zero if $\mu = \lambda$ (see Eq. (20)), i.e., (μ, λ)-ES does not use any heuristic information if $\mu = \lambda$.

Using a μ around $\frac{1}{2}\lambda$ may be a good choice for (μ, λ)-ES because it leads to a large IUR. When $\mu = \frac{1}{2}\lambda$, the information used by (μ, λ)-ES is the most. From the perspective of exploration and exploitation, we may come to a similar conclusion. If μ is too small (elitism), the information of the population is only used to select the best few solutions, and resultantly the diversity of the population may suffer quickly. If μ is too large (populism), the information of the population is only used to eliminate the worst few solutions, and resultantly the convergence speed may be too slow.

Different values of μ/λ are evaluated on the CEC 2013 benchmark suite containing 28 different test functions (see Table 1) which are considered as black-box problems [26]. The meta parameter is set to $\Delta\sigma = 0.5$. The algorithm using each set of parameters is run 20 times independently for each function. The dimensionality is $d = 5$, and the maximal number of function evaluations is $10000d$ for each run. For each fixed λ, the mean errors of 20 independent runs of each μ/λ are ranked. The rankings are averaged over 28 functions, shown in Fig. 2. $-\log\binom{\lambda}{\mu}/\lambda$ curves are translated along the vertical axis, also shown in Fig. 2.

According to the experimental results, μ/λ around 0.5 is a good choice, which accord with our expectation. Moreover, the tendency of the performance (average ranking) curve is generally identical to that of the IUR ($IUR_{(\mu,\lambda)-ES} \propto \log\binom{\lambda}{\mu}/\lambda$ when g is large). The experimental results indicate a positive correlation between the performance and the IUR: the parameter value with larger IUR is prone to perform better. Thus, the IUR can be used to guide the choice of parameters. After all, tuning the parameters by experiments is much more expensive than calculating the IURs.

Table 1. Test functions of CEC 2013 single objective optimization benchmark suite [26]

	No.	Name
Unimodal Functions	1	Sphere Function
	2	Rotated High Conditioned Elliptic Function
	3	Rotated Bent Cigar Function
	4	Rotated Discus Function
	5	Different Powers Function
Basic Multimodal Functions	6	Rotated Rosenbrock's Function
	7	Rotated Schaffers F7 Function
	8	Rotated Ackley's Function
	9	Rotated Weierstrass Function
	10	Rotated Griewank's Function
	11	Rastrigin's Function
	12	Rotated Rastrigin's Function
	13	Non-Continuous Rotated Rastrigin's Function
	14	Schwefel's Function
	15	Rotated Schwefel's Function
	16	Rotated Katsuura Function
	17	Lunacek Bi_Rastrigin Function
	18	Rotated Lunacek Bi_Rastrigin Function
	19	Expanded Griewank's plus Rosenbrock's Function
	20	Expanded Scaffer's F6 Function
Composition Functions	21	Composition Function 1 (Rotated)
	22	Composition Function 2 (Unrotated)
	23	Composition Function 3 (Rotated)
	24	Composition Function 4 (Rotated)
	25	Composition Function 5 (Rotated)
	26	Composition Function 6 (Rotated)
	27	Composition Function 7 (Rotated)
	28	Composition Function 8 (Rotated)

CMA-ES. Different from (μ, λ)-ES, CMA-ES adopts a rank-based weighted recombination instead of a selection operation, in which the rank information of the best μ individuals is utilized.

On the one hand, the rank-based weighted recombination achieves the largest IUR when $\mu = \lambda$ (see Eq. (21)). The larger μ is, the more information is used (because the rank information of the rest $\lambda - \mu$ individuals are wasted).

On the other hand, the rank-based weighted recombination also achieves the best performance when $\mu = \lambda$ [1]. This is also an evidence on the correlation between IUR and performance. However, the optimal weighted recombination requires the use of negative weights, which is somehow not adopted in CMA-ES [15]. The manner of information utilization and other conditions (termination criterion, performance measure, etc.) should also be taken into consideration when parameters are chosen. Therefore using $\mu = \lambda$ is probably not the best choice for CMA-ES even though it leads to a large IUR.

4.2 Algorithms in the Same Family

Usually different algorithms in the same family utilize information in similar manners, in which case we may compare their performances to show the correlation between IUR and performance. However, we need to be more cautious here because IUR is not the only factor as long as different algorithms in the same family do not utilize information in identical manners.

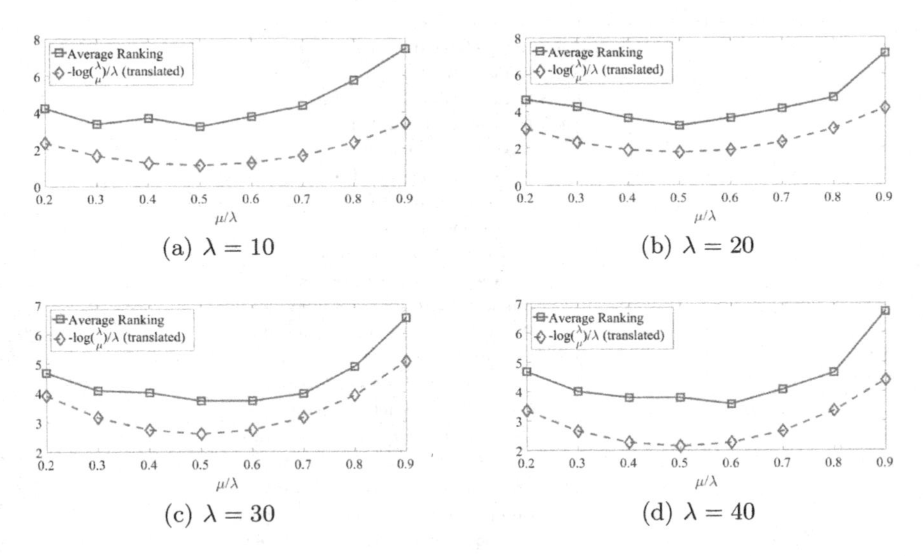

(a) $\lambda = 10$ (b) $\lambda = 20$

(c) $\lambda = 30$ (d) $\lambda = 40$

Fig. 2. The average rankings and the $-\log\binom{\lambda}{\mu}/\lambda$ curves of each value of λ.

As shown in Sect. 3, $IUR_{LJ} \geq IUR_{MC}$ and $IUR_{CMA\text{-}ES} \geq IUR_{(\mu,\lambda)-ES}$. LJ and CMA-ES are more sophisticated designed compared with the previous algorithms since they are able to utilize more information of the objective function. Naturally we would expect that LJ outperforms MC and CMA-ES outperforms (μ, λ)-ES.

The four algorithms are evaluated on the CEC 2013 benchmark suite. The parameter of LJ is set to $\gamma = 0.99$. The parameters of (μ, λ)-ES are set to $\lambda = 30, \mu = 15, \Delta\sigma = 0.5$. The parameters of CMA-ES are set to suggested values [15] except that $\sigma = 50$ because the radius of the search space is 100. The dimensionality is $d = 5$, and the maximal number of function evaluation is $10000d$ for each run. Each algorithm is run 20 times independently for each function. Their mean errors are shown in Table 2. The best mean errors are highlighted. Their mean errors are ranked on each function, and the average rankings (AR.) over 28 functions are also shown in Table 2.

Pair-wise Wilcoxon rank sum tests are also conducted between MC and LJ and between (μ, λ)-ES and CMA-ES. The p values are shown in the last two columns of Table 2. Significant results (with confidence level 95%) are underlined. The results of LJ are significantly better than MC on 14 functions, and

Table 2. Mean errors and average rankings of the four algorithms and p values

F.	MC	LJ	(μ, λ)-ES	CMA-ES	MC vs. LJ	(μ, λ)-ES vs. CMA-ES
1	2.18E+02	**0.00E+00**	5.91E−12	**0.00E+00**	8.01E−09	4.01E−02
2	4.25E+05	**0.00E+00**	3.49E+05	**0.00E+00**	8.01E−09	8.01E−09
3	8.02E+07	**0.00E+00**	2.18E+07	**0.00E+00**	1.13E−08	1.13E−08
4	4.23E+03	**0.00E+00**	2.20E+04	**0.00E+00**	1.13E−08	1.13E−08
5	8.00E+01	6.79E+01	1.95E−05	**0.00E+00**	1.33E−02	1.90E−04
6	9.94E+00	2.51E+01	**2.46E+00**	7.86E−01	4.17E−05	8.15E−06
7	2.02E+01	7.10E+01	1.66E+01	**5.66E+00**	1.99E−01	2.56E−03
8	**1.83E+01**	2.01E+01	2.03E+01	2.10E+01	3.42E−07	1.61E−04
9	2.53E+00	1.67E+00	2.37E+00	**1.08E+00**	1.48E−03	1.63E−03
10	2.30E+01	1.78E+00	1.30E+01	**4.16E−02**	6.80E−08	1.23E−07
11	2.22E+01	1.40E+01	6.67E+00	**6.57E+00**	3.04E−04	8.17E−01
12	2.10E+01	1.33E+01	1.20E+01	**7.36E+00**	1.12E−03	2.04E−02
13	2.21E+01	1.90E+01	1.87E+01	**1.28E+01**	1.20E−01	5.98E−01
14	3.78E+02	7.53E+02	**1.35E+02**	4.61E+02	1.10E−05	7.41E−05
15	**3.84E+02**	6.85E+02	5.27E+02	4.52E+02	3.99E−06	1.81E−01
16	7.43E−01	**5.34E−01**	8.27E−01	1.49E+00	1.93E−02	1.11E−01
17	3.25E+01	2.23E+01	**9.87E+00**	1.07E+01	1.78E−03	3.65E−01
18	3.43E+01	1.82E+01	**1.01E+01**	1.01E+01	2.60E−05	9.89E−01
19	4.08E+00	7.21E−01	5.45E−01	**4.82E−01**	9.17E−08	9.46E−01
20	**1.23E+00**	1.85E+00	2.50E+00	1.92E+00	1.10E−05	6.97E−06
21	3.23E+02	3.05E+02	**2.55E+02**	2.80E+02	1.94E−02	9.89E−01
22	5.91E+02	7.91E+02	**4.01E+02**	7.20E+02	2.56E−03	5.63E−04
23	**6.04E+02**	8.33E+02	7.01E+02	6.08E+02	8.29E−05	3.37E−01
24	**1.26E+02**	2.04E+02	1.99E+02	1.76E+02	6.80E−08	4.60E−04
25	**1.27E+02**	1.96E+02	1.98E+02	1.81E+02	1.60E−05	7.71E−03
26	**1.01E+02**	2.38E+02	1.67E+02	1.98E+02	1.43E−07	7.76E−01
27	3.57E+02	3.52E+02	3.65E+02	**3.27E+02**	4.25E−01	2.47E−04
28	3.05E+02	**3.00E+02**	3.25E+02	3.15E+02	8.59E−01	2.03E−01
AR	2.82	2.68	2.43	**1.82**	14 : 10	14 : 3

significantly worse on only 10 functions. While the results of CMA-ES are significantly better than (μ, λ)-ES on 14 functions, and significantly worse on only 3 functions. Generally speaking, the performance of LJ is better than MC and the performance of CMA-ES is better than (μ, λ)-ES. These experimental results imply that the extent of information utilization may be an important factor in the performance.

The algorithms in the same family utilize information in similar but different manners. In this case, the influence of IUR on the performance is crucial, but sometimes not deterministic. For example, in CMA-ES, there are several different mechanisms proposed to improve the performance. The improvement in IUR does not reflect all of them. The improvement related to IUR is the rank-based weighted recombination. It has significant impact on performance [1,16]. While other mechanisms such as adapting the covariance matrix and the step size are not related to IUR but also very important. These mechanisms are introduced as different information utilization manners, which help the algorithm to better fit the underlying distribution of objective functions. Similar comparisons can be made between PSO and SPSO and between DE and JADE (see appendix).

However, after all, an algorithm cannot perform very well if little information is used. Hence, just like LJ, CMA-ES, SPSO and JADE, the tendency of elevating the IUR is quite clear in various families of heuristics. Many mechanisms have been proposed to better preserve historical information for further utilization [8,31,38]. Many general methods (adaptive parameter control [12], estimation of distribution [23], fitness approximation [20], Bayesian approaches [29], Gaussian process models [6], hyper-heuristic [7]) have been proposed to elevate the IURs of heuristics. Not to mention these numerous specified mechanisms. In summary, the IUR provides an important and sensible perspective on the developments in this field.

4.3 Algorithms in Different Families

The correlation between the IUR and the performance of the algorithms in different families (such as LJ and (μ, λ)-ES) can be vaguer because the manners of information utilization are different, though the above experimental results accord with our expectation ((μ, λ)-ES performs better than LJ and $IUR_{LJ} \leq IUR_{(\mu,\lambda)-ES}$ unless $\mu = \lambda$). If algorithms utilize information in extremely different manners, the IUR may not be the deterministic factor. There are infinite manners to utilize information. It is difficult to judge which manner is better. Whether a manner is good or not depends on how it fits the underlying distribution of the objective functions, which is difficult to describe. A well designed algorithm with low IUR may outperform a poorly designed algorithm with high IUR because it utilizes information more efficiently and fits the underlying distribution better. Nonetheless, certainly the extent of information utilization is still of importance in this case because 1) the algorithms with larger IURs have greater potential 2) the IUR of the "best" algorithm (if any) must be very close to one and 3) an algorithm that uses little information cannot be a good algorithm.

The exact correlation between the IUR and the performance requires much more theoretical works on investigating the manners of information utilization and how they fit the underlying distributions, which are very difficult but not impossible.

5 Upper Bound for Comparison-Based Algorithms

Above examples have covered several approaches of information utilization in heuristic optimization algorithms. But the IURs of these algorithms are all not high because they are comparison-based algorithms, in which only the rank information is utilized.

Theorem 2 (Upper bound for comparison-based algorithms). *If the maximal number of evaluations is m, $y = f(x)$ are i.i.d, and algorithm $\mathscr{A}$ is a comparison-based optimization algorithm,*

$$IUR_{\mathscr{A}} \leq \frac{\log m}{H(f(x))}. \tag{22}$$

Proof. Suppose in a certain run, the actual evaluation number is $m' \leq m$. In this case, M is drawn from a set with cardinal number at most $m'!$ (with m' individuals all sorted), then the maximal information quantity is $H(M) \leq \log m'!$ for a comparison-based algorithm. Thus $IUR_{\mathscr{A}} \leq \frac{\log m'!}{m'H(f(x))}$. Note that the right hand side is a monotonically increasing function of m', and $\frac{\log m!}{mH(f(x))} \leq \frac{\log m}{H(f(x))}$.

Suppose $|\mathcal{Y}| = n$ and $f(x)$ obey uniform distribution, than $\frac{\log m}{H(f(x))} = log_n m$. Typically $m << n$, thus this upper bound is quite low. Most iterative algorithms do not allow the information in past iterations (because it requires a lot of memory space to do so), in which case the upper bound becomes $\frac{\log \lambda}{H(f(x))}$ where λ is the evaluation number in each generation. The IUR of CMA-ES is able to approach this bound when $\mu = \lambda$. That is, CMA-ES has almost the largest IUR in comparison-based algorithms without historical information.

There exist algorithms which use exact evaluation values in the searching process, such as genetic algorithm [19], ant colony optimization [10], estimation of distribution algorithms [23], invasive weed optimization [28], artificial bee colony [22], fireworks algorithm [36], etc. They can achieve higher IURs, even close to 1, because the cardinal number of the set from which M is drawn can be up to n^m. These algorithms have greater potential than comparison-based algorithms and can outperform them if well designed.

6 Conclusion

It is natural and often effective to utilize more heuristic information in optimization algorithms, which has been widely realized. However, there was no metric to reflect the extent of information utilization. In this paper, a metric called the information utilization ratio (IUR) is defined as the ratio of the utilized information quantity over the acquired information quantity. IUR can be an index to reflect how sophisticated and advanced an algorithm is designed. IUR proves to be well defined. Several examples are given to demonstrate the procedure of calculating IURs. Generally speaking, the IUR determines the upper bound of the performance of an optimization algorithm. To further indicate the importance of this metric, several experiments are conducted to show the correlation between the IUR and the performance. The experimental results imply that 1) for a certain algorithm, the parameter value with larger IUR has advantage; 2) for algorithms in the same family, the one with larger IUR is prone to be more efficient; 3) for algorithms in different families, the IUR is also an important factor. We also give the IUR's upper bound for comparison-based algorithms.

The IUR can be used to guide the choice of parameters, guide the design of new algorithms and guide the improvement of existing algorithms. For example, if you are inventing a new algorithm, or adapting an existing one, it is promising to include mechanisms that can enhance the information utilization in your algorithm. If you want to know which one among several algorithms is more likely efficient before you use them, it would be quite informative to compare their IURs to show which one is better designed and has greater potential.

Most works in the field of heuristic search or optimization focus on inventing new mechanisms or tricks, while few have considered the potential driver behind these works. We consider this work as a fundamental theory, which is surprisingly not easy. Hopefully the definition of IUR will lead to a more systematic manner of research about how mechanisms should be designed and how information should be utilized.

Extending this metric to other fields in artificial intelligence such as classification and time series prediction may be an interesting future work.

Acknowledgments. This work is supported by the National Natural Science Foundation of China (Grant No. 62076010), and partially supported by Science and Technology Innovation 2030 - "New Generation Artificial Intelligence" Major Project (Grant Nos.: 2018AAA0102301 and 2018AAA0100302).

Appendices

A Particle Swarm Algorithms

A.1 Particle Swarm Optimization

Particle swarm optimization (PSO) [11] is one of the most famous swarm and heuristic algorithms which is quite simple but surprisingly efficient in numerical optimization. In PSO, a fixed number (s) of particles moves in the search space to find the optimal solutions. The position of a particle is updated as follows. In generation g, for each particle i and each dimension j,

$$\begin{aligned} v_{ij}(g+1) \leftarrow & v_{ij}(g) + \phi_1 r_{1,ij}(pbest_{ij}(g) - x_{ij}(g)) \\ & + \phi_2 r_{2,ij}(gbest_j(g) - x_{ij}(g)), \end{aligned} \tag{23}$$

$$x_{ij}(g+1) \leftarrow x_{ij}(g) + v_{ij}(g+1), \tag{24}$$

where ϕ_1 and ϕ_2 are constant coefficients, r_1 and r_2 are random numbers, *pbest* is the best position in history found by this particle and *gbest* is the best position found by the entire swarm.

The output distribution in each generation is determined by $I(f(x_i(g)) < f(pbest_i(g-1)))$ and $\arg\min_i f(pbest_i(g))$. Although it is difficult to calculate $H(M)$, we have the lower and upper bounds:

$$s\sum_{i=1}^{g-1}\pi(i) \le H(M) \le \sum_{i=2}^{g} H(M_i) \le (g-1)\log s + s\sum_{i=1}^{g-1}\pi(i). \tag{25}$$

Proposition 5.

$$\frac{s\sum_{i=1}^{g-1}\pi(i)}{sgH(f(x))} \le IUR_{PSO}(g) \le \frac{(g-1)\log s + s\sum_{i=1}^{g-1}\pi(i)}{sgH(f(x))}. \tag{26}$$

A.2 Standard Particle Swarm Optimization

After years of development, many improvements and variants are proposed for PSO. In order to construct a common ground for further researches, a standard particle swarm optimization (SPSO) was defined [5]. Compared with original PSO, there are two main modifications: the local ring topology and the constricted update rule. The constricted update rule uses a new coefficient derived from ϕ_1 and ϕ_2 to constrict the velocity to guarantee convergence. In the local ring topology, the *gbest* in the velocity update equation is replaced with a *lbest*, which is the best position among this individual and its two neighbourhoods on the ring.

For each group (consisting of three particles), information with quantity at most $\log 3$ is needed to decide *lbest*.

Proposition 6.

$$\frac{s\sum_{i=1}^{g-1}\pi(i)}{sgH(f(x))} \leq IUR_{SPSO}(g) \leq \frac{s(g-1)\log 3 + s\sum_{i=1}^{g-1}\pi(i)}{sgH(f(x))}. \tag{27}$$

Usually $IUR_{PSO} \leq IUR_{SPSO}$ though their exact values are difficult to derive. It turns out that the information utilization ratio of the local model is larger than the global model because in local topology the particles interact with each other more frequently.

According to experimental results, SPSO significantly outperform PSO on a large range of test functions [5].

B Differential Evolution Algorithms

B.1 Differential Evolution

Differential evolution (DE) [34] is a powerful heuristic algorithm for numerical optimization. The number of individuals in DE is also fixed. The mutation is conducted as below (take DE/rand/1 as an example). For each x in the population, generate

$$z = x_{r1} + F(x_{r2} - x_{r3}), \tag{28}$$

where $r1, r2$ and $r3$ are random indexes and F is a constant coefficient. Then a crossover is conducted between z and x to generate a new candidate y, where there is a parameter CR to control the probability that a dimension of y is identical to that of z. If $f(y) < f(x)$, x is replaced with y, otherwise, x is kept.

In DE, the distribution of generating new offspring is determined by $I(f(y) < f(x))$ of each individual. So the IUR of DE is equal to that of LJ with the same g. However, they would be different with the same number of evaluation times.

Proposition 7.

$$IUR_{DE}(g) = \frac{s\sum_{i=1}^{g-1}\pi(i)}{sgH(f(x))}. \tag{29}$$

IURs of some other DE variants are given in Table 3.

B.2 JADE

JADE [38] is an important development of DE. There are three main adaptations proposed in JADE:

Table 3. IURs of other DE variants

	IUR
DE/best/1	$= IUR_{PSO}$
DE/current-to-best/1	$= IUR_{PSO}$
DE/rand/2	$= IUR_{DE}$
DE/best/2	$= IUR_{PSO}$

1. A DE/current-to-$p$best/1 mutation strategy. In JADE,
$$z_i = x_i + F_i(x^p_{best} - x_i) + F_i(x_{r1} - x_{r2}). \tag{30}$$
where x^p_{best} is a randomly chosen individual from the 100p% best individuals.
2. An optional external archive.
3. Adaptive mutation parameters.

External archive is a useful tool to improve information utilization. However, in JADE these individuals are just randomly chosen and randomly removed from the archive, where no information of the objective function is used. Compared to DE, JADE elevates IUR after all because the indexes of the best 100p% individuals are used. Note that the output distribution is determined only when all indexes of the best 100p% individuals are given.

Proposition 8.
$$\frac{s\sum_{i=1}^{g-1}\pi(i)}{sgH(f(x))} \le IUR_{JADE}(g) \le \frac{(g-1)\log\binom{s}{ps} + s\sum_{i=1}^{g-1}\pi(i)}{sgH(f(x))}. \tag{31}$$

According to experimental results, JADE significantly outperform DE on a large range of test functions [38].

References

1. Arnold, D.V.: Optimal weighted recombination. In: Foundations of Genetic Algorithms, International Workshop, Foga 2005, Aizu-Wakamatsu City, Japan, 5–9 January 2005, Revised Selected Papers, pp. 215–237 (2005)
2. Auger, A., Teytaud, O.: Continuous lunches are free plus the design of optimal optimization algorithms. Algorithmica **57**(1), 121–146 (2010)
3. Bäck, T., Hoffmeister, F., Schwefel, H.P.: A survey of evolution strategies. In: Proceedings of the Fourth International Conference on Genetic Algorithms (1991)

4. Beyer, H.: Evolution strategies. Scholarpedia **2**(8), 1965 (2007). revision #130731
5. Bratton, D., Kennedy, J.: Defining a standard for particle swarm optimization. In: Swarm Intelligence Symposium. SIS 2007, pp. 120–127. IEEE (2007)
6. Büche, D., Schraudolph, N.N., Koumoutsakos, P.: Accelerating evolutionary algorithms with gaussian process fitness function models. IEEE Trans. Syst. Man Cybern. Part C: Appl. Rev. **35**(2), 183–194 (2005)
7. Burke, E.K., Hyde, M., Kendall, G., Ochoa, G., Özcan, E., Woodward, J.R.: A classification of hyper-heuristic approaches. In: Handbook of metaheuristics, pp. 449–468. Springer (2010). https://doi.org/10.1007/978-1-4419-1665-5_15
8. Deb, K., Agrawal, S., Pratap, A., Meyarivan, T.: A fast elitist non-dominated sorting genetic algorithm for multi-objective optimization: NSGA-II. Springer, Berlin Heidelberg (2000). https://doi.org/10.1007/3-540-45356-3_83
9. Dorigo, M., Birattari, M., Stützle, T.: Ant colony optimization. Comput. Intell. Mag. IEEE **1**(4), 28–39 (2006)
10. Dorigo, M., Maniezzo, V., Colorni, A.: Ant system: optimization by a colony of cooperating agents. IEEE Trans. Syst. Man Cybern. Part B Cybern. **26**(1), 29–41 (1996)
11. Eberhart, R., Kennedy, J.: A new optimizer using particle swarm theory. In: Proceedings of the Sixth International Symposium on Micro Machine and Human Science. MHS 1995, pp. 39–43. IEEE (1995)
12. Eiben, A.E., Hinterding, R., Michalewicz, Z.: Parameter control in evolutionary algorithms. IEEE Trans. Evolut. Comput. **3**(2), 124–141 (1999)
13. English, T.M.: Some information theoretic results on evolutionary optimization. In: Proceedings of the Congress on Evolutionary Computation, CEC 99 (1999)
14. Everitt, T., Lattimore, T., Hutter, M.: Free lunch for optimisation under the universal distribution. In: 2014 IEEE Congress on Evolutionary Computation (CEC), pp. 167–174. IEEE (2014)
15. Hansen, N.: The CMA evolution strategy: A tutorial. Vu le 29 (2005)
16. Hansen, N., Kern, S.: Evaluating the CMA Evolution Strategy on Multimodal Test Functions. Springer, Berlin Heidelberg (2004). https://doi.org/10.1007/978-3-540-30217-9_29
17. Harik, G.R., Lobo, F.G., Goldberg, D.E.: The compact genetic algorithm. IEEE Trans. Evolut. Comput. **3**(4), 287–297 (1999)
18. He, J., Lin, G.: Average convergence rate of evolutionary algorithms. IEEE Trans. Evolut. Comput. **20**(2), 1 (2015)
19. Holland, J.H.: Adaptation in Natural and Artificial Systems: An Introductory Analysis with Applications to Biology, Control, and Artificial Intelligence. U Michigan Press, Ann Arbor (1975)
20. Jin, Y.: A comprehensive survey of fitness approximation in evolutionary computation. Soft. Comput. **9**(1), 3–12 (2005)
21. Jones, D.R.: A taxonomy of global optimization methods based on response surfaces. J. Global Optim. **21**(4), 345–383 (2001)
22. Karaboga, D., Basturk, B.: A powerful and efficient algorithm for numerical function optimization: artificial bee colony (ABC) algorithm. J. Global Optim. **39**(3), 459–471 (2007)
23. Larranaga, P.: A review on estimation of distribution algorithms. In: Estimation of Distribution Algorithms, pp. 57–100. Springer (2002). https://doi.org/10.1007/978-1-4615-1539-5_3
24. Lawler, E.L.: The traveling salesman problem: a guided tour of combinatorial optimization. Wiley-Interscience Series in Discrete Mathematics (1985)

25. Li, J., Zheng, S., Tan, Y.: The effect of information utilization: introducing a novel guiding spark in the fireworks algorithm. IEEE Trans. Evolut. Comput. (99), 1 (2016). https://doi.org/10.1109/TEVC.2016.2589821
26. Liang, J., Qu, B., Suganthan, P., Hernández-Díaz, A.G.: Problem definitions and evaluation criteria for the CEC 2013 special session on real-parameter optimization (2013)
27. Luus, R., Jaakola, T.: Optimization by direct search and systematic reduction of the size of search region. AIChE J. **19**(4), 760–766 (1973)
28. Mehrabian, A.R., Lucas, C.: A novel numerical optimization algorithm inspired from weed colonization. Ecol. Inform. **1**(4), 355–366 (2006)
29. Pelikan, M.: Bayesian optimization algorithm. In: Hierarchical Bayesian Optimization Algorithm, pp. 31–48. Springer (2005). https://doi.org/10.1007/978-3-540-32373-0_6
30. Quinlan, J.R.: Induction of decision trees. Mach. Learn. **1**(1), 81–106 (1986)
31. Reyes-Sierra, M., Coello Coello, C.A.: Multi-objective particle swarm optimizers: a survey of the state-of-the-art. Int. J. Comput. Intell. Res. **2**(3), 287–308 (2006)
32. Shannon, C.: A mathematical theory of communication. Bell Syst. Tech. J. **27**(4), 623–656 (1948). https://doi.org/10.1002/j.1538-7305.1948.tb00917.x
33. Srinivas, M., Patnaik, L.M.: Genetic algorithms: a survey. Computer **27**(6), 17–26 (1994)
34. Storn, R., Price, K.: Differential evolution-a simple and efficient heuristic for global optimization over continuous spaces. J. Global Optim. **11**(4), 341–359 (1997)
35. Streeter, M.J.: Two Broad Classes of Functions for Which a No Free Lunch Result Does Not Hold. Springer, Berlin Heidelberg (2003). https://doi.org/10.1007/3-540-45110-2_15
36. Tan, Y., Zhu, Y.: Fireworks algorithm for optimization. In: Advances in Swarm Intelligence, pp. 355–364. Springer (2010). https://doi.org/10.1007/978-3-642-13495-1_44
37. Wolpert, D.H., Macready, W.G.: No free lunch theorems for optimization. IEEE Trans. Evolut. Comput. **1**(1), 67–82 (1997)
38. Zhang, J., Sanderson, A.C.: JADE: adaptive differential evolution with optional external archive. IEEE Trans. Evolut. Comput. **13**(5), 945–958 (2009)

A General Framework for Intelligent Optimization Algorithms Based on Multilevel Evolutions

Chenchen Wang, Caifeng Chen, Ziru Lun, Zhanyu Ye, and Qunfeng Liu(✉)

School of Computer Science and Technology, Dongguan University of Technology, Dongguan 523808, China
liuqf@dgut.edu.cn

Abstract. Intelligent optimization is a kind of global optimization algorithms based on simulating biological intelligent behaviors such as evolution and foraging. Currently, there are numerous intelligent optimization algorithms have been proposed based on a large mount of animals' or plants' behaviors. This phenomenon shows the prosperity of this field, but bring issues about these algorithms' analysis and applications. We believe an extensive development stage has passed in the field of intelligent optimization, and more theoretical analysis and deep understanding about these algorithms become favorite. In this paper, we try to build a general framework for all population-based global optimization algorithms. This framework employs the idea of multilevel evolution, and therefore it can include not only the traditional bio-inspired evolution algorithms which often only evolute in a single level of search space, but also those population-based algorithms adopt data-driven strategies or cultural evolutions. By the help of the proposed framework, we can classify all population-based global optimization algorithms into three types, and improve the traditional algorithms. In this paper, this framework is then applied to the popular particle swarm optimization, and a modified particle swarm optimization with three-level of evolutions is proposed. Numerical results show that the modified algorithm improves the original one significantly.

Keywords: Intelligent optimization algorithms · Population-based optimization algorithms · General framework · Multilevel evolutions

1 Introduction

Intelligent optimization algorithm is a class of global optimization algorithms for finding the global optimal solution to optimization problems.

$$\min_{x \in R^n} f(x). \quad (1)$$

Supported by Guangdong Universities' Special Projects in Key Fields of Natural Science under Grant 2019KZDZX1005.

Y. Tan et al. (Eds.): ICSI 2022, LNCS 13344, pp. 23–35, 2022.
https://doi.org/10.1007/978-3-031-09677-8_2

These algorithms generally use the population composed of multiple individuals and rely on randomness to simulate behaviors with certain intelligence factors such as bio-inspired evolution and foraging. For example, Genetic Algorithm (GA), a well-known algorithm, imitates the mutation and recombination of genes in the process of bio-inspired evolution [4]. For another example, Particle Swarm Optimization Algorithm (PSO) simulates the foraging behavior of birds and fish [7] and Ant Colony Optimization Algorithm (ACO) simulates the foraging behavior of ants [2]. Of course, the intelligent optimization algorithm mentioned in this paper refers to a kind of algorithms in a broader sense, including population-based heuristic algorithms inspired by natural and social phenomena, such as Differential Evolution (DE) [16], and so on. Meanwhile, for ease of exposition, the term "bio-inspired evolution" is used in this paper to represent the simulation or inspire of genetic operation, biological behavior and all kind of phenomena can be beneficial for optimization.

Up to now, hundreds of intelligent optimization algorithms have been proposed, involving the intelligent behavior of a large number of animals and plants. So these phenomena have been nicknamed "Algorithmic Zoo" and "Algorithmic Botanical Garden" in academic community. Based on these bio-inspired evolution or heuristic algorithms, researchers have also considered some strategies such as data-driven [5], cultural evolution [14] and local search enhancement (Memetic) [12], resulting in an enormous number of improved algorithms and algorithm variants. These phenomena indicate that the field of intelligent optimization may have entered the late stage of booming development. Then, more theoretical analysis and deep understanding about these algorithms become favorite [9], so as to provide rigorous mathematical analysis and theoretical support for the algorithms' operation and performance.

In this paper, we attempt to develop a general framework for all population-based global optimization algorithms. This framework adopts the idea of multilevel evolution, that is, there exist multiple levels evolution, and the evolution at different levels can search different scales and use different evolution strategies. We prove that with the help of this idea, the various intelligent optimization algorithms mentioned above can be incorporated into a general framework, providing a basis and platform for subsequent improvement, analysis and theoretical support of algorithms. In addition, this framework is applied to the PSO and the PSO based on multilevel evolutions is proposed. The proposed method has been conducted on a lot of numerical tests. The experimental results show that the multilevel evolutions improve the performance of the original method.

The remainder of this paper is organised as follows. In Sect. 2, we first briefly introduce the traditional intelligent optimization algorithms and further explain the imitation or bio-inspired object of the algorithms. Some modern ideas are also described: data-driven, culture evolution and memetic strategy, etc. In Sect. 3, we present the proposed general framework based on multilevel evolutions in detail. This framework incorporates the algorithms mentioned in Sect. 2 and their improvements. Section 4 develops the PSO based on multilevel evolutions by applying the proposed framework. Then we provide and analyze the experimental results. Finally, we conclude the work with a summary as well as outlook for further work in Sect. 5.

2 Intelligent Optimization Algorithms

In this section, we describe briefly traditional heuristic or bio-inspired optimization algorithms. Then, some modern ideas of algorithm improvement are also presented, such as data-driven, culture evolution and memetic strategy, etc.

2.1 Traditional Heuristic or Bio-Inspired Optimization Algorithms

As mentioned in the introduction, a large number of intelligent behaviors such as evolution and foraging of animals and plants are imitated by traditional algorithms, resulting in "Algorithmic Zoo" and "Algorithmic Botanical Garden". Table 1 presents a portion of these algorithms and gives information on their corresponding names, objects and behaviors that are imitated or emulated, and when they were proposed. In addition, which multilevel evolution type (Sect. 3) these algorithms belong to is listed in the last column of the Table 1.

There were only algorithms of bio-inspired evolution at the beginning. Until the early 1990s, the algorithms based on cultural evolution have appeared. We also find that there are few algorithms based on multilevel cultural evolution.

2.2 Modern Ideas of Algorithm Improvement

In this part, we mainly provide a description to data-driven and knowledge-driven, cultural evolution and memetic strategy.

Data-Driven and Knowledge-Driven. In the context of the rise of big data, the concept of data-driven was proposed. In the field of optimization, it refers to the mining and utilization of various possible empirical data, especially the data generated during the execution of the algorithm. Then, knowledge to guide optimization is generated and can better guide the further optimization [5]. It can be found that data-driven and knowledge-driven are closely related. From data awareness to data-cognition to generating effective decisions, data-driven can guide the optimization and accelerate the algorithms' convergence speed or improve the quality of solutions.

Cultural Evolution. Cultural evolution, derived from the social sciences, is used to describe the evolution how human culture has evolved. Thanks to language and writing, cultural evolution opened up a more efficient path for the evolution of human civilization, in addition to biological or genetic evolution [14]. Under the scenario of optimization, cultural evolution mines and uses the data generated in biological or genetic evolution to obtain a certain culture or belief. Finally, it can accelerate the optimization process through the belief space.

Table 1. Some population-based global optimization algorithms. The algorithms based on bio-inspired evolutions refer to AB, the algorithms based on bio-inspired and cultural evolutions denote as ABC, and the algorithms based on bio-inspired and multilevel cultural evolutions denote as ABC^2.

Algorithm	Bio-inspired objects or behaviors	Year	Types of multilevel evolutions
Simulated Annealing Algorithm [8]	Natural algorithms that simulate the high-temperature annealing process of metallic materials	1953	AB
Genetic Algorithm [4]	Simulating Darwin's genetic selection and natural elimination of biological evolution	1975	AB
Ant Colony Optimization Algorithm [2]	Simulating the behavior of ant colonies such as foraging and nesting	1991	AB
Memetic Algorithm [12]	Combining biological-level evolution with social-level evolution	1992	ABC
DIRECT Algorithm [6]	Global optimization based on search space partition	1993	AB
Cultural Algorithm [14]	An algorithm based on the idea of cultural evolution of human society	1994	ABC
Particle Swarm Optimization [7]	A swarm intelligence optimization algorithm, simulating the foraging behavior of birds	1995	AB
Differential Evolution [16]	A genetic algorithm with special "mutation", "crossover" and "selection" formulas, using real codes	1995	AB
Social Evolutionary Algorithm [13]	Combining multi-intelligent systems and traditional genetic mechanisms	2009	ABC
Fireworks Algorithm [17]	Inspired by observing fireworks explosion	2010	AB
Wind Driven Optimization [1]	Based on the physical equations that govern atmospheric motion	2010	AB
Brain Storm Optimization Algorithm [15]	Simulation of human creative problem-solving brainstorming	2011	AB
Multilevel Robust DIRECT Algorithm [10]	Recursive invocation of DIRECT algorithm in different search spaces	2015	ABC^2
Bilevel-search particle swarm optimization [19]	Adding elite strategies to classical particle swarm optimization	2021	ABC

Memetic Strategy. The traditional memetic strategy combines intelligent optimization algorithms with local search in order to improve the optimization process [12]. It is mainly concerned with the selection of local search methods and the interaction, integration and balance between global and local search. This idea was later used for multitask learning (optimization) [20] and transfer learning (optimization) [18]. In this paper, we emphasize the information sharing and interaction between two different scales of optimization: global exploration and local exploitation.

3 A General Framework for Intelligent Optimization Algorithms Based on Multilevel Evolutions

3.1 From Bio-Inspired Evolution to Cultural Evolution

As reviewed in Sect. 2, the traditional heuristic or bio-inspired optimization algorithms generally perform population evolution in the original search space. However, the algorithm improvements, including data-driven, knowledge-driven, cultural evolution and memetic strategy, implicitly involve another optimization beyond the population evolution of the original search space, and the information generated by the latter improves the evolutionary path of the former. For example, in memetic strategy, the introduced local search method is a new process for optimizing. As for data-driven, knowledge-driven and cultural evolution, they hypothesize that historical data or experience can generate valid knowledge. This useful knowledge can affect subsequent optimization and improve the original evolution of population.

In that case, if information obtained from the optimization process or various empirical knowledge can be considered as "cultural information", the general term "cultural evolution" can be used as the generic term for improvements such as data-driven, knowledge-driven, cultural evolution and memetic strategy. In this way, we can find that intelligent optimization algorithms have gradually evolved from simulating bio-inspired evolution to true bio-inspired evolution and cultural evolution. In the field of optimization, the success of cultural evolution is based on the fact that it evolves faster than bio-inspired evolution, which can speed up the convergence of the algorithm.

3.2 A Framework Based on Multilevel Evolutions

From the previous discussion, it can be found that the original population optimization and various forms of improvement ideas can be unified by a framework based on multilevel evolutions. At each level, different search strategies can be evolved and different scales of search spaces can be defined. In general, in the original search space, population executes genetic operations or other bio-inspired operations based on biological behavior. Then, according to the process information, the useful knowledge or "cultural information" can be acquired for a more efficient local search in a new space with a new population or subpopulation. The search space can also be the original space or a smaller one. Finally, the information created at this stage is fed back into the original space to accelerate the evolutionary process. This mechanism of "bio-inspired evolution in the original space and cultural evolution in the new space" can be denoted as a framework "bio-inspired evolution $\oplus$ cultural evolution". Of course, more levels can also be created to further speed up the optimization process. In this paper, we define this multilevel evolutionary mechanism as a framework "bio-inspired evolution $\oplus$ cultural evolution $\oplus$ multilevel cultural evolution". The skeleton of this framework is shown in Fig. 1.

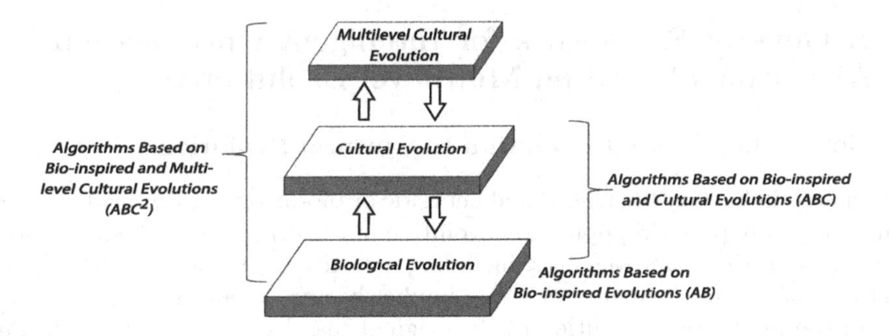

Fig. 1. The framework based on multilevel evolutions.

In this part, we will discuss that how all population-based optimization algorithms can be incorporated into this multilevel evolutionary framework. For ease of distinction, we state the formal definition of bio-inspired evolution, cultural evolution and multilevel cultural evolution.

First of all, bio-inspired evolution is to introduce the concept of population in the search space according to the characteristics of the optimization problem. It simulates the genetic manipulation or the intelligent behavior of organisms or various beneficial inspired phenomena to find the optimal solution through iterative evolution.

Secondly, according to the features of optimization problems, cultural evolution is the process of extracting rules and information using all possible prior knowledge and data produced in bio-inspired evolution. Then, the optimization process is improved by the algorithm itself or by other local search methods.

At last, based on the characteristics of the optimization problem, multilevel cultural evolution is the process of using all possible prior knowledge and data created in cultural evolution to further find rules and information. The optimization process is improved by the algorithm itself or other local search methods.

With the above formal definitions, all population-based optimization algorithms can be classified into three categories. The first type refers to the algorithms based on bio-inspired evolutions (AB), which only implement bio-inspired evolution in the original search space. The second type denotes as the algorithms based on bio-inspired and cultural evolutions (ABC), In addition to bio-inspired evolution, there is another space (may similar to the original space) where cultural evolution is implemented. The cultural information is derived from the prior knowledge of the problem or information in bio-inspired evolution. The third one is defined as the algorithms based on bio-inspired and multilevel cultural evolutions (ABC^2), having both biological and cultural evolution, and further extracting more effective subcultures for optimization in the new search space. This new space is generally a subset of the space in cultural evolution. In other words, a population-based optimization algorithm belongs to the algorithm based on bio-inspired evolutions if it searches only in the original space and does not utilize any prior knowledge or information in optimization. If an algorithm adopts this prior knowledge or information, it is an algorithm based on bio-inspired and cultural evolutions. On the basis of the second type, if there are also local searches in the space, it will become an algorithm based on bio-inspired and multilevel cultural evolutions.

The type of multilevel evolutions of some optimization algorithms is given in Table 1. It is clear that most of the algorithms are the AB type algorithms and cultural evolution has become mainstream in recent years. But the multilevel cultural evolution is relatively rare.

3.3 The Construction of Culture and Multilevel Culture

It can find that an increasing number of algorithms are moving toward cultural evolution and multilevel cultural evolution beyond bio-inspired evolution. In this part, we give some common ways in which cultures are constructed. Because the multilevel culture is also part of a culture, these constructing ways are applicable to both cultural and multilevel cultural constructions in principle.

Prior Experience-Based Cultural Construction. In the scenario of optimization problems, if there is some prior experience that can be used to guide the search for the optimal value, a suitable culture can be constructed directly based on prior experience. This newly constructed culture can lead to the process of cultural evolution. However, intelligent optimization algorithms are often applied to deal with black-box optimization problems, so the prior experience is generally difficult to obtain. With this in mind, constructing cultures based on prior experience generally rarely occurs.

Knowledge-Based Cultural Construction. This construction is derived from mining and learning from data produced by bio-inspired evolution, mainly by finding available knowledge to guide cultural evolution. Thus, those data-driven or knowledge-driven intelligent optimization algorithms adopt this way to construct cultures. In addition, most of the machine learning techniques can be used for data mining, which is a current hot research topic [5].

Geography-Based Cultural Construction. This type of constructed way is usually used for global optimization based on partition [6]. In optimization, the search space is continuously partitioned according to the individual's fitness value, so that the individual has a naturally "territory". For some "territories", different symbols can be assigned to mark the cultures, and multilevel evolution according to these symbols can be constructed flexibly [10].

Heuristic-Guided Cultural Construction. This way directly applies "successful cultures" to cultural evolution. This successful culture refers to the culture that has been effective in other similar issues. Transfer learning (optimization) and multitask learning (optimization) can be considered as such heuristic-guided cultural construction [18,20]. Additionally, other useful heuristics can be directly constructed as culture and applied to cultural evolution.

Local-Exploitation-Based Cultural Construction. As for the traditional memetic strategy, it is combined with the local search method to enhance the speed of convergence of bio-inspired evolution [12]. This is also a process of cultural construction, denoted as a local-exploitation-based cultural construction. Generally, local search works among the elite individuals of the population [19]. This is similar to giving elites more resources to make greater contributions to society, and is a manifestation of elite culture.

3.4 Sharing Information Among the Evolutions at Different Levels

In the framework based on multilevel evolutions, it is very important to share information between different evolutionary levels. They generally have the following characteristics.

Sharing Information Between the Evolutions at Adjacent Levels. Sharing information usually occurs in the evolution of neighboring levels, as shown in Fig. 1. In other words, information and feedback can be shared between bio-inspired evolution and cultural evolution, as well as between cultural evolution and multilevel cultural evolution. However, there is no direct information and feedback between bio-inspired evolution and multilevel cultural evolution.

Sharing Information in Different Directions with Different Roles. In the framework based on multilevel evolution, there are two different directions for sharing information. The first is bottom-top, which refers to the information sharing from bio-inspired evolution to cultural evolution to multilevel cultural evolution. Conversely, the second one is from top to bottom. Sharing information in these directions plays different roles. Specifically, bottom-up is a process of addressing or extracting information with the aim of finding usable knowledge or rules. In this case, this sharing information can give a guidance to the "top" evolution. The top-bottom is information feedback to send information back to the below level and accelerate its evolution.

Information in Different Levels Possibly with Different Dimensions. In the framework based on multilevel evolutions, the evolution of the upper level is usually intended to accelerate the evolution of the lower level. That is to say that the bottom level (bio-inspired evolution) is used to solve the optimization problem. The cultural evolution and multilevel evolution are only for accelerating bio-inspired evolution, not for solving the optimization problem itself. It indicates that the search spaces of cultural evolution and multilevel cultural evolution are usually smaller than the original space. Moreover, their dimension of the control variables can be decreased, which can better achieve the goal of accelerating bio-inspired evolution with low-cost (multilevel) cultural evolution. Therefore, bottom-top tends to lead to a decrease in the amount of information (reduced-dimension or smaller information space), while top-bottom leads to an increase in the amount of information (increased-dimension or larger information space).

4 Particle Swarm Optimization Algorithm Based on Multilevel Evolutions

In the previous section, we proposed the framework based on multilevel evolutions, which is able to cover all population-based intelligent optimization algorithms. Moreover, this framework also provides a direction for performance improvement of such algorithms: accelerating the bio-inspired evolution in the original search space by introducing cultural evolution and multilevel cultural evolutions. In this section, we are going to apply this framework to Particle Swarm Optimization Algorithm (PSO), and propose a new algorithm denoted as PSO based on three-level evolutions (PSO-3Level).

First of all, we use the standard PSO (SPSO2011) proposed by French professor Clerc at the level of bio-inspired evolution. This is an important implementation of classical PSO. Based on the SPSO2011, we introduce two types of elite culture. The first type is the ordinary elite, which selects the best 30% of individuals of the population to build the elite sub-population for cultural evolution. The second type is the top elite, which selects the best individual in the population and gives it the power of local search to carry out multilevel cultural evolution. In our paper, we adopt the well-known local search method BFGS. The pseudo-code of PSO based on three-level evolutions is as follows.

Algorithm 1: PSO based on three-level evolutions (PSO-3Level)

Input: Maximum computational cost $MaxNF$; population size $N = 100$;
Output: the found best function value.

1 Initialize the population P;
2 **while** *Stop condition is not satisfied* **do**
3   Bio-inspired evolution: execute the SPSO2011 and iterate 1 time;
4   Cultural evolution: select 30 best individuals from the P and build the sub-population, execute the SPSO2011 and iterate 1 time;
5   Multilevel cultural evolution: select the best individual, execute the BFGS and iterate 10 times;

The above parameters have not been specifically optimized, but simple set. In this paper, our main focus is to illustrate the application of the framework based on multilevel evolutions, not to design the best current variant of PSO. With the use of two elite cultures in the PSO-3Level, we have reason to believe that this method can improve the solution accuracy of the SPSO2011 and have a significant advantage with less computational cost.

To evaluate the performance of the PSO-3Level, we test out the proposed method on Hedar test suite [3]. Hedar test suite has a total of 68 functions (containing functions with different dimensions), with a maximum number of 48. By recording the function values during the evolution, a high-dimensional matrix H(MaxNF, 30, 68, 2) is eventually obtained. This matrix H indicates that two algorithms are tested (SPSO2011, PSO-3Level), each with 68 functions, and each function is solved 30 times independently to eliminate randomness as much as possible. MaxNF is the number of function evaluations in each test. With the help of matrix H, we can describe the decreasing trend of the function values in each test. Due to a large number of functions, data profile, proposed by Moré and Wild in 2009 [11], is used to analyze the experimental data in order to facilitate the presentation of comparative results. It uses the following inequalities as convergence condition for the algorithm to solve the problem.

$$f(x) \leq f_L + \tau \left(f\left(x_0\right) - f_L\right). \tag{2}$$

where x_0 is a point of the initial iteration and τ is the precision parameter (τ is set to 10^{-7} in this paper). At a specific computational cost, f_L is the minimum found by all tested optimization algorithms, and each function has its own f_L.

$t_{p,s}$ is denoted as the lowest computational cost that satisfies the Eq. 2 in solving the functions $p \in P$ by the algorithms $s \in S$. If Eq. 2 is not satisfied, $t_{p,s}$ is set to infinity. The sets S and P are the set of algorithms to be compared and the set of test functions, respectively. Based on this, the data profile curve of algorithm $s \in S$ can be defined as:

$$d_s(\alpha) = \frac{1}{|P|} \left| \left\{ p \in P : \frac{t_{p,s}}{n_p + 1} \leq \alpha \right\} \right|. \tag{3}$$

where n_p is the dimension of function p and $\frac{t_{p,s}}{n_p+1}$ roughly represents the relative computational cost in per dimension on average. Therefore, the data profile curve describes the proportion of problems that can be solved by algorithm s within α relative computational cost.

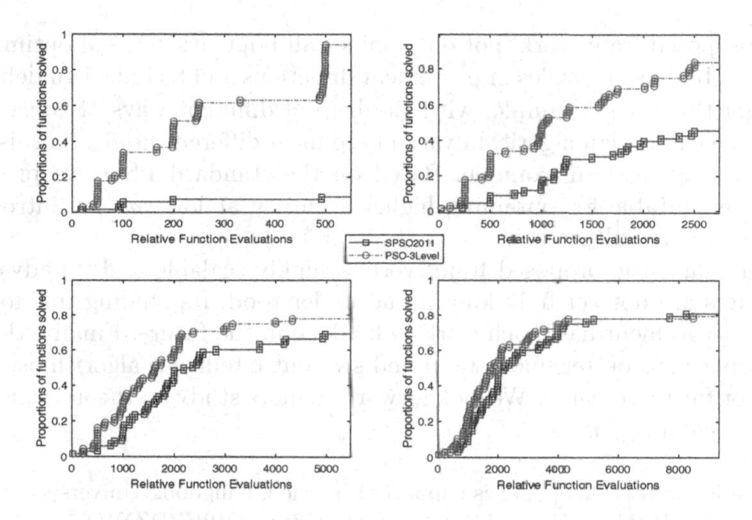

Fig. 2. Comparison of data profiles of SPSO2011 and PSO-3Level. The computational cost are 1000, 5000, 10000 and 20000, respectively.

Figure 2 presents the curve of the data profile, and the four subgraphs correspond to 1000, 5000, 10000 and 20000 function evaluations, respectively. It can be found that the lower the computational cost, the better our proposed algorithm PSO-3Level will be. When there are only 1000 function evaluations, the algorithm PSO-3Level solves more than 90% of the problems (according to Eq. 2). While SPSO2011 only solves less than 10% of problems. The difference in performance between them is more than 80%. As the computational cost increase, the difference of their performance begin to decline. When the computational cost increases to 20000, they perform almost the same. This is manly because the optimal solutions of most problems have been found, and the effect of elite culture has gradually become invalid.

5 Conclusion and Future Work

The field of intelligent optimization has proposed a large number of algorithms, which are jokingly referred to as "Algorithmic Zoo" and "Algorithmic Botanical Garden". With this in mind, we propose a general framework based on multilevel evolutions, which incorporates all population-based optimization algorithms. Our framework indicates that these algorithms are either single-level algorithms only based on bio-inspired evolution (AB), two-level algorithms based on bio-inspired and cultural evolutions (ABC), or three-level algorithms based on bio-inspired and multilevel cultural evolutions (ABC^2).

We also describe the way of constructing culture and multilevel cultures: prior experience-based, knowledge-base, geography-based, heuristic-guided and local exploitable-based cultural construction. Meanwhile, the analysis of sharing information among the evolutions at different levels is also presented.

The proposed framework, not only unifies all population-based optimization algorithms, but also provides improvement directions and technical guidelines for these algorithms. For example, with the help of different ways of constructing cultures, we can design algorithm variants to meet different needs. So this paper provides an application example. Based on the standard PSO, we propose a PSO-3Level suitable for obtaining higher accuracy at low cost by introducing two types of elite culture.

In summary, the proposed framework is highly scalable and its advantages and features are not yet fully known and understood. Its guiding role for algorithms' improvement has much work to be done in the future. Finally, the work in this paper is to better understand and sort out intelligent algorithms, paving the way for future research. We look forward to more study on theoretical studies and numerical analysis.

Acknowledgment. This paper is supported by the Guangdong Universities' Special Projects in Key Fields of Natural Science under Grant 2019KZDZX1005.

References

1. Bayraktar, Z., Komurcu, M., Werner, D.H.: Wind driven optimization (WDO): a novel nature-inspired optimization algorithm and its application to electromagnetics. In: 2010 IEEE Antennas and Propagation Society International Symposium, pp. 1–4. IEEE (2010)
2. Colorni, A., Dorigo, M., Maniezzo, V., et al.: Distributed optimization by ant colonies. In: Proceedings of the First European Conference on Artificial Life, vol. 142, pp. 134–142 (1991)
3. Hedar, A.R.: Test functions for unconstrained global optimization. http://www-optima.amp.i.kyoto-u.ac.jp/member/student/hedar/Hedar_files/TestGO_files/Page364.htm
4. Holland, J.H.: Adaptation in natural and artificial systems : an introductory analysis with applications to biology (1992)
5. Jin, Y., Wang, H., Chugh, T., Guo, D., Miettinen, K.: Data-driven evolutionary optimization: an overview and case studies. IEEE Trans. Evol. Comput. **23**(3), 442–458 (2018)
6. Jones, D.R., Perttunen, C.D., Stuckman, B.E.: Lipschitzian optimization without the Lipschitz constant. J. Optim. Theory Appl. **79**(1), 157–181 (1993)
7. Kennedy, J., Eberhart, R.: Particle swarm optimization. In: Proceedings of ICNN'95-International Conference on Neural Networks, vol. 4, pp. 1942–1948. IEEE (1995)
8. Kirkpatrick, S., Gelatt, C.D., Jr., Vecchi, M.P.: Optimization by simulated annealing. Science **220**(4598), 671–680 (1983)
9. Liu, B., Wang, L., Liu, Y., Wang, S.: A unified framework for population-based metaheuristics. Ann. Oper. Res. **186**(1), 231–262 (2011)
10. Liu, Q., Zeng, J., Yang, G.: MrDIRECT: a multilevel robust direct algorithm for global optimization problems. J. Global Optim. **62**(2), 205–227 (2015)
11. Moré, J.J., Wild, S.M.: Benchmarking derivative - free optimization algorithms. SIAM J. Optim. **20**(1), 172–191 (2009)

12. Moscato, P., Norman, M.G.: A memetic approach for the traveling salesman problem implementation of a computational ecology for combinatorial optimization on message-passing systems. Parallel Comput. Transput. Appl. **1**, 177–186 (1992)
13. Pan, X., Liu, f., Jiao, L.: Multiobjective social evolutionary algorithm based on multi-agent. J. Softw. **20**, 1703–1713 (2009)
14. Reynolds, R.G.: An introduction to cultural algorithms **24**, 131–139 (1994)
15. Shi, Y.: Brain storm optimization algorithm. In: Tan, Y., Shi, Y., Chai, Y., Wang, G. (eds.) ICSI 2011. LNCS, vol. 6728, pp. 303–309. Springer, Heidelberg (2011). https://doi.org/10.1007/978-3-642-21515-5_36
16. Storn, R., Price, K.: Differential evolution - a simple and efficient heuristic for global optimization over continuous spaces. J. Global Optim. **11**(4), 341–359 (1997)
17. Tan, Y., Zhu, Y.: Fireworks algorithm for optimization. In: Tan, Y., Shi, Y., Tan, K.C. (eds.) ICSI 2010. LNCS, vol. 6145, pp. 355–364. Springer, Heidelberg (2010). https://doi.org/10.1007/978-3-642-13495-1_44
18. Wu, K., Wang, C., Liu, J.: Evolutionary multitasking multilayer network reconstruction. IEEE Trans. Cybern. (2021, online). https://doi.org/10.1109/TCYB.2021.3090769
19. Yan, Y., Zhou, Q., Cheng, S., Liu, Q., Li, Y.: Bilevel-search particle swarm optimization for computationally expensive optimization problems. Soft. Comput. **25**(22), 14357–14374 (2021). https://doi.org/10.1007/s00500-021-06169-3
20. Zhang, F., Mei, Y., Nguyen, S., Zhang, M., Tan, K.C.: Surrogate-assisted evolutionary multitask genetic programming for dynamic flexible job shop scheduling. IEEE Trans. Evol. Comput. **25**(4), 651–665 (2021)

Swarm-Based Computing Algorithms for Optimization

Improved Hamming Diversity Measure for Set-Based Optimization Algorithms

Kyle Erwin[1(✉)] and Andries Engelbrecht[2]

[1] Computer Science Division, Stellenbosh University, Stellenbosh, South Africa
kyle.erwin24@gmail.com
[2] Department of Industrial Engineering, and Computer Science Division, Stellenbosh University, Stellenbosh, South Africa
engel@sun.ac.za

Abstract. Proper evaluation of the performance of a population-based algorithm often requires a good understanding of the search behaviour of the population over time. One approach towards understanding search behaviour is to measure the diversity of the population, which is an indicator of how similar the search agents in the population are to one another. Historically, diversity metrics have focused on populations with a continuous-valued, fixed-length solution encoding. Two diversity measures, Jaccard diversity and Hamming diversity, for populations with a set-based solution encoding were recently proposed. It was shown that Jaccard diversity accurately represents the diversity of sets over time while Hamming diversity under represents diversity in the same scenarios. This paper proposes a simple improvement to the Hamming diversity measure, that makes it equivalent to the Jaccard diversity measure, accurately quantifying set-based population diversity.

Keywords: Set-based meta-heuristics · Set-based diversity · Hamming distance · Jaccard distance

1 Introduction

Measures of population diversity are useful tools in a researcher's toolbox for better understanding algorithm performance. These measures give a researcher insight into the exploration and exploitation behaviour of population-based optimization algorithms. Diversity is at a maximum when the search agents in a population totally differ from one another. In contrast, diversity is a minimum, i.e. 0.0, when all the agents are the same. Relatively high diversity values are indicative of exploratory behaviour, while lower diversity values indicate exploitation behavior. By measuring the diversity of a population over time trends can be obtained for the exploratory and exploitative search behaviour of an algorithm. Ideally, large diversity values should be observed at the beginning of the search, and overtime, the diversity values should approach 0.0. If the diversity values

Y. Tan et al. (Eds.): ICSI 2022, LNCS 13344, pp. 39–47, 2022.
https://doi.org/10.1007/978-3-031-09677-8_3

are close to 0.0 by the end of the search, it can be concluded that the agents have converged on a solution.

There are a variety of diversity measures for population-based algorithms with a continuous-valued solution encoding [1,7]. These measures, which typically use Euclidean distance, are not suitable for set-based populations, because sets can differ in size and there is no concept of distance between elements in the sets. In light of this, Erwin and Englebrecht [2] proposed two set-based diversity measures, one based on the Jaccard distance (Jaccard diversity) [5] and the other based on the Hamming distance (Hamming diversity) [4].

The aforementioned set-based diversity measures have the following properties: 1) When the contents of sets become more similar, diversity values decrease. 2) Conversely, when the contents of sets become less similar, diversity values increase. 3) When the contents of sets are identical, the diversity values will be zero. Then, the Jaccard diversity measure has a fourth property that when sets are totally different, the Jaccard diversity value will be 1.0. In the same scenario, Hamming diversity will under represent the diversity of sets.

The purpose of this paper is to address the under representation of diversity caused by Hamming diversity. This paper proposes a simple modification to the way sets are converted into bit strings for the Hamming diversity measure. The result is that Hamming diversity produces diversity values equivalent to those of Jaccard diversity. Thus, Hamming diversity is now able to accurately represent diversity especially in cases where diversity is intuitively at a maximum.

This work is useful as it can help to further understand and develop set-based algorithms for combinatorial optimization problems. For example, algorithms like the set-based adaptation of particle swarm optimization by Langeveld and Engelbrecht [6] or various set-based genetic algorithms [3,8,9]. The focus and improvement of the Hamming diversity measure is particularly useful for problems where a variable-length binary solution encoding is needed, e.g. data compression problems.

The rest of the paper is organized as follows: Formal definitions for the Hamming and Jaccard distance measures are given in Sect. 2. Then, in Sect. 3, Hamming diversity and Jaccard diversity are introduced. Section 4 proposes a change to the Hamming diversity measure. Numerical examples demonstrating the difference between the proposed Hamming diversity measure and the original Hamming diversity measure are given in 5. Section 6 presents an experiment that mimics set-based agents converging to a solution and discusses the difference in diversity between the two Hamming diversity measures. Lastly, Sect. 7 concludes the paper.

2 Discrete-Valued Distance Measures

This section defines the Hamming distance measure in Sect. 2.1 and the Jaccard distance measure in Sect. 2.2. Both measures calculate a distance value that indicates how similar two discrete-valued representations are.

2.1 Hamming Distance

The Hamming distance is an indication of the minimum number of substitutions required to change one string into another string of equal length, by counting the number of positions where corresponding symbols differ [4]. This can generally be applied to vectors of equal length, so long as the vectors are bit vectors. The Hamming distance between vectors $\boldsymbol{u}$ and $\boldsymbol{v}$, each of length n, is calculated as

$$d_H(\boldsymbol{u}, \boldsymbol{v}) = \frac{d'_H(\boldsymbol{u}, \boldsymbol{v})}{n} \tag{1}$$

where

$$d'_H(\boldsymbol{u}, \boldsymbol{v}) = \sum_{j=1}^{n} \iota(u_j, v_j) \tag{2}$$

and

$$\iota(u, v) = \begin{cases} 1 \text{ if } u \neq v \\ 0 \text{ if } u = v \end{cases} . \tag{3}$$

The Hamming distance produces a maximum value of 1.0 when all components between $\boldsymbol{u}$ and $\boldsymbol{v}$ are in disagreement. The Hamming distance is 0.0 when all the components are in agreement.

2.2 Jaccard Distance

Jaccard distance measures the dissimilarity between two sets as the complement of the cardinality of their intersection divided by the cardinality of their union [5]. Formally, the Jaccard distance measure is

$$J(A, B) = 1 - \frac{|A \cap B|}{|A \cup B|} \tag{4}$$

where A and B are two sets.

The Jaccard distance is equal to 1.0 when A and B are disjoint. Moreover, when A and B are identical, d_J is equal to 0.0.

3 Diversity Measures for Sets

This section builds upon the previous section and formally introduces the Hamming and the Jaccard diversity measures for set-based population-based optimization algorithms. Hamming diversity is defined in Sect. 3.1 and Jaccard diversity is defined in Sect. 3.2.

3.1 Hamming Diversity

Hamming diversity is the average Hamming distance between all set pairs in S. Note that the Hamming distance is defined for vectors of the same length of discrete valued elements. In order to apply the Hamming distance to sets, set-based representations have to be converted to vector-based representations. This is done by applying a mapping function, ω, to the set which is defined as

$$\omega : S \subseteq U \rightarrow X^n \tag{5}$$

where $X = \{0, 1\}$.

Given $S = \{u_1, u_2, \ldots, u_n\}$, the corresponding bit vector is

$$\omega(S) = (V_1, V_2, \ldots, V_n) \tag{6}$$

where

$$V_j = \begin{cases} 1 \text{ if } U_j \in S \\ 0 \text{ otherwise} \end{cases} . \tag{7}$$

The following illustrates the use of ω to convert the set $\{1, 2, 5\}$ in the universe of the first five natural numbers to a bit vector. The length of the bit vector is the size of the universe. Each index in the bit vector corresponds to an element in the universe. For each element in the set $\{1, 2, 5\}$, the corresponding bit vector values are set to 1 and all other values are set to 0, since they are not present in the set. Therefore, the bit vector representation of set $\{1, 2, 5\}$ is

$$\omega(\{1, 2, 5\}) = (1, 1, 0, 0, 1) .$$

Diversity using Hamming distance, or simply, Hamming diversity is

$$\bar{d}_H = \frac{\sum_i^{n-1} \sum_{j=i+1}^{n} d_H(\omega(S_i), \omega(S_j))}{\sum_i^{n-1} \sum_{j=i+1}^{n} 1} . \tag{8}$$

It should also be noted that a downside to the Hamming diversity is that each set has to be converted to a larger $|U|$-dimensional vector where U is the set universe.

3.2 Jaccard Diversity

Jaccard diversity is the average Jaccard distance between all set pairs in a given population, S. Formally, Jaccard diversity is calculated as

$$\bar{d}_J = \frac{\sum_i^{n-1} \sum_{j=i+1}^{n} d_J(S_i, S_j)}{\sum_i^{n-1} \sum_{j=i+1}^{n} 1} \tag{9}$$

where n is the cardinality of S, and S_i and S_j are the sets corresponding to indices i and j, respectively.

4 Improved Hamming Diversity

The Hamming diversity measure under represents diversity because it takes into account the entire set universe. The poor performance is not necessarily a result of the Hamming diversity per se, but rather the mapping function ω used to convert a set into a bit vector. For example, consider the sets {1, 2, 3} and {4, 5} in the universe of the first 10 natural numbers, i.e. $\{1, 2, \ldots, 10\}$. The sets differ in that the elements they contain are different. However, the sets are similar in that they exclude common elements. To better illustrate this point, the bit vectors of the aforementioned sets are

$$\omega(\{1,2,3\}) = (1,1,1,0,0,0,0,0,0,0)$$

and

$$\omega(\{4,5\}) = (0,0,0,1,1,0,0,0,0,0) \; .$$

One can see that even though the sets are totally different from one another, the bit vectors bare some resemblance. This resemblance results in a Hamming diversity value of 0.5 when intuitively we would expect a value of 1.0. Furthermore, as the size of the set universe increase, the Hamming diversity values for totally different sets will decrease.

This paper proposes a simple adjustment to the mapping function ω that uses the union of the sets to determine the bit vectors. Formally, the proposed mapping function ω^* is

$$\omega^* : S \subseteq (A \cup B) \rightarrow B^n \tag{10}$$

where A an B are sets and $X = \{0, 1\}$.

Using ω^*, the bit vectors of {1, 2, 3} and {4, 5} are

$$\omega^*(\{1,2,3\}) = (1,1,1,0,0)$$

and

$$\omega^*(\{4,5\}) = (0,0,0,1,1) \; .$$

The bit vectors above are arguably more intuitive as they better represent what is the same and what is different between the two sets. The Hamming diversity measure using ω^*, denoted as $\bar{d}^*_H$, produces a value of 1.0 for {1, 2, 3} and {4, 5} - equal to the value produced by the Jaccard diversity measure for the same sets. This makes sense because $\bar{d}^*_H$, like Jaccard diversity, measures the elements of sets that differ with respect to the size of union of the sets. Another benefit of using ω^* is that each set no longer has to be converted to a larger $|U|$-dimensional vector.

Examples accompanied by discussions of results are given in the next section to showcase the usefulness of this change. It is expected that the change will retain the three properties described about the Hamming diversity measure in Sect. 1 while introducing the fourth property ascribed to Jaccard diversity.

5 Numerical Examples

The purpose of this section is to illustrate the application of the proposed change to the Hamming diversity measure on special cases. The examples also serve to highlight the differences in diversity values between $\bar{d}_H$ and $\bar{d}^*_H$. Jaccard diversity, which has been shown to accurately represent the diversity of sets, is included as a baseline diversity measure.

The examples show sets that start at maximum diversity progressively moving to a minimum diversity. Furthermore, the results of the examples are also discussed. The examples in this section are all based on the scenario in which three set-based agents attempt to converge to a known optimum of {1, 2} in a set universe that contains the first 10 natural numbers, i.e. $U = \{1, 2, \ldots, 10\}$. The search space consists of all permutations of set elements of all possible cardinalities. Each agent is initialized such that the diversity of the population S is at a maximum by ensuring the elements of each agent is unique. The agents then move progressively closer towards the optimal set by changing one element at a time to better resemble the optimal set. The purpose of these examples is to mimic a scenario wherein diverse agents move over a number of iterations to collapse on a target set, i.e. $\{1, 2\}$.

5.1 Example 1 - Sets at Maximum Diversity

The first example in this scenario uses

$$S_0 = (\{1, 3, 10\}, \{4, 6, 8\}, \{2, 5, 7, 9\})$$

where S_t is the collection of sets at time step t. Each set in S_t corresponds to an agent. Through inspection, the reader can deduce that the contents of each set differs from one another. The diversity measures are

$$\bar{d}_J(S_0) = 1.0$$

$$\bar{d}^*_H(S_0) = 1.0$$

$$\bar{d}_H(S_0) = 0.6667\ .$$

The $\bar{d}^*_H$ value, equal to $\bar{d}_J$, accurately represents the diversity of sets when maximum diversity is present. Thus, achieving the purpose set out in the beginning of the paper. By comparison, $\bar{d}_H$ under represents diversity for reasons discussed in Sect. 4.

5.2 Example 2 - Sets that Are Slightly Similar

This example shows the set-based agents from the previous example beginning to converge to the optimal set. Here, S_1 is $(\{1, 2, 10\}, \{1, 6\}, \{2, 5, 7\})$. The diversity measures are

$$\bar{d}_J(S_1) = 0.85$$

$$\bar{d}_H^*(S_0) = 0.85$$

$$\bar{d}_H(S_1) = 0.4 \; .$$

The $\bar{d}_H^*$ value has decreased as the sets have become more similar but is also twice that of $\bar{d}_H$, and reasonably so. There is some similarity between the sets, but enough to justify a value of 0.4.

5.3 Example 3 - Sets that Are Extremely Similar

Two of the set-based agents are shown to have converged on the optimal set, with $S_2 = (\{1,2\}, \{1,2\}, \{1,2,7\})$. The values for the diversity measures are

$$\bar{d}_J(S_2) = 0.2221$$

$$\bar{d}_H^*(S_0) = 0.2221$$

$$\bar{d}_H(S_2) = 0.0667 \; .$$

Both $\bar{d}_J$ and $\bar{d}_H^*$ indicate that there is some dissimilarity between the sets, while $\bar{d}_H$ suggests that the sets have almost converged. The extremely low $\bar{d}_H$ is a result of more components being in an agreement once the sets have been converted to 10-dimensional bit vectors.

5.4 Example 4 - Sets at Minimum Diversity

The last example shows that all the set-based agents have converged on the optimal set. Here, S_3 contains $(\{1,2\}, \{1,2\}, \{1,2\})$. The Jaccard diversity and the Hamming diversity values are

$$\bar{d}_J(S_3) = 0.0$$

$$\bar{d}_H^*(S_0) = 0.0$$

$$\bar{d}_H(S_3) = 0.0 \; .$$

All values are in agreement that S_3 has converged, as expected.

6 Computational Results

The previous section showed that Hamming diversity better represents the diversity of sets when using the updated mapping function. This section presents a computational process to give the reader a sense of how Hamming diversity values differ over time as a result of which mapping function is used. The computational process mimics set-based agents moving through a search space to a randomly generated optimal solution and is further explained in Sect. 6.1. Section 6.2 describes the settings for the computational process. Comments about the results are given in Sect. 6.3.

6.1 Process

This process simulates set-based agents converging on a randomly chosen optimal set. During the simulation, changes are made to the agents to increase the similarity between the agents. These changes are chosen at random by selecting one of two options: The first is to randomly remove an element not within the optimal set. The second option is to add a randomly chosen element from the optimal set to the agent if the element is not already present. If one of these options is not applicable, then the other is used. If neither of the options is applicable, then the agent is equal to the optimal set. Furthermore, the agents are randomly initialized such that the intersection between all sets is empty. In other words, the sets are initialized such that the diversity of the population is maximized. It is therefore expected that the diversity decreases from the maximum diversity to the minimum diversity.

6.2 Settings

Each agent was randomly initialised with a random (from a uniform distribution in [1,10]) number of elements. The sizes for the optimal set, were 1, 5 and 10. The optimal set size is denoted as O_n where n is the size. For each scenario, a population of 10 was used, and a universe of 100 elements, i.e. $U = \{1, 2, \ldots, 100\}$. The diversity values for each time step are recorded over 30 independent runs and are graphed.

6.3 Experimental Results

Readers familiar with [2] will notice that the graphs in Fig. 1 look similar to those comparing Jaccard and Hamming in the aforementioned work. This makes sense as $\bar{d}_H^*$ and Jaccard diversity produce equal values given the same input. The graphs in this paper show that $\bar{d}_H^*$ better represents the diversity of sets, which are initially totally different, progressively move toward a single solution. Given the results of this synthetic, but realistic, experiment, one can conclude that $\bar{d}_H^*$ can help researchers better understand the search behaviour of set-based optimization algorithms.

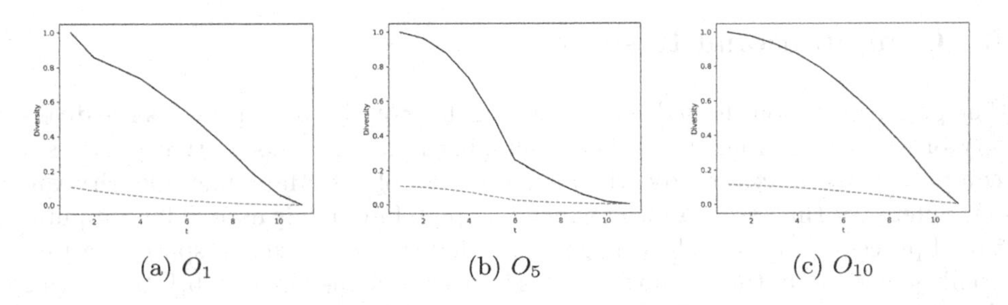

(a) O_1 (b) O_5 (c) O_{10}

Fig. 1. Comparison of $\bar{d}_H^*$ (full-line) and $\bar{d}_H$ (dashed-line) as set-based agents converge to variously sized optimal solutions

7 Conclusion

Diversity measures allow researchers to better understand the exploration and exploitation of population-based optimization algorithms. This paper focused on a set-based diversity measure, namely Hamming diversity. The Hamming diversity measure uses a function, ω, to convert a set to a bit vector so that the diversity of sets can be calculated using Hamming distance. However, ω takes into account the entire set-universe when converting sets to bit vectors. This results in an under representation of diversity. This paper investigated an alternative set-to-bit vector conversation process that results in Hamming diversity values that are equal to those of Jaccard diversity values - where Jaccard diversity has been shown to accurately represent the diversity of sets. This work is useful as researchers now have two set-based diversity measures to make use of, and depending on the context of the problem, it may be more suitable to use one over the other. In general, however, there is a cost to using Hamming diversity as it requires sets to be converted to bit vectors. Future work can investigate Hamming diversity for population that use a variable-length bit vector encoding.

References

1. Cheng, S., Shi, Y.: Diversity control in particle swarm optimization. In: IEEE Symposium on Swarm Intelligence, pp. 1–9 (2011). https://doi.org/10.1109/SIS.2011.5952581
2. Erwin, K., Engelbrecht, A.P.: Diversity measures for set-based meta-heuristics. In: Proceedings of the 7th International Conference on Soft Computing Machine Intelligence, pp. 45–50 (2020)
3. Gong, D., Sun, J., Miao, Z.: A set-based genetic algorithm for interval many-objective optimization problems. IEEE Trans. Evolut. Comput. **22**(1), 47–60 (2018). https://doi.org/10.1109/TEVC.2016.2634625, conference Name: IEEE Transactions on Evolutionary Computation
4. Hamming, R.W.: Error detecting and error correcting codes. Bell Syst. Tech. J. **29**(2), 147–160 (1950)
5. Jaccard, P.: Etude comparative de la distribution florale dans une portion des Alpes et des Jura. Bulletin de la Société vaudoise des sciences naturelles **37**, 547–579 (1901)
6. Langeveld, J., Engelbrecht, A.P.: Set-based particle swarm optimization applied to the multidimensional knapsack problem. Swarm Intell. **6**(4), 297–342 (2012)
7. Olorunda, O., Engelbrecht, A.P.: Measuring exploration/exploitation in particle swarms using swarm diversity. In: Proceedings of the IEEE Congress on Evolutionary Computation, pp. 1128–1134 (2008)
8. Ruiz-Torrubiano, R., Suarez, A.: Hybrid approaches and dimensionality reduction for portfolio selection with cardinality constraints. IEEE Comput. Intell. Mag. **5**(2), 92–107 (2010)
9. Woodside-Oriakhi, M., Lucas, C., Beasley, J.: Heuristic algorithms for the cardinality constrained efficient frontier. Eur. J. Oper. Res. **213**(3), 538–550 (2011)

A Mapping Solution Search Garden Balsam Optimization for Solving Invariant Point Problems

Xiaohui Wang[1] and Shengpu Li[2](✉)

[1] College of Computer, Pingdingshan University, Pingdingshan 467002, China
[2] College of Information Engineering, Pingdingshan University, Pingdingshan 467002, China
Lsp1519@163.com

Abstract. Mapping solution search (MSS), which maps current solution to a mapping solution, increases population diversity and promotes algorithm without the difficulty of premature convergence. This paper presents an MSS based garden balsam optimization (MGBO). This avoids the premature convergence of the algorithm, improves the convergence speed of the algorithm, and increases the possibility that the solution is closer to the global optimum. To evaluate the performance of MGBO, four complex invariant point problems are chosen from the literature. Experimental studies show that the MGBO can solve these problems with great precision compared with some state-of-the-art algorithms.

Keywords: Mapping solution search · Garden balsam optimization · Invariant point problem · MGBO

1 Introduction

In mathematics, an invariant point (also called as a fixed point) of a function is an element such that its function value is equal to itself. People have made extensive and in-depth research for the existence of invariant points for nonlinear system. In the past years, variety of methods have been proposed, and researchers have published abundant valuable invariant point problems through literature areas [1–6]. As the complexity of the problem increases, the traditional method may not find the ideal invariant points. To reduce the above difficulties, many evolutionary algorithms have been proposed. During research on evolutionary algorithms, researchers are often inspired by nature [7].

Garden balsam optimization (GBO) [8–10] is a population-based stochastic optimization algorithm inspired by the seed transmission mode of garden balsam. GBO has shown good performance in solving real problems, but it still suffer from local optima in strongly multi-modal problems. The drawbacks of GBO are slow convergence and premature convergence. In the process of swarm intelligence algorithm evolution, many excellent evolutionary techniques have been applied to improve algorithm performance.

Y. Tan et al. (Eds.): ICSI 2022, LNCS 13344, pp. 48–58, 2022.
https://doi.org/10.1007/978-3-031-09677-8_4

The main contributions of this paper are as follows:

1. This paper proposed a new evolutionary technique, namely mapping solution search (MSS).
2. Motivated by the advantages of GBO and MPS, we propose a new hybrid method that combines GBO with MSS (MGBO). MGBO is applied to solving four real invariant point problems, and the results obtained are compared with those using Artificial Bee Colony Algorithm(ABC), Particle Swarm Optimization (PSO), Teaching-Learning-Based Optimization (TLBO), Differential Evolution (DE), and GBO to evaluate the performance of proposed algorithm.

The rest of the paper is organized as follows: In Sect. 2, we discuss the background of GBO, MSS technique and invariant point problems. In Sect. 3, the MSS is applied to GBO. In Sect. 4 gives the experimental results and discussions on MGBO. Finally, some conclusions are summarized in Sect. 5.

2 Background

2.1 Invariant Point Problem

In this section, we will introduce the invariant point problems.In mathematics, an invariant point (also called as a fixed point) of a function is a point that is mapped to itself by the function. A nonlinear equation with periodic inputs can, in general, be described by the following equations:

$$f : x \to f(x) \tag{1}$$

where $x \in \mathbb{R}$ is a input value.

We denote the actual input values by x and the nominal values by $f(x)$. The purpose of steady-state solution is to obtain deviations $\Delta = f(x) - x \to 0$. When the actual parameter values are known, the dynamical systems can be identified and calibrated as the nominal values.

Generally, these design specifications define a set in the solution space called the feasible solution set R_c and can be defined as:

$$R_c = \{\mathrm{x} \in \mathbb{R} | f(x) = x\} \tag{2}$$

For example, let: f(x) be a real function on $\mathbb{R}$ and: $\mathrm{f(x)} = \mathrm{x}^2 - 2\mathrm{x} + 2$, then x = 1 and x = 2 are the invariant points of: f(x) because f(1) = 1 and f(2) = 2. It's worth noting that not all functions have invariant points. For example, f(x) = x does not have invariant points, because:f(x) is never equal to x, but some functions may have more than one invariant points, and the set of all these invariant points is called an invariant set.

An iterative method for invariant point equation $f(x) = x$ is the recursive relation $x_{r+1} = f(x_r)$, $r = 0, 1, 2, ..., iter_{max}$ with some initial guessx_0, where $iter_{max}$ represents the maximum number of iterations. The algorithm stops when the stopping criterion is met. The stopping criterion can be defined as: $r > iter_{max} or |x_{r+1} - x_r| < \varepsilon$.

Theorem 1. If f is continuous on [a, b] and f(x) ∈ [a, b] for all x ∈ [a, b], then f has an invariant point in [a, b].

Proof. (See Sect. 2 of [11]).

Theorem 2. If $f(x)$ and its derivatives are continuous, f(x) = x and $f'(x) < 1$, then there is an interval $I = [x - \delta, x + \delta], \delta \geq 0$, such that the iterative scheme $x_{r+1} = f(x_r)$ converges to x for every $x_0 \in I$. Further, if $f'(x) \neq 0$, then the convergence is linear. Alternatively, if $f'(x) = f''(x) = \cdots = f^{(p-1)}(x) = 0$ and $f^{p}(x) \neq 0$, then the convergence is of order p.

Proof. (See theorem 2.4 of [11]).

From Theorems 1 and 2, it follows that the iterative scheme is convergent if there exists $\delta > 0$, such that

$$\left|f'(x)\right| < 1, \forall x \in (x - \delta, x + \delta) \tag{3}$$

However, finding an interval $I = [x - \delta, x + \delta]$ is difficult.

Let $g(x) = f(x) - x$, if x' is a root of the function $g(x)$ then $f\left(x'\right) - x' = 0$.

Lemma *1*:f(x) has an invariant point at x' *iff* g(x) = f(x) − x has a root at x'.

According to the lemma1, the invariant point problem is transformed into an unconstrained optimization problem, and defined as:

Minimize $y = (f(x) - x)^2$, subject to $x \in \mathbb{R}$.

And the global minimum value of y is 0.

2.2 Mapping Solution Search

With the development of EAs, many new evolutionary techniques are proposed, such as oppositional-based learning(OBL) [12], space transformation search (STS) [13], Artificial Bee Colony algorithm [14] etc. Inspired by STS, Mapping solution search MSS is proposed to improve the robustness and efficiency of meta-heuristic algorithms. The idea behind of MSS is mapping current solution to a mapping solution. Then the fitness values of the two solutions were calculated and compared respectively, and the solutions with better fitness values were retained. MSS technology increases population diversity and promotes algorithm without the difficulty of premature convergence. If we are searching for x, and if we agree that searching the mapping solution could be beneficial, then calculating the mapping solution x^* is the first step.

Definition 1-Let x be a solution in the current search space, $x \in [a, b]$, where as x^* is the mapping solution, the relationship is defined as follows

$$x^* = \Delta - x \tag{4}$$

where $\Delta = a + b$ is a computable value, a and b represent the limits. It is obvious that x and x^* are on the symmetry of Δ.

Then, the mapping solution x^* is closer to the global optimum x^o than the current solution x only if the following conditions are met

$$\left|x^o - x^*\right| < \left|x^o - x\right| \tag{5}$$

Hence,

$$\left(x^{o} - x^{*}\right)^{2} - \left(x^{o} - x\right)^{2} < 0 => \left(\Delta - 2x^{o}\right)(\Delta - 2x) < 0 \tag{6}$$

It is obvious that the x^* is closer to the x^o than the x, when x^o and x are located on the different sides of $\frac{\Delta}{2}$.

Analogously, the mapping solution in a multidimensional case can be defined.

Definition-Let $X = (x_1, x_2, \ldots, x_D)$ be a solution in the D-dimensional search space, $x_j \in \left[a_j, b_j\right]$.

The mapping solution X^* is defined by its coordinates $x_1{}^*, x_2{}^*, \ldots, x_D{}^*$ where

$$x_j{}^* = \Delta_j - x_j,\ j = 1, 2, \cdots, D \tag{7}$$

where $\Delta_j = a_j + b_j$, a_j and b_j represent the limits of j-th direction.

Mapping solution search-Let $X = (x_1, x_2, \ldots, x_D)$ and $X^* = (x_1{}^*, x_2{}^*, \ldots, x_D{}^*)$ be the current solution and its mapping solution by MSS, respectively and (X) be a fitness function to measure the quality of a solution. If $f(X)$ is better than $f(X^*)$, is survived for the next generation, otherwise X is replaced by X^*.

2.3 Garden Balsam Optimization

The GBO algorithm is a new swarm intelligence algorithm, first posed by Li et.al. in 2020 [8]. GBO was inspired by the transmission of garden balsam seeds. After maturation, the seeds of garden balsam are scattered around the mother by their mechanical force, and the size of the mechanical cracking force is related to the growth environment of the plant. As the machinery spreads out, a portion of the seed changes its position again under the influence of external forces.

It can be seen that in each iteration of the algorithm, mechanical propagation operator, secondary propagation operator, and selection strategy are executed successively until the end condition is met. In the process of seed propagation, if seed transboundary behavior occurs, the mapping rule should be executed.

3 Proposed Hybrid Garden Balsam Optimization

The MGBO algorithm is implemented as follows:

3.1 Initialize a Population

Initially, MGBO generates a uniformly distributed initial population of N_{init} seeds where each seed $x_i (i = 1, 2, ..., N_{init})$ is a D-dimensional vector. Here D is the number of variables in the optimization problem and x_i represent the i-th seed in the population. Each seed corresponds to the potential solution of the problem under consideration. Each x_i is initialized as follows:

$$x_i^k = x_{LB}^k + U(0, 1) \times \left(x_{UB}^k - x_{LB}^k\right) \tag{8}$$

where, x_{LB}^kandx_{UB}^k are bounds of x_i in k-th direction, $k = 1, 2, \cdots, D$, and $U(0, 1)$ represents a random number uniformly distributed in the range [0,1].

3.2 Mechanical Transmission

The number of seeds produced by the parent x_i :

$$S_i = \frac{f_{max} - f(x_i)}{f_{max} - f_{min}} \times (S_{max} - S_{min}) + S_{min} \tag{9}$$

S_i represents the number of seeds produced by the i-th parent; $f(x_i)$ represents the fitness value of the i-th parent, f_{max} and f_{min} are the maximum and minimum values of the fitness function of the current population, respectively; S_{max} and S_{min} are the maximum and the minimum number of seeds produced by garden balsam, respectively.

Seed diffusion range is as follows:

$$A_i = (\frac{iter_{max} - iter}{iter_{max}})^n \times \frac{f_{max} - f(x_i)}{f_{max} - f_{min}} \times A_{\text{init}} \tag{10}$$

when $iter = iter_{max}$ or $f(x_i) = f_{max}$. Here $iter$ represents the current number of iterations, $iter_{max}$ represents the maximum number of iterations.

3.3 Second Transmission

In the real world, individual seeds will be randomly transmitted to other places by the influence of natural forces such as animals, running water, and wind to increase the population diversity.

Its manifestation is as follows:

$$x'_{i1} = x_B + F(x_{i2} - x_{i3}) \tag{11}$$

where, x_{i1} is the target individual, x_B is the optimal solution, F is the zoom factor, x_{i2} *and* x_{i3} are the solutions of two dissimilar individuals.

3.4 Mapping Solution by MSS

In this step, the new population is generated using MSS. After all new seeds have found their positions in the solution space, the new mapping solution is created using formula (7), and each member is evaluated.

3.5 Competitive Exclusion Rules

There is a maximum limit for population size within a specific region. When the population reaches its maximum (N_{max}), individuals with poor fitness will be eliminated in competition within the population.The rule is to rank all individuals in the current population according to the fitness value, retain individuals with good fitness values (elite solutions), randomly select the remaining individuals, and eliminate excess individuals. The number of elite solutions (N_{best}) is calculated according to formula (12) and rounded up to an integer.

$$N_{best} = \frac{iter}{iter_{max}} N_{max} \tag{12}$$

3.6 Termination Condition

The above five steps are repeated until the termination condition is met. In the process of transmission, seeds may fall outside the scope of feasible areas. Such kind of seeds is meaningless, and they must be pulled back to the feasible area according to certain rules. The MGBO handles this situation using random mapping rule. That is, the out-of-bounds seeds are mapped using formula (13), which guarantees that all individuals remain in the feasible space.

$$x_i^{k'} = x_{LB}^k + U(0,1) \times \left(x_{UB}^k - x_{LB}^k\right) \tag{13}$$

where, x_{UB}^k, x_{LB}^k and $U(0,1)$ are the same as in the Eq. (8).

4 Experimental Results and Discussions

An extensive empirical study has been conducted to verify the behavior and performance of MGBO. In the study, four intricate invariant point problems commonly used in research are chosen. Basic information ranges of variables and regarding definitions are shown in Table 1.

Table 1. Invariant point problems.

Function	Definition	Range
G1	$f_1(x) = x^2/4000 - \cos(x) + 1 = x$	[−20, 20]
G2	$f_2(x) = x^2 - 10\cos(2\pi x) + 10 = x$	[−20, 1]
G3	$f_3(x) = 20 + e - 20e^{-0.2\sqrt{x^2}} - e^{\cos(2\pi x)} = x$	[1, 22]
G4	$f_4(x) = 418.9829 - x\sin(\sqrt{x}) = x$	[400, 500]

4.1 Parameters Setting

To prove the efficiency of MGBO algorithm, it is compared with GBO algorithm, four state-of-art algorithms, namely PSO [15], DE [16], ABC [17], and TLBO [18]. The comparative algorithm selected in the experiment has been previously used by different people in attempts to solve various constrained optimization problems [19–25]. For the experiments, same stopping criteria and maximum number of function evaluations are used for the six algorithms.

The termination conditions are a maximum number of iterations (100) and a number of fitness evaluations (240000). The remaining parameters of comparison algorithms are considered to be the same as in the original work mentioned above. The parameters of GBO and MGBO in the experiment are shown in [9].

Table 2. The obtained statistical results of considered algorithms for four invariant point problems.

Fun		G1	G2	G3	G4
ABC	x*	1.3540E−14	1.0000	19.9187	490.0310
	Best	0.8334E−15	0.0000	1.3078E−13	7.2134E−13
	Mean	1.0995E−14	1.0428E−10	3.2026E−11	4.7434E−12
	Std	1.2608E−10	3.3387E−09	1.0447E−06	2.8323E−10
	Rank	4	5	3	4
PSO	x*	1.3540E−07	− 2.1491E−06	20.0982	490.0312
	Best	1.8334E−14	4.6227E−12	2.7032E−11	1.1157E−10
	Mean	1.0995E−10	1.8584E−09	2.8323E−10	3.4669E−08
	Std	1.2147E−10	3.9784E−09	5.0332E−10	7.3857E−08
	Rank	5	6	4	5
TLBO	x*	− 5.6791E−06	7.8456E−07	20.8259	490.0312
	Best	3.2252E−11	6.1534E−13	1.5609E−07	1.3141E−09
	Mean	3.7622E−09	4.1623E−10	1.2475E−06	5.6612E−07
	Std	4.1853E−09	5.2538E−10	1.5884E−06	6.6350E−07
	Rank	6	4	6	6
DE	x*	1.9380E−13	0.9950	20.8258	490.0312
	Best	3.7560E−26	1.8851E−14	1.8859E−10	3.2312E−25
	Mean	7.3610E−21	1.5942E−12	9.0160E−09	7.2716E−24
	Std	1.2669E−20	2.0479E−12	1.4106E−08	7.2716E−24
	Rank	2	3	5	1
GBO	x*	−1.2549E−10	1.0000	20.0983	490.0312
	Best	1.2782E−20	1.1907E−20	5.4308E−19	6.9917E−17
	Mean	2.6298E−18	6.0087E−17	7.1187E−18	1.0647E−16
	Std	2.7852E−18	2.9006E−16	5.2251E−18	2.5084E−16
	Rank	3	2	2	3
MGBO	x*	-7.1856E−23	1.0000	21.7043	490.0312
	Best	2.3341E−44	0.0000	0.0000	3.4239E−20
	Mean	5.7904E−39	4.7689E−39	6.7905E−29	3.0887E−17
	Std	2.5462E−38	2.7681E−38	4.9073E−29	2.7942E−17
	Rank	1	1	1	2

4.2 Numerical Analysis

In this study, the proposed WGBO algorithm was applied to four complex invariant point problems. The function G1 have invariant point at zero. The function G2 have invariant point at the vicinity of 490. The function G3 has invariant points at 0 and 1. The function G4 has more than one invariant point.

Targeting at the four invariant point problems, the results, consisting of an optimal solution (x), the mean value ('Mean'), best value ('Best') and standard deviation ('Std') for 100 independent runs, are showed in Table 2 and are compared with other algorithms.

From the Table 2, WGBO found the exact invariant points for the problems G2 and G3.No other algorithm found an exact invariant point. For the problems G1 and G4, WGBO found better approximations of the invariant points than the other algorithms.

Furthermore, to prove the overall efficiency of WGBO, the other algorithms are ranked with WGBO. The algorithm with the optimal mean value is set to 1, and the next best performing algorithm is set to 2, and so on. From the Table 3,the best performing algorithm is WGBO, as it achieves the lowest value for Rank.

To test the convergence speed of WGBO, the convergence characteristics for the given functions for all algorithms are plotted in Fig. 1. These figures indicate that WGBO is faster than the other algorithms and is quite stable in approaching fixed points for all functions.

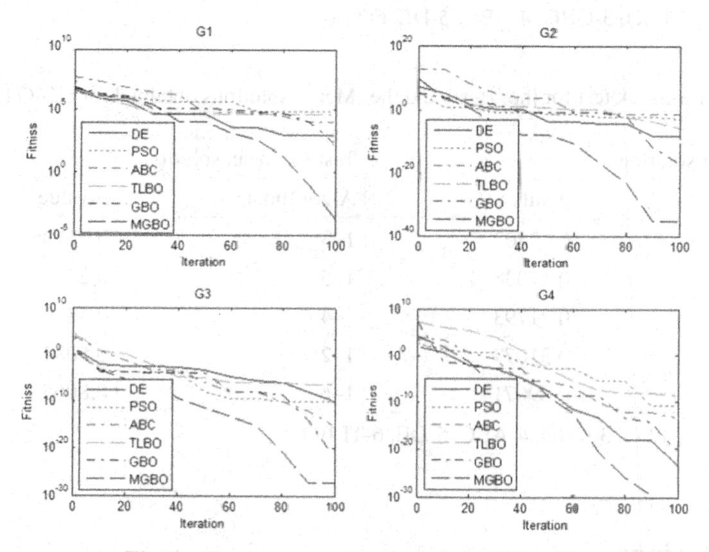

Fig. 1. Convergence characteristics of G1–G4.

It can be seen from the results in Table 2 that MGBO algorithm outperforms other competitive algorithms in performance. However, a t – test with a significance level of 0.05 is necessary to prove the significance of the proposed algorithm.The test reports are given in Table 3,which contains p-values and *h*. When h = 1, the MGBO performed better than contrast algorithms; when h = 0, it works the other way around. Furthermore, if the p-value is less than 0.05, then H0 is rejected so the MGBO performed better than contrast algorithms; otherwise, H0 is accepted. From Table 4, it can be seen that the proposed algorithm performed better than contrast algorithms.

In addition, Holm-Sidak test is necessary to prove the significance of the proposed algorithm. Holm-Sidak test as a post-hoc test method can be used to determine statistical differences between algorithms. Table 4 shows the Holm-Sidak test results in which the 'best' and 'mean' solutions are obtained on G1–G4 function. The p-values obtained by all the algorithms from Holm-Sidak test show the statistical difference between the proposed garden balsam optimization algorithm and other algorithms.

Table 3. Comparative results for four invariant point problems using t-test.

Algorithms	G1		G2		G3		G4	
	p-value	*h*	*p*-value	*h*	*p*-value	*h*	*p*-value	*h*
1–2	1.1883E−06	1	6.7992E−03	1	1.3683E−03	1	6.5408E−03	1
1–3	2.0614E−07	1	8.9306E−03	1	2.1547E−03	1	2.7036E−04	1
1–4	1.3664E−07	1	8.5067E−07	1	1.9084E−10	1	5.1453E−08	1
1–5	9.1782E−04	1	1.8437E−05	1	2.0117E−04	1	2.2794E−05	1
1–6	1.3408E−06	1	1.3954E−05	1	1.5906E−05	1	3.5227E−06	1

* a 1-MGBO, 2-PSO, 3-GBO, 4-ABC, 5-DE, 6-TLBO

Table 4. Holm-Sidak test for the 'Best' and the 'Mean' solutions obtained for G1–G4 functions.

Test for best solution		Test for mean solution	
Algorithma	p-value	Algorithma	p-value
1–3	0.02407	1–3	0.07105
1–5	0.17038	1–5	0.27082
1–4	0.21793	1–4	0.48108
1–2	0.31077	1–2	0.50871
1–6	0.89071	1–6	0.80653

* a 1-MGBO, 2-PSO, 3-GBO, 4-ABC, 5-DE, 6-TLBO

5 Conclusions

In the paper, we have proposed the new evolutionary technique MSS, which maps current solution to a mapping solution. The MSS method has been applied to garden balsam optimization, and proposed the novel hybrid algorithm MGBO to solve complex invariant point problems. Experimental studies show that MGBO behaves well on four selected specific invariant point problems with great precision compared with some classical intelligence algorithms. The performance of this algorithm in solving other optimization problems needs to be verified by subsequent experiments.

References

1. Hirstoaga, S.A.: Iterative selection methods for common fixed point problems. J. Math. Anal. Appl. **324**(2), 1020–1035 (2020)
2. Pakkaranang, N., Kumam, P., Cho, Y.J.: Proximal point algorithms for solving convex minimization problem and common fixed points problem of asymptotically quasi-nonexpansive mappings in cat(0) spaces with convergence analysis. Numer. Algorithms **78**(3), 827–845 (2018)
3. Abed, S.S., Hasan, Z.: Convergence comparison of two schemes for common fixed points with an application. Ibn AL-Haitham J. Pure Appl. Sci. **32**(2), 81(2019)
4. Chidume, C.E., Romanus, O.M., Nnyaba, U.V.: An iterative algorithm for solving split equality fixed point problems for a class of nonexpansive-type mappings in banach spaces. Numer. Algorithms (4), 1–21 (2019)
5. Ogbuisi, F.U., Mewomo, O.T.: On split generalised mixed equilibrium problems and fixed-point problems with no prior knowledge of operator norm. J. Fixed Point Theory Appl. **19**(3), 2109–2128 (2017). https://doi.org/10.1007/s11784-016-0397-6
6. Spaces, B.R.H.: Weak and strong convergence theorems for the multiple-set split equality common fixed-point problems of demicontractive mappings. J. Func. Spaces **2017**(2), 1–11 (2017)
7. Berdahl, A., Torney, C.J., Ioannou, C.C., Faria, J.J., Couzin, I.D.: Emergent sensing of complex environments by mobile animal groups. Science **339**(6119), 574–576 (2018)
8. Li, S., Sun, Y.: A novel numerical optimization algorithm inspired from garden balsam. Neural Comput. Appl. **32**(22), 16783–16794 (2018). https://doi.org/10.1007/s00521-018-3905-3
9. Li, S., Sun, Y.: Garden balsam optimization algorithm. Concurr. Comput. Pract. Exper. **32**(2), e5456 (2020)
10. Li, S., Sun, Y.: Predicting ink transfer rate of 3D additive printing using EGBO optimized least squares support vector machine model. Math. Probl. Eng. (2020)
11. Watson, G.A. (ed.): Numerical Analysis. LNM, vol. 773. Springer, Heidelberg (1980). https://doi.org/10.1007/BFb0094158
12. Tizhoosh H.R.: Opposition-based learning: a new scheme for machine intelligence. In: Proceedings international conference on computational intelligence for modelling control and automation, CIMCA2005, vol. 1, pp. 695–701. Vienna, Austria (2005)
13. Naidu, Y.R., Ojha, A.K.: A space transformational invasive weed optimization for solving fixed-point problems. Appl. Intell. **48**(4), 942–952 (2017). https://doi.org/10.1007/s10489-017-1021-1
14. Yu, W., Wang, J.: A new method to solve optimization problems via fixed point of firefly algorithm. Int. J. Bio-Inspired Comput. **11**(4), 249–256 (2018)
15. Kennedy, J., Eberhart, R.C.: Particle swarm optimization. In: IEEE International Conference on Neural Networks, Piscataway, NJ, vol. 4, pp. 1942–1948 (1995)
16. Storn, R., Price, K.: Differential evolution-a simple and ecient heuristic for global optimization over continuous spaces. J. Global Optim **11**(4), 341–359 (1997)
17. Karaboga D., Basturk B.: A powerful and ecient algorithm for numerical function optimization: artificial bee colony (ABC) algorithm. J. Glob. Optim. **39**(3), 459–471(2007)
18. Rao, R.V., Savsani, V.J., Vakharia, D.P.: Teaching-learning-based optimization : a novel method for constrained mechanical design optimization problems. Comput. Aided Des. **43**(3), 303–315 (2011)
19. Parsajoo, M., Armaghani, D.J., Asteris, P.G.:. A precise neuro-fuzzy model en-hanced by artificial bee colony techniques for assessment of rock brittleness in-dex. Neural Comput. Appl. **34**(4), 3263–3281 (2022)

20. Tien Bui, D., et al.: Spatial prediction of rainfall-induced landslides for the Lao Cai area (Vietnam) using a hybrid intelligent approach of least squares support vector machines inference model and artificial bee colony optimization. Landslides **14**(2), 447–458 (2016). https://doi.org/10.1007/s10346-016-0711-9
21. Boudane, F., Berrichi, A.: Multi-objective artificial bee colony algorithm for parameter-free neighborhood-based clustering. Int. J. Swarm Intell. Res. (IJSIR), **12**(4), 186–204 (2021)
22. Wu, H. , Huang, Y., Chen, L. , Zhu, Y., Li, H.: Shape optimization of egg-shaped sewer pipes based on the nondominated sorting genetic algorithm (nsga-ii). Environ. Res. **204**, 111999 (2022)
23. Li, M., Wang, L., Wang, Y., Chen, Z.: Sizing optimization and energy management strategy for hybrid energy storage system using multi-objective optimization and random forests. IEEE Trans. Power Electron. **36**(10), 11421–11430 (2021)
24. Yang, G., Cao, Y., Tao, H.: A method for multi-objective optimization and application in automobile impact. J. Phys. Confer. Ser. **1802**(3), 032129 (5pp) (2021)
25. Chogueur, O., Bentouba, S., Bourouis, M.: Modeling and optimal control applying the flower pollination algorithm to doubly fed induction generators on a wind farm in a hot arid climate. J. Sol. Energy Eng. **143**(4), 1–26 (2021)

Parallel Symbiotic Lion Swarm Optimization Algorithm Based on Latin Hypercube Distribution

Zongxin Han and Mingyan Jiang(✉)

School of Information Science and Engineering, Shandong University, Qingdao 266237, China
jiangmingyan@sdu.edu.cn

Abstract. In order to optimize the search process of lion swarm optimization algorithm, improve the accuracy of the result and save the time, we propose the parallel symbiotic lion swarm algorithm based on Latin Hypercube. First, Latin Hypercube is used to initial population. Secondly, the mutualistic symbiosis mechanism is proposed to increase the communications between agents. Then, reverse learning strategy of dimensional keyhole imaging is added to increase diversity of the optimization process. Finally, on the premise of ensuring the quality of the solution, the parallel computation is combined with the improved lion swarm optimization algorithm to improve the convergence rate. Several benchmark functions are used to evaluate the performance of the improved lion swarm algorithm, and the running time of the parallel algorithm is compared. Experimental results show that the efficiency, stability and convergence rate of the improved lion swarm algorithm are greatly improved.

Keywords: Lion swarm optimization algorithm · Parallel computing · Latin hypercube distribution · Mutualism · Dimensional keyhole imaging reverse learning

1 Introduction

In recent years, with the expansion of engineering field, optimization problems also increase gradually. Using as few resources as possible to solve various problems in the best way is a hot topic in engineering fields.

To solve the optimization problem, the researchers invented evolutionary computing. Rechenberg Schwefel proposed Evolution Strategies (ES) [1]. With the continuous development of evolutionary computing, swarm intelligence algorithms gradually become the main branch of evolutionary computing. Kennedy proposed Particle Swarm Optimization Algorithm (PSO) by simulating birds' behavior [2]. The Whale Optimization Algorithm (WOA) proposed by Mirjalili, uses random individuals or optimal individuals to simulate the hunting behavior of whales [3]. Artificial Bee Colony Algorithm (ABC) is an algorithm inspired by bee colony behavior, which was proposed by Karaboga [4]. In addition, the Lion Swarm Optimization Algorithm (LSO) with self-organization and robustness is one of the better algorithms proposed in recent years.

Y. Tan et al. (Eds.): ICSI 2022, LNCS 13344, pp. 59–69, 2022.
https://doi.org/10.1007/978-3-031-09677-8_5

In the field of LSO, there have been some research progresses. For example, Wang B proposed a lion swarm optimization on the idea of the genetic algorithm [5]. LSO is an emerging swarm intelligence optimization algorithm proposed by Shengjian Liu in recent years [6]. Improving the theoretical framework of LSO makes it have a good optimization mechanism and efficiency, reliability, stability and robustness, which is the current required research work.

As the scale of the heuristic algorithm increases, the running time of the algorithm gradually increases, so the parallel evolutionary computation begins to appear [7]. In the process of development, parallel computing is divided into two models: island model and master-slave model. The research on the island model mainly has two directions. The first is to gradually extend parallel computing to broader evolutionary algorithms [8]. The second is to deploy the parallel algorithm to different hardware architectures and explore the influence of different hardware architectures on the algorithm, including multi-core CPU, GPU and FPGA.

2 Parallel Computing

2.1 Coarse-Grained Parallel Computing and Fine-Grained Parallel Computing

Coarse-grained parallel computing can use a few of processors to divide a large population into several groups. Each sub-population contains individuals with different division, and each sub-population is assigned to a processor, which can independently perform serial computing in parallel. After several iterations, the local optimum in the population is transferred to all the other neighboring sub-populations. In the exchange of local optimum, two processes of sending and receiving are carried out in parallel [9].

When the number of processors in the parallel system is large enough, the parallel system can allocate one processor to each agent. But above situation is ideal. Each sub-population is assigned one processor, and these subgroups are guaranteed to have a lot of frequent communications, ensuring that the optimal fitness is transmitted to each processor [10].

2.2 Master-Slave Parallel Computing and Island Parallel Computing

The master-slave model and island model is shown in Fig. 1. In coarse-grained parallel model, there is a master processor that stores the entire population, and assigns population individuals to multiple slave processors for evaluation. The master-slave parallelization method is to leave the global operations to the processor that controls the algorithm process. In this form of parallel intelligent optimization algorithm, there is only one population.

In island model, each sub-population is an island that contains multiple individuals. The process of the algorithm is as follows: each island runs a serial algorithm independently of each other, and an agent exchange takes place between the islands at a specific interval [11]. It is found that the accuracy of the island model is higher than that of the original evolutionary algorithm in specific application scenarios, so the paper studies the parallel algorithm of multiple sub-populations. Traditionally, each process is associated with one CPU core [12].

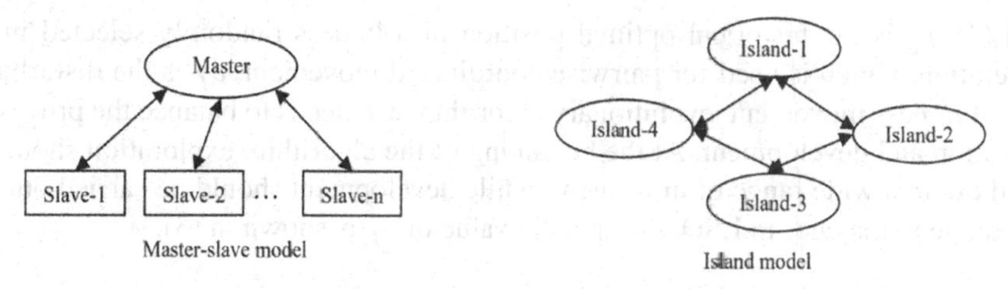

Fig. 1. Master-slave model and island model.

3 Research on Lion Swarm Optimization Algorithm

3.1 Lion Swarm Optimization Algorithm

Shengjian Liu proposed LSO by modeling the hunting behaviors of lion swarm in nature. Funston, P. J proposed the specific way of lion swarm hunting behavior through a large number of observations: lion swarm in nature consists of adult male lion, lionesses and lion cubs [13]. Male lion accounts for a small proportion of the lion swarm, and their hunting behaviors are less. Lionesses account for a large number of lions and are the main force of hunting. Due to small size of adult lions, lion cubs usually wander randomly or hunt with adult lions.

The individual lion is abstracted as the independent variable, and the prey is abstracted as the fitness of the function. According to the fitness of the lion, LSO divides the lion population into three types: male lion, lionesses and lion cubs. Among them, there is only one male lion, which represents the global optimum. The male lion and lionesses are adult lions, and the proportion of adult lions in the population is η, while the proportion of lion cubs is $1 - \eta$.

Take solving the minimum value of the function: let the dimension of the function be D, and there are SN lions in the population, where the position of the *i-th* lion in the *t-th* iteration is $x_i^t = (x_{i1}, x_{i2}, \ldots, x_{iD})$, $1 \leq i \leq SN$. The specific movements of lions can be divided into three categories:

The Male Lion Moves Slightly. According to the behavior of the male lion, we set the male lion to make small movement near the global optimal, as shown in (1):

$$x_{opt}^{t+1} = \left(1 + \gamma p_i^t - g^t\right) g^t \tag{1}$$

where, γ is a random number generated according to the standard Gaussian distribution $N(0, 1)$; g^t represents the optimal position in the whole population at the *t-th* iteration; p_i^t represents the historical best position of the *i-th* individual in the *t-th* iteration.

The Lionesses Move in Coordination. In the lion swarm algorithm, a group of individuals with lower fitness form lioness swarm, and move in pairs, as shown in (2):

$$x_i^{(t+1)} = \frac{\left(1 + \gamma \alpha_f\right)\left(p_i^t + p_c^t\right)}{2} \tag{2}$$

In (2), p_c^t is the historical optimal position of a lioness randomly selected in the *t-th* iteration, which is used for pairwise coordinated movement. α_f is the disturbance factor of lioness movement, evolutionary algorithm also needs to balance the process of exploration and development. At the beginning of the algorithm, exploration should be carried out in a wide range of movement, while development should be carried out in a small range at the end. In LSO, the specific value of α_f is shown in (3):

$$\alpha_f = step \times \exp\left(\frac{-30t}{T}\right)^{10} \tag{3}$$

In (3), T is the maximum number of iteration; the specific value of step is $step = 0.1(\overline{high} - \overline{low})$, where $\overline{high}$ and $\overline{low}$ respectively represent the mean of the maximum and minimum of the value range of each dimension.

The Lion Cubs Imitate Movement. In LSO, lion cubs are the individual with large fitness, which are mainly used for global optimization. In order to increase the randomness of the lion cubs' behaviors, the movement mode of the lion cubs is randomly selected among the three movement modes: following the male lion, following the lionesses and wandering randomly, as shown in (4):

$$x_i^{t+1} = \begin{cases} \frac{(1+\gamma\alpha_c)(p_i^t+g^t)}{2}, & q \le \frac{1}{3} \\ \frac{(1+\gamma\alpha_c)(p_i^t+p_m^t)}{2}, & \frac{1}{3} < q \le \frac{2}{3} \\ \frac{(1+\gamma\alpha_c)\left(p_i^t+\overline{g^t}\right)}{2}, & \frac{2}{3} < q \le 1 \end{cases} \tag{4}$$

In (4), α_c is the disturbance factor of lion cubs movement, which is similar to lionesses, and decreases with the increase of iteration. In the algorithm, $\alpha_c = step \times (T - t)/T$, q is a random number generated by uniformly distributed $U(0, 1)$. If $q \le \frac{1}{3}$, lion cubs move to the current global optimum. If $\frac{1}{3} < q \le \frac{2}{3}$, lion cubs move with the lionesses, p_m^t represents the historical optimal position of the lionesses in the *t-th* iteration; and $\overline{g^t} = \overline{high} + \overline{low} - g^t$ if $\frac{2}{3} < q \le 1$. The authors suggest that the value of η should be between 0.2 and 0.3, and the values of α_c and α_f are related to the problem to be optimized [6].

3.2 Latin Hypercube Distribution

The Latin Hypercube (LHC) method is a stratified stochastic process proposed by M.D. McKay [14]. In the LHC, to generate n points, the spatial distribution of each variable is divided into n equal intervals, within which the LHC randomly generates a point. These points can be generated as (5):

$$x_i = \frac{1}{nr} + \frac{i-1}{n} \tag{5}$$

In (5), r is a random number from 0 to 1, x_i is the *i-th* point generated in the *i-th* interval. And LHC remembers formation interval of the point that can help to get a representative set of points, increasing the diversity of the population.

3.3 Symbiotic Search Algorithm

Symbiotic Organisms Search Algorithm (SOS) is a heuristic search algorithm based on biological symbiosis, which simulates the interaction between individuals [15]. Common symbiotic relationships include mutually beneficial symbiotic relationships, partial benefit symbiotic relationships and parasitic relationships. This paper guides the excellent symbiosis mechanism in SOS, strengthens the information interaction of the agents in the lion swarm algorithm.

A new individual is obtained through the interaction between two individuals, and the interaction formulas are shown as (6), (7), (8):

$$x_i^{t+1} = x_i^t + rand(0, 1) \times \left(x_{best}^t - R_{MV} * bf1\right) \tag{6}$$

$$x_j^{t+1} = x_j^t + rand(0, 1) \times \left(x_{best}^t - R_{MV} * bf2\right) \tag{7}$$

$$R_{MV} = \left(x_i^t + x_j^t\right)/2 \tag{8}$$

In (6), (7), R_{MV} represents the interaction relationship between two agents, $rand(0, 1)$ is a random number between 0 and 1, x_{best}^t is the optimal individual. $bf1$ and $bf2$ are benefit factors, representing the benefit level obtained by the individual from the relationship. $bf1$ and $bf2$ can be set to 1 or 2, indicating partial or full benefits.

An agent is randomly selected to interact with the current agent, so that only the current agent can benefit, the formula is as follows (9):

$$x_i^{t+1} = x_j^t + rand(-1, 1) \times \left(x_{best}^t - x_j^t\right) \tag{9}$$

Copy and mutate the current agent. If the fitness of the mutated agent is better than that of the mutated agent, the formula is as follows (10):

$$R_{PV}(pick) = rand(1, length(pick)) * (ub(pick) - lb(pick)) + lb(pick) \tag{10}$$

In (10), pick is the variant, ub is the upper bound of search, lb is the lower bound of search.

3.4 Dimensional Keyhole Imaging Reverse Learning

In view of the problem that the algorithm falls into local optimum in the process of optimization, we use reverse learning strategy in the improvement of the intelligent optimization algorithm, which makes the results more accurate [16]. The formulas created by the reverse learning method based on the creation of keyhole imaging are as follows (11), (12):

$$x_i^{t+1} = \frac{\left(\overline{ub} + \overline{lb}\right)}{2} + \frac{\left(\overline{ub} + \overline{lb}\right)}{2k} - \frac{x_i^t}{k} \tag{11}$$

$$k = \left(1 + \left(\frac{t}{T}\right)^{\frac{1}{2}}\right)^{10} \tag{12}$$

In the (11), (12), x_i^t is the component of the current individual in the $i\text{-}th$ dimension, x_i^t is the current iteration, and x_i^t is the maximum iteration.

3.5 Ideas and Formulas for Algorithm Improvement

Improvement Way of Thinking and Counter Plan. The stagnation prevention mechanism and disturbance mechanism are added to avoid falling into local optimum. Hybrid algorithms are designed to combine two or multiple algorithms to enhance the overall performance.

The Latin Hypercube strategy is introduced through (5) to change the initial population value. In the process of location updating, lionesses updating formula (2) is combined with symbiosis mechanism, so that lionesses updating formula as (13):

$$x_i^{t+1} = \begin{cases} p_i^t + rand(0,1) \times \left(g^t - \frac{(p_i^t + p_c^t)}{2}\right), q \le \frac{1}{2} \\ p_i^t + rand(0,1) \times \left(g^t - (p_i^t + p_c^t)\right), \frac{1}{2} < q \le 1 \end{cases} \quad (13)$$

The adjustment factor λ and the reverse learning strategy of keyhole imaging are introduced, and the improved lion cubs' formulas are as follows (14), (15):

$$x_i^{t+1} = \begin{cases} \frac{(1+\gamma\alpha_c)(p_i^t + g^t)}{2}, q \le \frac{\lambda}{2} \\ \frac{(1+\gamma\alpha_c)(p_i^t + p_m^t)}{2}, \frac{\lambda}{2} < q \le \lambda \\ \frac{\left(\overline{high} + \overline{low}\right)}{2} + \frac{\left(\overline{high} + \overline{low}\right)}{2} - \frac{x_i^t}{k}, \lambda < q \le 1 \end{cases} \quad (14)$$

$$\lambda = \frac{T}{5t + T} \quad (15)$$

$p_i^t g^t, p_m^t, p_i^t, \overline{high}, \overline{low}, k, \gamma, \alpha_c$ have the same meaning as formulas (1) (2) (3) (4) (12).

In the early of the algorithm, the lion cubs perform more local optimization; more lion cubs perform the reverse learning strategy in the end. And if the proportion of adult lions change, there will still be a certain number of lion cubs to carry out the reverse learning strategy in the end, which improves the robustness of the algorithm and ensures the stability of the algorithm.

Pseudocode. The steps of the improved lion swarm optimization algorithm (ILSO) are as follows:

Step1: Initialization, set the maximum number of iterations T and the adult lion proportion η; randomly generate *SN* lion individuals. Then, the lions are divided into male lion, lionesses and lion cubs according to the fitness. Finally, the initial historical optimal position of the individual is set as the current position, and the initial global optimal position is set as the male lion position.
Step2: According to formulas (1), (13) and (14), the movement of male lion, lionesses and lion cubs is completed respectively.
Step3: Lion swarm update, update the fitness of all individuals; update p_i^t and g^t based on the fitness.
Step4: Algorithm termination, repeat Step2 through Step3 until $t = T$, the global optimal individual is obtained and its fitness is calculated.
Step5: Output lion position and optimal fitness.

3.6 Computational Results and Conclusion

PSO is an uncertain algorithm that reflects natural organisms. It has more opportunities to solve the global optimum, and also has self-organization and evolution. WOA is a new type of intelligent optimization algorithm. Its local search uses the shrinkage and encirclement mechanism, which has good universality and has a good effect on some test functions. In order to verify the effect of ILSO proposed in this paper, the ILSO, LSO, PSO and WOA are simulated for some basic test functions. The number of populations is 60, the number of experimental iterations is 500, the dimension of the test function is $D = 15$.

The fitness curve after running the eight benchmark functions with the four algorithms is shown in Fig. 2. The optimal fitness of the simulation is shown in Table 1. From Fig. 2 and Table 1, it can be seen that PSO converges quickly in the early stage, but the convergence rate is slow and the results are poor because it cannot jump out of the local optimum; WOA has slow convergence rate, low precision and it is easy to fall into local optimum; although LSO can obtain the optimum, the accuracy of the result also needs to be improved. Compared with other algorithms, ILSO can reach the theoretical optimal fitness in multiple test functions, and after adding the symbiosis strategy and dimension by dimensional keyhole imaging strategy, ILSO can improve the result accuracy and convergence rate. We can conclude from the simulation experiment that ILSO is effective and helpful to the optimization process and the final result.

Table 1. Results of algorithm.

Function		LSO	ILSO	WOA	PSO
F1	Bent Cigar	1.00E+01	**1.00E+01**	1.88E+08	4.09E+08
F2	Sum of different power	1.00E+01	**1.00E+01**	3.80E+23	3.69E+25
F3	Zakharov	3.42E+02	**1.00E+01**	7.05E+01	4.49E+06
F4	Rosenbrock	1.10E+01	**1.00E+01**	1.27E+04	9.60E+06
F5	Rastrigin	1.00E+01	**1.00E+01**	1.54E+02	2.81E+02
F6	Expanded Schaffer	1.00E+01	**1.00E+01**	1.29E+01	1.51E+01
F7	Levy function	1.19E+01	**1.18E+01**	4.91E+01	1.18E+02
F8	Ackley	9.16E-38	**0.00E+00**	1.24E+04	1.31E+04

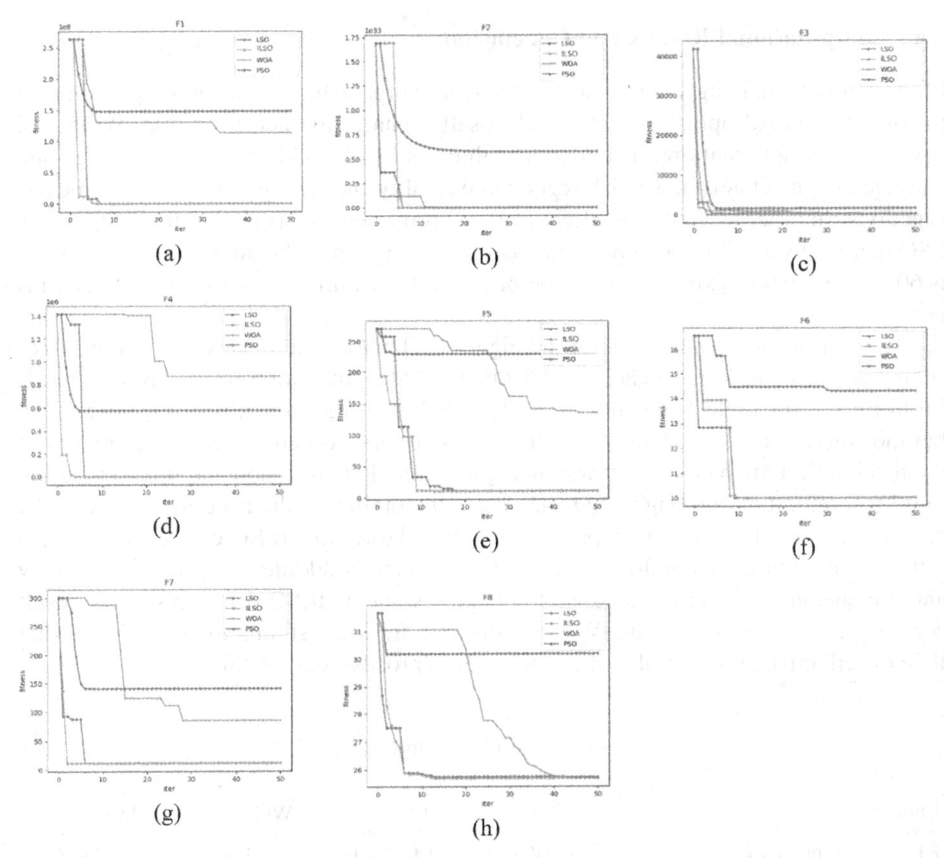

Fig. 2. Comparison charts of test function results of optimization algorithms: (a) F1 (b) F2 (c) F3 (d) F4 (e) F5 (f) F6 (g) F7 (h) F8.

4 Parallel Lion Swarm Optimization Algorithm

4.1 Ideas and Pseudocode

The idea of the parallel lion swarm algorithm is as follows. The size of the total population is set as SN. The parallel algorithm based on the island model divides the total population into N sub-populations on average, and the size of each sub-population is denoted as SN_{sub}. According to the formula (16), we can obtain the value of SN_{sub}:

$$SN_{sub} = \frac{SN}{N} \tag{16}$$

Then each sub-population independently runs LSO. After a specific number of iterations, each sub-population performs male lion exchange until the algorithm terminates. The optimal fitness in all sub-populations is the final solution of the algorithm. The connection between islands is called the topology, and the topologies include Star topology, Ring topology, and Fully connected topology [17].

Another important aspect of island model-based parallel LSO is parallel strategy. Topology is a coarse-grained parallel computing and information communication, and the parallel strategy determines the communication between sub-populations. The parallel strategy needs to be determined in three aspects: the interval of iterations between two migrations, which is called migration cycle R; the number of agents that determines each migration is called migration ratio; migration rules that determine which agents migrate out and which are replaced. When the migration cycle is met, the optimal individual in each sub-population is migrated to the adjacent population, and the worst individual in the adjacent population is replaced. Using the parallel strategy above, suppose that there are N sub-populations, and the communication time between each sub-population can be expressed by formula (17):

$$comm = \frac{N \times T}{R} \tag{17}$$

Table 2 shows the pseudocode of ILSO after the addition of parallel computing:

Table 2. Pseudocode for parallel lion swarm optimization algorithm.

Parallel Symbiotic Lion Swarm Optimization Algorithm Based on Latin Hypercube Distribution
Input: population quantity SN; adult lion proportion η; maximum number of iterations T; sub-population number N; the migration cycle R
1. t=0
2. parallel **for** n=1: N **do**
3. **if** t=0 **then**
4. In sub-population generate initial population;
5. Calculate the fitness;
6. Sort population and distribution;
7. Update the optimal fitness in each sub-population;
8. **end if**
9. **for** i=1: SN_{sub} **do**
10. Update individuals in each sub-population;
11. Next generation optimal fitness is updated;
12. **end for**
13. **if** t % R=0 **then:**
14. The optimal fitness is sent to the adjacent sub-population
15. Replace the worst fitness with the accepted best value
16. **end if**
17. t=t+1
18.until all sub-populations are satisfied t=T
Output: the optimal fitness for all sub-populations $g^t = \min(g_1^T, g_2^T, g_3^T, \ldots, g_N^T)$.

4.2 Experimental Results and Analysis

In this experiment, parallel computing is added to ILSO. The experimental environment is Window10, 32G RMA and 3.0GHz CPU, the simulation platform is Python 3.8. The number of agents in the population is 96, and the number of experimental maximum iterations is 500, the dimension of the test function is $D = 1000$, the number of iterations reaches 300, the migration period is $R = 10$ and other parameters remain unchanged. Experimental results are shown in Table 3.

From Table 3, we can conclude that after adding parallel computing to the LSO, the parallel lion swarm optimization algorithm has the same precision as the LSO, but as the process increases, the optimization ability decreases. This is because the number of agents in the sub-population decreases. It shows that when the total number of agents in the population is constant, as the number of sub-populations increases, the accuracy of the algorithm will decrease. From the above analysis, the parallel lion swarm optimization algorithm can effectively reduce the running time of the algorithm under the condition of ensuring the same accuracy as LSO. Moreover, parallel algorithms have good portability and can be extended to other swarm intelligence optimization algorithms.

Table 3. Results of algorithm.

F	Serial computing		Parallel two processes		Parallel three processes		Parallel four processes	
	Mean	Time/s	Mean	Time/s	Mean	Time/s	Mean	Time/s
1	1.00E+01	46.7	1.00E+01	23.2	1.22E+01	20.6	1.25E+01	16.5
2	1.00E+01	60.9	1.00E+01	33.5	1.18E+01	18.0	1.26E+01	17.6
3	1.01E+01	54.6	1.00E+01	30.7	1.11E+02	25.4	1.14E+01	17.1
4	1.10E+01	70.2	1.00E+01	37.8	1.31E+01	29.7	1.91E+01	18.1
5	1.00E+01	79.0	1.00E+01	40.4	1.11E+01	19.6	1.16E+01	22.2

5 Conclusion

In order to improve the performance of LSO, this paper introduces the Latin Hypercube strategy to make the initial population evenly distributed and increase the diversity of the population. Then the paper introduces the symbiosis mechanism from the Symbiotic Search Algorithm, which increases the collaboration and communication between agents and accelerates the convergence rate of the algorithm. Inspired by the principle of optics, the reverse learning strategy of dimensional keyhole imaging is introduced to improve the ability of the algorithm to jump out of local optimum. Finally, parallel computing is introduced to speed up the calculation speed of the algorithm. After multiple test function solving experiments, the stability of the improved algorithm is verified by the fitness curve, the time spent, and the obtained optimal fitness.

References

1. Dujardin, J., Kahl, A., Lehning, M.: Synergistic optimization of renewable energy installations through evolution strategy. Environ. Res. Lett. **16**(6), 064016 (2021)
2. Jakubik, J., Binding, A., Feuerriegel, S.: Directed particle swarm optimization with Gaussian-process-based function forecasting. Eur. J. Oper. Res. **295**(1), 157–169 (2021)
3. Mirjalili, S., Lewis, A.: The whale optimization algorithm. Adv. Eng. Softw. **95**, 51–67 (2016)
4. Mingyan, J., Dongfeng, Y.: Artificial Bee Colony Algorithm and Its Application. Science Press, BeiJing (2014)
5. Ji, F., Jiang, M.: Lion swarm optimization by reinforcement pattern search. In: Tan, Y., Shi, Y. (eds.) ICSI 2021. LNCS, vol. 12689, pp. 119–129. Springer, Cham (2021). https://doi.org/10.1007/978-3-030-78743-1_11
6. Falei, J., Mingyan, J.: Tabu annealing lion swarm optimization algorithm. In: International Conference on Computer Engineering and Artificial Intelligence 2021, pp. 119–129. IEEE, Shanghai (2021)
7. Song, Y., et al.: MPPCEDE: multi-population parallel co-evolutionary differential evolution for parameter optimization. Energy Convers. Manag. **228**(2), 113661 (2021)
8. Prakash, A., Lal, R.K.: Floorplanning for area optimization using parallel particle swarm optimization and sequence pair. Wirel. Pers. Commun. **118**(1), 323–342 (2021). https://doi.org/10.1007/s11277-020-08015-5
9. Wu, Z., Zhao, C., Liu, B.: Polygonal approximation based on coarse-grained parallel genetic algorithm. J. Vis. Commun. Image Represent. **71**, 102717 (2020)
10. Tasoulas, Z.G., Anagnostopoulos, I.: Kernel-based resource allocation for improving GPU throughput while minimizing the activity divergence of SMs. IEEE Trans. Circuits Syst. I Regul. Pap. **67**, 428–440 (2019)
11. Gozali, A.A., Kurniawan, B., Weng, W., Fujimura, S.: Solving university course timetabling problem using localized island model genetic algorithm with dual dynamic migration policy. IEEJ Trans. Electr. Electron. Eng. **15**(3), 389–400 (2020)
12. Zhang, Y., Tao, L., Wang, C., Ye, L., Sun, S.: Numerical study of icebreaking process with two different bow shapes based on developed particle method in parallel scheme. Appl. Ocean Res. **114**, 102777 (2021)
13. Daoqing, Z., Mingyan, J.: Parallel discrete lion swarm optimization algorithm for solving traveling salesman problem. J. Syst. Eng. Electron. **31**(4), 751–760 (2020)
14. Rosli, S.J., Rahim, H.A., Abdul Rani, K.N., et al.: A hybrid modified method of the sine cosine algorithm using latin hypercube sampling with the Cuckoo search algorithm for optimization problems. Electronics **9**(11), 1786 (2020)
15. Bramerdorfer, G.: Tolerance analysis for electric machine design optimization: classification, modeling and evaluation, and example. IEEE Trans. Magn. **55**(8), 1–9 (2019)
16. Xu, H., Wang, Y.: Whale optimization algorithm for embedded circle mapping and one dimensional oppositional learning based small hole imaging. Control Decis. **36**(5), 1173–1180 (2021)
17. Lalwani, S., Sharma, H., Satapathy, S.C., Deep, K., Bansal, J.C.: A survey on parallel particle swarm optimization algorithms. Arab. J. Sci. Eng. **44**(4), 2899–2923 (2019)

Research on Spectrum Allocation Algorithm Based on Quantum Lion Swarm Optimization

Keqin Jiang and Mingyan Jiang(✉)

School of Information Science and Engineering, Shandong University, Qingdao 266237, China
jiangmingyan@sdu.edu.cn

Abstract. As a famous representative of the NP-Hard problem, the optimization of cognitive radio spectrum allocation has attracted the attention of many scholars. In this paper, a quantum lion swarm optimization (QLSO) algorithm is proposed to solve the problem of spectrum allocation. Firstly, we introduce the basic lion swarm optimization algorithm and cognitive radio network model. Secondly, we introduce quantum coding and order some operators in the QLSO algorithm. Finally, we select several common swarm intelligence algorithms as a comparison and conduct simulation experiments. The experiments on randomly generated spectrum allocation models with different topologies show that the QLSO algorithm has higher solution quality and convergence performance than the other algorithms, such as discrete particle swarm optimization (DPSO) algorithm, genetic algorithm (GA), and binary lion swarm optimization (BLSO) algorithm.

Keywords: Quantum lion swarm optimization (QLSO) algorithm · Cognitive radio · Spectrum allocation · Particle swarm optimization (DPSO) algorithm · Genetic algorithm · Binary lion swarm optimization (BLSO) algorithm

1 Introduction

In recent years, with the vigorous development of wireless communication technology, the number of wireless terminal devices has also increased sharply, resulting in the problem of a lack of spectrum resources. To improve the utilization of spectrum resources, cognitive radio (CR) [1] can be adopted. Cognitive radio can adaptively adjust its internal parameters, and achieve the best spectrum utilization based on reliable network communication. Spectrum allocation refers to allocating the available spectrum to one or more designated users according to the number of users who need to access the spectrum and their service requirements. Its main purpose is to make rational and effective use of spectrum resources and at the same time avoid interference caused by sharing spectrum between cognitive users and authorized users.

At present, scholars like to use intelligent optimization algorithms to solve NP-hard problems such as spectrum allocation, and have achieved good research results in this field. In the field of artificial fish swarm algorithm, literature [2] gives a method to solve the spectrum allocation problem based on the artificial fish swarm algorithm and studies the influence of its internal mechanism on the algorithm convergence. Literature [3]

Y. Tan et al. (Eds.): ICSI 2022, LNCS 13344, pp. 70–81, 2022.
https://doi.org/10.1007/978-3-031-09677-8_6

introduces the concrete process of applying an artificial bee colony algorithm to solve the spectrum allocation problem of cognitive radio, and proves the effectiveness and stability of the algorithm.

Lion Swarm Optimization (LSO) [4] algorithm is a new swarm intelligence optimization algorithm proposed by Liu et al. in 2018, which has the advantages of fast convergence, high optimization precision, and good stability [5]. However, the basic LSO algorithm is mainly used for continuous function optimization. To make the basic LSO algorithm solve the combinatorial optimization problem, some scholars introduced binary coding based on the basic lion swarm optimization algorithm and proposed a BLSO algorithm for solving the 0–1 knapsack problem [6].

However, when the BLSO algorithm is used to solve the spectrum allocation problem of cognitive radio, the convergence speed is slow. Literature [7] proposed a spectrum allocation model based on a quantum genetic algorithm, which effectively reduced the number of iterations and improved the convergence of the algorithm.

Based on the above research, this paper proposes a quantum lion swarm optimization (QLSO) algorithm to solve the spectrum allocation problem of cognitive radio.

2 Fundamental Knowledge

2.1 LSO Algorithm

Lions swarm optimization algorithm is a swarm intelligence algorithm that simulates the behavior of lions [5]. Suppose in the D-dimensional search space, N lions from a group and the position of the jth lion in the D-dimensional search space is denoted as $x_j = (x_{j1}, x_{j2}, \cdots, x_{jD})$.

The lion king updates the position according to formula (1).

$$x_j^{t+1} = g^t(1 + \gamma||p_j^t - g^t||) \tag{1}$$

The lionesses work together to update their positions according to formula (2).

$$x_j^{t+1} = \frac{p_j^t + p_c^t}{2}(1 + \alpha_f \gamma) \tag{2}$$

The cubs update their positions according to formula (3).

$$x_j^{t+1} = \begin{cases} \frac{g^t+p_j^t}{2}(1+\alpha_c\gamma), & 0 < q \le \frac{1}{3} \\ \frac{p_m^t+p_j^t}{2}(1+\alpha_c\gamma), & \frac{1}{3} < q \le \frac{2}{3} \\ \frac{\overline{g}^t+p_j^t}{2}(1+\alpha_c\gamma), & \frac{2}{3} < q < 1 \end{cases} \tag{3}$$

where g^t is the optimal position of the tth generation population; γ is a random number generated according to the normal distribution N(0,1); p_j^t is the historical optimal position of the jth lion in the tth generation; p_c^t is the historical best position of a hunting partner randomly selected from the tth generation of lioness population; α_f is the disturbance factor of female lion's moving range; $\overline{g}^t = \overline{LOW} + \overline{HIGHT} - g^t$ is the jth cub being driven away within the hunting range position, $\overline{LOW}$ and $\overline{HIGHT}$ are the minimum mean and maximum mean of each dimension in the lion's range space respectively; p_m^t is the tth historical best position of the cub following the lioness; α_c is the disturbance factor of the juvenile's moving range; the probability factor q is in accordance with the uniform distribution Uniform random value generated by U[0,1].

The algorithm flow is as follows [8]:

Step 1. Initialization: Initialize parameters T. Randomly generate N lion swarm individuals x_i^k. Firstly, sort according to the fitness value of the individual, and then determine the initial positions of the lion king, lioness, and cubs.

Step 2. Lions move: According to (1), (2), and (3), complete the movement of the lion king, lioness and lion cubs.

Step 3. Lions update: Update the fitness value; update p_j^t and g^t.

Step 4. Algorithm termination: Repeat steps 2 to 3 until k = T; record the optimal fitness value and the optimal individual of the population.

2.2 Cognitive Radio Network Model

Figure 1 is the topology of a cognitive radio network, in which ①–⑤ represents five secondary users, I–IV represents four primary users, and there are three frequency bands available in the cognitive radio network, which are indicated by A, B, and C respectively.

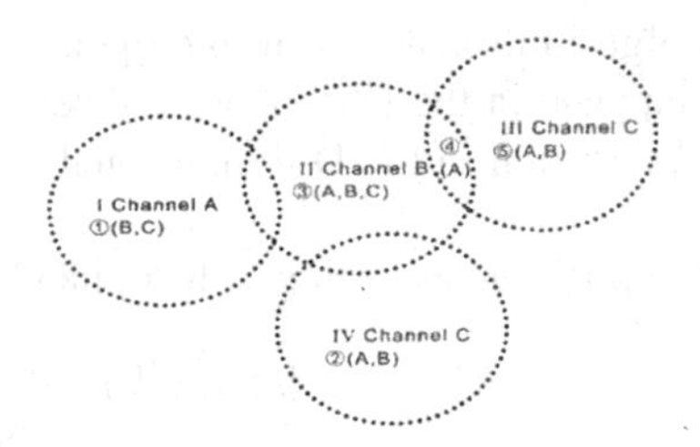

Fig. 1. Topology of cognitive radio network

Users I to I–IV use frequency bands A, B, C, and C respectively. Secondary users of cognitive radio should use frequency bands that are not currently used by primary users to access. If the channel has been assigned to the primary user, nearby secondary users cannot use the channel to avoid interference with the primary user. The circle in Fig. 1 represents the coverage of each major user. If secondary users happen to fall into this circle, then secondary users cannot use this channel. For example, if secondary user ① falls into the circle of primary user 1, secondary user ① can't use channel A, but can only use channels B and C.

2.3 Mathematical Description of Cognitive Radio Graph Theory Model

Assuming that N secondary users are competing for M channels in the cognitive radio network, the availability matrix, reward matrix, interference matrix, and assignment matrix can be defined as follows:

$$L = \{l_{n,m} | l_{n,m} \in \{0, 1\}\}_{N*M} \tag{4}$$

$$B = \{b_{n,m}\}_{N*M} \tag{5}$$

$$C = \{c_{n,k,m} | c_{n,k,m} \in \{0, 1\}\}_{N*N*M} \tag{6}$$

$$A = \{a_{n,m} | a_{n,m} \in \{0, 1\}\}_{N*M} \tag{7}$$

where $l_{n,m} = 1$ indicates that user n can use channel m, and $l_{n,m} = 0$ means that user n can't use channel m. $b_{n,m}$ indicates the benefit obtained by user n when using channel m. $c_{n,k,m} = 1$ indicates that interference will occur when users n and users k use channels m at the same time, and $c_{n,k,m} = 0$ means no interference. If $c = k$, then $c_{n,n,m} = 1 - l_{n,m}$. $a_{n,m} = 1$ indicates that the channel m has been assigned to the user n, and $a_{n,m} = 0$ means that the channel m is not assigned to the user n. The assignment matrix must meet the interference constraint conditions defined by the interference matrix C according to formula (8).

$$a_{n,m} + a_{k,m} \leq 1, c_{n,k,m} = 1, \forall n, k < N, m < M \tag{8}$$

According to the assignment matrix A, the total benefit of the whole cognitive radio network are as follows:

$$U(R) = \sum_{n=1}^{N} \beta_n = \sum_{n=1}^{N} \sum_{m=1}^{M} a_{n,m} \cdot b_{n,m} \tag{9}$$

where β_n is the total benefit obtained by user n.

The purpose of spectrum allocation in cognitive radio is to maximize the total benefit of all users in the system according to the formula (10). While fairness among participating users should be considered an important factor according to formula (11).

$$f1 = \max(U(R)) = \max\left(\sum_{n=1}^{N} \sum_{m=1}^{M} a_{n,m} \cdot b_{n,m}\right) \tag{10}$$

$$f2 = \max\left(U_{fair}\right) = \max\left(\prod_{n=1}^{N} \left(\beta_n + 10^{-4}\right)\right)^{\frac{1}{N}} \tag{11}$$

In order to ensure that the parameter value is not 0, for each β_n, plus a tiny positive number10^{-4}.

The optimization benefit maximization objective function f_1 and fairness objective function f_2 are weighted and summed to solve the maximum function f:

$$f = \omega_1 * f_1 + \omega_2 * f_2, \ \omega_1 + \omega_2 = 1 \tag{12}$$

3 Proposed Method

3.1 Coding Mode of Quantum Lion Swarm

In the QLSO algorithm, the coding of lions is realized by using a quantum bit (qubit), and a qubit can be "0" or "1" or the superposition of both. A qubit can be expressed as follows:

$$|\psi\rangle = \alpha|0\rangle + \beta|1\rangle \tag{13}$$

where α and β represent the probability amplitudes of state $|0\rangle$ and state $|1\rangle$. The probability of a qubit in the "0" state is $|\alpha|^2$, and the probability of a qubit in the "1" state is $|\beta|^2$. The normalization conditions can be expressed as $|\alpha|^2 + |\beta|^2 = 1$.

With qubit coding, the QLSO population with the population of N can be expressed as $x(t) = \{x_1^t, x_2^t, x_3^t, \cdots, x_j^t, \cdots, x_N^t\}$. Where t is the current iteration number, and x_j^t is the j lion in the t generation population, which can be expressed as follows [9]:

$$x_j^t = \begin{bmatrix} \alpha_{j1}^t \Big| \alpha_{j2}^t \Big| \alpha_{j3}^t \Big| \cdots \Big| \alpha_{ji}^t \Big| \cdots \Big| \alpha_{jD}^t \\ \beta_{j1}^t \Big| \beta_{j2}^t \Big| \beta_{j3}^t \Big| \cdots \Big| \beta_{ji}^t \Big| \cdots \Big| \beta_{jD}^t \end{bmatrix}, \left|\alpha_{ji}^t\right|^2 + \left|\beta_{ji}^t\right|^2 = 1 (i = 1, 2, \cdots, D) \tag{14}$$

3.2 Quantum Measurement of Population

The measurement will change the state of the qubit, and make it collapse from the superposition state of state $|0\rangle$ and state $|1\rangle$ to a specific state. A group of states $P(t)$ will be obtained by measuring each individual in the population $x(t)$ once. $P(t)$ is a set of binary solutions. The value of "0" or "1" for each bit is determined according to the value of $\left|\alpha_{ji}^t\right|^2$ or $\left|\beta_{ji}^t\right|^2$ in x_j^t. In a quantum computer, measuring the quantum state will make the system collapse into a single state. However, when the QLSO algorithm runs in a classic computer, the collapse will not happen, so we can choose to randomly generate some [0,1]. If it is greater than $\left|\alpha_{ji}^t\right|^2$, take "1"; otherwise, take "0".

3.3 Renewal of Quantum Population

When the population is initialized, both α_{ji}^1 and β_{ji}^1 in x_j^1 are initialized to $\frac{1}{\sqrt{2}}$, which means that all states are superimposed with the same probability in the initial search. In quantum theory, the transition between states is realized by the quantum gate transformation matrix, so the rotation angle of the quantum revolving door can also represent the variation of quantum chromosomes, and then the information of the best individual can be added to the variation to accelerate the convergence of the algorithm. In QLSO, the revolving door is used as the Q-gate, and the matrix expression is as follows:

$$U(\Delta\theta_i) = \begin{bmatrix} \cos(\Delta\theta_i) & -\sin(\Delta\theta_i) \\ \sin(\Delta\theta_i) & \cos(\Delta\theta_i) \end{bmatrix} \tag{15}$$

Table 1. Adjustment strategy of quantum revolving door

				$S(\alpha_i, \beta_i)$			
x_i	b_i	$f(x_i) \geq f(b_i)$	$\Delta\theta_i$	$\alpha_i\beta_i > 0$	$\alpha_i\beta_i > 0$	$\alpha_i = 0$	$\beta_i = 0$
0	0	F	0	-	-	-	-
0	0	T	0	-	-	-	-
0	1	F	Delta	+1	−1	0	∓1
0	1	T	Delta	−1	+1	∓1	0
1	0	F	Delta	−1	+1	∓1	0
1	0	T	Delta	+1	+1	0	∓1
1	1	F	0	-	-	-	-
1	1	T	0	-	-	-	-

where $\Delta\theta_i(i = 1, 2, \cdots, D)$ is the angle at which each qubit rotates to "0" or "1" (Table 1).

Among them, x_i is the ith dimension of the current particle, b_i is the ith dimension of the target particle, $f(x)$ is the fitness function, $\Delta\theta_i$ is the size of the rotation angle, which controls the convergence speed of the algorithm and $s(\alpha_i, \beta_i)$ is the direction of the rotation angle to ensure the convergence of the algorithm.

The operation of the quantum revolving door can be defined as the operator $\Theta\left(x_j^t, b_k^t\right)$, where x_j^t represents the jth lion in the tth generation population, and b_k^t represents the kth lion selected in the tth generation population.

The quantum NOT gate is a 1-bit quantum gate, and the qubit $|1\rangle$ becomes $|0\rangle$ after passing through the NOT gate, and the qubit $|0\rangle$ becomes $|1\rangle$ after passing through the Not gate. The matrix is expressed as:

$$U = \begin{bmatrix} 0\,1 \\ 1\,0 \end{bmatrix} \tag{16}$$

The quantum Not gate is defined as the operator $\varphi(x_j^t, i)$, where x_j^t represents the jth lion in the tth generation population and i represents the ith dimension of the jth lion. Operator φ means that only x_{ji}^t t is inverted, and the rest remains unchanged.

3.4 QLSO Algorithm

In the process of optimization, the movement of the lion king, lioness and cubs are completed according to

$$x_j^{t+1} = \varphi(g^t, pos) \tag{17}$$

$$x_j^{t+1} = \Theta\left(x_j^t, p_c^t\right) \tag{18}$$

$$x_j^{t+1} = \begin{cases} \Theta(x_j^t, g^t) & 0 < rand(0,1) \le \frac{1}{3} \\ \Theta(x_j^t, p_m^t) & \frac{1}{3} < rand(0,1) \le \frac{2}{3} \\ \Theta(x_j^t, \overline{g}^t) & \frac{2}{3} < rand(0,1) \le 1 \end{cases} \tag{19}$$

where the meanings of the symbol g^t, p_c^t, p_m^t are the same as those in formula (1), (2) and (3). $\overline{g}^t = 1 - g^t$, which is to invert every bit of g^t by bit, which is a typical elite reverse learning thought [10]. φ and Θ are the operators defined before.

The pseudo-code of the QLSO algorithm is as follows:

Algorithm1: QLSO algorithm

Input Population size N; percentage of adult lions β; maximum iteration number T.

1. t←0
2. **while**(t < T)**do**
3. **If** t=0 **then**
4. Randomly generate quantum coding lion swarm $x_j^t (j = 1,2,\cdots,N)$
5. Sort by fitness value $f(x_j^t)$, divide swarm into lion king, lioness and lion cubs.
6. $p_j^t \leftarrow x_j^t$; $g^t \leftarrow arg\max(f(x))$
7. **end if**
8. **for** i = 1 : N **do**
9. **if** x_j^t is the lion king **then**
10. Generate x_j^{t+1} according to (17)
11. **else if** x_j^t is the lioness **then**
12. Generate x_j^{t+1} according to (18)
13. **else if** x_j^t is the lion cubs **then**
14. Generate x_j^{t+1} according to (19)
15. **end if**
16. $p_j^{t+1} \leftarrow \text{argmax}(f(x))$.
17. **end for**
18. $g^t \leftarrow \text{argmax}(f(x))$.
19. t←t+1
20. **end**
21. **Output** Optimal individual of population g^t

3.5 Spectrum Allocation Algorithm Based on QLSO

When the QLSO algorithm is applied to solve the spectrum allocation problem in the cognitive radio system, the position of each lion corresponds to a spectrum allocation solution. For a cognitive radio system with N sub-users and M channels, the location of the lion should correspond to an N*M matrix. With the increase of the number of users N and the number of channels M, the size of the location matrix rapidly increases. To solve this problem, it is necessary to encode the position variable of the lions.

In the previous mathematical description of the spectrum allocation model, if $l_{n,m} = 0$, channel M cannot be allocated to user N, so $a_{n,m} = 0$. In this way, only the elements in the distribution matrix A corresponding to the positions of the elements with a value of 1 in the availability matrix L need to be taken out to form the position vector of the lion group, and the remaining elements with a value of 0 are ignored. A specific codec demonstration diagram [11] is shown in Fig. 2 (N = 5, M = 6), and x is the position vector of the lion needed in the algorithm.

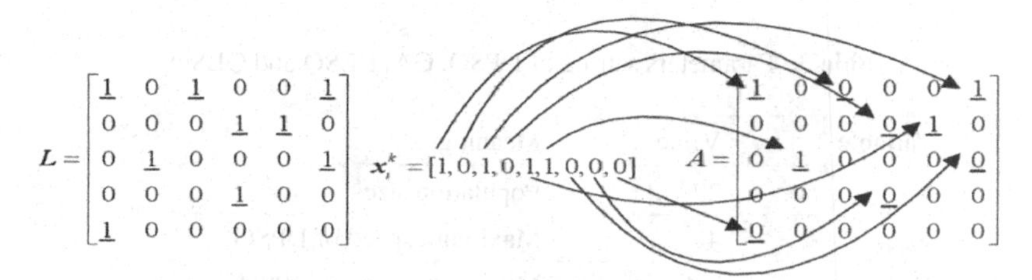

Fig. 2. Schematic diagram of coding

4 Experiments and Results

4.1 Simulation Experiment Parameter Setting

The simulation is a noiseless and static cognitive radio system. Given an area of 10 × 10, k primary users and n secondary users are randomly set, and there are m available channels. Each primary user randomly selects one channel as his authorized channel, and the power coverage of the primary user is d_p. The power coverage of the secondary user is D, which is set between $[d_{min}, d_{max}]$. The values of above parameter are shown in Table 2. Parameters of available matrix L, reward matrix B, and interference matrix C are generated by referring to the pseudo-code in Appendix 1 of reference [12].

The values of algorithms parameter are shown in Table 3. In order to compare the performance of the algorithms in this paper, the algorithms in this paper take the total system bandwidth revenue ($\omega_1 = 1, \omega_2 = 0$), the fairness of cognitive users' access ($\omega_1 = 0, \omega_2 = 1$) and the overall system performance ($\omega_1 = 0.4, \omega_2 = 0.6$) as evaluation functions, and get different algorithms under three evaluation functions according to the formula (12).

Table 2. Parameters setting of simulation

Variable	Value	Meaning
K	20	Primary user number
N	10	Secondary user number
M	10	Free channel number
d_p	2	Primary user's power coverage
d_{min}	1	Minimum secondary user's power coverage
d_{max}	4	Maximum secondary user's power coverage

Table 3. Parameters setting of DPSO, GA, BLSO and QLSO

Variable	Value	Meaning
N	20	Population size
V_{max}	4	Maximum speed of DPSO
V_{max}	−4	Minimum speed of DPSO
c1, c2	2	Learning factor of DPSO
P_c	0.8	Cross probability of GA
P_m	0.01	Variation probability of GA
T	1000	Maximum iteration number

4.2 Experimental Results and Analysis

To verify the performance of spectrum allocation in cognitive radio based on the QLSO algorithm, the DPSO algorithm, GA, and the BLSO algorithm, and QLSO algorithm will be applied to solve the spectrum allocation problem in cognitive radio at the same time, and their experiences will be compared.

Firstly, 10 topological structures are randomly generated based on the cognitive radio parameters set above. Then, the DPSO algorithm, GA, the BLSO algorithm, and the QLSO algorithm are used for spectrum allocation respectively, and their convergence speed is observed.

From Figs. 3, 4, and 5, we can see that the running results of BLSO algorithms and QLSO algorithms are better than those of DPSO algorithms and GA. In 10 different topologies, the running result of the QLSO algorithm is slightly better than that of the BLSO algorithm. From Figs. 6, 7, and 8, it can be found that the convergence speed of the DPSO algorithm and QLSO algorithm is faster than that of the BLSO algorithm and GA. The convergence results of the QLSO algorithm and BLSO algorithm are similar, and the iterative results of these two algorithms are better than those of the DPSO algorithm and GA.

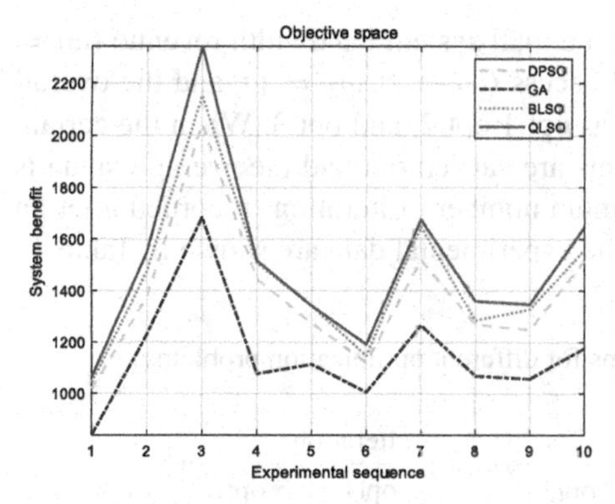

Fig. 3. Comparison chart of system benefit

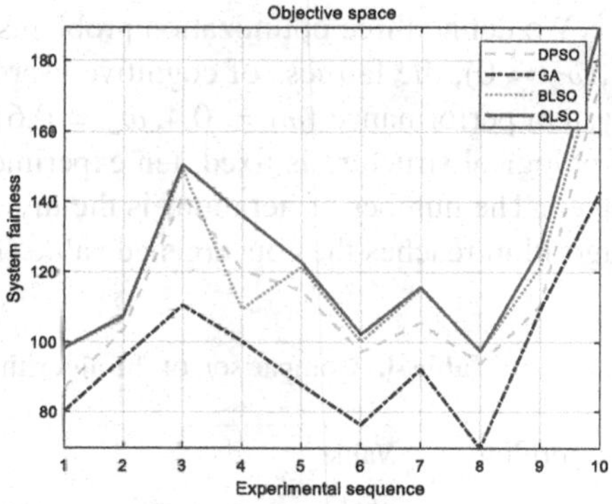

Fig. 4. Comparison chart of system fairness

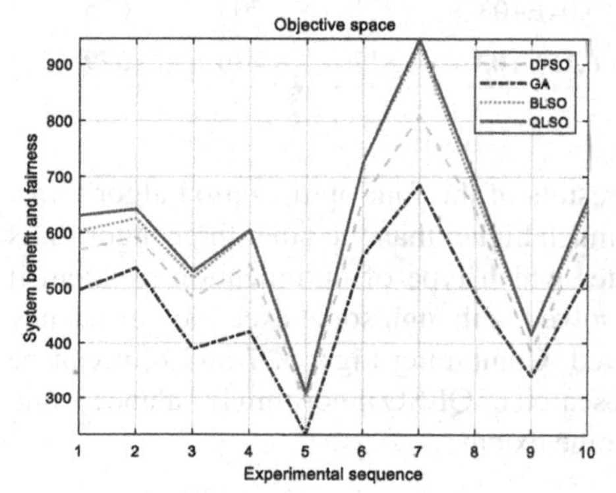

Fig. 5. Comparison chart of system with both benefit and fairness

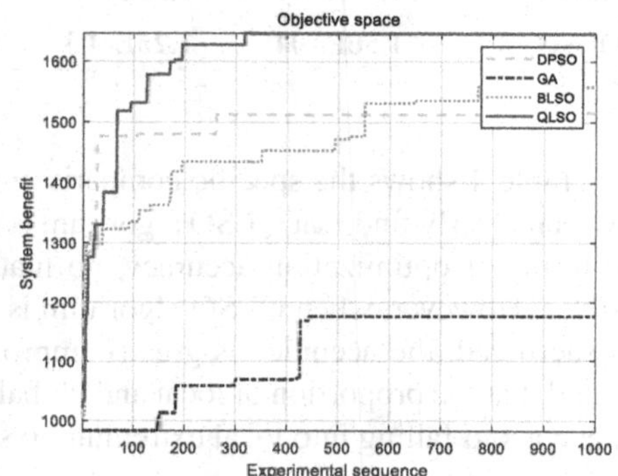

Fig. 6. Iterative process of system benefit

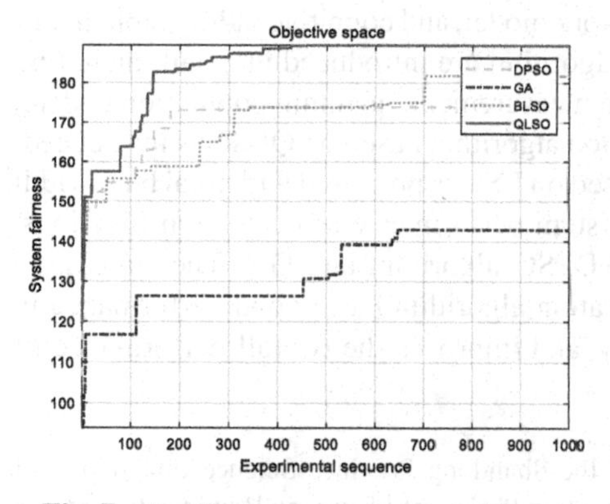

Fig. 7. Iterative process of system fairness

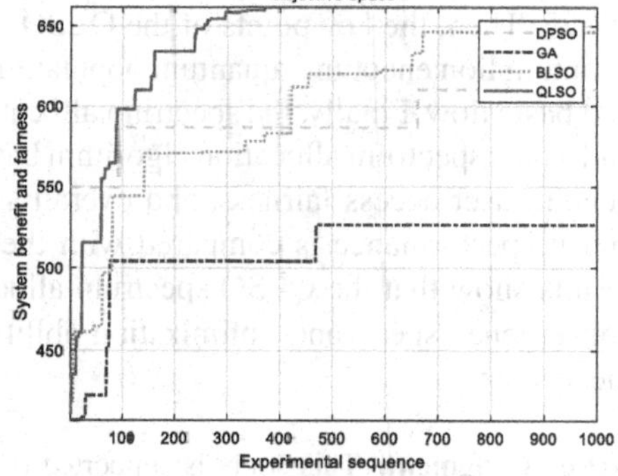

Fig. 8. Iterative process of system with both benefit and fairness

We define three optimization problems: the total system bandwidth revenue ($\omega_1 = 1, \omega_2 = 0$), the fairness of cognitive users' access ($\omega_1 = 0, \omega_2 = 1$), and the overall system performance ($\omega_1 = 0.4, \omega_2 = 0.6$) as opt 1 opt 2, and opt 3. When the certain topological structure is fixed, ten experiments are carried out and the average value is taken. The number of iterations is the minimum number of iterations recorded after an algorithm reaches the convergence value. The experimental data are shown in Table 4.

Table 4. Comparison of the algorithms for different optimization problems

Algorithm	Value			Iteration		
	opt1	opt2	opt3	opt1	opt2	opt3
GA	1.15E+04	9.64E+02	4.67E+03	450	647	469
DPSO	1.38E+04	1.15E+03	5.58E+03	**262**	**186**	662
BLSO	1.45E+04	1.20E+03	6.04E+03	771	705	675
QLSO	**1.50E+04**	**1.25E+03**	**6.15E+03**	317	436	**379**

Table 4 shows the specific comparison results of the four optimization algorithms. We can clearly find that QLSO algorithm is much higher than the other three algorithms in terms of optimization accuracy, no matter which type of optimization problem it solves. However, when QLSO algorithm is solved, although some execution efficiency is sacrificed, the accuracy is greatly improved. Combining Figs. 6, 7 and 8, it can be found that the proportion of local and global search of QLSO algorithm is balanced, and it can avoid falling into local extremum to some extent.

5 Conclusion

Firstly, this paper introduces the mathematical descriptions of the standard lion swarm optimization algorithm, cognitive radio network model, and cognitive radio graph theory model. Then, the key points of the QLSO algorithm are introduced in detail, including quantum lion encoding, quantum population measurement, quantum population update, and basic flow. Finally, the spectrum allocation algorithm based on QLSO is introduced, and a new spectrum allocation algorithm based on LSO is proposed with total bandwidth benefit, user access fairness, and overall system performance as evaluation functions, and its performance is compared with the DPSO algorithm and GA. The simulation results show that the QLSO spectrum allocation algorithm has a good performance in convergence speed and optimization ability, and improves the overall performance of the system.

Acknowledgment. This study is supported by the Shandong Province Science Foundation of China (Grant No. ZR2020MF153) and Key Innovation Project of Shandong Province (Grant No. 2019JZZY010111).

References

1. Mitola, J., Maguire, G.Q.: Cognitive radio: making software radios more personal. IEEE Pers. Commun. **6**(4), 13–18 (1999)
2. Jiang, M., Yuan, D.: Artificial Fish School Algorithm and Its Application. Science Press, Beijing (2012)
3. Jiang, M., Yuan, D.: Artificial Bee Colony Algorithm and Its Application. Science Press, Beijing (2014)
4. Liu, S., Yang, Y., Zhou, Y.: A swarm intelligence algorithm-Lion swarm algorithm. IEEE Pers. Pattern Recogn. Artif. Intell. **31**(5), 431–441 (2018)
5. Guo, Y., Jiang, M.: Job-shop scheduling problem with improved lion swarm optimization. In: Meng, H., Lei, T., Li, M., Li, K., Xiong, N., Wang, L. (eds.) ICNC-FSKD 2020. LNDECT, vol. 88, pp. 661–669. Springer, Cham (2021). https://doi.org/10.1007/978-3-030-70665-4_72
6. Liu, S., Yang, Y., Zhou, Y.: Binary lion swarm algorithm for solving 0-1 Knapsack problem. Comput. Eng. Sci. **41**(11), 2079–2087 (2019)
7. Zhao, Z., Peng, Z., Zheng, S.: Spectrum allocation of cognitive radio based on quantum genetic algorithm, **58**(2), 1358–1363 (2009)
8. Zhang, D., Jiang, M.: Parallel discrete lion swarm optimization algorithm for solving traveling salesman problem. J. Syst. Eng. Electron. **31**(4), 751–760 (2020)
9. Xu, M., Lu, Y., Zhou, J.: An elite quantum wolves algorithm for cognitive radio spectrum allocation. Mod. Electron. Technol. **44**(14), 33–38 (2021)
10. Zhou, X.: Elite opposition-based particle swarm optimization. Acta Electron. Sin. **41**(8), 1647–1652 (2013)
11. Peng, Z., Zhao, Z., Zheng, S.: Spectrum allocation of cognitive radio based on hybrid Shuffled Frog Leading Algorithm. Comput. Eng. **11**, 2079–2087 (2019)
12. Peng, C., et al.: Utilization and fairness in spectrum assignment for opportunistic spectrum access. Mob. Netw. Appl. **11**(4), 555–576 (2006)

Bean Optimization Algorithm Based on Differential Evolution

Yongqiang Hu[1,2(✉)], Ying Li[1,2], Tingjuan Li[1], Jiaqing Xu[4], Hang Liu[3], and Changshun Zhang[1,2]

[1] Qinghai Institute of Science and Technology Information, Xining 810008, China
975462415@qq.com
[2] Qinghai Normal University, Xining 810016, China
[3] Anhui University, Hefei 230601, China
[4] Guilin University of Technology, Guilin 541006, China

Abstract. Inspired by the evolution of natural plant distributions, bean optimization algorithm (BOA) is proposed and become an efficient swarm intelligence algorithm. Aiming at the disadvantage of low efficiency of fine search in BOA, an algorithm (DBOA) is proposed by integrating the mutation and selection operators of differential evolution into BOA. The mutation operator enriches the population diversity and improves the local optimization speed of the algorithm. The selection operator further ensures the evolution direction and enhances the optimization accuracy of DBOA. The proposed DBOA has been tested on a set of well-known benchmark problems and compared with other typical swarm intelligence algorithms. The experimental results show that DBOA effectively improves the accuracy and speed of the BOA and has better performance in solving complex optimization problems.

Keywords: Bean optimization algorithm · Differential evolution · Swarm intelligence · Mutation operator · Fine search

1 Introduction

Swarm intelligence refers to the decentralized and self-organization behavior at the collective level. It is a process in which individuals compete and interact with each other to form an optimal solution. Swarm intelligence algorithm is an effective method to solve complex optimization problems without local information and model [1].

Inspired by natural biological clusters, many excellent swarm intelligence optimization algorithms have been constructed, including classical particle swarm optimization particle swarm optimization (PSO) [2], ant colony optimization (ACO) [3], differential evolution (DE) [4], bean optimization algorithm (BOA) [5], whale optimization algorithm (WOA) [6], etc. Among them, the BOA is inspired by the distribution and evolution of plant population in nature. It has the advantages of simple structure, easy implementation, strong global optimization ability and fast convergence speed, but there are still some deficiencies in the speed of local fine optimization.

Y. Tan et al. (Eds.): ICSI 2022, LNCS 13344, pp. 82–94, 2022.
https://doi.org/10.1007/978-3-031-09677-8_7

BOA has been improved by many scholars and applied to solve some problems since it was proposed. For example, in 2008, the bean optimization algorithm [7] was first proposed, an evolutionary model of population distribution based on piecewise function is proposed. and the simulation experiments of typical optimization problems proved that the algorithm has good optimization performance. In 2010, Wang [8] applied the bean optimization algorithm to the multi-to-multi disaster relief material scheduling model, and obtained a feasible solution with less time and cost. In 2010, Sun [9] applied BOA algorithm to the optimization model based on fuzzy preference relation, and obtained the ranking value of the post-disaster reconstruction and restoration system, which conformed to the principle of the post-earth reconstruction planning in China. In 2011, Zhang [10] constructed an adaptive selection mechanism by using the knowledge action idea based on inverse reasoning induction. The experimental results show that the optimized adaptive selection mechanism has better performance than the basic algorithm. In 2012, Zhang [11] overcome the disadvantage that BOA with continuous distribution function cannot be used to solve discrete optimization problems by increasing population migration and information cross-sharing, the algorithm is applied to the TSP problem, and its feasibility is proved. In 2013, Zhang [12] constructed a Markov model of BOA algorithm. By studying the randomness and convergence of BOA, it is proved that the algorithm meets the global convergence criterion of random search algorithm. Two mechanisms of population migration and cross sharing of information were added, which overcome the disadvantage that BOA with continuous distribution function cannot be used to solve discrete optimization problems. In 2015, Feng [13] applied the population distribution model based on negative binomial distribution to BOA algorithm. In 2017, Zhang [14] introduced chaos theory into BOA algorithm to improve the population distribution of offspring and improve the global search performance of BOA. In 2020, Ali [15] aiming at the lognormal distribution in the population abundance pattern, a lognormal bean optimization algorithm is proposed, which is proved to be globally convergent. In 2021, Liu [16] improved the spatial exploration ability and bean individual distribution diversity of BOA by constructing a population distribution model based on Cauchy distribution in BOA.

At present, the improvement of the algorithm is mostly from the evolution of population distribution. There is no significant improvement in the global optimization performance and fine search ability of the algorithm. The research combined with other swarm intelligence optimization algorithms will provide new ideas for the performance improvement of BOA. DE algorithm has the advantages of simple principle, less parameter control and strong global search ability. Therefore, the mutation operator and selection operator of DE are introduced into BOA to improve the optimization performance and convergence accuracy of BOA algorithm.

2 Bean Optimization Algorithm with Operators of Differential Evolution

2.1 Related Algorithm Theory

Bean optimization algorithm is a kind of swarm intelligence optimization algorithm, which simulates the process of plant propagation and bean propagation in nature. In order to simulate the reproduction process of natural plants, the algorithm always divides the population into two parts, one is the parent part and the other is the bean part.

In order to simulate the reproduction process of natural plants, the algorithm always divides the population into two parts, the parent part and the children part. In each evolution, the population will finally select some beans with the best fitness as the parent species for the next generation of the population to reproduce. The number of parent species can be set reasonably, and the algorithm will ensure that the geographic locations of multiple parent species are sufficiently dispersed by judging the Euclidean distance between the parent species:

$$M = m(ub - lb) - n \quad (1)$$

$$M < pdist(X_n(t), X_{n+1}(t)) \quad (2)$$

Among them, *ub*, *lb* represent the upper and lower boundaries of the search space, respectively, *m* and *n* are threshold parameters. *M* is the distance threshold, *pdist*() is a function of Euclidean distance, $X_n(t)$, $X_{n+1}(t)$ are all individuals of the parent species. Such threshold selection can make the distance threshold associated with the search space of the solution, so that BOA can select the distance threshold that fits each optimization function. At the same time, with the increase of the number of evolutions, the distance threshold will gradually decrease. Eventually, it will be reduced to 0 in a certain generation at a later stage of evolution. At this time, the iteration has entered the late stage, and the population has converged enough. At this time, the focus of the search should be placed on the local accurate search, and the setting of the distance threshold will affect the convergence of the population in the later stage of evolution.

In order to simulate the propagation mode of natural plant beans, the specific update method of the algorithm is not unique. The more classic population distribution models include the piecewise function model, the normal distribution model, and the negative binomial distribution model. Among them, the normal distribution model is more classic and efficient. The BOA population update method under the normal distribution model is as follows:

$$\mathrm{X(t+1)} = normrnd\ (\mathrm{X}_n(t), \alpha) \quad (3)$$

Among them, in formula (3), $X(t+1)$ is the position of the individual after the population update normrnd (x,y) is a function that generates a normal distribution, x, y *are the mean and standard deviation, respectively,* $X_n(t)$ is the position of n the parent individual, α is the adaptive decreasing variance. The variance is set to be related to the number of iterations and the size of search space, and the variance will gradually decrease as the number of evolutions increases. This ensures that the population in the early stage of evolution can be distributed more divergently around the parent species, and the population in the later stage can gather towards the parent species.

In order to give the population a certain ability to jump out of the local optimum, the individuals of the population have a very small probability to perform mutation operations by regenerating positions in the search space:

$$X(t+1) = lb + (ub - lb)rand(X(t)) \quad (4)$$

Among them, $X(t + 1)$ is the position of the individual after mutation, *ub* and *lb* respectively represent the upper and lower boundaries of the search space, and *rand* (*X*(*t*)) is multiplied by a random number of [0, 1] at the bean position $X(t)$.

Differential Evolution (DE) is an evolutionary algorithm based on population differences proposed by Rainer Storn and Kenneth Price. It is widely used due to its advantages of simple structure, less parameter adjustment and strong robustness. The basic idea of differential evolution is to first mutate between the parent individuals to form a mutant individual, and then according to a certain probability, the parent individual and the mutated individual are crossed to form a new intermediate transition individual, and then through the fitness value of the parent individual and the fitness value between the excess individuals are selected for optimal operation, and the better individual is retained to realize the evolution of the population. The DE algorithm mainly solves the optimal value through three operation operators: mutation, crossover and selection.

(1) Mutation operation

After selecting an individual, the algorithm completes the mutation by adding the weighted difference of the two individuals to the individual. In the early stage of the algorithm iteration, the individual differences in the population are large, and such mutation operation will make the algorithm have a strong global search ability; in the later stage of the iteration, when it tends to converge, the individual differences in the population are small, which also makes the algorithm have a strong local search ability.

(2) Cross operation

The basic principle of crossover operation is to exchange some elements between the individual to be mutated and the new individual generated after the mutation operation to increase the diversity of the population.

(3) Select operation

The differential evolution algorithm uses a greedy mechanism to select individuals entering the next generation to ensure the evolution direction of the population, that is, when the newly generated offspring individuals are better than the parent individuals, retain the offspring individuals, otherwise retain the parent individuals, and then enter the next generation one cycle.

2.2 The Idea of DBOA

There are many excellent operator in the differential evolution algorithm (DE), and its unique mutation mechanism makes good use of the differences within the population [17], which can well promote intraspecific communication in the process of population evolution. The selection mechanism can well guarantee the direction of population evolution and the convergence speed and accuracy of the population. In order to further explore and develop a better bean optimization algorithm, it is considered to combine these two excellent mechanisms in the bean optimization algorithm and the differential evolution algorithm. A new composite algorithm with better performance.

The mutation operator can be expressed as:

$$V_i(t) = X_{r1}(t) + F_r \times (X_{r2}(t) - X_{r3}(t)) \tag{5}$$

The three individuals randomly selected in the mutation operator are sorted from good to bad, $X_b(t)$, $X_m(t)$, $X_w(t)$ and the corresponding fitness is $f_b, f_m, f_{w,}$, and the adaptive scaling factor F_r is:

$$Fr = Fl + (Fu - Fl)fm{-}{-}f\,b/fw - fb \tag{6}$$

where, $F_l = 0.1$, $F_u = 0.9$.

The selection operator is expressed as:

$$\begin{aligned} X_i(t+1) = V_i(t)\ f(V_i(t) < f(X_i(t))) \\ X_i(t)\ \text{otherwise} \end{aligned} \tag{7}$$

The selection operator in the differential evolution algorithm can well guarantee the evolution direction of the population, and this operator can ensure that the better individuals in the population are left. Improving the mutation operator of BOA can only promote the information exchange within the population, but is not enough to greatly improve the overall optimization performance of the algorithm. In order to explore a bean optimization algorithm with better optimization performance, it is considered to introduce the selection operator into BOA. If the fitness of the current individual is better than that of the mutant individual, choose to retain the current individual, otherwise choose to retain the mutant individual.

2.3 Algorithm Flow

The specific steps of bean optimization algorithm based on differential evolution (DBOA) are as follows:

Step 1: initialize the population,

Step 2: determine the distance threshold and select the parent specie,

Step 3: update the population according to the normal distribution model,

Step 4: carry out the mutation operation, combine the new bean population with the mutation operation through formula (4) to update the population position,

Step 5: carry out the selection operation, compare the population fitness updated according to the normal distribution with the population fitness after mutation operation, and select the position with better fitness as the position of the bean according to formula (6),

Step 6: calculate the individual fitness value,

Step 7: determine whether the iteration termination condition is reached, if yes, end the loop and output the value of the parent species step 1, otherwise return to step 2.

3 Experiment and Result Analysis

3.1 Experimental Setup and Test Function

In order to verify the optimization performance of DBOA, DBOA is used in the optimization of CEC2013 test function [18] set, f_1 ~ f_5 they are a unimodal function, and its main function is to investigate the optimization accuracy of the algorithm, f_6~ f_{20} based on multimodal functions, these functions have a large number of local extreme points. In the process of optimization, the algorithm is easy to fall into local optimization, so it can be mainly used to detect whether the algorithm has the ability to avoid premature and jump out of local optimization. f_{21}~ f_{28} they are compound functions, which are composed of multiple basic functions, showing the characteristics of multimodal functions, so these functions are more complex and comprehensive. And the optimization results are compared with the optimization results of particle swarm optimization (PSO), differential evolution algorithm (DE) and basic bean optimization algorithm (BOA). Comparative analysis. Among them, the PSO and DE algorithms are both classic optimization algorithms, and WOA is a relatively novel optimization algorithm proposed in recent years. The basic BOA can more intuitively see the optimization performance of the improved algorithm.

The operating environment of all experiments in this paper is Intel(R) Core (TM) i5-8265U CPU @ 1.60 GHz 1.80 GHz, memory 16.00 GB, Windows 10 system, and the running software is Matlab R2020a.

In order to ensure the fairness of the test results, in this experiment, it is ensured that the population size and the maximum number of iterations of DBOA and PSO, DE, and WOA are consistent with the basic BOA settings. The population size of each algorithm is uniformly set to 30, and the individual dimension is D. It is set to 30, the maximum number of iterations is set to 100D, and the parameter settings of each algorithm are shown in Table 1. These parameters are set to ensure that the optimization effect of the respective algorithm is the best and can be stably converged to the value obtained.

Table 1. Parameter setting of each algorithm.

Algorithm	Parameter setting
PSO	The learning factors C1 and C2 are both 2, and the inertia weight W = 0.5 + RAND / 2
DE	The mutation probability is 0.5 and the crossover probability is 0.2
WOA	The initial value of convergence factor A is 2 and decreases linearly to 0
BOA	The variation probability is 0.2, the number of parent species is 3, the population number of parent species 1 is 0.5 * n, the population number of parent species 2 is 0.3n, and the population number of parent species 3 is 0.2 * n
DBOA	The probability of differential variation is 0.5, the number of parent species is 3, the population number of parent species 1 is 0.5 * n, the population number of parent species 2 is 0.3n, and the population number of parent species 3 is 0.2 * n

3.2 Analysis of Simulation Experiment Results

The mean and standard deviation of 30 optimization calculations performed on the CEC2013 test function set for 5 algorithms are selected as the evaluation indicators of the optimization performance of each algorithm. Among them, Table 2, Table 3 and Table 4 respectively show the results obtained by 30 optimizations on 28 test functions for 6 algorithms, and mark the optimal mean of the optimization results in black.

Table 2. Optimization results of each algorithm on unimodal function.

Type	Fun	PSO	DE	WOA	BOA	DBOA
		Mean	Mean	Mean	Mean	Mean
Unimodal function	*f1*	−1.31e + 03	**−1.40e + 03**	−1.37e + 03	**−1.40e + 03**	**−1.40e + 03**
	f2	**1.09e + 06**	8.81e + 07	6.42e + 07	2.87e + 06	2.02e + 06
	f3	8.38e + 09	2.74e + 08	**2.43e + 10**	4.70e + 08	3.54e + 08
	f4	417.89	6.30e + 04	8.32e + 04	−232.75	**−1.03e + 03**
	f5	−818.44	**−1.00e + 03**	−735.84	−999.96	−999.96

Table 3. Optimization results of each algorithm on unimodal function.

Type	Fun	PSO	DE	WOA	BOA	DBOA
		Std	Std	Std	Std	Std
Unimodal function	*f1*	323.45	0	20.05	0.01	5.97e−04
	f2	8.56e + 05	1.84e + 07	2.94e + 07	1.04e + 06	4.41e + 05
	f3	1.49e + 10	2.01e + 08	1.25e + 10	5.91e + 08	4.92e + 08
	f4	1.51e + 03	1.31e + 04	3.13e + 04	659.13	37.27
	f5	566.65	0	157.12	0.01	0.01

Table 2, Table 3 shows the optimization results of each algorithm on the unimodal function f_1, DE, BOA and DBOA can basically converge to the theoretical optimal value. For function f_2, the average optimization value of DBOA is better than that of BOA and DE. For function f_3, the optimization effect of WOA is the best. For function f_4, the optimization mean of DBOA is the best. It can be seen that it is better than the optimization effect of BOA and DE, which is very close to the understanding of the optimal value. At the same time, it can be seen from its standard deviation that its optimization stability is also relatively high. For function f_5, the optimization effects of BOA and DBOA are basically the same.

Table 4. Optimization results of each algorithm on multimodal function.

Type	Fun	PSO	DE	WOA	BOA	DBOA
		Mean	Mean	Mean	Mean	Mean
Multimodal function	*f6*	−827.79	−857.10	−751.77	−851.84	**−859.99**
	f7	−622.61	**−729.52**	1.25e + 04	−703.93	−691.18
	f8	−679.03	−679.01	−679.03	**−679.04**	−679.02
	f9	−563.49	−563.79	−561.50	−573.51	**−575.87**
	f10	−387.29	−462.86	−276.63	−499.68	**−499.87**
	f11	60.34	**−399.21**	127.57	−228.85	−277.01
	f12	140.60	−98.70	302.15	−112.22	**−159.50**
	f13	330.27	**6.95**	331.50	48.38	21.74
	f14	4.17e + 03	**422.46**	4.94e + 03	3.75e + 03	3.90e + 03
	f15	4.44e + 03	7.72e + 03	6.40e + 03	**3.99e + 03**	4.14e + 03
	f16	201.38	202.73	201.98	200.56	**200.44**
	f17	680.07	**351.29**	923.03	471.06	448.74
	f18	835.01	626.03	1.05e + 03	572.53	**530.33**
	f19	591.29	507.17	569.23	509.16	**507.13**
	f20	614.70	614.97	614.87	**613.45**	614.95

Table 5. Optimization results of each algorithm on multimodal function.

Type	Fun	PSO	DE	WOA	BOA	DBOA
		Std	Std	Std	Std	Std
Multimodal function	*f6*	33.26	7.32	40.82	30.89	**27.91**
	f7	58.07	**13.52**	2.88e + 04	23.26	29.34
	f8	0.05	0.07	0.06	**0.04**	0.04
	f9	3.51	1.44	3.05	4.22	**4.24**
	f10	108.39	23.88	78.77	0.30	**0.02**
	f11	116.48	**2.43**	110.62	49.36	38.96
	f12	136.65	10.10	142.09	59.37	**45.93**
	f13	84.49	**14.35**	94.28	46.03	39.44
	f14	895.10	**149.08**	1.14e + 03	586.30	434.05
	F15	543.44	184.40	944.99	624.67	469.98

(*continued*)

Table 5. (*continued*)

Type	Fun	PSO	DE	WOA	BOA	DBOA
		Std	Std	Std	Std	Std
	f16	0.29	0.28	0.47	0.3840	0.21
	f17	64.12	**3.28**	80.73	28.64	29.34
	f18	86.57	12.14	126.13	40.11	**20.21**
	f19	270.55	0.71	36.72	3.56	**1.79**
	f20	0.25	0.10	0.28	1.87	0.15

Table 6. Optimization results of each algorithm on composite function.

Type	Fun	PSO	DE	WOA	BOA	DBOA
		Mean	Mean	Mean	Mean	Mean
Composite function	*f21*	1.00e + 03	**973.93**	1.11e + 03	1.00e + 03	1.00e + 03
	f22	5.86e + 03	**3.12e + 03**	7.52e + 03	5.77e + 03	5.37e + 03
	f23	6.69e + 03	8.78e + 03	7.94e + 03	5.5746e + 03	**5.22e + 03**
	f24	1.34e + 03	1.28e + 03	1.31e + 03	1.27e + 03	**1.26e + 03**
	f25	1.48e + 03	1.39e + 03	1.42e + 03	1.39e + 03	**1.38e + 03**
	f26	1.54e + 03	1.42e + 03	1.52e + 03	1.53e + 03	**1.40e + 03**
	f27	2.58e + 03	2.50e + 03	2.66e + 03	2.33e + 03	**1.53e + 03**
	f28	5.44e + 03	1.**70e** + 03	6.37e + 03	1.88e + 03	1.80e + 03

Table 7. Optimization results of each algorithm on composite function.

Type	Fun	PSO	DE	WOA	BOA	DBOA
		Std	Std	Std	Std	Std
Composite function	*f21*	101.22	44.78	45.60	80.15	68.90
	f22	840.19	265.94	1.00e + 03	1.06e + 03	1.19e + 03
	f23	1.12e + 03	276.58	906.116	860.97	1.25e + 03
	f24	31.41	3.41	9.57	12.33	10.51
	f25	31.70	4.83	9.19	8.16	15.76
	f26	89.66	46.04	100.33	71.92	26.50
	f27	182.55	38.20	57.61	88.70	71.99
	f28	428.39	2.57e−13	895.96	496.70	400.45

Table 4, Table 5, Table 6, Table 7 show the optimization results of each algorithm on multimodal function and complex composite function, where DE is f_7, f_{11}, f_{13}, f_{14}, f_{17}, f_{21}, f_{22}, f_{28} there are 8 best search functions in total DBOA in f_6, f_9, f_{10}, f_{12}, f_{16}, f_{18}, f_{19}, f_{23}, f_{24}, f_{25}, f_{26}, f_{27}, the optimization effect is the best on 12 test functions. Among the 28 test functions in Table 4, Table 5, Table 6 and Table 7, 23 optimization results of DBOA are better than BOA, and 19 optimization results of DBOA are better than DE. On the whole, the optimization performance of DBOA has been significantly improved compared with that of BOA. At the same time, DBOA is also superior to DE. DBOA absorbs the advantages of BOA and DE at the same time. Because DBOA has two populations: the normal distribution population of BOA and the variant population of DE, the better final population is finally determined through the selection mechanism, so this is the reason for the excellent optimization performance of DBOA.

3.3 Wilcoxon Rank-Sum Test

In order to test the performance of the algorithm more comprehensively [19], in this paper, the Wilcoxon rank-sum test method [20] is introduced to test the significance of the optimal results of DBOA and PSO, DE, WOA, BOA under 30 independent operations, test whether there is significant difference and judge the reliability. The original assumption of H_0 is that there is no significant difference between the data of the two algorithms. The alternative hypothesis for H_1 is that the data populations of the two algorithms differ significantly. Using the test result p to determine the value of h, when $p > 0.05$, the corresponding value of h is 0, which indicates the acceptance assumption of H_0, which shows that the two algorithms have the same performance of searching. When $p < 0.05$, the corresponding value of h is 1, this paper presents the acceptance hypothesis H_1, which indicates that there is a big difference between the two algorithms.

Table 8. Wilcoxon signed rank-sum test results

Fun	DBOA-PSO		DBOA-DE		DBOA-WOA		DBOA-BOA	
	p	h	p	h	p	h	p	h
f_1	1.8267e–04	1	6.3864e–05	1	1.8267e–04	1	0.0091	1
f_2	0.0312	1	1.8267e–04	1	1.8267e–04	1	0.2730	0
f_3	0.0017	1	0.3847	0	1.8267e–04	1	0.0376	1
f_4	1.8267e–04	1	1.8267e–04	1	1.8267e–04	1	1.8267e–04	1
f_5	1.8267e–04	1	8.7450e–05	1	1.8267e–04	1	0.3447	0
f_6	0.0173	1	0.5205	0	1.8267e–04	1	0.8501	0
f_7	0.7337	0	0.6776	0	0.0757	0	0.2413	0
f_8	0.7337	0	0.3075	0	0.2123	0	0.4727	0
f_9	4.3964e–04	1	3.2984e–04	1	3.2984e–04	1	0.9698	0

(continued)

Table 8. *(continued)*

Fun	DBOA-PSO		DBOA-DE		DBOA-WOA		DBOA-BOA	
	p	h	p	h	p	h	p	h
f_{10}	1.8267e–04	1	1.8267e–04	1	1.8267e–04	1	0.5205	0
f_{11}	1.8267e–04	1	1.8267e–04	1	1.8267e–04	1	0.6776	0
f_{12}	1.8267e–04	1	0.0010	1	1.8267e–04	1	0.2730	1
f_{13}	1.8267e–04	1	0.2413	0	1.8267e–04	1	0.1405	0
f_{14}	0.0091	1	1.8267e–04	1	4.3964e–04	1	0.3075	0
f_{15}	0.5708	0	1.8267e–04	1	1.8267e–04	1	0.9698	0
f_{16}	1.8267e–04	1	1.8267e–04	1	1.8267e–04	1	1	0
f_{17}	1.8267e–04	1	1.8267e–04	1	1.8267e–04	1	0.0312	1
f_{18}	1.8267e–04	1	1.8267e–04	1	1.8267e–04	1	0.0376	1
f_{19}	1.8267e–04	1	0.3447	0	1.8267e–04	1	0.0312	1
f_{20}	0.0565	0	0.0017	1	0.0565	0	0.0013	1
f_{21}	0.3075	0	0.0072	1	0.0257	1	0.1859	0
f_{22}	0.1620	0	1.8267e–04	1	0.0058	1	0.7913	0
f_{23}	0.0757	0	1.8267e–04	1	0.0017	1	0.3847	0
f_{24}	2.4613e–04	1	0.0312	1	1.8267e–04	1	0.1620	0
f_{25}	1.8267e–04	1	0.0640	0	0.0022	1	0.0640	0
f_{26}	0.1405,	0	0.3075	0	0.0539	0	0.0022	1
f_{27}	1.8267e–04	1	0.0028	1	1.8267e–04	1	0.1405	1
f_{28}	1.8267e–04	1	6.3864e–05	1	1.8267e–04	1	0.5708	0
+/=/–	19/8/1		11/8/9		24/4/0		8/18/2	

Table 8 shows the rank-sum test results of DBOA and the four algorithms, showing that the number of times that DBOA outperforms, equals, and outperforms the comparison algorithm on 28 test functions. First, we observe the rank-sum test results of DBOA and Wilcoxon. It is found that the optimal performance of DBOA on 8 functions is better than BOA, and the optimal performance of only 2 functions is weaker than BOA, furthermore, the improvement of DBOA is effective. The optimization performance of DBOA on 19 functions is better than that of PSO, and on 24 functions is better than that of WOA. The experimental results in Sect. 3.2 are further verified. Finally, the optimization performance of DBOA on 8 functions is equivalent to that of DE, and the optimization performance of DBOA on 11 functions is better than that of DE, combining the results of mean and standard deviation in Sect. 3.2, it can be concluded that the optimal performance of DBOA with BOA and De is better than that of DE. Through the Wilcoxon rank-sum test, the excellent performance of DBOA is further shown.

4 Summary

This paper proposes a bean optimization algorithm based on differential evolution to solve the problem that the local fine optimization of the bean optimization algorithm is slow. The algorithm first uses the basic BOA population distribution model to update the population position, so that it can be updated in a shorter time. Obtain faster convergence speed and search performance, and then introduce the core mutation and selection operators in differential evolution to further improve the optimization performance of the algorithm for complex optimization problems, realize the complementary advantages of the two algorithms, and obtain a global search capability. Efficient hybrid optimization algorithm with local search capability. Later, through experiments on the CEC2013 test function set, it was found that DBOA has better optimization performance, faster convergence speed and higher universality, and is more suitable for solving various function optimization problems than DE and basic BOA.

Acknowledgement. This research was funded by Qinghai Science Foundation under grant number 2020-ZJ-913, Special project of scientific and technological achievements transformation in Qinghai province number 2021-GX-114, Scientific research project of graduate students in Anhui universities number YJS20210087.

References

1. Yang, X.S., Deb, S., Fong, S., et al.: From swarm intelligence to metaheuristics: nature-inspired optimization algorithms. Computer **49**(9), 52–59 (2016)
2. Li, A.D., Xue, B., Zhang, M.: A forward search inspired particle swarm optimization algorithm for feature selection in classification. In: 2021 IEEE Congress on Evolutionary Computation (CEC). IEEE, pp. 786–793 (2021)
3. Wu, W., Wei, Y.: Guiding unmanned aerial vehicle path planning design based on improved ant colony algorithm. Mechatronic Syst. Control Учредители: Acta Press **49**(1), 48–54 (2021)
4. Brest, J., Maučec, M.S., Bošković, B.: Differential evolution algorithm for single objective bound-constrained optimization: algorithm. In: IEEE Congress on Evolutionary Computation (CEC). IEEE, pp. 1–8 (2020)
5. Zhang, X., Sun, B., Mei, T., et al.: Post disaster restoration based on fuzzy preference relation and bean optimization algorithm. In: 2010 IEEE Youth Conference on Information, Computing and Telecommunications. IEEE, pp. 271–274 (2010)
6. Sun, Y., Wang, X., Chen, Y., et al.: A modified whale optimization algorithm for large-scale global optimization problems. Expert Syst. Appl. **114**, 563–577 (2018)
7. Zhang, X.-M., Wang, R.-J., Song, L.-T.: A novel evolutionary algorithm-sees optimization algorithm. PR&AI, **21**(05), 677–681 (2008)
8. Wang, P., Chen, Y.: The optimization alogorithm of SOA during the dispatching of disaster relief supplied. Econ. Res. Guide, (08), 252–253 (2010)
9. Zhang, X.: Research on a Novel Swarm Intelligence Algorithm Inspired by Beans Dispersal. University of Science and Technology of China (2011)
10. Zhang, X., Sun, B., Mei, T., Wang, R.: Post-disaster restoration based on fuzzy preference relation and Bean Optimization Algorithm. In: 2010 IEEE Youth Conference on Information, Computing and Telecommunications, pp. 271–274 (2010)

11. Zhang, X., Wang, H., Sun, B., et al.: The Markov model of bean optimization algorithm and its convergence analysis. Int. J. Comput. Intell. Syst. **6**(6), 609–615 (2013)
12. Zhang, X., Jiang, K., Wang, H., et al.: An improved bean optimization algorithm for solving TSP. In: International Conference on Advances in Swarm Intelligence, pp. 261–267 (2012)
13. Feng, T., Xie, Q., Hu, H., et al.: Bean optimization algorithm based on negative binomial distribution. Lect. Notes Comput. Sci. **9140**, 82–88 (2015)
14. Feng, T.: Study and Application of Bean Optimization Algorithm on Complex Problem. Master's thesis, University of Science and Technology of China, Hefei, China (2017)
15. Mohsin, A.: Research on Bean Optimization Algorithm Based on Abundance Distribution Patterns. Anhui Agricultural University (2020)
16. Liu, H., Zhang, X., Wang, C.: Bean optimization algorithm based on cauchy distribution and parent rotation mechanism. Patt. Recogn. Artif. Intell. **34**(07), 581–591 (2021)
17. Jedrzejowicz, P., Skakovski, A.: Improving performance of the differential evolution algorithm using cyclic decloning and changeable population size. J. Univers. Comput. Sci. **22**(6), 874–893 (2016)
18. Liang, J.J., Qu, B.Y., Suganthan, P.N., et al.: Problem definitions and evaluation criteria for the CEC 2013 special session on real-parameter optimization. Comput. Intell. Lab. Zhengzhou Univ. Zhengzhou, China Nanyang Technol. Univ. Singapore, Tech. Rep. **201212**(34), 281–295 (2013)
19. Zhang, X.-M., Jiang, Y., Liu, S.-W.: Hybird coyote optimization with grey wolf optimizer and its application to clustering optimization. Acta Automatica Sinica, 1–17 (2022)
20. Derrac, J., García, S., Molina, D., Herrera, F.: A practical tutorial on the use of nonparametric statistical tests as a methodology for comparing evolutionary and swarm intelligence algorithms. Swarm Evol. Comput. **1**(1), 3–18 (2011)

A Backbone Whale Optimization Algorithm Based on Cross-stage Evolution

Xin Yang, Limin Wang, Zhiqi Zhang, Xuming Han(✉), and Lin Yue

Jinan University, Guangzhou, China
hanxuming@jnu.edu.cn

Abstract. The swarm intelligent algorithms (SIs) are effective and widely used, while the balance between exploitation and exploration directly affects the accuracy and efficiency of algorithms. To cope with this issue, a backbone whale optimization algorithm based on cross-stage evolution (BWOACS) is proposed. BWOACS is mainly composed of three parts: (1) adopts the density peak clustering (DPC) method to actively divide the population into several sub-populations, generates the backbone representatives (BR) during backbone construction stage; (2) determines the deviation placement (DP) by constructing the co-evolution operators (CE), the search space expansion operators (SE) and the guided transfer operators (GT) during bionic evolution strategy stage; (3) realises the bionic optimisation through DP during backbone representatives guiding co-evolution stage. To verify the accuracy and performance of BWOACS, we compare BWOACS with other variants on 9 IEEE CEC 2017 benchmark problems. Experimental results indicate that BWOACS has better accuracy and convergence speed than other algorithms.

Keywords: Whale optimization algorithm · Density peak clustering · Bionic evolution strategy

1 Introduction

In swarm intelligent algorithms (SIs), exploration, a collaboration of exploration and exploitation (called collaboration in next) and exploitation are independent and mutually constrained processes. In the exploration phase [14], each individual searches the solution space randomly to find the global optimal solution; in the collaboration phase [18], the relevant cooperative strategies are used to achieve cooperation during the two processes validly. In the exploitation phase [15], individuals have relative independence, make more use of local optimal information to discover the local optimal solution as soon as possible.

According to the characteristics of creatures, various SIs and variants have been proposed to achieve exploitation and exploration capabilities. Whale optimization algorithm (WOA) [8] adopted alternating shrinking encircling, spiral

Y. Tan et al. (Eds.): ICSI 2022, LNCS 13344, pp. 95–104, 2022.
https://doi.org/10.1007/978-3-031-09677-8_8

updating, and random search to reach the balance between the three processes. Particle swarm optimization (PSO) [4] realized the balance through the coordination of direction and velocity. Saeed et al. [16] proposed QWOA, using the quantum revolving gate operator as the variant operator to achieve three phases. CWOA [10] combines the characteristics of the standard WOA with chaos mapping actively. Chaos theory is used to generate parameter ratios; a dynamic neighborhood learning (DNL) strategy is proposed in DNLGSA [17], each particle can learn search information from the historical best experience (gbest) of the whole population. Direction Learning Strategy and Elite Learning Strategy are raised in LLABC [1] to form a machine that can complement each other. A weighting strategy upon Sigmoid function is introduced in AWPSO [6], adaptively adjusting the control parameters and the acceleration coefficients for three stages. There are widely used in practical problems, such as Automatic Control [7,9], Wireless Communication [5,12], and Task Scheduling [2,11].

Specific SIs described above are only for the best individuals in population to learn, which are easy to fall into the local optimal. In this paper, we present a backbone whale optimization algorithm based on cross-stage evolution (BWOACS), which effectively solves the problem of striking a balance between three phases. The contributions of this paper are mainly in the following three aspects:

1) We propose BWOACS. BWOACS provides a new method which can make most individuals in WOA obtain more effective information.
2) The concept of backbone representatives (BR) are proposed. BR are individuals that can reflect the commonality and characteristics of each subpopulation.
3) Bionic evolution strategy is proposed. The strategy designs different operators during different stages: co-evolution operator (CE), search space expansion operator (SE) and guiding transfer operator (GT).

2 Related Work

WOA is a new SI algorithm, which has the ability to quickly find the global optimal solution. It mainly includes three stages: shrinking encircling, spiral updating, and search for prey. Where shrinking encircling and spiral updating are carried out at the same time with the probability p ($p = 0.5$) in bubble-net attacking.

During the phase of shrinking encircling, the whale position is updated as Eq. 1.

$$X(t+1) = X^*(t) - A \cdot D_1 \tag{1}$$

where $D_1 = |K \cdot X^*(t) - X(t)|$. $A = 2a \cdot \lambda - a$, $K = 2 \cdot \lambda$. $X^*(t)$ is the position vector of the best solution obtained so far. a is linearly decreased from 2 to 0 throughout iterations, λ is a random vector in [0, 1].

During the spiral updating stage, position is updated as shown in Eq. 2.

$$X\left(t+1\right)=X^{*}\left(t\right)+D_{2}\cdot e^{bl}\cdot cos\left(2\pi l\right) \tag{2}$$

where $D_2 = |X^*(t) - X(t)|$. b is a constant for defining the shape of the logarithmic spiral, l is a random number in $[-1, 1]$.

During the phase of search for prey,when $|A| \leq \mathbf{1}$, the whale updates its position according to Eq. 1. When $|A| > 1$, the whale carries out a random search according to Eq. 3.

$$X\left(t+1\right)=X_{rand}\left(t\right)-A\cdot D_{3} \tag{3}$$

where $D_3 = |K \cdot X_{rand}(t) - X(t)|$. $X_{rand}(t)$ is a random position vector (a random whale) chosen from the current population.

3 A Backbone Whale Optimization Algorithm Based on Cross-stage Evolution

BWOACS consists of three main components: backbone construction stage, bionic evolution strategy stage and backbone representatives guiding co-evolution stage, details in Algorithm 1.

3.1 Backbone Construction Stage

Based on the analysis of the SIs, individuals in relatively close locations have a strong similarity to each other. To simplify the computational complexity of mutual learning between individuals, we divide sub-populations by DPC clustering [13], where the local optimal solution within each sub-population is defined as the backbone representative(BR). The definition of BRs is in Eq. 4.

$$x_{c*}=argminf(x),\ x\in X_{c} \tag{4}$$

where x_{c*} is the BR of the sub-population and X_c is the set of individuals of the sub-population, $f(x)$ is the fitness of individual.

This paper proposes the double power-set operator (DS) to generate the backbone representatives of each sub-population. With the iteration increasing, the individuals tend to be more stable. The frequency of clustering gradually decreases. DS is given in Eq. 5.

$$t=m^{x}-n \tag{5}$$

where t is the number of iteration, x is the number of clustering. m and n are constants, which set by maximum number of iterations.In this paper, $m = n = 2$.

Algorithm 1. BWOACS

Require: N: the population size; S: the number of sub-populations; T_{max}: the maximum number of iterations
Ensure: X_f : the final population; P_f : the final optimal result;
1: Randomly initialise X_i, DP_i
2: Calculate the fitness of each search agent
3: **while** $t < Tmax$ **do**
4: **if** $t = 2^x - 2$ **then**
5: Cluster the population by DPC to generate S sub-populations
6: Calculate the number of search agents Nc in each sub-swarm
7: **end if**
8: **for** $c = 1 \rightarrow S$ **do**
9: Find out the best search agent x_{c*} in each sub-population
10: **for** $j = 1$ to N_c **do**
11: Update a, A, K, l, p, $cf1$ and $cf2$
12: **if** $p < 0.5$ and $|A| \leq 1$ **then**
13: **if** $x = x_{c*}$ **then**
14: Update the position of the current search agent by (9)(10)
15: **else**
16: Update the position of the current search agent by (15)(16)
17: **end if**
18: **else if** $p \geq 0.5$ **then**
19: **if** $x = x_{c*}$ **then**
20: Update the position of the current search agent by (11)(12)
21: **else**
22: Update the position of the current search agent by (17)(18)
23: **end if**
24: **else** $p < 0.5$ and $|A| > 1$
25: Update the position of the current search agent by (13)(14)
26: **end if**
27: **end for**
28: **end for**
29: Check if any search agent goes beyond the search space and amend it
30: Calculate the fitness of each search agent
31: Update P_f if there is a better solution
32: $t = t + 1$
33: **end while**
34: **return** X_f, P_f

3.2 Bionic Evolution Strategy Stage

Exploration. BR are the individuals most likely to find the global optimum. BR can learn more search information from the mutual exchange process of effective information among sub-populations. CE is proposed to guide BR in updating.as defined in Eq. 6.

$$CE = \begin{cases} A|K \cdot \frac{1}{s}\sum_{c=1}^{s} x_{c*}^{t} - x_{c*}^{t}| & p < 0.5 \\ e^{bl} \cdot cos\left(2\pi l\right)|\frac{1}{s}\sum_{c=1}^{s} x_{c*}^{t} - x_{c*}^{t}| & p \geq 0.5 \end{cases} \tag{6}$$

where x_{c*}^{t} is the position of BR in the t-th iteration of c-th sub-population. S is the number of sub-populations.

Collaboration. To avoid occur premature convergence, let small number of individuals enter other regions of the solution space to search. We propose the search space expansion operator SE, whose definition is in Eq. 7.

$$SE = A|K \cdot x_{crand}^{t} - x_{ca}^{t}| \tag{7}$$

where x^t_{crand} is the position of the randomly selected individual in the c-th sub-population, x^t_{ca} is the position of individuals who will enter other regions.

Exploitation. BWOACS directs the update of ordinary individuals within this sub-population through BR, and the ordinary individuals within each sub-swarm are constrained to refine exploitation in their specific sub-population. We propose the guided transfer operators (GT) to update ordinary individuals.

$$GT = \begin{cases} A|K \cdot x^t_{c*} - x^t_{ci}| & p < 0.5 \\ e^{bl} \cdot \cos(2\pi l)\,|x^t_{c*} - x^t_{ci}| & p \geq 0.5 \end{cases} \tag{8}$$

where x^t_{ci} is the position of the i-th individual in the t-th iteration and the c-th sub-population.

3.3 Backbone Representatives Guiding Co-evolution Stage

With an analysis of the current and previous stages, the concept of deviation placement DP is proposed. DP is determined by CE SE GT during different stages, respectively.

In exploration, when $p < 0.5$,DP^{t+1}_{c*} is in Eq. 9, $p \geq 0.5$, DP^{t+1}_{c*} is in Eq. 11;

$$DP^{t+1}_{c*} = \left(w \cdot DP^t_{c*} - CE\right)/cf1 = \left(w \cdot DP^t_{c*} - A|K \cdot \frac{1}{s}\sum_{c=1}^{s} x^t_{c*} - x^t_{c*}|\right)/cf1 \tag{9}$$

$$x^{t+1}_{c*} = \left(x^t_{c*} + DP^{t+1}_{c*}\right)/cf1 \tag{10}$$

$$DP^{t+1}_{c*} = w \cdot DP^t_{c*} + CE = w \cdot DP^t_{c*} + e^{bl} \cdot cos\,(2\pi l)\,|\frac{1}{s}\sum_{c=1}^{s} x^t_{c*} - x^t_{c*}| \tag{11}$$

$$x^{t+1}_{c*} = \left(x^t_{c*} + DP^{t+1}_{c*}\right)/cf2 \tag{12}$$

where DP^{t+1}_{c*} is the DP of BR in the t-th iteration and c-th sub-population. $w = 1/(1+t)$, $cf1 = 2 + 1/(t+1)$, $cf2 = 1 + 1/(t+1)$.

In collaboration, DP^{t+1}_{ca} is in Eq. 13;

$$DP^{t+1}_{ca} = \left(w \cdot DP^t_{ca} - SE\right)/cf1 = \left(w \cdot DP^t_{ca} - A|K \cdot x^t_{crand} - x^t_{ca}|\right)/cf1 \tag{13}$$

$$x^{t+1}_{ca} = \left(x^t_{crand} + DP^{t+1}_{ca}\right)/cf2 \tag{14}$$

where DP^{t+1}_{ca} is the DP of individuals who will enter other regions in the t-th iteration and c-th sub-population.

In exploitation, when $p < 0.5$, DP^{t+1}_{ci} is in Eq. 15, $p \geq 0.5$, DP^{t+1}_{ci} is in Eq. 17.

$$DP^{t+1}_{ci} = \left(w \cdot DP^t_{ci} - GT\right)/cf1 = \left(w \cdot DP^t_{ci} - A|K \cdot x^t_{c*} - x^t_{ci}|\right)/cf1 \tag{15}$$

$$x^{t+1}_{ci} = \left(x^t_{c*} + DP^{t+1}_{ci}\right)/cf1 \tag{16}$$

$$DP_{ci}^{t+1} = w \cdot DP_{ci}^{t} + GT = w \cdot DP_{ci}^{t} + e^{bl} \cdot cos\left(2\pi l\right) \left|x_{c*}^{t} - x_{i}^{t}\right| \quad (17)$$

$$x_{ci}^{t+1} = \left(x_{c*}^{t} + DP_{ci}^{t+1}\right) / cf2 \quad (18)$$

where DP_{ci}^{t+1} is the DP of the i-th individual in the t-th iteration and c-th sub-population.

3.4 Computational Complexity

Suppose the population number is N, the maximum iteration number is $T_{\max}$, and the computational complexity of BWOACS is calculated as follows:

The computational complexity for initialising and assessing of population fitness both are $O\left(N\right)$; ranking the populations and finding the best individuals are $O\left(N^2\right)$; in the iterative process, clustering of the population is $O\left(N^2\right)$, clustering number is $\lfloor \log_2 T_{\max} \rfloor$; the updating of individuals in each sub-population and calculating of fitness values are $O\left(2N\right)$; sorting the fitness of individuals in each sub-population and finding out BR are $O\left(\sum_{c=1}^{S} N_c^2\right)$; ranking BR to find the best individuals is $O\left(S^2\right)$. In a word, the time complexity of BWOACS for T_{max} cycles is in Eq. 19:

$$O(2N) + O(N^2) + T_{\max}(O(2N) + O(\sum_{c=1}^{S} N_c^2) + O(S^2)) + \lfloor \log_2 T_{\max} \rfloor \cdot O(N^2) \quad (19)$$

From [3], the time complexity of WOA are in Eq. 20:

$$O(2N) + O(N^2) + T_{\max}(O(2N) + O(N^2)) \quad (20)$$

We can conclude that the computational complexity of BWOACS and WOA are in the same order by analysing Eq. 19 and 20.

4 Experiment

4.1 Functions and Parameter Settings

Nine widely used benchmark functions were selected from IEEE CEC 2017, details in Table 1. To verify the performance of BWOACS, BWOACS was compared with DNLGSA [17], CWOA [10], LLABC [1] and AWPSO [6].

Besides, We set the iteration number and the population size as 500 and 30. Each function was run with 50 independent replications. The best results were in boldface. All the proposed algorithms were coded in MATLAB 2016b. The computation was conducted on a personal computer with an Intel Core i7-3770, 3.40 GHz CPU, 8 GB RAM.

Table 1. Benchmark functions.

Group	Function	D	Search range	Global optimum	Name
Unimodal	f_1	30	(−100, 100)	0	Sphere
	f_2	30	(−10, 10)	0	Schwefel's P2.22
	f_3	30	(−100, 100)	0	Schwefel's P1.2
Multimodal	f_4	30	(−500, 500)	−418.9829n	Schwefel's P2.26
	f_5	30	(−5.12, 5.12)	0	Rastrigin
	f_6	30	(−32, 32)	0	Ackley
Fixed-dimension Multimodal	f_7	2	(−65.536, 65.536)	0.998	Hartman
	f_8	4	(−5, 5)	0.0003075	Kowalik
	f_9	2	lb = [−5, 0]; ub = [10, 15]	0.398	Shekel

4.2 Comparison of Solution Accuracies

Table 2 showed the experimental results compared with 4 algorithms. It can be seen from the results that BWOACS performed better than others in average values. For unimodal and multimodal functions, BWOACS derived nice average values in all. For fixed-dimension multimodal functions, the best average values in BWOACS were 2, while DNLGSA, CWOA, LLABC, AWPSO were 1, 0, 1 and 2, respectively. BWOACS was robust in terms of the standard deviation. For 9 functions, there were 5 functions whose standard deviation values had been 0. For most functions, the standard deviation values of BWOACS were much lower than DNLGSA, CWOA, LLABC, AWPSO. It was proved that BWOACS had better stability than other algorithms.

Table 2. Results of 5 algorithms on 9 functions

Function		DNLGSA	CWOA	LLABC	AWPSO	BWOACS
f1	Ave	2.56E+04	5.94E+03	2.22E+03	5.96E+00	**5.67E−180**
	Std	4.01E+07	3.35E+08	6.15E+05	5.24E+00	**0**
f2	Ave	3.48E+01	1.82E+10	1.86E+01	1.28E+01	**1.46E−98**
	Std	4.15E+02	4.57E+21	1.17E+02	1.10E+01	**2.39E−195**
f3	Ave	5.35E+04	6.95E+04	2.01E+03	4.02E+02	**1.53E−63**
	Std	1.10E+09	2.05E+09	1.04E+06	2.12E+04	**4.57E−125**
f4	Ave	−4.60E+03	−9.76E+03	−1.16E+04	−2.07E+03	**−1.25E+04**
	Std	2.29E+05	5.56E+06	1.74E+05	2.81E+05	**1.41E+04**
f5	Ave	1.18E+02	1.07E+02	1.56E+02	1.41E+02	**0**
	Std	5.85E+02	3.63E+04	1.80E+03	6.26E+02	**0**
f6	Ave	1.83E+01	3.02E+00	1.20E+01	4.63E+00	**8.88E−16**
	Std	9.09E−01	5.46E+01	2.66E+00	3.69E−01	**0**
f7	Ave	**−3.86E+00**	−3.83E+00	**−3.86E+00**	**−3.86E+00**	**−3.86E+00**
	Std	3.92E−30	1.13E−02	**3.93E−30**	4.29E−09	2.20E−05
f8	Ave	8.26E−04	1.84E−02	**8.02E−04**	9.05E−04	8.69E−04
	Std	1.31E−07	2.14E−03	**3.13E−08**	7.27E−08	2.73E−07
f9	Ave	−5.38E+00	−7.26E+00	−9.29E+00	**−9.38E+00**	−8.40E+00
	Std	7.17E+00	6.30E+00	**3.32E+00**	5.33E+00	9.87E+00

4.3 Convergence Speed Analysis

Fig. 1 manifested the convergence curves of 5 algorithms in 9 functions. For f1–f3, f5 and f6, the rapid decline of the curve indicated that BWOACS had a faster convergence speed in the early evolution. The search strategy could complete the fast search effectively. For f4 and f7–f9, the convergence curve of BWOACS represented a shape with some twists and turns because it needed to determine the global optimal value through *BR*. In addition, it was discovered that BWOACS had a strong ability to jump out of the local optimal value because the slope of the curves increasing became larger gradually with the iteration number increasing. In the aspect of the convergence curve, the effectiveness of BWOACS is proved.

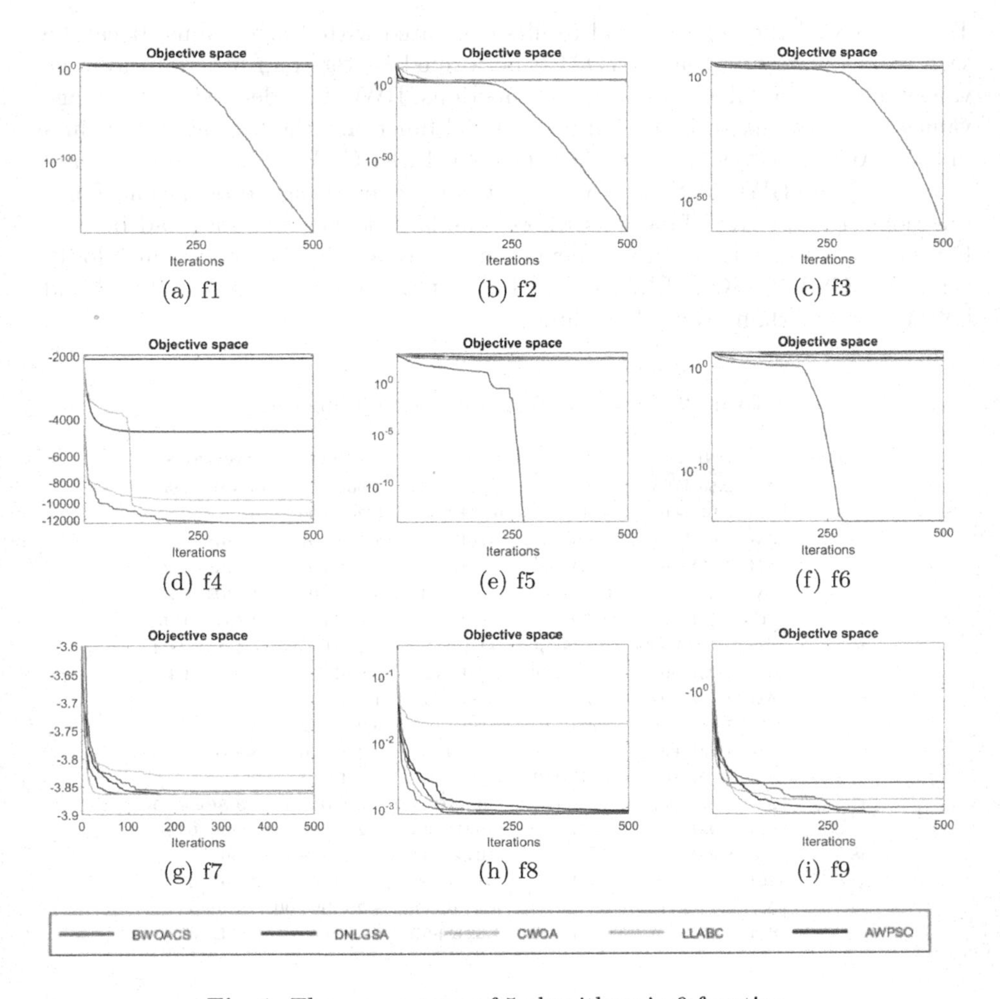

Fig. 1. The convergence of 5 algorithms in 9 function

5 Conclusion

By analyzing the advantages and characteristics of SIs, BWOACS is proposed. BWOACS proposes BR and DS to coordinate three stages, and effectively arrange CE, SE, GT for three stages, ensuring the constrained allocation of exploration capabilities and exploitation capabilities in each stage. In order to verify the effectiveness of BWOACS, BWOACS is compared with the most advanced algorithms DNLGSA, CWOA, LLABC and AWPSO. The results show that BWOACS is superior to them. In future work, we will achieve a common framework by this strategy and apply other algorithms.

References

1. Gao, W., Sheng, H., Wang, J., Wang, S.: Artificial bee colony algorithm based on novel mechanism for fuzzy portfolio selection. IEEE Trans. Fuzzy Syst. **27**(5), 966–978 (2018)
2. García-Nieto, J., Alba, E., Olivera, A.C.: Swarm intelligence for traffic light scheduling: application to real urban areas. Eng. Appl. Artif. Intell. **25**(2), 274–283 (2012)
3. Jiang, R., Yang, M., Wang, S., Chao, T.: An improved whale optimization algorithm with armed force program and strategic adjustment. Appl. Math. Model. **81**, 603–623 (2020)
4. Kennedy, J., Eberhart, R.: Particle swarm optimization. In: Proceedings of ICNN 1995-International Conference on Neural Networks, vol. 4, pp. 1942–1948. IEEE (1995)
5. Lim, H., Hwang, T.: User-centric energy efficiency optimization for miso wireless powered communications. IEEE Trans. Wireless Commun. **18**(2), 864–878 (2018)
6. Liu, W., Wang, Z., Yuan, Y., Zeng, N., Hone, K., Liu, X.: A novel sigmoid-function-based adaptive weighted particle swarm optimizer. IEEE Trans. Cybern. **51**, 1085–1093 (2019)
7. Luo, H., Krueger, M., Koenings, T., Ding, S.X., Dominic, S., Yang, X.: Real-time optimization of automatic control systems with application to BLDC motor test rig. IEEE Trans. Industr. Electron. **64**(5), 4306–4314 (2016)
8. Mirjalili, S., Lewis, A.: The whale optimization algorithm. Adv. Eng. Softw. **95**, 51–67 (2016)
9. Olaru, S., Dumur, D.: Avoiding constraints redundancy in predictive control optimization routines. IEEE Trans. Autom. Control **50**(9), 1459–1465 (2005)
10. Oliva, D., Abd El Aziz, M., Hassanien, A.E.: Parameter estimation of photovoltaic cells using an improved chaotic whale optimization algorithm. Appl. Energy **200**, 141–154 (2017)
11. Pan, Q.K.: An effective co-evolutionary artificial bee colony algorithm for steelmaking-continuous casting scheduling. Eur. J. Oper. Res. **250**(3), 702–714 (2016)
12. Pham, Q.V., Mirjalili, S., Kumar, N., Alazab, M., Hwang, W.J.: Whale optimization algorithm with applications to resource allocation in wireless networks. IEEE Trans. Veh. Technol. **69**(4), 4285–4297 (2020)
13. Rodriguez, A., Laio, A.: Clustering by fast search and find of density peaks. Science **344**(6191), 1492–1496 (2014)

14. Tian, Y., Zhang, X., Wang, C., Jin, Y.: An evolutionary algorithm for large-scale sparse multiobjective optimization problems. IEEE Trans. Evol. Comput. **24**(2), 380–393 (2019)
15. Wang, Z.J., et al.: Dynamic group learning distributed particle swarm optimization for large-scale optimization and its application in cloud workflow scheduling. IEEE Trans. Cybern. **50**(6), 2715–2729 (2019)
16. Yan, Z., Zhang, J., Zeng, J., Tang, J.: Nature-inspired approach: an enhanced whale optimization algorithm for global optimization. Math. Comput. Simul. **185**, 17–46 (2021)
17. Zhang, A., Sun, G., Ren, J., Li, X., Wang, Z., Jia, X.: A dynamic neighborhood learning-based gravitational search algorithm. IEEE Trans. Cybern. **48**(1), 436–447 (2016)
18. Zhang, X., Tian, Y., Cheng, R., Jin, Y.: A decision variable clustering-based evolutionary algorithm for large-scale many-objective optimization. IEEE Trans. Evol. Comput. **22**(1), 97–112 (2016)

An Improved Hunger Games Search Algorithm for Global Optimization

Shaolang Li, Xiaobo Li(✉), HuiChen, Yuxin Zhao, and Junwei Dong

College of Mathematics and Computer Science, Zhejiang Normal University, Jinhua 321004, China

{shaolang1997,lxb}@zjnu.edu.cn

Abstract. This paper proposes an improved version of the hunger games search algorithm (HGS) based on the opposition-based learning (OBL) and evolutionary population dynamics (EPD) strategies called OEHGS for solving global optimization tasks. The proposed OEHGS algorithm consists of three stages: the first stage generates an initial population and its opposite using OBL strategy; the second stage uses the EPD to prevent premature convergence and stagnation; and the third stage uses the OBL as an additional phase to update the HGS population at each iteration. The opposition-based learning approach is incorporated into HGS with a selecting rate, which can jump out of the local optimum without increasing the computational complexity. The performance of our proposed algorithm was tested through a set of experimental series. The experiments revealed that the proposed algorithm is superior to those of state-of-the-art algorithms in this domain for solving optimization problems.

Keywords: Hunger games search · Opposition-based learning · Evolutionary population dynamics

1 Introduction

Hunger games search (HGS) [1] is a new and efficient meta-heuristic algorithm inspired by the behavior of animals in nature when they are hungry. Compared with other algorithms, HGS has high performance in processing numerical values and other optimization tasks. However, when solving the optimization problem, it will quickly converge on the local optimum rather than global optimal solution, and this behavior is called immature convergence, which is a widely existing problem of the meta-heuristic algorithms [2, 3]. Another problem is the stagnation in local search. In fact, the success of meta-heuristic methods to solve this problem depends on balance of exploration and exploitation. HGS fails to achieve a balance between exploration and exploitation, which can significantly to degrade the quality of the solution.

In order to overcome the limitation of HGS, the opposition-based learning (OBL) is used in HGS. The concept of OBL was introduced by Tizhoosh in [4]. It has been widely used to accelerate learning in neural network. The idea has also been used in heuristic algorithms, for example, OBL is used in DE algorithm to generate new offspring

Y. Tan et al. (Eds.): ICSI 2022, LNCS 13344, pp. 105–116, 2022.
https://doi.org/10.1007/978-3-031-09677-8_9

during the evolution [5], and the elite OBL strategy and simplex method are introduced into GWO to improve global search ability [6]. In addition, the concept of evolutionary population dynamics (EPD) [7] is applied for the optimization process to update the poor solution on the population and reduce the impact on the poor solution to the population quality.

The performance of our proposed algorithm has been tested through a set of experimental series, including 45 benchmark functions [8, 9]. The experimental results compared with those of other similar algorithms revealed that the OEHGS obtains superior results in terms of performance and efficacy.

The remainder of this paper is organized as follows: the basic concept of the HGS, OBL strategy and EPD are introduced in Sect. 2. The improved HGS based on OBL and EPD is presented in detail in Sect. 3. Then, the OEHGS is validated, and the results and discussion are given in Sect. 4. Finally, we discuss relevant issues and conclude the paper in the Sect. 5.

2 Background

2.1 Hungry Games Search (HGS)

The HGS algorithm was proposed by Yang et al. In 2021 [10]. The HGS algorithm is inspired from the behaviors of animals in the state of starvation. Similar to other metaheuristic algorithms, *N* individuals are randomly initialized according to the given search space in HGS. During the iteration process, each ainimal is assigned to a state, cooperative or non-cooperative. Cooperative and Non-cooperative phases of HGS are described below:

2.1.1 Non-cooperative Phase

Non-cooperative:

$$x_i^{G+1} = x_i^G \times (1 + randn(1)), \quad rand < L \tag{1}$$

where *rand* is a random number in the range of [0, 1], *randn* is the Gaussian distribution, and *L* is a parameter to control the behaviors of each individual.

2.1.2 Cooperative Phase

$$\text{Cooperative: } x_i^{G+1} = \begin{cases} w_1 \times x_{best}^G + w_2 \times R \times |x_{best}^G - x_i^G|, & rand > L, rand > E \\ w_1 \times x_{best}^G - w_2 \times R \times |x_{best}^G - x_i^G|, & rand > L, rand < E \end{cases} \tag{2}$$

where w_1 and w_2 are hunger weight, *R* is a ranging controller, and *E* represents the variable number that controls the global location. The formulations of w_1, w_2, *R*, and *E* are described below:

$$w_1^i = \begin{cases} hungry_i \times \dfrac{N}{SHungry} \times r3, & rand < L \\ 1, & rand < L \end{cases} \tag{3}$$

$$w_2^i = (1 - e^{-|hungry_i - SHungry|}) \times rand \times 2 \quad (4)$$

$$hungry_i = \begin{cases} 0, & fit_i = BF \\ hungry_i + H, & fit_i \neq BF \end{cases} \quad (5)$$

$$H = \begin{cases} LH \times (1 + rand), & TH < LH \\ TH, & TH \geq LH \end{cases} \quad (6)$$

$$TH = \frac{fit_i - BF}{WF - BF} \times rand \times 2 \times (ub_i - lb_i) \quad (7)$$

$$R = 2 \times shrink \times rank - shrink \; shrink = 2 \times \left(1 - \frac{t}{T}\right) \quad (8)$$

$$E = \frac{2}{e^{fit_i - BF} + e^{BF - fit_i}} \quad (9)$$

where LH is a limited parameter, fit_i is the fitness of ith individual, BF and WF are the best fitness and worst fitness in the last iteration, respectively, ub_i and lb_i are the upper and lower bounds, respectively.

2.2 Opposed-Based Learning

Opposed-based learning [11] is used in MH algorithms for improving their performance through exploring the global solution to a selected problem, it uses the value of the fitness function to determine whether the opposite are better than the current solution. The basic definition of the OBL can be calculated as follows:

$$\overline{x} = u_i + l_i - x_i \quad (10)$$

where $\overline{x} \in R^k$ is the opposite vector from the real vector $x \in R^k$, u_i and l_i are the upper bound and lower bound of the search space, respectively. Through the optimization process, the two solutions are compared, the best fitness of $\overline{x}$ and x are stored, and the worst fitness is replaced. For example, if $fit(x) \leq fit(\overline{x})$, then x is stored; otherwise,$\overline{x}$ is saved.

2.3 Evolutionary Population Dynamics

The EPD strategy is basically based on the theory of self-organized criticality (SOC) [12, 13], which indicates that the local changes in the population may influence the entire population without the intervention of any external [14, 15]. The EPD aims to improve the quality of the solutions by removing the worst solutions from the current iteration of the solutions and replacing the worst solutions by generating the new individuals around the best solution.

3 The Improved HGS with OBL and EPD Stategies

In this section, the structure of the proposed algorithm is explained. The HGS is modified by combining its original structure with OBL and EPD strategies to introduce the ability to deeply explore the search domain and rapidly reach the optimal value. The proposed algorithm is called OEHGS, and the detailed procedures of OEHGS are shown in Algorithm 1.

In general, the proposed OEHGS consists of the following stages: 1) the initial stage, 2) Update population with EPD, 3) Apply OBL in the population and 4) termination criterion.

Initial Stage: In this stage, the OEHGS algorithm begins by determining the initial parameter values of the HGS; then, the OEHGS randomly creates a population of size N in dimension *Dim*. Each individual is updated, and evaluated using the fitness function. Subsequently, the best solution and fitness value are saved.

Update Population with EPD: In this stage, EPD eliminates the worst solutions in the population and generates new populations in the neighborhood of the best agent. EPD is a simple and efficient operator based on the population technique, which is applied it to the proposed algorithm. Its calculation formula is as follows:

$$x_w = rand \times (ub_k - lb_k) + lb_k, \quad w = 1, 2, ..., N_w \tag{11}$$

where ub_k and lb_k represent the boundaries of the k-neighborhood of the best agent x_{best}. *rand* is a random value in the interval (0,1). The N_w represents the number of the worst agents and it is defined as follows:

$$N_w = round(N \times (rand \times (c_1 - c_2))), \quad c_1 = 0.1, c_2 = 0.9, r \in [0, 1] \tag{12}$$

where *rand* is a random value in the interval (0,1), *round*() is a function used to convert real number to integer [16].

Apply OBL in the Population: In this stage, the OBL technique receives the updated solutions from the HGS and calculates the opposite population of this part. The OBL is used to generate opposite population with a select rate S_r. We randomly generate a number in the range of 0 and 1. If the number is smaller than S_r, we adopt OBL to generate new solutions based on the current solutions. The result is re-evaluated by the fitness function; if the fitness value is better than the current value, the OEHGS updates the population with this value.

Termination Criterion: The second and third stages are repeated until the termination criterion is met. In this paper, the value of the maximum number of iterations is used as the stop criterion to evaluate the ability of OEHGS to find the optimal solution in a specific number of iterations. When OEHGS reaches the stop condition, the global optimal solution will be returned.

Algorithm 1 The proposed OEHGS algorithm

Step 1: Parameter initialization. Initialize the swarm size of the population x, the number of the dimensions of the search space D, sum of hungry feelings of all individuals *SHungry*, the maximum of iterations *itermax*.

Step 2: While (stopping condition is not met) do

Calculate the fitness of each individual fit_i;
Update the BF, WF, X_{best};
Use Eq. (5) to calculate *hungry* value
Use Eq. (3) to calculate w_1;
Use Eq. (4) to calculate w_2;
For each individual:
  Use Eq. (9) to calculate E;
  Use Eq. (8) to calculate R;
  Use Eq. (1) and Eq. (2) to update the positions;
End For
Apply the EPD strategy to update the individuals by Eq. (11) and Eq. (12);
Using OBL to generate another solution $\bar{x}$ by Eq. (10).
If ($random<S_r$)
  For(i=0;i<N;i++)
    For(d=0;d<D;d++)
$\bar{x} = l_i + u_i - x_i$
End for
  End for
  Evaluate x based on the fitness function
  Select N fitness solutions from $\{\bar{x}, x\}$ as the current individuals
End while
Return the best solution

Let the population size of individuals to be N, the dimension of the problem to be D, the total iteration to be *itermax*. The time complexity of the conventional HGS is $O(N*(1 + itermax*N*(2 + logN + 2*D)))$. The computational complexity of fitness evaluation and hunger update are both $O(N)$, the computational complexity of sorting requires $O(NlogN)$, the computational complexity of OBL and EPD is $O(N*D)$, the computational complexity of weight and location update is O(N*D). The total computational complexity of the entire phase is $O(N*(1 + itermax*N*(2 + logN + 4*D)))$. Therefore, the time complexity of the proposed version is consistent with the conventional HGS and does not increase the computational complexity [17].

4 Experiments and Analysis

4.1 Function Description

The proposed algorithm was compared with eight meta-heuristic algorithms, including SSA [18], PSO [19], WOA [20], MFO [21], SCA [22], DE [14], BBO [23] and FA [24]. Forty-five benchmark testing functions were selected from 23 well-known benchmarks (F5-F15), CEC2014 (F20-F30) and CEC2017 (F1-F20). The benchmark testing functions include six unimodal functions, twelve multimodal functions, seventeen fix-dimension functions and ten combine functions. The numerical results of these algorithms in terms of the average value (AVG) and standard deviation (STD) of the function error rates were obtained to assess the potentials of associated techniques, and the best result of each task is marked in **boldface**. Furthermore, the non-parametric statistical test of Wilcoxon rank-sum was performed at a significance level of 5% [25] is used to estimate the statistically significant difference between the proposed method and other competitors, and the algorithms were ranked based on the minimum mean values.

4.2 Parameter Settings

For fair comparisons, the parameters of involved algorithms are all the same as the original paper. The dimension D of the search space was set to be 30, the swarm size was set to be 30, S_r is set to be 0.5, and the maximum number of iterations was set to be 1000. All the algorithms were tested under the same condition and operating system. The initial parameters were set based on the acclaimed settings in prior papers. It is noteworthy that each algorithm of the experiments was executed according to the average results over 30 runs to reduce stochastic error in this paper.

4.3 Comparison with OEHGS and Other Well-Known Algorithms

We compared the OEHGS with the eight well-known algorithms for the 45 benchmark problems using means, standard deviation and average rank. The **boldface** shows the winner. In Table 1, it can be seen that the performance of OEHGS is the best in dealing with F1-F7, F10-F12, F14, F16, F17, F26, F27, F31, F34 and F38-F43. OEHGS ranks first in 10 algorithms, DE ranks second in 45 benchmark functions, and the original HGS ranks 4th in the 45 benchmark functions. The convergence curves of some different benchmark functions are shown in Fig. 1. It can be seen that OEHGS converges faster than the other algorithms in the majority of cases. For function F2, F38, F42 and F43, there is a close competition between OEHGS and HGS, both of them have obtained the best solutions. It can be seen that the average rank of our algorithm is only 2.22, which is much smaller than other algorithms. Compared with the fourth-ranked HGS, the average rank of OEHGS is about half of the HGS. Accordingly, it can be concluded that the performance of OEHGS is superior to the other counterparts. Table 2 shows the consequences of the Wilcoxon sign-rank test performed by OEHGS and other algorithms. Most of the p-values are less than 0.05. Even in SCA, all p-values are less than 0.05. This fact further indicates that OEHGS has a strong statistical significance compared to the other methods. In some benchmark functions, some algorithms and OEHGS reach the optimal values simultaneously, so the difference between the proposed algorithm and other competitors is not statistically significant.

The results of Fig. 1 shows that the convergence rate of OEHGS is fast. From F1, F4, F6, F7 and F27, it can be seen that OEHGS converges the fastest among all the algorithms, other algorithms converge quite slowly, and some of them even fall into local optimum. the convergence curves of F16, F26, F27 and F34 indicate that OEHGS has high accuracy in solving problems and can quickly find the global optimum at the beginning of the iteration. Although some algorithm's convergence speed is competitive in some stages, the accuracy of the solution to those methods is not as high as that of OEHGS, and the solution found by OEHGS has a higher quality. By observing the performance algorithms on F11, F12, F14, F31 and F41, it can be concluded that OEHGS has a strong ability for global exploration. As shown in these figures, it can be observed that the opposed-based learning and evolutionary population dynamics can guarantee a balance between exploration and exploitation.

Table 1. Comparison results of OEHGS and other well-known algorithms.

F		OEHGS	HGS	SSA	PSO	WOA	MFO	SCA	DE	BBO	FA
F1	AVG/	2.95E-19±	2.18E-13±	1.01E-08±	8.50E-09±	7.53E+02±	1.66E+03±	4.53E+00±	3.45E-13±	1.64E+01±	2.62E-03±
	STD	0.00E+00	0.00E+00	2.17E-09	2.63E-08	9.16E-02	3.72E+03	2.74E-01	2.57E-13	6.67E+00	8.71E-04
	RANK	1	2	5	4	9	10	7	3	8	6
F2	AVG/	1.11E-04±	1.31E-04±	5.38E-02±	3.90E+00	1.47E+01±	2.77E+00±	4.11E-02±	1.95E-02±	6.20E-01±	4.65E-01±
	STD	0.00E+00	0.00E+00	2.22E-02	±4.09E+00	1.84E-03	5.60E+00	4.89E-02	5.22E-03	2.40E-01	1.71E-01
	RANK	1	2	5	9	10	8	4	3	7	6
F3	AVG/	-1.26E+04±	-8.03E+03±	-7.67E+03±	-5.74E+03±	-6.85E+03±	-8.48E+03±	-3.85E+03±	-6.66E+03±	-1.25E+04±	-5.52E+03±
	STD	2.11E-01	6.36E+02	7.06E+02	1.17E+03	1.77E+03	9.36E+02	3.31E+02	4.91E+02	1.25E+01	3.58E+03
	RANK	1	5	6	9	7	4	10	8	3	2
F4	AVG/	0.00E+00±	5.68E-14±	4.73E+01±	9.14E+01±	2.92E+02±	1.63E+02±	1.23E+01±	1.60E+02±	5.92E+00±	7.88E+01±
	STD	0.00E+00	0.00E+00	1.84E+01	2.91E+01	1.42E-14	3.51E+01	2.07E+01	1.13E+01	1.33E+00	2.00E+01
	RANK	1	2	5	7	10	9	4	8	3	6
F5	AVG/	4.44E-16±	4.44E-16±	1.66E+00±	9.83E-05±	1.99E+01±	1.84E+01±	1.73E+01±	2.25E-07±	2.02E+00±	1.91E-01±
	STD	0.00E+00	0.00E+00	8.33E-01	2.63E-04	2.32E-15	4.97E+00	6.51E+00	9.04E-08	2.64E-01	4.12E-01
	RANK	1	1	6	4	10	9	8	3	7	5
F6	AVG/	0.00E+00±0.	1.11E-16±	1.05E-02±	5.42E-03±	4.29E+02±	6.05E+00±	2.77E-01±	2.47E-04±	1.14E+00±	1.01E-02±
	STD	00E+00	0.00E+00	1.37E-02	6.56E-03	0.00E+00	2.26E+01	2.11E-01	1.33E-03	7.79E-02	6.83E-03
	RANK	1	2	6	4	10	9	7	3	8	5
F7	AVG/	1.31E-18±	6.54E-04±	4.40E+00±	3.46E-03±	2.94E+01±	1.71E+07±	2.84E+00±	3.46E-03±	6.48E-02	1.04E-02±
	STD	0.00E+00	0.00E+00	3.14E+00	1.86E-02	4.92E-03	6.39E+07	5.25E+00	1.86E-02	±5.06E-02	3.11E-02
	RANK	1	2	8	4	9	10	7	3	6	5
F8	AVG/	6.81E-01±	2.49E-01±	8.34E-03±	3.63E-03±	7.63E+00±	1.13E-01±	3.57E+03±	5.32E-02±	6.91E-01±	2.03E-03±
	STD	8.60E-01	5.10E-01	1.89E-02	5.81E-03	1.47E-01	4.54E-01	1.90E+04	2.87E-01	2.55E-01	4.13E-03
	RANK	7	6	3	2	9	5	10	4	8	1
F9	AVG/	1.06E+00±	2.24E+00±	9.98E-01±	2.38E+00±	1.05E+01±	1.95E+00±	1.40E+00±	1.13E+00±	1.09E+00±	1.06E+00±
	STD	5.20E-01	4.05E+00	2.61E-16	1.60E+00	1.85E+00	1.52E+00	7.92E-01	5.56E-01	4.31E-01	2.48E-01
	RANK	2	8	1	9	10	7	6	5	4	3
F10	AVG/	5.21E-04±	5.53E-04±	7.96E-04±	7.48E-03±	3.65E-03±	9.63E-04±	1.05E-03±	1.20E-03±	1.34E-02±	9.40E-04±
	STD	0.00E+00	0.00E+00	2.71E-04	9.51E-03	3.92E-04	3.75E-04	3.73E-04	3.57E-03	1.29E-02	1.36E-04
	RANK	1	2	3	9	8	5	6	7	10	4
F11	AVG/	-1.03E+00±	-1.03E+00±	-1.03E+00±	-1.03E+00±	-9.77E-01±	-1.03E+00±	-1.03E+00±	-1.03E+00±	-1.03E+00±	-1.03E+00±
	STD	0.00E+00	0.00E+00	7.32E-15	6.60E-16	7.22E-11	6.66E-16	2.40E-05	6.66E-16	5.60E-03	2.58E-09
	RANK	1	1	1	1	10	1	8	1	9	7
F12	AVG/	3.98E-01±	3.98E-01±	3.98E-01±	4.75E-01±	3.98E-01±	3.98E-01±	3.99E-01±	3.98E-01±	4.51E-01±	3.98E-01±
	STD	0.00E+00	0.00E+00	1.01E-14	4.14E-01	1.15E-06	0.00E+00	9.59E-04	2.17E-03	9.31E-02	2.93E-09
	RANK	1	1	4	10	6	1	8	7	9	5
F13	AVG/	3.00E+00±	3.00E+00±	3.00E+00±	3.00E+00±	3.02E+00±	3.00E+00±	3.00E+00±	3.00E+00±	1.29E+01±	3.00E+00±
	STD	0.00E+00	0.00E+00	6.13E-14	1.11E-15	2.26E-05	1.56E-15	2.20E-05	2.32E-15	1.38E+01	1.23E-08
	RANK	2	6	5	2	9	2	8	1	10	7
F14	AVG/	-3.86E+00±	-3.86E+00±	-3.86E+00±	-3.86E+00±	-3.85E+00±	-3.86E+00±	-3.86E+00±	-3.86E+00±	-3.86E+00±	-3.86E+00±
	STD	0.00E+00	0.00E+00	1.42E-14	2.36E-03	3.41E-03	2.66E-15	2.23E-03	2.66E-15	5.18E-04	5.70E-10
	RANK	1	6	4	8	10	1	9	1	7	5
F15	AVG/	-3.25E+00±	-3.21E+00±	-3.22E+00±	-3.20E+00±	-2.97E+00±	-3.21E+00±	-2.98E+00±	-3.24E+00±	-3.26E+00±	-3.27E+00±
	STD	6.00E-02	8.00E-02	4.45E-02	3.06E-01	8.56E-02	4.75E-02	1.80E-01	6.01E-02	5.96E-02	5.91E-02
	RANK	3	6	5	8	10	7	9	4	2	1
F16	AVG/	3.84E+03±	1.31E+04±	4.23E+03±	4.59E+08±	1.40E+10±	1.40E+11±	1.83E+11±	7.23E+03±	2.65E+08±	5.63E+04±
	STD	4.88E+03	2.66E+04	5.94E+03	2.47E+09	7.08E+09	7.55E+10	3.20E+10	6.03E+03	1.13E+08	2.92E+04
	RANK	1	4	2	7	8	9	10	3	6	5
F17	AVG/	1.06E+04±	2.86E+04±	1.42E+04±	1.33E+04±	2.60E+05±	1.43E+05±	6.46E+04±	1.37E+05±	1.63E+05±	1.23E+05±
	STD	3.53E+03	8.74E+03	5.90E+03	5.83E+03	6.57E+04	5.31E+04	1.35E+04	2.71E+04	3.30E+04	5.46E+04
	RANK	1	4	3	2	10	8	5	7	9	6
F18	AVG/	4.84E+02±	4.91E+02±	5.05E+02±	4.93E+02±	8.21E+02±	1.29E+03±	2.51E+03±	4.95E+02±	5.43E+02±	4.40E+02±
	STD	3.69E+01	2.94E+01	3.04E+01	2.16E+01	1.33E+02	6.56E+02	7.92E+02	1.12E+01	3.50E+01	2.54E+01
	RANK	2	3	6	4	8	9	10	5	7	1
F19	AVG/	6.69E+02±	7.15E+02±	6.46E+02±	6.83E+02±	8.44E+02±	6.92E+02±	8.21E+02±	7.04E+02±	5.89E+0±	5.76E+02±
	STD	3.85E+01	4.39E+01	3.51E+01	2.86E+01	6.65E+01	5.15E+01	2.55E+01	1.02E+01	21.97E+01	2.06E+01
	RANK	4	8	3	5	10	6	9	7	2	1
F20	AVG/	6.50E+02±	6.74E+02±	6.50E+02±	6.58E+02±	6.99E+02±	6.52E+02±	6.76E+02±	6.00E+02±	6.32E+02±	6.06E+02±
	STD	1.15E+01	1.14E+01	1.86E+01	7.00E+00	1.60E+01	1.34E+01	6.45E+00	2.53E-01	1.23E+01	3.01E+00
	RANK	4	8	5	7	10	6	9	1	3	2

(continued)

Table 1. *(continued)*

F21	AVG/	1.05E+03±	1.10E+03±	8.97E+02±	8.84E+02±	1.27E+03±	1.11E+03±	1.22E+03±	9.38E+02±	8.82E+02±	**8.08E+02±**
	STD	5.82E+01	9.69E+01	6.04E+01	4.21E+01	8.64E+01	2.10E+02	6.00E+01	**1.27E+01**	2.29E+01	2.36E+01
	RANK	6	7	4	3	10	8	9	5	2	**1**
F22	AVG/	9.47E+02±	9.68E+02±	9.53E+02±	9.33E+02±	1.04E+03±	1.04E+03±	1.08E+03±	1.00E+03±	8.88E+02±	**8.85E+02±**
	STD	1.93E+01	3.50E+01	4.31E+01	2.73E+01	5.65E+01	4.73E+01	2.65E+01	**8.86E+00**	1.91E+01	2.52E+01
	RANK	4	6	5	3	8	9	10	7	2	**1**
F23	AVG/	4.34E+03±	5.49E+03±	5.00E+03±	4.39E+03±	1.05E+04±	6.52E+03±	7.52E+03±	**9.00E+02±**	1.74E+03±	9.64E+02±
	STD	9.61E+02	1.03E+03	1.38E+03	1.25E+03	3.41E+03	2.10E+03	1.47E+03	**1.93E-01**	5.25E+02	2.24E+02
	RANK	4	7	6	5	10	8	9	**1**	3	2
F24	AVG/	5.30E+03±	5.48E+03±	5.12E+03±	4.60E+03±	6.88E+03±	5.59E+03±	8.69E+03±	8.45E+03±	**4.07E+03±**	4.33E+03±
	STD	5.86E+02	7.44E+02	7.01E+02	5.93E+02	7.37E+02	5.98E+02	3.84E+02	**2.74E+02**	4.15E+02	5.65E+02
	RANK	5	6	4	3	8	7	10	9	**1**	2
F25	AVG/	1.27E+03±	1.36E+03±	1.31E+03±	**1.23E+03±**	6.41E+03±	7.15E+03±	3.26E+03±	1.24E+03±	8.53E+03±	1.72E+03±
	STD	7.63E+01	8.02E+01	9.07E+01	3.53E+01	1.95E+03	8.36E+03	1.01E+03	**2.15E+01**	6.08E+03	2.92E+02
	RANK	3	5	4	**1**	8	9	7	2	10	6
F26	AVG/	**6.13E+05±**	1.07E+07±	2.17E+08±	1.21E+06±	2.31E+09±	1.18E+09±	1.42E+10±	3.18E+06±	3.58E+07±	1.75E+08±
	STD	**5.85E+05**	9.17E+06	1.18E+08	3.78E+06	1.23E+09	2.04E+09	4.29E+09	5.50E+06	2.12E+07	1.38E+08
	RANK	**1**	4	7	2	9	8	10	3	5	6
F27	AVG/	**2.93E+04±**	1.19E+05±	1.26E+05±	2.54E+07±	2.70E+07±	1.45E+08±	8.33E+09±	4.29E+04±	1.01E+08±	1.09E+06±
	STD	**2.99E+04**	8.55E+04	7.61E+04	1.29E+08	3.75E+07	2.86E+08	3.47E+09	6.00E+04	1.42E+08	1.38E+06
	RANK	**1**	3	4	6	7	9	10	2	8	5
F28	AVG/	3.62E+04±	7.81E+04±	4.07E+04±	3.41E+04±	2.13E+06±	4.14E+05±	5.45E+05±	**2.03E+03±**	2.35E+06±	1.36E+05±
	STD	3.12E+04	7.86E+04	3.50E+04	4.33E+04	2.33E+06	8.51E+05	3.52E+05	**1.24E+03**	2.39E+06	1.34E+05
	RANK	3	5	4	2	9	7	8	**1**	10	6
F29	AVG/	2.47E+04±	3.07E+04±	6.74E+04±	9.22E+03±	1.70E+07±	8.33E+04±	3.72E+08±	3.20E+03±	2.86E+07±	5.76E+05±
	STD	3.11E+04	2.05E+04	5.82E+04	9.18E+03	2.60E+07	9.35E+04	2.94E+08	1.94E+03	2.52E+07	6.42E+05
	RANK	3	4	5	2	8	6	10	1	9	7
F30	AVG/	2.75E+03±	2.96E+03±	2.86E+03±	2.79E+03±	4.17E+03±6	3.06E+03±2	4.04E+03±	3.20E+03±	2.77E+03±	**2.66E+03±**
	STD	3.10E+02	3.30E+02	3.65E+02	2.86E+02	.26E+02	.83E+02	2.62E+02	**1.77E+02**	3.56E+02	3.18E+02
	RANK	2	6	5	4	10	7	9	8	3	1
F31	AVG/	**2.23E+03±**	2.47E+03±	2.28E+03±	2.44E+03±	2.73E+03±	2.50E+03±	2.72E+03±	2.25E+03±	2.38E+03±	2.26E+03±
	STD	1.96E+02	2.29E+02	2.07E+02	2.47E+02	2.26E+02	2.90E+02	1.93E+02	1.44E+02	1.58E+02	2.29E+02
	RANK	**1**	7	4	6	10	8	9	2	5	3
F32	AVG/	6.64E+05±	1.19E+06±	7.69E+05±	**3.01E+05±**	9.42E+06±	4.20E+06±	9.41E+06±	3.73E+05±	5.08E+06±	9.90E+05±
	STD	6.25E+05	1.26E+06	5.84E+05	2.18E+05	9.82E+06	9.46E+06	4.95E+06	**2.10E+05**	5.12E+06	6.42E+05
	RANK	3	6	4	**1**	10	7	9	2	8	5
F33	AVG/	2.90E+04±4.	2.34E+04±2.	1.89E+07±1	6.55E+03±5	5.95E+07±7	1.19E+08±3	5.79E+08±3	**6.22E+03±7**	1.55E+07±1	1.65E+07±1
	STD	53E+04	75E+04	.43E+07	.25E+03	.62E+07	.52E+08	.12E+08	**.82E+03**	.19E+07	.08E+07
	RANK	4	3	7	2	8	9	10	**1**	5	6
F34	AVG/	**2.44E+03±**	2.77E+03±	2.58E+03±	2.63E+03±1	2.80E+03±2	2.71E+03±2	2.75E+03±1	2.52E+03±2	2.59E+03±2	2.54E+03±1
	STD	**1.41E+02**	2.01E+02	1.45E+02	.99E+02	.03E+02	.45E+02	.67E+02	.03E+02	.32E+02	.98E+02
	RANK	**1**	9	4	6	10	7	8	2	5	3
F35	AVG/	1.03E+04±	2.24E+04±	2.79E+04±	1.75E+04±7	1.18E+05±9	7.71E+04±3	4.39E+04±1	**2.93E+03±1**	5.87E+04±2	7.61E+04±4
	STD	5.40E+03	1.10E+04	1.69E+04	.71E+03	.22E+04	.99E+04	.85E+04	**.38E+03**	.72E+04	.84E+04
	RANK	2	4	5	3	10	9	6	**1**	7	8
F36	AVG/	3.42E+05±	4.73E+05±	2.73E+05±	2.12E+05±	8.58E+06±	1.01E+06±	4.08E+06±	**3.79E+04±**	2.87E+06±	5.91E+05±
	STD	2.49E+05	3.82E+05	2.76E+05	1.65E+05	7.55E+06	7.51E+05	2.88E+06	**7.06E+04**	1.74E+06	5.20E+05
	RANK	4	5	3	2	10	7	9	**1**	8	6
F37	AVG/	2.68E+03±	2.97E+03±	2.75E+03±	2.94E+03±	3.11E+03±	2.96E+03±	3.29E+03±	2.71E+03±	2.91E+03±	2.62E+03±
	STD	1.51E+02	2.30E+02	2.00E+02	1.81E+02	2.53E+02	2.43E+02	1.68E+02	2.01E+02	2.65E+02	2.03E+02
	RANK	2	8	4	6	9	7	10	3	5	1
F38	AVG/	**2.50E+03±**	**2.50E+03±**	2.63E+03±	2.62E+03±	2.69E+03±	2.67E+03±	2.71E+03±	2.62E+03±	2.62E+03±	2.64E+03±
	STD	**0.00E+00**	**0.00E+00**	7.29E+00	1.50E+00	4.03E+01	3.78E+01	2.73E+01	0.00E+00	2.14E+00	1.68E+01
	RANK	**1**	**1**	6	4	9	8	10	3	5	7
F39	AVG/	**2.60E+03±**	**2.60E+03±**	2.64E+03±	2.62E+03±	2.61E+03±	2.69E+03±	2.61E+03±	2.63E+03±	2.64E+03±	2.64E+03±
	STD	**0.00E+00**	**0.00E+00**	1.06E+01	6.80E+00	4.74E+00	3.26E+01	1.52E+01	3.93E+00	5.79E+00	5.69E+00
	RANK	**1**	**1**	8	5	3	10	4	6	7	9
F40	AVG/	**2.70E+03±**	**2.70E+03±**	2.72E+03±	2.72E+03±	2.71E+03±	2.72E+03±	2.74E+03±	2.72E+03±	2.72E+03±	2.71E+03±
	STD	**0.00E+00**	**0.00E+00**	4.27E+00	6.50E+00	1.68E+01	9.79E+00	1.18E+01	3.27E+00	2.78E+00	2.33E+00
	RANK	**1**	**1**	7	8	4	9	10	5	5	3
F41	AVG/	**2.70E+03±**	2.73E+03±	2.70E+03±	2.78E+03±	2.71E+03±	2.70E+03±	2.70E+03±	2.71E+03±	2.72E+03±	2.72E+03±
	STD	**1.00E-01**	4.69E+01	1.20E-01	4.02E+01	3.38E+01	1.04E+00	3.30E-01	3.66E+01	4.26E+01	5.45E+01
	RANK	**1**	9	2	10	6	3	4	5	8	7
F42	AVG/	**2.90E+03±**	**2.90E+03±**	3.53E+03±1	3.40E+03±	3.88E+03±	3.59E+03±	3.80E+03±	3.10E+03±	3.47E+03±	3.12E+03±
	STD	**0.00E+00**	**0.00E+00**	.38E+02	3.21E+02	3.39E+02	2.05E+02	3.21E+02	6.95E+01	1.71E+02	5.03E+01
	RANK	**1**	**1**	7	5	10	8	9	3	6	4
F43	AVG/	**3.00E+03±**	**3.00E+03±**	4.14E+03±	6.85E+03±	5.47E+03±	3.85E+03±	5.54E+03±	3.67E+03±	3.99E+03±	3.26E+03±
	STD	**0.00E+00**	**0.00E+00**	3.23E+02	5.07E+02	7.27E+02	1.12E+02	4.23E+02	3.79E+01	2.73E+02	7.29E+01
	RANK	**1**	**1**	7	10	8	5	9	4	6	3
F44	AVG/	2.56E+05±	2.85E+05±	4.63E+06±	2.25E+06±	1.34E+07±	3.12E+06±	3.33E+07±	1.65E+06±	1.04E+04±	**3.12E+03±**
	STD	3.23E+05	1.51E+06	7.37E+06	5.35E+06	1.23E+07	3.74E+06	1,39E+07	3.70E+06	3.91E+03	**1.62E+01**
	RANK	3	4	8	6	9	7	10	5	2	**1**
F45	AVG/	7.92E+03±	1.11E+04±	3.29E+04±	7.16E+03±	2.67E+05±	4.07E+04±	4.92E+05±	6.04E+03±	2.16E+04±	**4.12E+03±**
	STD	1.56E+03	8.91E+03	1.66E+04	3.60E+03	1.53E+05	3.20E+04	2.01E+05	1.11E+03	9.46E+03	**3.78E+02**
	RANK	4	5	7	3	9	8	10	2	6	**1**
Sum of ranks		**100**	195	217	223	395	316	373	168	269	188
Average rank		**2.22**	4.33	4.82	4.96	8.78	7.02	8.29	3.73	5.98	4,18
Overall rank		**1**	4	5	6	10	8	9	2	7	3

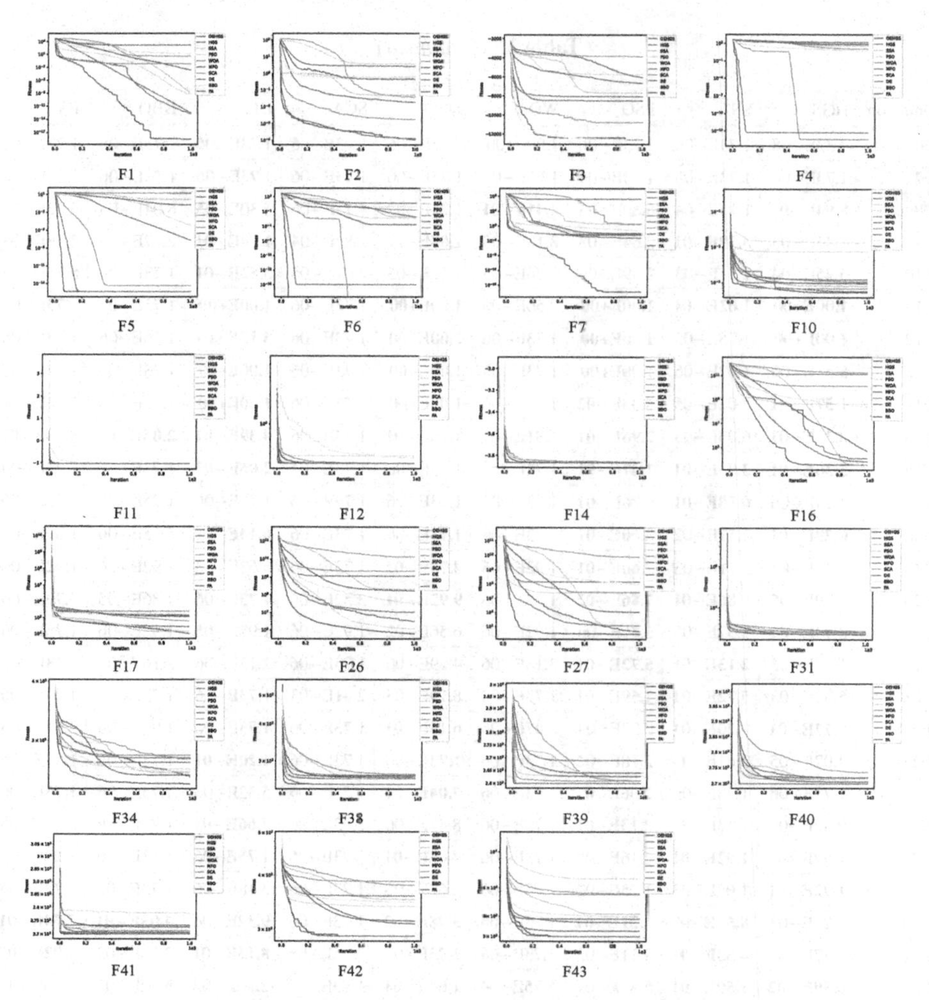

Fig. 1. Convergence trends curves for OEHGS versus other optimize

Table 2. The p−value of the Wilcoxon test obtained from comparison with traditional algorithms.

Function	HGS	SSA	PSO	WOA	MFO	SCA	DE	BBO	FA
F1	7.95E−07	1.73E−06	7.95E−07	1.73E−06	7.95E−07	1.73E−06	1.73E−06	7.95E−07	7.95E−07
F2	**4.65E−01**	1.73E−06	1.73E−06	1.25E−04	1.73E−06	1.73E−06	1.73E−06	1.73E−06	1.73E−06
F3	1.73E−06	1.73E−06	1.73E−06	1.73E−06	1.73E−06	1.73E−06	1.73E−06	1.73E−06	1.73E−06
F4	4.32E−08	1.73E−06	1.73E−06	**1.00E+00**	1.73E−06	1.73E−06	1.73E−06	1.73E−06	1.73E−06
F5	**1.00E+00**	1.73E−06	1.73E−06	7.90E−05	1.73E−06	1.73E−06	1.73E−06	1.73E−06	1.73E−06

(*continued*)

Table 2. (*continued*)

Function	HGS	SSA	PSO	WOA	MFO	SCA	DE	BBO	FA
F6	4.32E−08	1.73E−06	1.73E−06	1.00E+00	1.73E−06	1.73E−06	1.73E−06	1.73E−06	1.73E−06
F7	1.73E−06	1.73E−06	1.73E−06	1.73E−06	1.73E−06	1.73E−06	1.73E−06	1.73E−06	1.73E−06
F8	**8.94E−01**	1.15E−04	6.89E−05	**4.41E−01**	1.25E−02	1.73E−06	5.30E−05	6.04E−03	6.32E−05
F9	1.45E−03	**3.39E−01**	2.64E−03	8.18E−05	**1.08E−01**	3.88E−04	**1.94E−01**	2.77E−03	2.58E−03
F10	5.25E−04	1.11E−03	2.59E−06	2.26E−03	1.12E−05	4.07E−05	**3.82E−01**	1.73E−06	5.75E−06
F11	**1.00E+00**	**1.02E−01**	**1.00E+00**	2.56E−06	**1.00E+00**	1.73E−06	**1.00E+00**	1.73E−06	1.73E−06
F12	**1.00E+00**	1.75E−05	**1.00E+00**	1.73E−06	**1.00E+00**	1.73E−06	**3.17E−01**	1.73E−06	1.73E−06
F13	5.61E−06	1.67E−06	**1.00E+00**	1.73E−06	**1.00E+00**	1.73E−06	**1.00E+00**	1.73E−06	1.73E−06
F14	**1.57E−01**	3.02E−05	**8.33E−02**	1.73E−06	**1.00E+00**	1.73E−06	**1.00E+00**	1.73E−06	1.73E−06
F15	**8.33E−01**	6.03E−03	**2.96E−01**	**7.81E−01**	**1.12E−01**	1.73E−06	**4.39E−01**	**2.06E−01**	**8.13E−01**
F16	**5.30E−01**	**3.93E−01**	**1.47E−01**	1.73E−06	1.73E−06	1.73E−06	**2.06E−01**	1.73E−06	1.73E−06
F17	4.29E−06	**6.58E−01**	**9.26E−01**	1.73E−06	1.73E−06	1.73E−06	1.73E−06	1.73E−06	1.73E−06
F18	**6.29E−01**	1.57E−02	**3.60E−01**	1.73E−06	1.73E−06	1.73E−06	**6.14E−01**	5.75E−06	1.89E−04
F19	7.27E−03	1.71E−03	**7.66E−01**	1.92E−06	**4.78E−01**	1.73E−06	5.67E−03	1.92E−06	1.92E−06
F20	5.79E−05	**5.86E−01**	**7.86E−02**	1.73E−06	**9.92E−01**	3.52E−06	1.73E−06	4.86E−05	1.73E−06
F21	8.94E−04	2.37E−05	2.35E−06	1.92E−06	6.56E−02	1.92E−06	1.97E−05	1.92E−06	1.73E−06
F22	1.20E−03	**2.13E−01**	**5.72E−01**	3.18E−06	4.29E−06	1.73E−06	2.13E−06	2.16E−05	1.13E−05
F23	5.32E−03	**5.72E−01**	**1.59E−01**	1.73E−06	8.19E−05	2.41E−03	1.73E−06	1.73E−06	1.73E−06
F24	**8.77E−01**	**1.20E−01**	2.22E−04	8.47E−06	**6.14E−01**	1.73E−06	1.73E−06	1.92E−06	3.52E−06
F25	4.07E−05	3.61E−03	2.18E−02	1.73E−06	1.73E−06	1.73E−06	**1.20E−01**	1.73E−06	1.92E−06
F26	1.73E−06	1.73E−06	**2.06E−01**	1.73E−06	**7.04E−01**	1.73E−06	5.32E−03	1.73E−06	1.73E−06
F27	9.32E−06	1.13E−05	**2.13E−01**	1.73E−06	8.47E−06	1.73E−06	**7.66E−01**	1.73E−06	1.73E−06
F28	**1.99E−01**	**1.92E−01**	3.16E−02	1.73E−06	**4.05E−01**	1.73E−06	1.73E−06	1.73E−06	9.27E−03
F29	**1.92E−01**	1.04E−03	1.25E−02	1.73E−06	2.60E−06	1.73E−06	3.18E−06	1.73E−06	1.73E−06
F30	9.27E−03	**8.59E−02**	**2.21E−01**	1.73E−06	**5.98E−02**	1.73E−06	9.32E−06	**4.05E−01**	**7.34E−01**
F31	2.77E−03	**4.53E−01**	2.11E−03	1.73E−06	1.73E−06	2.35E−06	**8.13E−01**	2.70E−02	**9.92E−01**
F32	**8.59E−02**	**1.59E−01**	5.32E−03	5.75E−06	1.04E−03	1.73E−06	2.30E−02	3.72E−05	3.00E−02
F33	**6.88E−01**	1.73E−06	1.40E−02	1.73E−06	1.92E−06	1.73E−06	4.99E−03	1.73E−06	1.73E−06
F34	1.36E−05	3.88E−04	6.16E−04	1.36E−05	1.73E−06	1.13E−05	7.52E−02	4.68E−03	2.18E−02
F35	1.04E−03	1.74E−04	7.16E−04	1.73E−06	1.73E−06	1.73E−06	3.88E−06	2.13E−06	2.60E−06
F36	**7.86E−02**	**2.13E−01**	2.18E−02	1.92E−06	1.73E−06	1.73E−06	1.73E−06	1.73E−06	1.96E−02
F37	4.86E−05	**8.59E−02**	9.71E−05	5.22E−06	**7.97E−01**	1.73E−06	**2.29E−01**	2.41E−03	**4.17E−01**
F38	**1.00E+00**	1.73E−06	1.73E−06	2.56E−06	1.73E−06	1.73E−06	1.73E−06	1.73E−06	1.73E−06
F39	**1.00E+00**	1.73E−06	1.73E−06	1.73E−06	1.73E−06	1.73E−06	1.73E−06	1.73E−06	1.73E−06
F40	**1.00E+00**	1.73E−06	1.73E−06	9.82E−04	1.73E−06	1.73E−06	1.73E−06	1.73E−06	1.73E−06

(*continued*)

Table 2. (*continued*)

Function	HGS	SSA	PSO	WOA	MFO	SCA	DE	BBO	FA
F41	1.60E−04	6.04E−03	2.35E−06	1.85E−01	1.73E−06	1.73E−06	**3.39E−01**	**1.97E−05**	**2.13E−01**
F42	**1.00E+00**	1.73E−06	1.73E−06	1.73E−06	1.73E−06	1.73E−06	1.73E−06	1.73E−06	1.73E−06
F43	**1.00E+00**	1.73E−06	1.73E−06	1.73E−06	1.72E−06	1.73E−06	1.73E−06	1.73E−06	1.73E−06
F44	**8.64E−01**	1.25E−02	**1.06E−01**	2.60E−06	**6.73E−01**	1.73E−06	1.57E−02	6.73E−01	**6.73E−01**
F45	**6.88E−01**	1.73E−06	6.84E−03	1.73E−06	6.32E−05	1.73E−06	6.89E−05	4.53E−04	2.13E−06

5 Conclusions

In this paper, we propose an improved version of the original HGS algorithm using OBL and EPD strategies. These strategies focus on increasing the convergence rate of meta-heuristic algorithms by calculating the opposite solution to the current solution and reducing the impact on the poor solution to the population quality, to balance the exploration and exploitation in the algorithm. To evaluate the performance of the proposed algorithm, a set of experiment series was performed. For the unconstrained continuous optimization problem, 45 benchmark functions were chosen to compare OEHGS with other state-of-the-art algorithms including SSA, PSO, WOA, MFO, SCA and DE. The results verify the accuracy and convergence speed of OEHGS. The comparison results indicate that the performance of OEHGS is better than that of many other methods.

In the future study, the OEHGS can be applied to several fields, such as image segmentation, feature selection, and multi objective optimization algorithms. Moreover, the proposed algorithm can be applied to the machine learning model such as the parameter optimization of SVM model, and a prediction method of new energy problems.

Acknowledgment. This work was supported by the National Natural Science Foundation of China under Grant No. 61373057.

References

1. Li, S., Li, X., Chen, H., Zhao, Y., Dong, J.: A Novel Hybrid Hunger Games Search Algorithm With Differential Evolution for Improving the Behaviors of Non-Cooperative Animals. IEEE Access **9**, 164188–164205 (2021)
2. Abdel-Basset, M., Hessin, A.-N., Abdel-Fatah, L.: A comprehensive study of cuckoo-inspired algorithms. Neural Comput. Appl. **29**(2), 345–361 (2016). https://doi.org/10.1007/s00521-016-2464-8
3. Jordehi, A.: Rezaee, Enhanced leader PSO (ELPSO): A new PSO variant for solving global optimisation problems. Appl. Soft Comput. J. **26**, 401–417 (2015)
4. Gupta, S., Deep, K., Heidari, A.A., Moayedi, H., Wang, M.: Opposition-based learning Harris hawks optimization with advanced transition rules: principles and analysis. Exp. Syst. Appl. **158**, 113510 (2020)
5. Rahnamayan, S., Tizhoosh, H.R., Salama, M.M.: Opposition-based differential evolution. EEE Trans. Evol. Comput. **12**, 64–79 (2008)

6. Zhang, S., Luo, Q., Zhou, Y.: Applications, Hybrid grey wolf optimizer using elite opposition-based learning strategy and simplex method. Int. J. Comput. Intell. Appl. **16**, 1750012 (2017)
7. Cushing, J.M.: Difference equations as models of evolutionary population dynamics. J. Biol. Dyn. **13**, 103–127 (2019)
8. Liang, J.J., Qu, B.Y., Suganthan, P.N.J.C.I.L.: Zhengzhou University, Zhengzhou China, N.T.U. Technical Report, Singapore. In: Problem Definitions and Evaluation Criteria for the CEC 2014 Special Session and Competition on Single Objective Real-Parameter Numerical Optimization, vol. 635, p. 490 (2013)
9. Latorre, A., Pena, J.M.: A comparison of three large-scale global optimizers on the CEC 2017 single objective real parameter numerical optimization benchmark. In: Evolutionary Computation (2017)
10. Yang, Y., Chen, H., Heidari, A.A., Gandomi, A.H.: Hunger games search: visions, conception, implementation, deep analysis, perspectives, and towards performance shifts. Expert Syst. Appl. **177**, 114864 (2021)
11. Tizhoosh, H.R.: Opposition-based learning: a new scheme for machine intelligence. In: International Conference on Computational Intelligence for Modelling, Control and Automation and International Conference on Intelligent Agents, Web Technologies and Internet Commerce (CIMCA-IAWTIC 2006), pp. 695–701. IEEE (2005)
12. E. Goles, Self-Organized Critically: An Explanation of 1/f Noise, Ann.inst.h.poincaré Phys.théor, 56 (1992) 75–90
13. Bak, P., Tang, C., Wiesenfeld, K.: Self-organized criticality: an explanation of 1/f noise. Phys. Rev. Lett. **71**, 364–374 (1987)
14. Storn, R.: Differential evolution-a simple and efficient heuristic for global optimization over continuous space. J. Global Optim. **11**, 341–359 (1997)
15. Yong, J., He, F., Li, H., Zhou, W.: A novel bat algorithm based on collaborative and dynamic learning of opposite population. In: 2018 IEEE 22nd International Conference on Computer Supported Cooperative Work in Design (CSCWD) (2018)
16. Abd Elaziz, M., Yousri, D., Mirjalili, S.: A hybrid Harris hawks-moth-flame optimization algorithm including fractional-order chaos maps and evolutionary population dynamics. Adv. Eng. Software **154**, 102973 (2021)
17. Chen, R., Yang, B., Li, S., Wang, S., Cheng, Q.: An effective multi-population grey wolf optimizer based on reinforcement learning for flow shop scheduling problem with multi-machine collaboration. Comput. Ind. Eng. **162**, 107738 (2021)
18. Mirjalili, S., Gandomi, A.H., Mirjalili, S.Z., Saremi, S., Faris, H., Mirjalili, S.M.: Salp Swarm Algorithm: a bio-inspired optimizer for engineering design problems. Adv. Eng. Softw. **114**, 163–191 (2017)
19. Chen, H., Xu, Y., Wang, M., Zhao, X.: A balanced whale optimization algorithm for constrained engineering design problems. Appl. Math. Model. **71**, 45–59 (2019)
20. Mirjalili, S., Lewis, A.: The whale optimization algorithm. Adv. Eng. Software **95**, 51–67 (2016)
21. Mirjalili, S.: Moth-flame optimization algorithm: a novel nature-inspired heuristic paradigm. Knowl.-Based Syst. **89**, 228–249 (2015)
22. Mirjalili, S.: SCA: a sine cosine algorithm for solving optimization problems. Knowl.-Based Syst. **96**, 120–133 (2016)
23. Simon, D.: Biogeography-based optimization. IEEE Trans. Evol. Comput. **12**, 702–713 (2008)
24. Yang, X.-S.: Firefly algorithm, Lévy flights and global optimization. In: Bramer, M., Ellis, R., Petridis, M. (eds.) Research and Development in Intelligent Systems XXVI, pp. 209–218. Springer, London (2010). https://doi.org/10.1007/978-1-84882-983-1_15
25. Cuzick, J.: A wilcoxon-type test for trend. Stat. Med. **14**, 445–446 (1995)

An Improved Hydrologic Cycle Optimization Algorithm for Solving Engineering Optimization Problems

Haiyun Qiu[1], Bowen Xue[1], Ben Niu[1,2], Tianwei Zhou[1,2], and Junrui Lu[1,2](✉)

[1] College of Management, Shenzhen University, Shenzhen 518060, China
luxunrui2021@email.szu.edu.cn
[2] Greater Bay Area International Institute for Innovation, Shenzhen University, Shenzhen 518060, China

Abstract. This paper proposes an improved hydrologic cycle optimization algorithm (IHCO) for solving real-world constrained engineering optimization problems. In the improved algorithm, a new flow strategy is carried out by utilizing the empirical knowledge of the population. Meanwhile, in order to balance exploration and exploitation, evaporation and precipitation operator in basic hydrologic cycle optimization is redesigned and an adaptive Gaussian mutation method is introduced. The standard deviation of the Gaussian distribution decreases linearly as the algorithm proceeds. Compared with several metaheuristic algorithms, the superiority of IHCO is validated through thirteen engineering optimization problems. The experimental results demonstrate that IHCO outperforms the basic algorithm, and it has a satisfactory capability to enhance performance.

Keywords: Improved hydrologic cycle optimization · Metaheuristic algorithm · Engineering optimization problem · Adaptive gaussian mutation

1 Introduction

Optimization is the essence of human decision-making. It attempts to maximize or minimize a predefined objective function. In science and engineering field, based on some design standards and safety rules, most of the optimization problems are highly nonlinear and have complex constraints. Such problems are called constrained optimization problems, which can be formulated as a D-dimensional minimization problem. The general form can be defined as formula (1):

$$\begin{aligned} & minimize\, f(x),\ x = (x_1, x_2 ..., x_D),\ x_i^{min} \le x_i \le x_i^{max} \\ & subject\ to: g_j(x) \le 0, j = 1, ..., n, \\ & \qquad h_j(x) = 0, j = n + 1, ..., m. \end{aligned} \tag{1}$$

where x is a feasible solution and each decision variable x_i is within the specified range $\left[x_i^{min}, x_i^{max}\right]$. $f(\cdot)$ is the predefined objective function. The feasible region set is defined by a set of n inequality constraints $g(x)$ and a set of $m - n$ equality constraints $h(x)$.

Y. Tan et al. (Eds.): ICSI 2022, LNCS 13344, pp. 117–127, 2022.
https://doi.org/10.1007/978-3-031-09677-8_10

To solve constrained engineering problems, various types of optimization methods are available. Traditional optimization methods mostly proceed based on derivative or gradient, such as steepest descent method [1] and Newton method [2]. They are proved to be suitable for solving continuous and smooth problems [3]. However, they are sensitive to the choice of the initial starting point. Besides, in the real world, engineering optimization problems are often nonlinear, non-differentiable and multi-modal, which means that it might be impossible to calculate the gradient of many objective functions. In addition, as the complexity of the problem increases, higher computational cost is required to obtain the optimal solution.

These drawbacks of traditional optimization methods encourage the birth and development of metaheuristic methods. By combining some classic heuristic methods with rule-based theories such as swarm intelligence and evolution, a large number of metaheuristic methods have been proposed. Genetic algorithms (GA) [4], particle swarm optimization (PSO) [5], differential evolution (DE) [6], grey wolf optimizer (GWO) [7] are some popular metaheuristic methods and have been used to solve constrained engineering design problems [8].

It is impossible for any algorithm to solve all optimization problems according to No Free Lunch (NFL) theorem [9]. Numerous researchers strive to develop new optimization algorithms based on natural phenomena. Besides, it is also a trend to modify the operators of existing methods to enhance their performance.

Recently, a hydrologic cycle optimization (HCO) [10] has been put forward. It introduces flow, infiltration, evaporation and precipitation operators to simulate the hydrological cycle process. It is confirmed that the HCO is a competitive approach for solving numerical and data clustering optimization problems [10]. In this paper, an improved hydrologic cycle optimization(IHCO) algorithm is proposed to boost the performance of the basic HCO. To demonstrate the superiority of IHCO, it is applied to solve engineering optimization problems. Compared to other famous metaheuristic algorithms, the experimental results show that our proposed algorithm provides better performance.

The rest of this paper is organized as follows. The basic hydrologic cycle optimization is described in Sect. 2. Details of the proposed algorithm are in Sect. 3. Parameter settings and experimental results of the applied algorithms for solving engineering problems are shown in Sect. 4. Finally, this paper is concluded in Sect. 5.

2 Basic Hydrologic Cycle Optimization Algorithm

In this section, the basic hydrologic cycle optimization algorithm(HCO) in [10] is briefly introduced.

Inspired by the phenomenon of the hydrological cycle in nature, HCO was proposed. The process of the hydrological cycle is briefly described as follows. Under the influence of solar radiation, gravity and other factors, water molecules circulate and exchange between the atmosphere, ocean and land through some physical actions, such as evaporation, precipitation, surface water runoff, groundwater runoff and infiltration. In this process, some of the water on the land would pool into rivers and finally flow into the oceans. The above phenomenon can be abstracted as the process of searching for the optimal solution of an objective function. By simulating the hydrological cycle process, three operators are introduced. They are flow operator, infiltration operator, evaporation and precipitation operator. These three operators of HCO are briefly described in Subsect. 2.1, Subsect. 2.2 and Subsect. 2.3, respectively.

2.1 Flow Operator

To simulate the flowing downhill phenomenon, a flow operator is conceived. For each individual X_i, another individual X_j with better fitness is randomly selected to generate a new candidate solution X_n. If the new location is better, the individual will flow to the new position and then flow again in the same direction until the maximum number of flow times is reached. The maximum number of flow times is predetermined to avoid premature convergence. For the current best individual, another individual is completely randomly chosen. The process is shown as formula (2), where $rand(\cdot)$ is a uniformly distributed number in the interval [0, 1] and D is the dimension of the problem.

$$X_n = X_i + (X_j - X_i). * rand(1, D). \tag{2}$$

2.2 Infiltration Operator

Infiltration plays a major role in increasing the diversity of the population. For each individual X_i, they randomly select another individual X_j to guide their infiltration actions. Several dimensions of the original individual X_i are randomly selected and eventually updated by moving closer or further away from the target individual. The number of the selected dimensions may be different for each individual. All individuals will accept the updated location, regardless of whether the new location is better. The process is shown as formula (3), where sd is the sd^{th} dimension of the randomly selected dimensions and $rand_{sd}$ is a random number generated for the sd^{th} dimension.

$$X_i^{sd} = X_i^{sd} + (X_i^{sd} - X_j^{sd}). * 2. * (rand_{sd} - 0.5). \tag{3}$$

2.3 Evaporation and Precipitation Operator

To enhance the global search capability, the evaporation and precipitation operator is introduced. A parameter P_e is set to control the probability of this action. When an individual is vaporized, there are two equal-probability ways to relocate. The first way is to randomly generate position in the decision space. The other way is to generate a new solution near the current optimal solution by applying Gaussian mutation.

3 Improved Hydrologic Cycle Optimization Algorithm

In this section, our improved hydrologic cycle optimization algorithm(IHCO) is described in detail. We attempt to redesign the flow operator as well as the evaporation and precipitation operator to enhance the performance of the basic HCO. In Subsect. 3.1, we propose a new strategy to generate candidate solution for the global optimal individual. In Subsect. 3.2, the improvement of the evaporation and precipitation operator is described and an adaptive Gaussian mutation strategy is introduced here.

3.1 Improvement of the Flow Operator

In the flow operator of the basic HCO, in addition to the global optimal individual, other individuals will randomly select another individual with better fitness, and the greedy selection is then applied. By continuous learning from exemplars, individuals can greedily flow to better new location. However, the global optimal individual has to learn from a randomly selected individual, as currently there is not any better solution. When an individual with extremely poor fitness becomes the exemplar, the current global optimal individual will gradually move away from the potential area. That is, learning from a single individual is often difficult to strike a better direction.

Based on the above thinking, we improve the flow strategy for the current best individual. First of all, a variable number of individuals are randomly selected to form a group. Then, by calculating the average of the current positions of all individuals in the group, the center of the group X_G is determined. At last, the current best individual X_{best} can make full use of this information to generate its candidate solution X_{new}. This strategy can be executed according to formula (4), where $rand_1$ is a uniformly distributed random number between 0 and 1. N_k is the number of individuals which are randomly selected to form a group, and t is the tth individual in the group. For other individuals that are not the current best solution, formula (2) is still used to generate their candidate solutions.

$$X_{new} = X_{best} + rand_1 .* (X_G - X_{best}).$$
$$X_G = \frac{\sum_t^{N_k} X_t}{N_k}. \tag{4}$$

3.2 Improvement of the Evaporation and Precipitation Operator

Two evaporation and precipitation strategies in HCO have been described in Subsect. 2.3. It is noteworthy that the two mutation strategies could be selected with the same probability. The advantage is that these strategies can help individuals to relocate or quickly approach the global optimum, which helps to enhance both global and neighborhood searching abilities. However, there are some deficiencies. Each individual randomly chooses a strategy without utilizing their empirical knowledge. This means that individuals may be randomly relocated in the decision space, even though it has strong exploitability in the current region. This will be detrimental to the development of the entire population. To make matters worse, if individuals frequently generate new positions randomly, it will prevent the population from convergence and local search.

These drawbacks encourage us towards new strategies. First, the parameter P_e is reserved to control the evaporation probability. Then, before precipitation, individuals are ranked according to their fitness values. Finally, if an individual ranks in the bottom five, it would be randomly relocated in the search space. Otherwise, it would move its current position to the current best position and then perform adaptive Gaussian mutation strategy on several selected dimensions to generate a new position. In this paper, adaptive Gaussian mutation is adjusted by the standard deviation σ of the Gaussian distribution. σ shows a decreasing trend with the process of iteration. Noted that σ helps high jump when it is close to 1 and helps low jump when it is close to 0.1. The adaptive Gaussian mutation strategy contributes to balance exploration and exploitation. The relocation strategy is shown as formula (5), while the adaptive Gaussian mutation is given by formula (6).

$$X_i^d = LB_d + (UB_d - LB_d). * rand_d, \text{ if } rand_2 < P_e \text{ and } rank(i) = 1. \quad (5)$$

$$X_i^{sd} = X_{Best}^{sd} + X_{Best}^{sd}. * Gaussian(\mu, \sigma^2), \text{ if } rand_2 < P_e \text{ and } rank(i) = 0.$$
$$\sigma = 1 - 0.9 * \frac{Iter}{MaxIter}. \quad (6)$$

$rand_d$ and $rand_2$ are uniformly distributed numbers between 0 and 1. $rank(\cdot)$ can be either 0 or 1. When it is equivalent to 1, it means that the individual ranks in the bottom five. P_e is the parameter to control the probability of evaporation. In formula (5), X_i^d is the dth dimension of the individual X_i. UB_d and LB_d are the upper and lower bounds of the dth decision variable, respectively. In formula (6), X_i^{sd} and X_{Best}^{sd} are the sd^{th} randomly selected dimension of the individual X_i and the current best individual with the best position X_{Best}, respectively. $Gaussian(\cdot)$ is the Gaussian distribution with mean μ and standard deviation σ. *Iter* and *MaxIter* represent the current iteration and the maximum number of iterations, respectively.

4 IHCO for Engineering Problems

In this section, the proposed algorithm is applied to solve thirteen engineering design problems. Some basic information of these problems is given in Subsect. 4.1. Several well-known metaheuristic algorithms for comparison and parameter settings are given in Subsect. 4.2. In Subsect. 4.3, experimental results and analysis are presented here.

4.1 Engineering Design Optimization Problems

Table 1. Basic information of thirteen engineering design optimization problems.

No	Name	DV	NC	Optimum results
F1	Speed reducer	7	11	2994.4244658
F2	Tension/compression spring design	3	4	0.012665232788
F3	Pressure vessel design	4	4	6059.714335048436
F4	Three-bar truss design problem	2	3	263.89584338
F5	Design of gear train	4	0	2.70085714e-12
F6	Cantilever beam	5	1	1.3399576

(*continued*)

Table 1. (*continued*)

No	Name	DV	NC	Optimum results
F7	Optimal Design of I-Shaped Beam	4	2	0.0130741
F8	Tubular column design	2	6	26.486361473
F9	Piston lever	4	4	8.41269832311
F10	Corrugated bulkhead design	4	6	6.8429580100808
F11	Car side impact design	10	10	22.84296954
F12	Design of welded beam	4	7	1.724852308597366
F13	Reinforced concrete beam design	3	2	359.2080

To verify the performance of the proposed IHCO, thirteen standard engineering optimization problems are employed, some basic information is shown in Table 1. The abbreviations "DV" and "NC" represent the number of design variables and constraints, respectively.

Among the 13 engineering design problems, except F5, the rest are constrained optimization problems. They are highly nonlinear problems with some complex constraints. When minimizing the objective functions, precise handling of design constraints must be considered. In this paper, a simple penalty approach is applied as the constraint handling method. More details on these engineering design problems mentioned above can be obtained in [8, 11].

4.2 Parameter Settings

Some settings are the same in all algorithms to ensure that the comparison is fair. In the trials, the population size of all algorithms is 50. Besides, the maximum number of function evaluations (FEs) is used for more equitable comparison to some degree. FEs is set to 10×10^4 in this paper. At the same time, in order to reduce the effect of randomness, on each engineering design problems, each algorithm runs 30 times independently.

Several well-known metaheuristic algorithms are implemented for comparison. They are basic HCO, PSO, DE, GA and water cycle algorithm(WCA). The algorithms and their parameter settings are shown in Table 2.

There are three main parameters in IHCO. *maxNF* represents the maximum flow time in the flow operator. P_e is used to control the evaporation probability. The settings of these two parameters in IHCO are consistent with the settings in HCO. σ is used for the adaptive Gaussian mutation strategy. It decreases linearly from 1 to 0.1 as the algorithm proceeds.

Table 2. Parameter settings of the algorithms

Algorithm	Parameter settings
IHCO (present study)	Maximum flow times *maxNF* increases from 1 to 3, The control parameter P_e decreases from 0.9 to 0.1, The standard deviation σ decreases from 1 to 0.1
HCO [10]	Maximum flow times *maxNF* increases from 1 to 3, The control parameter P_e decreases from 0.9 to 0.1
PSO [12]	Inertia weight ω decreases from 0.9 to 0.4, Learning rate $c_1 = c_2 = 2$
DE [6]	Step size $F\_weight = 0.4$, Crossover probability $F\_CR = 0.1$, DE/rand/1 strategy is used
GA [4]	Mutation rate $F\_mu = 0.1$, Crossover rate $F\_cr = 0.95$
WCA [13]	The number of rivers and sea $N_{sr} = 4$, A default value(close to zero) $d_{max} = 1e-3$

4.3 Experimental Results and Analysis

Table 3 presents the experimental results. In the second column, "MN" and is "SD" are the mean and standard deviation of all the optimal solutions obtained in each run, respectively. "RK" is the rank of the algorithm among all the applied algorithms.

In Table 3, the best mean, the best standard deviation and the highest ranking are highlighted in bold. "b/e/w" represents the total number of problems that IHCO performed better than, the same as, or worse than another algorithm, respectively. Besides, "AR/FR" are average ranking and final ranking for each algorithm.

As it could be clearly seen in Table 3, IHCO performs optimally on eight of the thirteen problems and it eventually ranks first.

These experimental results demonstrate that IHCO outperforms PSO and GA on all problems. It implies that the populations of PSO and GA are more likely to trap in local optimal solutions when solving engineering optimization problems. The results also show that our proposed algorithm IHCO has a better global searching ability to obtain more satisfactory solutions.

In addition, compared to the classic HCO, IHCO shows better performance on eleven problems and the superiority is especially significant on F3 and F9. On F1, F2, F4-F10, F12 and F13, IHCO obtains better mean values and standard deviations of the optimization results than HCO, which shows its advantage both in accuracy as well as stability. It can be concluded that IHCO has a satisfactory ability to boost the performance of the basic HCO for solving engineering problems.

Meanwhile, we can see that on F4, F6, F8, F10 and F11, DE obtains better results than IHCO. The optimal solutions of DE on F4, F6, F8, F10 and F11 are only 0.000038%, 0.014923%, 0.002265%, 0.027758% and 0.955449% lower than the results of IHCO, respectively, indicating that the performance gap between IHCO and DE on these five problems is small. However, the optimal results of DE on F2, F3, F9 and F12 are 414.1732%, 2.3054%, 335.8095% and 5.7786% higher than the optimal solutions of IHCO, respectively. These statistical results demonstrate that DE is not stable enough to solve all engineering optimization problems. They also confirm the robustness of IHCO.

Table 3. Comparison of experimental results between IHCO and other algorithms

	ST	IHCO	HCO	PSO	DE	GA	WCA
F1	MN	**2994.4378**	2994.4928	3025.8539	2994.8280	3053.9152	2996.3237
	SD	**0.0077**	0.0526	89.0968	1.4557	19.2577	4.3012
	RK	**1**	2	5	3	6	4
F2	MN	**0.0127**	0.0130	0.0815	0.0653	0.0213	0.0136
	SD	**6.2093E−05**	2.5156E−04	0.2653	0.2870	0.0049	0.0014
	RK	**1**	2	5	6	4	3
F3	MN	**6130.9076**	6240.7704	7786.3519	6272.2504	6791.8197	6523.5465
	SD	166.6928	**161.9452**	878.8256	349.9473	416.4013	531.6386
	RK	**1**	2	6	3	5	4
F4	MN	263.8959	263.8961	263.8962	**263.8958**	263.9158	263.8959
	SD	6.1377E−05	1.3218E−04	3.4397E−04	**6.7249E−08**	0.0128	1.3878E−04
	RK	2	4	5	**1**	6	2
F5	MN	**1.3691E−11**	1.7366E−11	1.0230E−08	7.2091E−10	7.5766E−10	3.5968E−09
	SD	**2.1309E−11**	2.9009E−11	8.7495E−09	8.1236E−10	8.0573E−10	4.4485E−09
	RK	**1**	2	6	3	4	5
F6	MN	1.3402	1.3406	1.6629	**1.3400**	1.4576	1.3407
	SD	1.2010E−04	3.3176E−04	0.2258	**1.6879E−05**	0.0802	1.0631E−03
	RK	2	3	6	**1**	5	4
F7	MN	**0.013074**	0.013107	0.02049	**0.013074**	0.013661	**0.013074**
	SD	**2.0379E−07**	3.6772E−05	0.0406	5.2935E−07	1.9697E−04	9.7390E−07
	RK	**1**	4	5	**1**	6	**1**
F8	MN	26.4870	26.4877	26.4953	**26.4864**	26.6074	**26.4864**
	SD	4.8827E−04	9.2592E−04	5.0604E−03	9.3592E−09	0.0977	**2.2749E−09**
	RK	3	4	5	**1**	6	**1**
F9	MN	**19.0279**	41.2442	282.0075	82.9254	65.7363	67.1484
	SD	**40.3864**	66.4895	152.9064	135.5672	126.3210	78.5440
	RK	**1**	2	6	5	3	4

(continued)

Table 3. (*continued*)

	ST	IHCO	HCO	PSO	DE	GA	WCA
F10	MN	6.8449	6.8457	6.9676	**6.8430**	7.0311	6.8439
	SD	8.8605E−04	1.2367E−03	0.2107	**1.0497E−04**	0.0852	4.3395E−03
	RK	3	4	5	**1**	6	2
F11	MN	23.0677	22.9883	23.5978	**22.8473**	24.2392	23.1481
	SD	0.2196	0.1596	0.4302	**6.6239E−03**	0.3560	0.2677
	RK	3	2	5	**1**	6	4
F12	MN	**1.7288**	1.7603	2.0343	1.8287	2.2869	1.7631
	SD	**1.8556E−03**	0.0218	0.1875	0.0952	0.4934	0.0581
	RK	**1**	2	5	4	6	3
F13	MN	**359.2080**	**359.2080**	360.4507	359.2095	359.2790	359.4130
	SD	**3.5416E−08**	1.5778E−07	1.5505	0.0071	0.0971	0.7803
	RK	**1**	**1**	6	3	4	5
b/e/w		/	11/1/1	13/0/0	7/1/5	13/0/0	9/2/2
AR/FR		1.6154/1	2.6154/3	5.3846/6	2.5384/2	5.1538/5	3.2307/4

The convergence curves of IHCO, HCO, PSO, DE, GA and WCA on F2, F3, F9 and F12 are given in Fig. 1. It is clear that IHCO shows faster convergence speed and better accuracy. In science and engineering field, people always highly focus on precision and accuracy. It is of great significance to improve the accuracy and precision of engineering optimization problems within a reasonable time. Besides, the stability of algorithms for solving engineering problems also needs to be paid attention to. According to these experimental results, IHCO has produced more stable solutions and has obtained the optimal results on eight problems. It has noticeable capability in solving constrained engineering optimization problems.

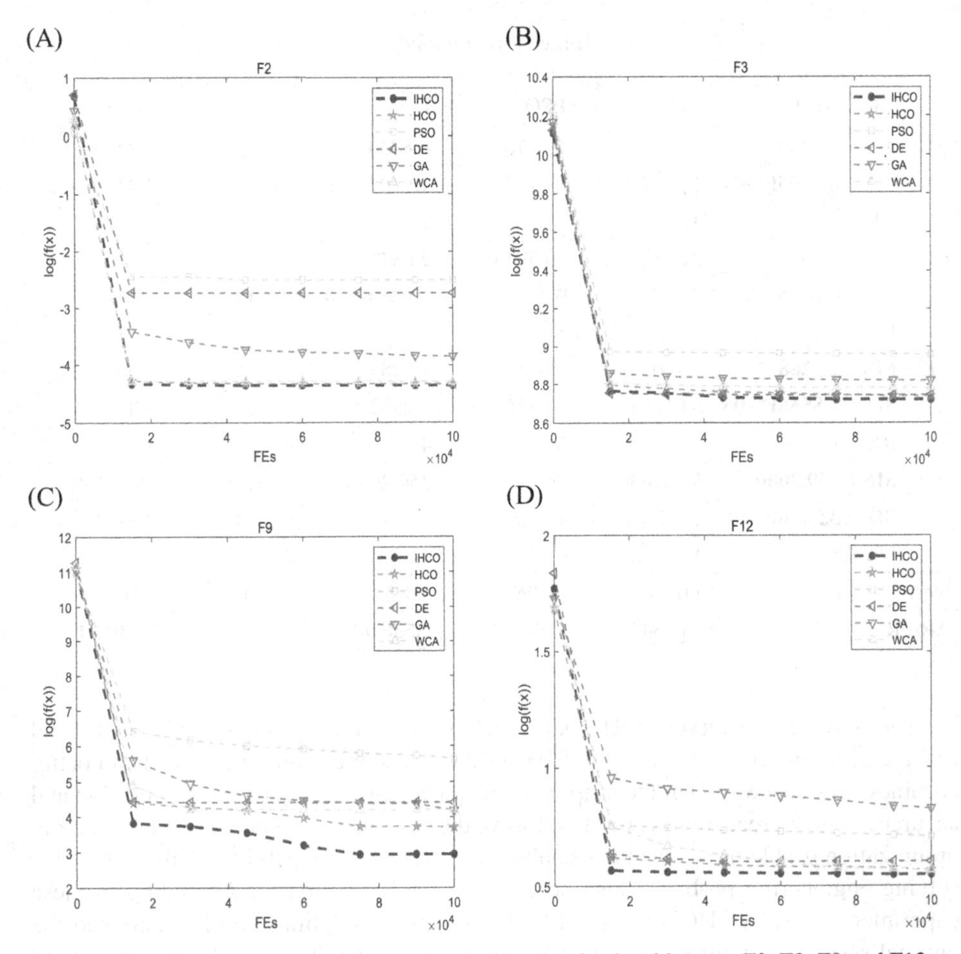

Fig. 1. The mean convergence results obtained by the six algorithms on F2, F3, F9 and F12

5 Conclusions

In this study, an improved hydrologic cycle optimization algorithm is proposed to enhance the performance of the basic algorithm.

We introduce a new flow strategy for the current global optimal individual, making great use of empirical knowledge generated by some individuals of the population. Moreover, the evaporation and precipitation operator is also redesigned to balance exploration and exploitation. In details, before precipitation, individuals are ranked according to their fitness values. Based on the probability of evaporation as well as its ranking, each individual may relocate in the search space or perform adaptive Gaussian mutation strategy. Finally, thirteen engineering design problems are conducted to evaluate the optimization ability of IHCO. The experimental results show that it is competitive for IHCO to solve real-world engineering design problems. Compared to other famous metaheuristic algorithms, IHCO performs well in the aspects of accuracy and robustness.

In the future, IHCO can be further tested in other real-world problems, such as business optimization problems and intelligent control problems. In addition, other enhanced strategies could also be introduced to develop a more robust and well-performed version of IHCO.

Acknowledgement. The study is supported by The National Natural Science Foundation of China (Nos. 71971143, 62103286), Major Project of Natural Science Foundation of China (No. 71790615), Integrated Project of Natural Science Foundation of China (No. 91846301), Social Science Youth Foundation of Ministry of Education of China (No. 21YJC630181), Key Research Foundation of Higher Education of Guangdong Provincial Education Bureau (No. 2019KZDXM030), Natural Science Foundation of Guangdong Province (Nos.2020A1515010749, 2020A1515010752), Guangdong Basic and Applied Basic Research Foundation (No. 2019A1515110401), Natural Science Foundation of Shenzhen City (No. JCYJ20190808145011259), and Guangdong Province Innovation Team (No. 2021WCXTD002).

References

1. Zhu, L.M., Ding, H., Xiong, Y.L.: A steepest descent algorithm for circularity evaluation. Comput. Aided Des. **35**(3), 255–265 (2003)
2. Sun, D.I., Ashley, B., Brewer, B., Hughes, A., Tinney, W.F.: Optimal power flow by newton approach. IEEE Trans. Power Appar. Syst. **103**(10), 2864–2880 (1984)
3. Zilinskas, A.: Practical mathematical optimization: an introduction to basic optimization theory and classical and new gradient-based algorithms. Interfaces **36**(6), 613–615 (2006)
4. Holland, J.H.: Genetic algorithms. Sci. Am. **267**(1), 66–72 (1992)
5. Kennedy, J., Eberhart, R.: Particle swarm optimization. In: Proceedings of the ICNN 1995-International Conference on Neural Networks, vol. 4, pp. 1942–1948, Perth, Australia. IEEE (1995)
6. Storn, R., Price, K.: Differential evolution - a simple and efficient heuristic for global optimization over continuous spaces. J. Global Optim. **11**(4), 341–359 (1997)
7. Mirjalili, S., Mirjalili, S.M., Lewis, A.: Grey wolf optimizer. Adv. Eng. Softw. **69**, 46–61 (2014)
8. Sheikhi Azqandi, M., Delavar, M., Arjmand, M.: An enhanced time evolutionary optimization for solving engineering design problems. Eng. Comput. **36**(2), 763–781 (2019). https://doi.org/10.1007/s00366-019-00729-w
9. Wolpert, D.H., Macready, W.G.: No free lunch theorems for optimization. IEEE Trans. Evol. Comput. **1**(1), 67–82 (1997)
10. Yan, X., Niu, B., Chai, Y., Zhang, Z., Zhang, L.: An adaptive hydrologic cycle optimization algorithm for numerical optimization and data clustering. Int. J. Intell. Syst. 1–29 (2022). https://doi.org/10.1002/int.22836
11. Azizi, M., Talatahari, S., Giaralis, A.: Optimization of engineering design problems using atomic orbital search algorithm. IEEE Access **9**, 102497–102519 (2021)
12. Shi, Y., Eberhart, R.C.: Empirical study of particle swarm optimization. In: Proceedings of the 1999 Congress on Evolutionary Computation-CEC 1999, pp. 1945–1950 (1999)
13. Eskandar, H., Sadollah, A., Bahreininejad, A., Hamdi, M.: Water cycle algorithm - a novel metaheuristic optimization method for solving constrained engineering optimization problems. Comput. Struct. **110**, 151–166 (2012)

Image Denoising Method of Auto-evolving PCNN Model Based on Quantum Selfish Herd Algorithm

Hongyuan Gao, Haijun Zhao(✉), and Shicong Chen

Harbin Engineering University, Harbin, People's Republic of China
zhj@hrbeu.edu.cn

Abstract. Pulse coupled neural network (PCNN) is used by predecessors to locate noise points, which overcomes the disadvantage that a large amount of image details will be lost in the traditional method of filtering the whole image. However, it is still a problem to find the optimal parameters of PCNN system. Therefore, the purpose of this paper is to improve the quality of image filtering and the flexibility of PCNN. This paper established a pulse coupled neural network model (PCNN-QSHA) based on quantum selfish herd algorithm. The PCNN's optimal parameters can be obtained by the quantum selfish herd algorithm without manual estimation parameter. The experimental results show that, compared with the previous methods, the proposed method has excellent performance and efficiency in image filtering. By comparing the proposed algorithm with GACS, PSO and SFLA, nine CEC benchmark functions are simulated, and the results show that QSHA has better convergence performance.

Keywords: Pulse coupled neural network · Image denoising · Optimal parameters · Quantum selfish herd algorithm

1 Introduction

It is widely known that image denoising is an important research direction of image processing. In reality, images will be affected by the interference of imaging equipment and external noise in the transmission process, and the image will be polluted by noise. The purpose of image denoising is to filter out the noise in the image under the condition of retaining the image's key information as much as possible. At present, there are many classical image denoising methods, such as median filtering, Wiener filtering, etc., but these methods are all denoising the whole image, so a lot of key details of the image will be lost.

In [1], Vorhies. JT et al. proposed A novel method is described for adaptive filtering of light fields to enhance objects at a given depth. In [2], Jialin Tang et al. improved the adaptive median filtering method. In [3], LINA JIA et al. proposed an effective image denoising algorithm with improved dictionaries. All of them make use of adaptive method to improve the applicability of this method and can denoise more noisy images.

Y. Tan et al. (Eds.): ICSI 2022, LNCS 13344, pp. 128–138, 2022.
https://doi.org/10.1007/978-3-031-09677-8_11

But they still denoised the whole image. In [4], A. Senthil Selvi et al. proposed de-noising of images from salt and pepper noise using hybrid filter, fuzzy logic noise detector and genetic algorithm. An experimental result shows the hybrid filter, fuzzy logic noise detector and genetic optimization algorithm rectifies the drawbacks of exiting filters and increases the visual quality of the image by increasing the PSNR value.

In recent years, image processing technology based on pulse-coupled neural network (PCNN) has attracted more and more experts' attention. PCNN is a neural network model proposed by Eckhorn in the 1990s. PCNN simulates the synchronous oscillation phenomenon in the visual cortex neurons of mammals [5]. The gray value of noise points in the image differs greatly from the gray value of surrounding pixels, while PCNN can compensate local micro-discontinuous values through the correlation between pixels, which is an advantage unmatched by traditional image filtering methods. Therefore, this paper chooses PCNN model for image filtering. In [6], Dong et al. proposed a new image fusion algorithm based on memristor-based PCNN. The algorithm is applied to multi-focus image fusion with improved multi-channel configuration. In [7], Yin et al. used the parameter adaptive pulse coupled neural network (PA-PCNN) model to fuse the high frequency band, in which all PCNN parameters could be estimated adaptively through the input frequency band. In [8], Yang et al. proposed a new heterogeneous simplified pulse coupled neural network (HSPCNN) model for image segmentation. HSPCNN is constructed from several simplified pulse-coupled neural network (SPCNN) models with different parameters corresponding to different neurons. In [9], Cheng et al. proposed variable step matrix of the simplified PCNN method. This method has achieved good results in Gaussian noise filtering.

To sum up, the existing literature focuses on the parameter setting of PCNN when using PCNN for image processing. It indicates that some key parameters of PCNN-based image processing model need to be set reasonably to obtain the optimal performance. However, the existing methods are difficult to obtain the optimal parameters of PCNN, which will reduce its performance. Moreover, the method of manual parameter estimation is not only inefficient but also poor in performance of PCNN. Therefore, in order to solve the problem of PCNN parameter setting and losing details in the process of image denoising in existing literatures, this paper established a pulse coupled neural network model (PCNN-QSHA) based on quantum selfish herd algorithm. The key parameters of the model are optimized by using the quantum mechanism combined with the evolutionary principle of selfish herd, and the peak signal-to-noise ratio as the fitness function. In this way, the denoising effect of the model is enhanced and its practicability is improved.

2 Image Filtering Model

This section introduces the image filtering model [10], which is divided into two parts: improved simplified PCNN model, and peak signal-to-noise ratio.

2.1 Improved Simplified PCNN Model

Adaptive window is used in the PCNN model, so it is also a pre-processing operation, and its specific process is as follow: The simplified PCNN is combined with structural similarity, and the filter window of a certain size $N \times N$ is applied to PCNN. If the function value of structural similarity after filtering of this window size is the maximum, the window size is output. The simplified structural similarity is used in this paper, which is as follow

$$S_{sim}(x, y) = \frac{(2 \cdot \mu_x \cdot \mu_y) \cdot (2 \cdot \sigma_{xy} + C_2)}{(\mu_x^2 + \mu_y^2 + C_1) \cdot (\sigma_x^2 + \sigma_y^2 + C_2)}. \tag{1}$$

where μ_x, μ_y represent the mean values of the original image $\boldsymbol{x}$ and the filtered image $\boldsymbol{y}$ respectively. σ_x, σ_y represent the standard deviation of $\boldsymbol{x}$ and $\boldsymbol{y}$. σ_{xy} is the covariance of $\boldsymbol{x}$ and $\boldsymbol{y}$.

In this paper, the PCNN model is improved to change the previous fixed size window into an adaptive size window. The number of iterations is K_1. In the model, each pixel corresponds to a neuron, so the feedback input of pixels in row i and column j of the n-th generation is

$$F_{ij}(n) = a_{ij}. \tag{2}$$

where a_{ij} represents the pixel value of row i and column j in the image. The linear connection input of pixels in row i and column j of the n-th generation is

$$L_{ij}(n-1) = \sum_{\iota=i-1}^{i+1} \sum_{l=j-1}^{j+1} \omega_{ij,\iota l} \cdot Y_{\iota l}(n-1). \tag{3}$$

where $\omega_{ij,\iota l}$ is the connection weight matrix's element. $Y_{\iota l}(n-1)$ is of output matrix's element. The internal activity item of the n-th generation is

$$U_{ij}(n) = F_{ij}(n) \cdot \left(1 + \beta_L \cdot L_{ij}(n)\right). \tag{4}$$

where β_L is coefficient of connection strength. Then, the dynamic threshold of generating pulse output is

$$t_{ij}(n) = \mathrm{e}^{-\alpha_t} \cdot t_{ij}(n-1) + v_t \cdot Y_{ij}(n). \tag{5}$$

where v_t is amplitude coefficient. α_t is attenuation coefficient. The pulse output of PCNN is

$$Y_{ij}(n) = \begin{cases} 1, U_{ij}(n) > t_{ij}(n-1) \\ 0, U_{ij}(n) \le t_{ij}(n-1) \end{cases}. \tag{6}$$

where $Y_{ij}(n)$ is the pulse output matrix's row i and column j element. $t_{ij}(n-1)$ is the dynamic threshold of $Y_{ij}(n)$. The pulse output indicates that when the internal activity term of the neuron is greater than the corresponding threshold value of the neuron, the output term is 1. After putting it into the formula (5), the threshold value will increase (the increase extent depends on the amplitude coefficient), and then it will prevent the neuron from firing continuously in the next iteration. In addition, when the pulse output element is 1, if it is not the first iteration, this neuron's corresponding position in the ignition matrix will fire this neuron in the iteration and record its iterations in the ignition matrix.

The ignition matrix is used to judge noise points. The judgment method is as follow: when more than half of the surrounding neurons in the filtering window fire before or after the neuron, the pixel point corresponding to the neuron is the noise point. For the noise points, the average filter is used to filter the image inside the window. For pixels, the grayscale value is returned directly. Such filtering for noise points rather than global filtering requires strict control of the parameters in PCNN. Because the expressions in these parameters are coupled, changing the value of one parameter will affect the filtering effect of the image. Therefore, QSHA with better convergence performance is selected to optimize parameters.

2.2 Peak Signal-to-Noise Ratio

The peak signal-to-noise ratio is the logarithm of the mean variance between the original image and the filtered image relative to the square of the maximum gray value. Taking it as the objective function, the expression is as follows:

$$f = 10 \cdot \log_{10}(\frac{\overline{m}}{M}). \tag{7}$$

where $\overline{m}$ denotes the maximum grayscale of image points, M is variance of the mean, which can be written as:

$$M = \frac{1}{\overline{M} \cdot \overline{N}} \sum_{i=1}^{\overline{M}} \sum_{j=1}^{\overline{N}} (x(i,j) - y(i,j))^2. \tag{8}$$

where $\overline{M}$ and $\overline{N}$ represent image dimensions. $x(i,j)$ and $y(i,j)$ respectively represent gray value of the original image and the filtered image in row i and column j.

3 Image Denoising Based on Quantum Herd Mechanism

The selfish herd algorithm is a global optimization algorithm based on the selfish behavior of group animals published by Fernando Fausto et al. in 2017 [11]. In this paper, this algorithm is combined with the theory of quantum swarm intelligence to obtain a quantum selfish herd algorithm. This algorithm is used to obtain the optimal parameters of PCNN in this paper.

3.1 Principles of Quantum Selfish Herd Algorithms

Firstly, the population size is set as K_2, the number of herds is G, the number of predators is $g = K_2 - G$, and the maximum number of iterations is K_3. In the k-th iteration, the quantum position of the i-th herd quantity is simply defined as $\boldsymbol{q}_i^k = [q_{i,1}^k, q_{i,2}^k, \cdots, q_{i,s}^k]$, $q_{i,s}^k \in [0, 1]$, $i = 1, 2, \cdots, G$, and the quantum position of predators is simply defined as $\boldsymbol{p}_i^k = [p_{i,1}^k, p_{i,2}^k, \cdots, p_{i,s}^k]$, $p_{i,s}^k \in [0, 1]$, $i = 1, 2, \cdots, g$. The quantum position $\boldsymbol{q}_i^k$ is mapped to position $\overline{\boldsymbol{q}}_i^k$ by the i-th herd individual according to the mapping interval, and the quantum position $\boldsymbol{p}_i^k$ is mapped to position $\overline{\boldsymbol{p}}_i^k$ by the i-th predator individual. The mapping rule is

$$\overline{q}_{i,d}^k = q_{i,d}^k(\xi_d^{\max} - \xi_d^{\min}) + \xi_d^{\min}. \tag{9}$$

$$\overline{p}_{i,d}^k = p_{i,d}^k(\xi_d^{\max} - \xi_d^{\min}) + \xi_d^{\min}. \tag{10}$$

where $d = 1, 2, \cdots, s$. $\xi_d^{\max}$ is the upper bound of the d-th dimension of the search space. $\xi_d^{\min}$ is the lower bound of the d-th dimension of the search space.

The fitness value of the i-th herd individual's position in the k-th iteration is $f(\overline{\boldsymbol{q}}_i^k)$, and predator individual's fitness is $f(\overline{\boldsymbol{p}}_i^k)$. According to the PCNN parameters corresponding to the position of the i-th herd or predator individual in the k-th iteration, PCNN is activated for image filtering. In QSHA, a survival value function was introduced to evaluate the survival of herd individuals. The survival value function is Formula (11), which is as follow

$$H_i^k = \frac{f(\overline{\boldsymbol{q}}_i^k) - f(\overline{\boldsymbol{q}}_{worst}^k)}{f(\overline{\boldsymbol{q}}_{best}^k) - f(\overline{\boldsymbol{q}}_{worst}^k)}. \tag{11}$$

where $f(\overline{\boldsymbol{q}}_{best}^k)$ is the optimal fitness of the herd in the k-th iteration. $f(\overline{\boldsymbol{q}}_{worst}^k)$ is the worst fitness of the herd in the k-th iteration. $f(\overline{\boldsymbol{q}}_i^k)$ is the fitness value of the i-th individual in the k-th iteration.

The predation risk of each individual in a herd is affected by the quantum position distance between the individual and the predator. And the pursuit probability $P_{i,j}^k$ mainly depends on the distance between the individuals and the survival value of the individuals. The pursuit probability is as follow

$$P_{i,j}^k = \frac{W_{i,j}^k}{\sum_{\tau=1}^{G} W_{i,\tau}^k}. \tag{12}$$

where $i = 1, 2, \cdots, g$, $j = 1, 2, \cdots, G$. $W_{i,j}^k$ is the attraction of $\boldsymbol{q}_j^k$ to $\boldsymbol{p}_i^k$, which is as follow

$$W_{i,j}^k = \left(1 - H_j^k\right) \cdot \mathrm{e}^{-\left\|\mathbf{p}_i^k - \mathbf{q}_j^k\right\|^2}. \tag{13}$$

It can be concluded from the formula (13), the smaller the survival value of individuals in the herd, the more attractive they are to predators. And the higher the pursuit probability, the more likely the herd individual is to be preyed upon. Pursuit probabilities are used to calculate cumulative probabilities, and the roulette wheel mechanism is used to catch prey. The cumulative probability mentioned is the same as the calculation method of the cumulative probability in the genetic algorithm, so it will not be described here. The prey that is hunted is put into set $\boldsymbol{O}$, and its quantum position is emptied. $\boldsymbol{O}$ is an empty set when the predator doesn't get the prey, and then the remaining individuals are used for crossover operation to re-allocate the quantum position to this individual. The specific steps of assignment: First, Crossover candidates $\overline{\boldsymbol{O}} = \{\boldsymbol{q}_i^k \notin \boldsymbol{O}\}$ are selected, and $i = 1, 2, \cdots, G$. Second, the mating probability is calculated by

$$\overline{P}_j^k = \frac{H_j^k}{\sum_{i=1}^{G} H_i^k}. \tag{14}$$

where $\forall i \in \{i = 1, 2, \cdots, G | \boldsymbol{q}_i^k \in \overline{\boldsymbol{O}}\}$, $\overline{P}_j^k$ is the j-th individual's mating probability.

Then, the cumulative probability is calculated by the mating probability. The quantum positions are allocated to the new generation of individuals, and the specific steps are as follows: for each dimension of the quantum position, the roulette wheel mechanism is adopted to select the corresponding dimension of the surviving individual's quantum position. And the quantum position dimension is directly assigned to the individual in the set $\boldsymbol{O}$. Then, quantum positions of all newborn individuals are mapped and their positions are obtained. And the fitness and survival value of new generation of individuals were calculated.

This paper identifies the herd leader $\boldsymbol{q}_{best}^k$ according to the maximum fitness value. It represents the individual with greatest fitness value in the k-th iteration. These herd individuals were divided into herd followers $\tilde{\boldsymbol{q}}_D^k$ and herd deserters $\tilde{\boldsymbol{q}}_F^k$, which is as follow

$$\tilde{\boldsymbol{q}}_D^k = \left\{ \boldsymbol{q}_i^k | H_i^k \geq H_{ave}^k \right\}. \tag{15}$$

$$\tilde{\boldsymbol{q}}_F^k = \left\{ \boldsymbol{q}_i^k | H_i^k < H_{ave}^k \right\}. \tag{16}$$

where $i = 1, 2, \cdots, G$. H_{ave}^k is the herd's average survival value in the k-th iteration. Then, the adjacent individual of each individual in the population is found, which is as follow

$$\tilde{\boldsymbol{q}}_{ci}^k = \left\{ \boldsymbol{q}_j^k \neq \boldsymbol{q}_i^k, \boldsymbol{q}_j^k \neq \boldsymbol{q}_{best}^k | H_i^k > H_j^k, r_{i,j}^k = \min_{j \in 1,2,\cdots,G} \left(\left\| \boldsymbol{q}_i^k - \boldsymbol{q}_j^k \right\| \right) \right\}. \tag{17}$$

where $i = 1, 2, \cdots, G$. $r_{i,j}^k$ is the Euclidean distance between $\boldsymbol{q}_i^k$ and $\boldsymbol{q}_j^k$. $\tilde{\boldsymbol{q}}_{ci}^k$ refers to the j-th herd individual as the i-th herd individual's adjacent individual.

The herd's center $\tilde{\boldsymbol{q}}_m^k$ is calculated by

$$\tilde{\boldsymbol{q}}_m^k = \frac{\sum_{i=1}^{G} H_i^k \cdot \boldsymbol{q}_i^k}{\sum_{i=1}^{G} H_i^k}. \tag{18}$$

3.2 Quantum Evolutionary of Quantum Herd Algorithm

Then, this paper uses a quantum revolving door to update the quantum positions of herd followers, herd deserters, and predators. In the herd members, the quantum rotation angle's updating formula of herd followers and herd deserters is different, so the movement pattern is different. Then, the i-th herd individual's the d-th dimensional quantum rotation angle is

$$\varphi_{i,d}^{k+1} = \begin{cases} \sigma \cdot H_{c,i}^{k} \cdot \left(\tilde{q}_{ci,d}^{k} - q_{i,d}^{k}\right) + \zeta \cdot H_{best}^{k} \cdot \left(q_{best,d}^{k} - q_{i,d}^{k}\right), \boldsymbol{q}_i^k \in \tilde{\boldsymbol{q}}_D^k \\ \sigma \cdot H_{qm}^{k} \cdot \left(\tilde{q}_{qm,d}^{k} - q_{i,d}^{k}\right) + \zeta \cdot \left(q_{best,d}^{k} - q_{i,d}^{k}\right), \boldsymbol{q}_i^k \in \tilde{\boldsymbol{q}}_F^k \end{cases}. \tag{19}$$

where $i = 1, 2, \cdots, G$, $d = 1, 2, \cdots, s$. s is the quantum position vector's maximum dimension. H_{best}^{k} is the survival value of herd leader in the k-th iteration. H_i^k, $H_{c,i}^k$ and H_{qm}^k represent the survival value of the i-th herd individual, the i-th herd individual's adjacent individual, and the herd's center respectively. σ and ζ are random numbers from 0 to 1. Then, quantum revolving gates are used to update the quantum position of herd individuals, which is as follow

$$q_{i,d}^{k+1} = \left| q_{i,d}^{k} \cdot \cos\left(\varphi_{i,d}^{k+1}\right) + \sqrt{1 - \left(q_{i,d}^{k}\right)^2} \cdot \sin\left(\varphi_{i,d}^{k+1}\right) \right|. \tag{20}$$

where $i = 1, 2, \cdots, G$, $q_{i,d}^{k}$ represents the d-th dimensional variable of the i-th herd individual's quantum position. $\varphi_{i,d}^{k+1}$ represents the d-th dimensional variable of the i-th herd individual's quantum rotation angle.

The d-th dimensional quantum rotation angle of the i-th predator is

$$\varphi_{i,d}^{k+1} = \sigma \cdot \left(q_{j,d}^{k} - p_{i,d}^{k}\right) + \rho \cdot \lambda_d. \tag{21}$$

where $i = 1, 2, \cdots, g$, ρ is a random number satisfying a gaussian distribution with a mean of 0 and a variance of 1. λ_d represents the d-th dimensional variable of a random unit vector in space. $q_{j,d}^{k}$ is the d-th dimensional variable of the preyed individual. Then, quantum revolving gates are used to update the quantum position of predator individuals, which is as follow

$$p_{i,d}^{k+1} = \left| p_{i,d}^{k} \cdot \cos\left(\varphi_{i,d}^{k+1}\right) + \sqrt{1 - \left(p_{i,d}^{k}\right)^2} \cdot \sin\left(\varphi_{i,d}^{k+1}\right) \right|. \tag{22}$$

where $i = 1, 2, \cdots, g$, $p_{i,d}^{k}$ represents the d-th dimensional variable of the i-th predator's quantum position in the k-th iteration.

When the algorithm reaches the maximum number of iterations, the herd leader is compared with the best individual in the predator population, and the better position is output as the PCNN model's optimal parameter. It can be seen that the optimal solution is possible for both the herd and the predator. In conclusion, there are two groups in QSHA: predator and herd, and the update strategies of the two groups are completely different. The two update strategies can prevent the optimization algorithm from falling into local convergence and improve the optimization rate. In Sect. 4, the superior convergence of QSHA is verified.

4 The Experimental Simulation

In this paper, three parameters are optimized by using quantum herd mechanism, which are amplitude coefficient v_t, attenuation coefficient α_t and connection strength coefficient β_L. The value of the three parameters ranges from 0 to 1000, and the accuracy is 10^{-2}. Even if three parameters are optimized in this paper, the optimization problem is difficult. Therefore, it is necessary to invent a new optimization algorithm to optimize the parameters of PCNN image filtering model. In the simulation, when filtering different noisy images, the optimal parameters corresponding to PCNN are different, so this is the advantage compared with some PCNN fixed parameter filtering methods. Table 1 gives the PCNN-QSHA's parameters required for the experiment, which is as follows.

Table 1. Simulation parameters.

Parameter	Value
C_1	6.502
C_2	58.522
K_1	8
K_2	40
K_3	50
G	32
$N \times N$	$3 \times 3, 5 \times 5, 7 \times 7, 9 \times 9, 11 \times 11$
$\xi_d^{\max}$	1000
$\xi_d^{\min}$	0

Matlab software 2016 is used for simulation in this paper. In Fig. 1, the optimal parameters obtained by QSHA are $v_t = 248.43$, $\alpha_t = 186.76$ and $\beta_L = 0.0054$. Figure a shows adding the pepper and salt density noise with a density of 0.1. Figure b shows the filtering effect of the method in this paper. Figure c is the filtering effect of median filter, and figure d is the filtering effect of Wiener filter.

a

b

c

d

Fig. 1. The effect comparison diagram of Image denoising.

In Fig. 2, the optimal parameters obtained by QSHA are $v_t = 839.12$, $\alpha_t = 121.66$ and $\beta_L = 0.34$. Figure a shows adding the pepper and salt density noise with a density of 0.3. Figure b shows the filtering effect of the method in this paper. Figure c is the filtering effect of median filter, and figure d is the filtering effect of Wiener filter.

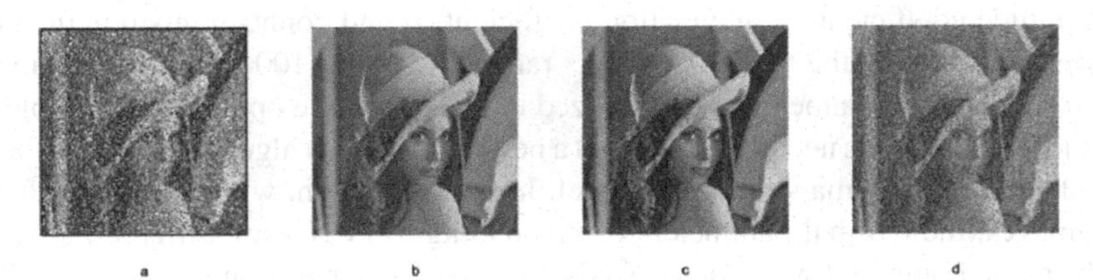

Fig. 2. The effect comparison diagram of Image denoising.

In Fig. 3, the optimal parameters obtained by QSHA are $v_t = 282.14$, $\alpha_t = 0.074$ and $\beta_L = 0.073$. Figure a shows adding the Gaussian noise with the mean is 0 and the variance is 0.005. Figure b shows the filtering effect of the method in this paper. Figure c is the filtering effect of median filter, and figure d is the filtering effect of Wiener filter. As can be seen from the above three figures, PCNN-QSHA is superior to other methods in image denoising without losing image details.

Fig. 3. The effect comparison diagram of Image denoising.

In order to explain the reason of choosing QSHA to optimize PCNN's parameters, instead of other classical algorithms, this paper selects shuffled frog leaping algorithm (SFLA) [12], Classical genetic algorithm (GA), and Particle swarm optimization [13] (PSO) algorithm to compare the proposed algorithm. These algorithm parameters are the same as those in the references. In Fig. 4, In order to test the performance of QSHO, nine CEC benchmark functions were selected, with dimension $s = 100$. Among the test functions: Ackeley function interval is $[-30, 30]$, Sphere function interval is $[-5.12, 5.12]$, Girewank function interval is $[-600, 600]$, Rastrigin function interval is $[-5.12, 5.12]$, Zakharov function interval is $[-5, 10]$, Sum squares function interval is $[-100, 100]$, Rosenbrock function interval is $[-30, 30]$, Schwefel2.21 function interval is $[-100, 100]$, Levy function interval is $[-10, 10]$. These algorithms have a population size of 40, the number of iterations is 500, and 200 independent replicates were performed. The simulation curves in the figure are the mean value of the fitness function of the corresponding test function. It can be seen that the convergence performance and efficiency of QSHO are better than other algorithms.

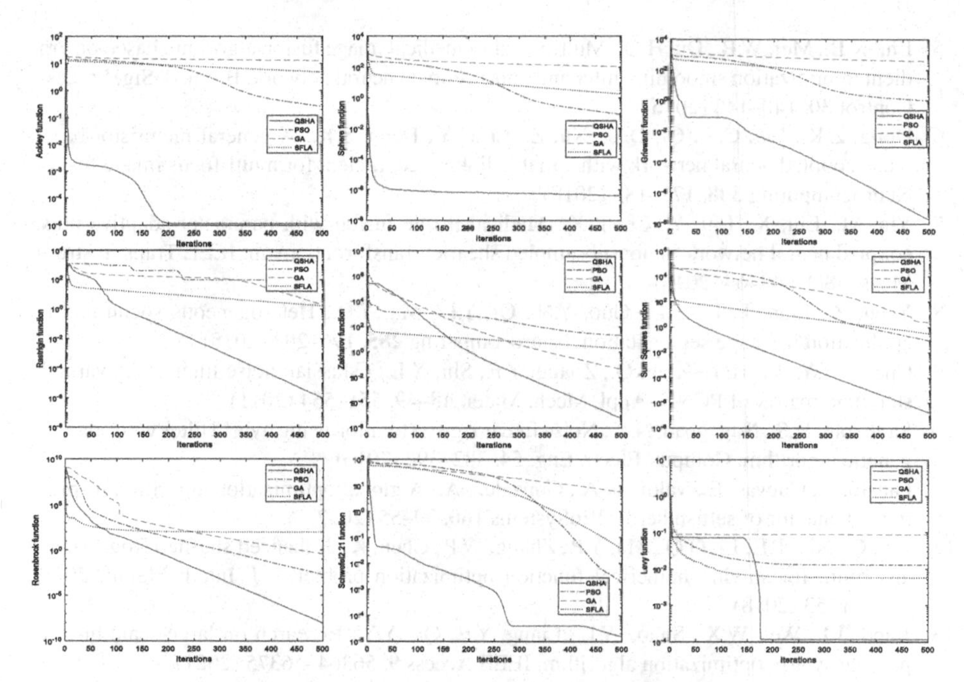

Fig. 4. Average convergence analysis of these test functions.

5 Conclusion

In this paper, an adaptive window PCNN model image filter is designed, and then the quantum herd mechanism is used to optimize the key parameters of the PCNN model. Different noise images correspond to different parameters. Compared with other methods, an excellent performance has been achieved in image noise filtering. In the future, quantum selfish herd algorithm will be extended to other fields, such as image segmentation, image fusion, face recognition, etc.

References

1. Vorhies, J.T., Hoover, A.P., Madanayake, A.: Adaptive filtering of 4-D light field images for depth-based image enhancement. IEEE Trans. Circuits Syst. II-Express Briefs **68**(2), 787–791 (2021)
2. Tang, J.L., Wang, Y.K., Cao, W., Yang, J.Q.: Improved adaptive median filtering for structured light image denoising. In: 7th International Conference on Information, Communication and Networks, Macau, pp. 146–149 (2019)
3. Jia, L.N., et al.: Image denoising via sparse representation over grouped dictionaries with adaptive atom size. IEEE Access **5**, 22512–22529 (2017)
4. Senthil Selvi, A., Pradeep Mohan Kumar, K., Dhanasekeran, S., Uma Maheswari, P., Senthil Pandi, S., Pandi. S.: De-noising of images from salt and pepper noise using hybrid filter, fuzzy logic noise detector and genetic optimization algorithm. Multimedia Tools Appl. **79**, 4115–4131 (2020)

5. Liu, X.B., Mei, W.B., Du, H.Q.: Multimodality medical image fusion algorithm based on gradient minimization smoothing filter and pulse coupled neural network. Biomed. Sig. Process. Control **30**, 140–148 (2016)
6. Dong, Z.K., Lai, C.S., Qi, D.L., Xu, Z., Li, C.Y., Duan, S.K.: A general memristor-based pulse coupled neural network with variable linking coefficient for multi-focus image fusion. Neurocomputing **308**, 172–183 (2018)
7. Yin, M., Liu, X., Liu, Y., Chen, X.: Medical image fusion with parameter-adaptive pulse coupled neural network in nonsubsampled shearlet transform domain. IEEE Trans. Instrum. Meas. **68**(1), 49–64 (2018)
8. Yang, Z., Lian, J., Li, S.L., Guo, Y.N., Qi, Y.L., Ma, Y.D.: Heterogeneous spcnn and its application in image segmentation. Neurocomputing **285**, 196–203 (2018)
9. Cheng, Y.Y., Li, H.Y., Xiao, Q., Zhang, Y.F., Shi, X.L.: Gaussian noise filter using variable step time matrix of PCNN. Appl. Mech. Mater. **48–49**, 551–554 (2011)
10. Sankaran, K.S., Nagappan, N.V.: Noise free image restoration using hybrid filter with adaptive genetic algorithm. Comput. Electr. Eng. **54**, 382–392 (2016)
11. Fausto, F., Cuevas, E., Valdivia, A., González, A.: A global optimization algorithm inspired in the behavior of selfish herds. BioSystems **160**, 39–55 (2017)
12. Liu, C., Niu, P.F., Li, G.Q., Ma, Y.P., Zhang, W.P., Chen, K.: Enhanced shuffled frog-leaping algorithm for solving numerical function optimization problems. J. Intell. Manuf. **29**(5), 1133–1153 (2018)
13. Jiand, J.J., Wei, W.X., Shao, W.L., Liang, Y.F., Qu, Y.Y.: Research on large-scale bi-level particle swarm optimization algorithm. IEEE Access **9**, 56364–56375 (2021)

Particle Swarm Optimization

Dynamic Multi-swarm Particle Swarm Optimization with Center Learning Strategy

Zijian Zhu[1], Tian Zhong[1], Chenhan Wu[1](✉), and Bowen Xue[2]

[1] Shenzhen College of International Education, Shenzhen 518048, China
nichols_wu@outlook.com

[2] College of Management, Shenzhen University, Shenzhen 518060, China

Abstract. In this paper, we propose a novel variant of particle swarm optimization, called dynamic multi-swarm particle swarm optimization with center learning strategy (DMPSOC). In DMPSOC, all particles are divided into several sub-swarms. Then, a center-learning strategy is designed, in which each particle within the sub-swarms will learn from the historical optimal position of a particle or the center position in a sub-swarm. Also, an alternative learning factor is given to determine the particle learning strategy, which can be classified as center-learning or optimum-learning. Four benchmark functions are used in order to compare the performance of DMPSOC algorithm with the standard particle swarm optimization (SPSO). Experiments conducted illustrate that the proposed algorithm outperform SPSO in terms of convergence rate and solution accuracy.

Keywords: Particle swarm optimization · Multi-swarm · Center-learning strategy

1 Introduction

Particle Swarm Optimization (PSO) developed by Kennedy et al. in 1995 [1] is an optimization method that modeled on social intelligence of animals such as birds. It relies on particles that follow their historical optimal position as well as the global optimal position, which can be easily implemented, has few parameters to adjust and quick convergence speed compared to other optimization algorithms.

However, the original PSO algorithm has problems of premature convergence and great loss of diversity. There are various methods of improving the performance of the PSO algorithm, such as novel topology structure [2–8] and parameters adjustments [9–13].

Although these improvements enhance the performance of the PSO algorithm, there are still many disadvantages of them. For instance, improvements on coefficients such as social learning factor, personal learning factor and constriction factor, cannot improve the performance of the PSO algorithm fundamentally. Meanwhile, basic enhancements on topology structure mainly focus on novel communication strategies within a swarm, and these simple interactions are less frequently observed within more organized animals compared to multi-swarm communication structure. Additionally, scholars paid more

Y. Tan et al. (Eds.): ICSI 2022, LNCS 13344, pp. 141–147, 2022.
https://doi.org/10.1007/978-3-031-09677-8_12

attention to interacting with the optimal particles, ignoring the other particles which are also worth learning from. Therefore, we present an evolutionary PSO algorithm by combining a dynamic multi-swarm topology structure and a center learning scheme [14] with an alternative learning strategy. In the proposed algorithm, the central particle of multi-swarms and the personal best particle are randomly used for the velocity updating equation.

The rest of this paper is organized as follows. Section 2 conducts a short overview of the original particle swarm optimization algorithm. Section 3 gives detailed information of the multi-swarm particle swarm optimization with center communication as well as alternative learning strategy. Experimental results and comparisons between the SPSO algorithm are presented in Sect. 4, ands further conclusions are given in Sect. 5.

2 An Overview of the Particle Swarm Optimization

The original PSO algorithm is derived from observations towards swarm intelligence of animals in the nature, for example, the hunting behaviors of a flock of birds. They follow a certain pattern of velocity as well as a position updating strategy which inspired Eberhart et al. to develop the algorithm.

For each particle in the original PSO algorithm, it demonstrates a potential solution of the problem within a regulated search range, and updates its velocity every iteration based on three factors: the inertia velocity, the optimum particle position it has found already, the best particle position in the whole swarm. The velocity as well as position updating equations of the i th particle can be represented by formulas (1) and (2):

$$V_i = V_i + c_1 * r_1 * (PBest - X_i) + c_2 * r_2 * (GBest - X_i) \tag{1}$$

$$X_i = X_i + V_i \tag{2}$$

where $i = 1, 2, \ldots, s$, s is the swarm size. c_1, c_2 are learning parameters of the personal learning behavior and societal learning behavior respectively. r_1, r_2 are two random number generated from 0 to 1, *PBest* is the historical optimal position , while *GBest* is the global optimal position in the current iteration. V_i stands for the velocity of the i th particle, and X_i is the position of the i th particle.

3 The Dynamic Multi-Swarm Particle Swarm Optimization with Center Learning Strategy

3.1 Dynamic Multi-Swarm Design

From the observation of diverse behaviors of animals in the nature such as an animal invading other animals' territories as well as the protecting its own land, the behaviors of particles can vary in this program. Thus, we proposed different strategies for velocity updating of particles. In order to utilize distinctive velocity updating equations to update the velocity of particles, an alternative learning factor $c_{alternativelearning}$ is presented to alternate between several velocity updating formulas. A random number is generated from 0 to 1 and compared with it to decide which velocity updating equation to be used. In this algorithm, we set $c_{alternativelearning}$ as 0.5 to balance the utilization of several velocity updating equations, since there will be equal possibility for them to be used to update the velocity of particles, this is good to test the performance of our algorithm as well. Thus, different behaviors of particles are balanced, the convergence speed and accuracy of the result will be obtained.

In the nature, animals with higher level of intelligence generally live in groups. For instance, the Sumpter [15] proposed that animals in a large group will self-organize into small groups in order to live better, and this phenomenon is not only found in animal society, but also modern human society: people form small groups to work more effectively. Therefore, we divide all the particles into several sub-swarms, each particle in sub-swarms utilizes the same equation to update their velocities as well as their positions.

3.2 Center Learning Strategy and Optimum Learning Strategy

Collective behaviors within a large group of animals are common. For example, the study proposed by Landeau et al. [16] indicates that predators are more likely to hunt animals far away from the center of a swarm. Thus, animals of a swarm are willing to move to the center based on the behavior of hunters depicted above. This inspiration can be presented to the PSO algorithm by applying a center learning strategy to particles of the swarm, which can be represented by Fig. 1 as they work in parallel within one iteration.

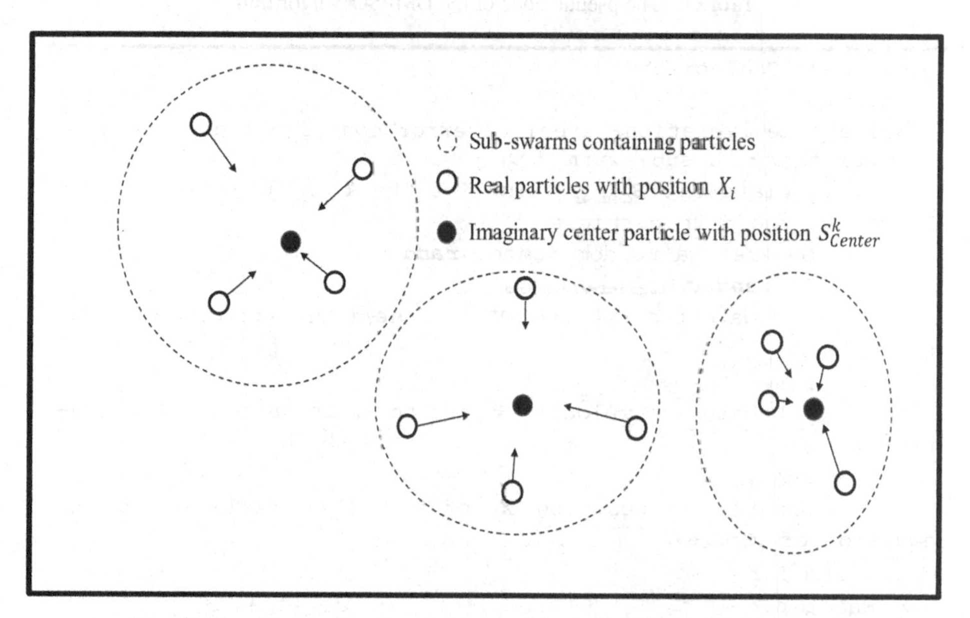

Fig. 1. Communication strategy for particles in the DMPSOC algorithm.

Then the velocity equation is:

$$V_i = w*V_i+c_2*r_2*(S^k_{Center} - X_i) \tag{3}$$

where S^k_{Center} is the "center" of the k th sub-swarm, which is the mean values in every dimension for every particle calculated by the formula:

$$S_{Center}=\frac{1}{ss}\sum_{i=1}^{ss}X_i \tag{4}$$

meanwhile, w is the linearly decreasing inertia weight of the equation that is updated by the following equation:

$$w = 0.9 - 0.5 * (\frac{iternow}{itertotal}) \tag{5}$$

where *iternow* is the current iterations and *itertotal* is the total iterations.

However, predators are more likely to move to other places based on their past experience. Meanwhile, decisions made by people in the society can be also influenced by their experiences. They inspire us to apply a strategy that particles move towards the position where it obtained its optimum fitness value and velocities of them are updated by:

$$V_i = w{*}V_i{+}c_1{*}r_1{*}(PBest{-}X_i) \tag{6}$$

Therefore, the pseudo code of DMPSOC can be presented as follows (see Table 1):

Table 1. The pseudo code of the DMPSOC algorithm

```
The DMPSOC algorithm
Begin
While (the iteration number or error goal isn't achieved)
  For the k th sub-swarm
    Calculate the S^k_Center value for the k th sub-swarm
    For the i th particle
      Generate a random number rand
      If rand>C_alternativelearning
        Update the velocity Vi for the i th particle by equa-
tion (3)
      Else
        Update the velocity Vi for the i th particle by equa-
tion (6)
      End if
      Update the position Xi of the i th particle by the
equation of Xi=Xi+Vi
    End for
  End for
End while
End
```

4 Experiment Results and Discussions

In order to compare the performance of DMPSOC algorithm and the SPSO algorithm, four benchmark functions (Sphere, Rastrigin, Ackley, Griewank, are respectively f_1, f_2, f_3, f_4) are used to test them. These functions and their search ranges are shown in Table 2.

Table 2. Results for two algorithms

Functions	*Algorithms*	*Best Value*	*Worst Value*	*Mean Value*	*Std*
f_1	PSO	**0.0000E+00**	**0.0000E+00**	**0.0000E+00**	**0.0000E+00**
	DMPSOC	5.7030E-41	8.3770E-35	1.2911E-40	1.0374E-37
f_2	PSO	7.4000E-03	**7.1300E-02**	3.2100E-02	1.4100E-02
	DMPSOC	**0.0000E+00**	4.2070E-01	**5.3801E-14**	**3.8811E-10**
f_3	PSO	**0.0000E+00**	**2.2204E-15**	1.3323e-15	1.0866E-15
	DMPSOC	**0.0000E+00**	2.6645E-14	**2.2204E-16**	**7.0217E-16**
f_4	PSO	1.0300E-02	7.1300E-02	3.8700E-02	2.5300E-02
	DMPSOC	**0.0000E+00**	**6.9400E-02**	**1.3800E-02**	**2.4000E-03**

In our experiment, the variable dimension is 10, the total swarm size of the DMPSOC algorithm is 80, consists of 4 sub-swarms with 20 particles per sub-swarm, for the SPSO algorithm, the swarm size is 80. Two algorithms both run 20 times with 15000 iterations in maximum. For parameters settings, the inertia weight for both algorithms both linearly decline from 0.9 to 0.4, and the c_1, c_2 values are both 2.0.

The results on four 10-dimentional benchmark functions of the SPSO algorithm as well as the DMPSOC algorithm are illustrated in Table 3 and Fig 2. Best values, worst values, mean values and standard deviations of two algorithms demonstrating the performances of algorithms are displayed. Bold numbers in Table 3 are the optimum values.

As presented in Table 3 and Fig 2, the DMPSOC algorithm consistently obtains minimum values for different objective functions. It can generally attain better results on these objective functions, and the result of the Rastrigin function is especially satisfactory. The reason for the less satisfactory result in the Sphere function of DMPSOC is the learning strategy is useless in single-modal objective functions. Overall, it outperforms the standard PSO algorithm.

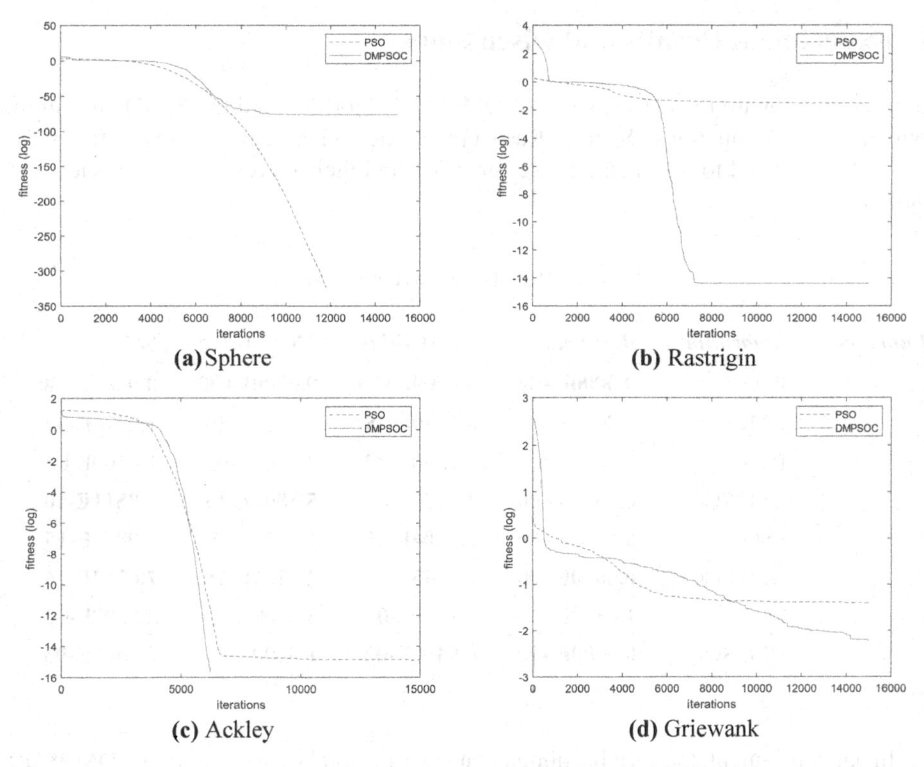

Fig. 2. Convergence curves of DMPSOC and PSO.

5 Summary and Further Work

A novel dynamic multi-swarm particle swarm optimization algorithm with center learning strategy is proposed in this paper. Distinctive to the standard particle swarm optimizer and other PSO variants, our algorithm innovatively combines the center learning strategy, the dynamic multi-swarm structure and alternative learning strategy to improve the performance of this algorithm, from the aspects of alleviating falling into local optimum and more accurate results.

Although it has relatively satisfactory results on several objective functions, it's still not the best choice for all optimizing problems. Various complex test functions can be used to examine the effectiveness of this presented algorithm, and several applications can be implemented with this proposed algorithm, improvements of the performance on unimodal objective functions could be made as well, which can be main focuses of further work.

Acknowledgment. This study is supported by Natural Science Foundation of Guangdong (2022A1515012077).

References

1. Eberhart, R., Kennedy, J.: Particle swarm optimization. In: Proceedings of the IEEE International Conference on Neural Networks, vol. 4, pp. 1942--1948 (1995)
2. Lin, A., Sun, W., Yu, H., Wu, G., Tang, H.: Global genetic learning particle swarm optimization with diversity enhancement by ring topology. Swarm Evol. Comput. **44**, 571–583 (2019)
3. Wang, Y.X., Xiang, Q.L.: Particle swarms with dynamic ring topology. In: 2008 IEEE Congress on Evolutionary Computation (IEEE World Congress on Computational Intelligence), pp. 419--423. IEEE (2008)
4. Yue, C., Qu, B., Liang, J.: A multiobjective particle swarm optimizer using ring topology for solving multimodal multiobjective problems. IEEE Trans. Evol. Comput. **22**(5), 805–817 (2017)
5. Yu, Y., Xue, B., Chen, Z., Qian, Z.: Cluster tree topology construction method based on PSO algorithm to prolong the lifetime of zigbee wireless sensor networks. EURASIP J. Wirel. Commun. Netw. **2019**(1), 1–13 (2019)
6. Niu, B., Zhu, Y., He, X., Wu, H.: MCPSO: a multi-swarm cooperative particle swarm optimizer. Appl. Math. Comput. **185**(2), 1050–1062 (2007)
7. Liang, J.J., Suganthan, P.N.: Dynamic multi-swarm particle swarm optimizer. In: Proceedings 2005 IEEE Swarm Intelligence Symposium, 2005. SIS 2005, pp. 124--129. IEEE (2005)
8. Ye, W., Feng, W., Fan, S.: A novel multi-swarm particle swarm optimization with dynamic learning strategy. Appl. Soft Comput. **61**, 832–843 (2017)
9. Shi, Y., Eberhart, R.C.: Parameter selection in particle swarm optimization. In: Porto, V.W., Saravanan, N., Waagen, D., Eiben, A.E. (eds.) EP 1998. LNCS, vol. 1447, pp. 591–600. Springer, Heidelberg (1998). https://doi.org/10.1007/BFb0040810
10. Nickabadi, A., Ebadzadeh, M.M., Safabakhsh, R.: A novel particle swarm optimization algorithm with adaptive inertia weight. Appl. Soft Comput. **11**(4), 3658–3670 (2011)
11. Li, L., Xue, B., Niu, B., Chai, Y., Wu, J.: The novel non-linear strategy of inertia weight in particle swarm optimization. In: 2009 Fourth International on Conference on Bio-Inspired Computing, pp. 1--5. IEEE (2009)
12. Cai, X., Cui, Y., Tan, Y.: Predicted modified PSO with time-varying accelerator coefficients. Int. J. Bio-Inspired Comput. **1**(1–2), 50–60 (2009)
13. Clerc, M.: The swarm and the queen: towards a deterministic and adaptive particle swarm optimization. In: Proceedings of the 1999 Congress on Evolutionary Computation-CEC99 (Cat. No. 99TH8406), vol. 3, pp. 1951--1957. IEEE (1999)
14. Yang, X., Jiao, Q., Liu, L.: Center particle swarm optimization algorithm. In: 2019 IEEE 3rd Information Technology, Networking, Electronic and Automation Control Conference (ITNEC), pp. 2084--2087. IEEE (2019)
15. Sumpter, D.: The principles of collective animal behaviour. Philosophical Trans. Royal Soc. B: Biol. Sci. **361**(1465), 5–22 (2006)
16. Landeau, L., Terborgh, J.: Oddity and the 'confusion effect' in predation. Anim. Behav. **34**(5), 1372–1380 (1986)

Alternative Learning Particle Swarm Optimization for Aircraft Maintenance Technician Scheduling

Tian Zhong[1,3], Chenhan Wu[1,3], Zijian Zhu[1,3](✉), and Zhenzhen Zhu[2,3]

[1] Shenzhen College of International Education, Shenzhen, China
MrZijianZhu@outlook.com
[2] College of Management, Shenzhen University, Shenzhen 518060, China
[3] Greater Bay Area International Institute for Innovation, Shenzhen University, Shenzhen 518060, China

Abstract. In the particle swarm optimization (PSO), each particle updates its velocity depending on its own experience and the best location that the swarm has approached so far, but this also means that it has a high tendency to fall into local optimum. To solve this problem, a new algorithm called alternative learning particle swarm optimization (ALPSO) based on the center-learning mechanism is proposed in this paper, consisting of three alternative learning strategies, like random learning strategy (ALPSO-RLS), central learning strategy (ALPSO-CLS), and mixed learning strategy (ALPSO-MLS). In the experimental part, four benchmark functions are chosen to test the performance of ALPSO, compared with the standard PSO. Finally, ALPSO is applied to tackle the aircraft maintenance technician scheduling problem. The results show that both ALPSO-RLS and ALPSO-CLS perform better than PSO in this test scenario.

Keywords: Alternative learning · Particle swarm optimization · Aircraft maintenance technician scheduling

1 Introduction

Particle swarm optimization (PSO), first developed by Kennedy and Eberhart [1], is inspired by the social behaviors of animals like birds and fish. The algorithm was based on a population of particles, each following its best and global best particle in the swarm to reach the best solution. However, PSO also faces problems like premature convergence, loss of diversity and stagnation in local optimum, causing many researchers to focus on the improvement of this algorithm, such as improvements of topology structure [2–6] and new learning strategy design [7–11].

However, most researchers tend to use only one method to achieve the best solution, easily leading to a stuck when searching for the optimal solution. In this paper, the alternative learning particle swarm optimization (ALPSO) is designed to tackle this problem, including random learning strategy (ALPSO-RLS), central learning strategy (ALPSO-CLS), and mixed learning strategy (ALPSO-MLS). In the random learning

Y. Tan et al. (Eds.): ICSI 2022, LNCS 13344, pp. 148–159, 2022.
https://doi.org/10.1007/978-3-031-09677-8_13

strategy, particles randomly choose to learn from their historical optimal solution or the swarm central solution. In the central learning strategy, particles update the velocity referring to both their historical optimal solution and the swarm central solution. In the mixed learning strategy, particles consult their historical optimal solution, the global optimal solution, and the swarm central solution to determine the velocity. These three strategies could effectively avoid the problem of local optimum by taking swarm central solution into consideration.

In addition, intelligent algorithms are suitable for solving NP-hard problems such as manpower scheduling and task scheduling. Therefore, to test the practicability of ALPSO, it is applied to solve a simple aircraft maintenance technician scheduling model referring to [13].

The rest of the paper is organized as the follows. Section 2 introduces the standard particle swarm optimization. Section 3 presents the alternative learning particle swarm optimization in detail and give the experimental results on four benchmark functions. The solution of aircraft maintenance technician scheduling based on ALPSO is given in Sect. 4. Section 5 draws the conclusion.

2 An Overview of Particle Swarm Optimization

Particle swarm optimization (PSO) was derived from the swarm intelligence of animals like flocks of birds or fish. Individuals in PSO are called particles, and the particles as a whole are called a swarm. Each particle has a current position vector $x_j(j = 1, 2, \ldots ss)$, current velocity vector v_j, the best position that the particle has encountered $pbest_j$, and the best position that the swarm has discovered $gbest_j$. ss represents swarm size, and x_j represents the j th particle in the swarm. The velocity is updated by Eq. (1).

$$v_{j+1} = w * v_j + c_1 * r_1 * \left(pbest_j - x_j\right) + c_2 * r_2 * \left(gbest_j - x_j\right) \quad (1)$$

where w represents the inertia weight of the particle, deciding how much the past velocity will affect the new velocity, c_1 and c_2 are the learning coefficients, and r_1 and r_2 are random numbers from 0 to 1. Then, x_j is updated by Eq. (2).

$$x_{j+1} = x_j + v_j \quad (2)$$

From the equations, we can see that the particles in the swarm of PSO only considered their historical best position and the location of the best particle. If all particles are stuck in local optimum, it will be hard for the particles to get out. The ALPSO algorithm could avoid this by considering all particles' experiences in a swarm. This will be more thoroughly explained in the following section.

3 Alternative Learning Particle Swarm Optimization

Collective intelligence is the intelligence that grows out of group, emerging from the collaboration at an aggregate level. In the swarm, the collective intelligence integrate the experience of all particles. Considering all experience of each particle is better than only considering the experience of one individual since if that individual is stuck, the rest of the population will be stuck with it. Considering all individual's experience can help the population to avoid trapping into local minima to some extent. For example, if a pack of wolves consider the experience of every wolf, they can avoid many traps.

In PSO, the center of the swarm is like the consideration of the experience of all particles. However, the center of the swarm is affected by the current location of particles. If each particle is around its best place it has found, the center of the swarm can be a better reflection of the experience of all particles, representing the collective intelligence of the swarm. So, if we let the particle to aim its best location or the center of the swarm, we can let the particles to only search for the best solution on its own or consider the swarm's experience to update its location. In addition, in order to increase the diversity of searching process, the random learning strategy is also introduced.

Therefore, in ALPSO, three alternative learning strategies are proposed, including random learning strategy, central learning strategy, and the mixed learning strategy.

3.1 Random Learning Strategy

In the random learning strategy, particles randomly learn from the center of the swarm or its own experience of the best location to decide their velocity (see Fig. 1).

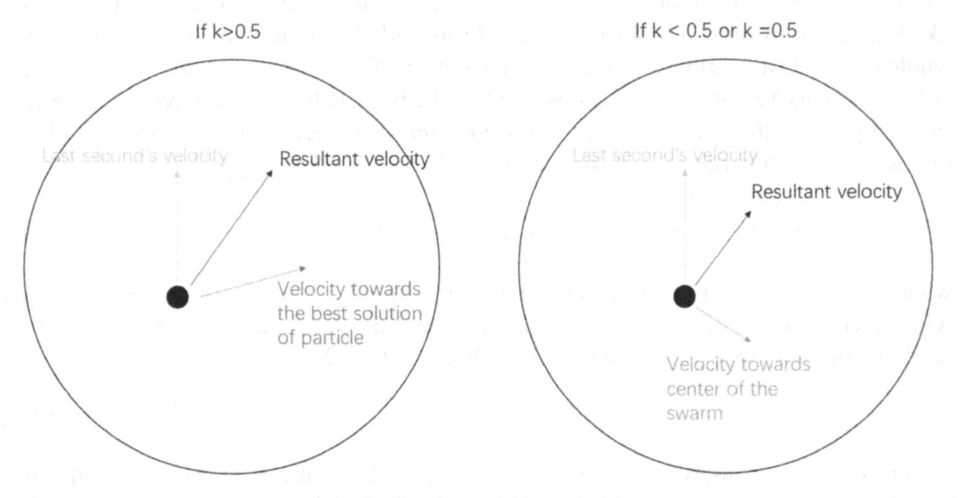

Fig. 1. Random learning strategy.

The velocity is updated by Eq. (3).

$$\begin{cases} v_{j+1} = w * v_j + c_1 * r_1 * \left(pbest_j - x_j\right), \textit{if} \quad k > 0.5 \\ v_{j+1} = w * v_j + c_2 * r_2 * \left(gmean_j - x_j\right), \textit{if} \quad k \le 0.5 \end{cases} \tag{3}$$

where $gmean_j (j = 1, 2, \ldots, ss)$ represents the center of the swarm, c_1 and c_2 are the learning parameters, and r_1, r_2, and k are the random numbers.

3.2 Central Learning Strategy

The second strategy is the central learning strategy referred in [12], in which the particles learn from both the center of the swarm and the historical experience of themselves (see Fig. 2).

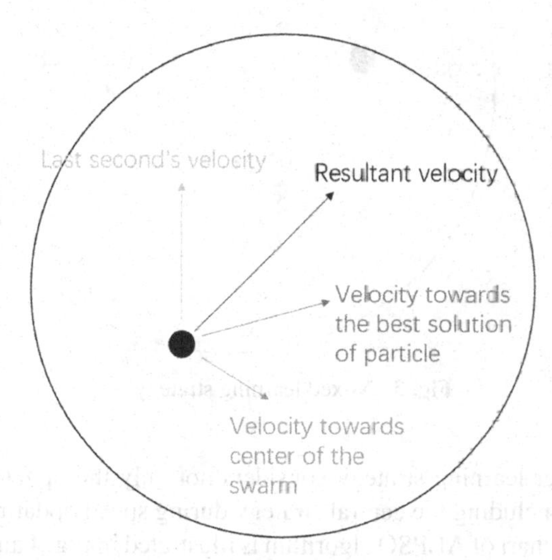

Fig. 2. Central learning strategy.

In this strategy, the velocity is updated by Eq. (4).

$$v_{j+1} = w * v_j + c_3 * r_3 * \left(pbest_j - x_j\right) + c_4 * r_4 * (gmean_j - x_j) \tag{4}$$

where c_3 and c_4 are the learning coefficients, and r_3 and r_4 are the randomly generated numbers.

Because PSO algorithm only learns from the local optimum and the global optimum, the central learning strategy emphasizes learning from the center of the swarm and the historical experience of themselves. Setting the parameters (r_3 and r_4) increases the randomness of the search in the future.

3.3 Mixed Learning Strategy

In the mixed learning strategy, particles aim to learn from the best particle, the center of the swarm as well as the experience of the particles themselves (see Fig. 3).

Velocity of this strategy is updated by Eq. (5).

$$v_{j+1} = w * v_j + c_5 * r_5 * \left(pbest_j - v_j\right) + c_6 * r_6 * \left(gbest_j - v_j\right) + c_7 * r_7 * \left(gmean_j - v_j\right) \tag{5}$$

where c_5, c_6 and c_7 are learning coefficients and r_5, r_6 and r_7 are numbers randomly generated.

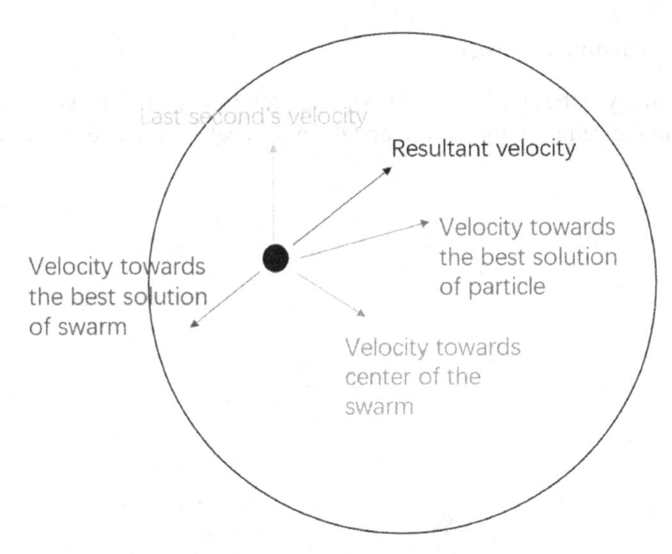

Fig. 3. Mixed learning strategy.

The mixed center learning strategy considers not only the update method of PSO algorithm, but also including the central strategy during speed updating.

Finally, the flowchart of ALPSO algorithm is illustrated in Fig. 4 and the pseudocode of ALPSO algorithm is showed in Table 1.

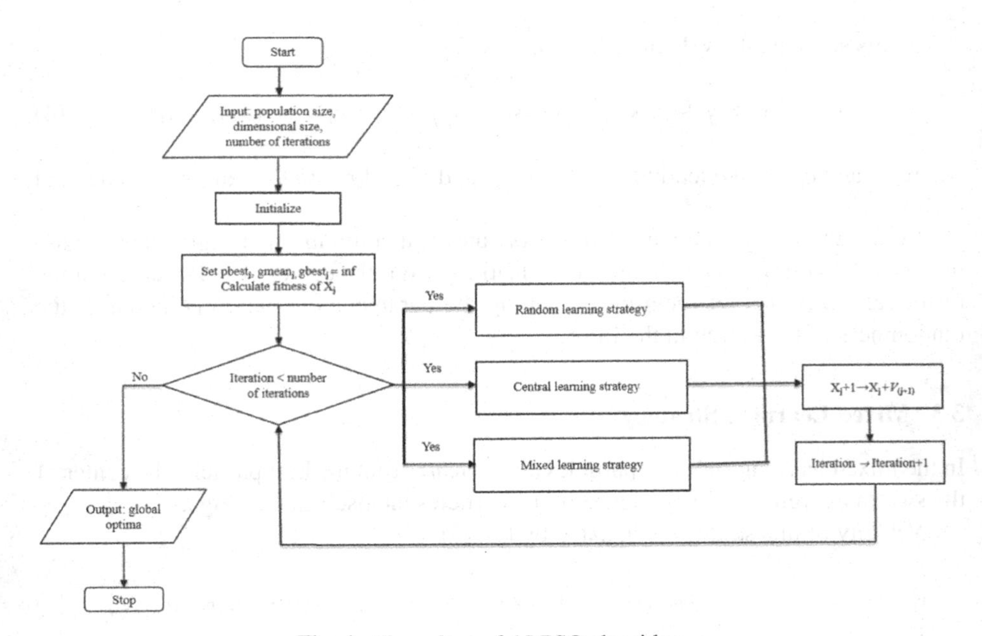

Fig. 4. Flow chart of ALPSO algorithm.

In the flow chart, we can see the overall description of the three strategies. First, the algorithm will initialize all parameters, including population size, dimension, iterations and so on, then set the iteration criteria, which named the learning strategy, using the three learning strategies proposed in the paper to process the data, and finally select the best performing learning strategy according to the end results of these algorithms. Instead of being three unrelated algorithms, they were tied under the same conditions and were chosen to reach the final best result, which is unique compared to former algorithms.

Table 1. Pseudocode of ALPSO.

Pseudocode of ALPSO
1: Alternative Learning Particle Swarm Optimization
2: **for** reach particle i
3: Initialize velocity V_i and position X_i for particle
4: Evaluate particle i and set $pBest_i = X_i$
5: **end for**
6: $gBest = \min\{pBest_i\}$
7: **while** not stop
8: **for** i =1 to N
9: Update the velocity and position of particle i
10: Update the velocity V_i for the i th particle by equation (3) or (4) or (5)
11: Evaluate particle i
12: **if** fit (X_i) < fit ($pBest_i$)
13: $pBest_i = X_i$;
14: **if** fit ($pBest_i$) < fit (gBest)
15: $gBest = pBest_i$;
16: **end for**
17: **end while**
18: compare three strategies, choose the best.
19: print gBest
20: **end procedure**

3.4 Experiments and Discussions

To evaluate the performance of ALPSO algorithm, we compare ALPSO with the original PSO on four functions—Rosenbrock, Griewank, Rastrigin and Alpine shown in Table 3. ALPSO is composed of ALPSO-RLS, ALPSO-CLS and ALPSO-MLS. In both two algorithms, the inertia weights decrease linearly from 0.9 to 0.4 [9]. We used general parameters to regulate functions, the variable dimension is 10 and the number of iterations is 5000 except Alpine, whose iteration number is 10000. The learning factors of ALPSO has different impacts. c_1, c_3 and c_5 are the learning factors of ALPSO-RLS, ALPSO-CLS and ALPSO-MLS. Their values decide how much the new velocity will be influenced by the particle's past experience. c_2,c_4 and c_6 are the learning coefficients of ALPSO-RLS, ALPSO-CLS and ALPSO-MLS respectively. The impact of the experience of the swarm on the new velocity of the particle depends on them. c_7 is the learning factor of ALPSO-MLS. It decides how much the best location experience of the swarm will impact its new velocity. The coefficients are set as shown in Table 2 in different functions. All experiments were done using Matlab. By comparing many sets of factors, we found out that under these sets of factors, ALPSO shows better results in functions and outperforms PSO. Table 4 and Fig. 5 presents the results of the four functions, and the best results are bolded.

Table 2. Parameters for experiments

No.	*Algorithms*	*Parameters*
1	Rosenbrock	*Swarmsize* = 300, run for 20 times, c_1=c_3=c_5=1.2, c_2=c_4=c_6=1.8, c_7=1.5, *inertia weight* = 0.9 to 0.4, *iteration* = 5000–1, *dimension* = 10
2	Griewank	*Swarmsize* = 300, run for 20 times, c_1=c_2=c_3=c_4=c_5=c_6=c_7=1.5, *inertia weight* = 0.9 to 0.4, *iteration* = 5000–1, *dimension* = 10
3	Rastrigin	*Swarmsize* = 300, run for 20 times, c_1=c_3=c_5=1.0, c_2=c_4=c_6=c_7=1.8, *inertia weight* = 0.9 to 0.4, *iteration* = 5000–1, *dimension* = 10
4	Alpine	*Swarmsize* = 300, run for 20 times, c_1=c_3=c_5=1.9, c_2=c_4=c_6=c_7=2.0, *inertia weight* = 0.9 to 0.4, *iteration* = 10000–1, *dimension* = 10

From Table 4, we can see that most of the time ALPSO-CLS and ALPSO-RLS has results smaller than PSO. Overall, the result of ALPSO is better than PSO. Also, as we can see from Fig. 5, ALPSO found results that are smaller in the four functions. In particular, ALPSO-RLS did well on functions Rosenbrock and Alpine, while ALPSO-C8 LS achieved better results on Griewank and Rastrigin. ALPSO-MLS only achieved better results compared to original PSO in Rosenbrock, but its convergence rate is higher than the other two strategies The strategy could do better than original PSO was due to the properties of functions. Alpine, Griewank, Rosenbrock and Rastrigin have many local stigma. The strong ability of ALPSO-CLS and ALPSO-RLS to avoid the problem of being stuck in local optimum let them to perform better in the four functions. Different

Table 3. Four benchmark functions.

Function	Mathematical representation	Search range
Rosenbrock (f_1)	$f_1(x) = \sum_{i=1}^{n-1}((x_i - 1)^2 + 100(x_{i+1} - x_i^2)^2)$	[–2.048,2.048]
Griewank (f_2)	$f_2(x) = 1 + \frac{1}{4000}\sum_{i=1}^{n} x_i^2 - \prod_{i=1}^{n}\cos(\frac{x_i}{\sqrt{i}})$	[–600,600]
Rastrigin (f_3)	$f_3(x) = \sum_{i=1}^{n}(10 - 10\cos(2\pi x_i) + x_i^2)$	[–5,12,5.12]
Alpine (f_4)	$f_4(x) = \sum_{i=1}^{d}\lvert x_i\sin(x_i) + 0.1x_i\rvert$	[0,10]

form them, ALPSO-MLS considered the best particle more, resulting in a higher convergence rate. So overall, ALPSO outperforms PSO on its final best solutions and results on the four functions.

Table 4. Results of ALPSO-RLS, ALPSO-CLS, ALPSO-MLS, and PSO on four functions.

Functions	Algorithms	Max value	Min value	Mean value	Standard deviation
Rosenbrock	ALPSO-RLS	**1.3804E – 05**	**0.00E + 00**	**8.00E – 07**	**9.43E-12**
	ALPSO-CLS	1.94E – 05	1.64E – 07	5.54E – 06	2.50E – 11
	ALPSO-MLS	1.72E – 01	**0.00E + 00**	2.38E – 02	1.40E – 03
	PSO	5.91E – 02	7.40E – 03	3.30E – 02	2.00E – 04
Griewank	ALPSO-RLS	5.69E – 02	7.80E – 03	2.24E – 02	1.2463E – 04
	ALPSO-CLS	**1.52E – 02**	4.89E – 04	**4.80E – 03**	**1.33E – 05**
	ALPSO-MLS	6.20E – 01	6.20E – 01	4.19E – 01	1.45E – 02
	PSO	6.64E – 02	**0.00E + 00**	3.91E – 02	3.37E – 04
Rastrigin	ALPSO-RLS	1.69E – 01	3.9790E – 13	3.40E – 02	2.00E – 03
	ALPSO-CLS	**9.30E – 03**	**0.00E + 00**	**7.9832E – 04**	**4.3223E – 06**
	ALPSO-MLS	6.43E + 01	4.37E + 01	5.52E + 01	3.43E + 01
	PSO	9.95E – 01	**0.00E + 00**	2.32E – 01	1.83E – 01
Alpine	ALPSO-RLS	**4.70E + 00**	2.60E + 00	**3.62E + 00**	**3.82E – 01**
	ALPSO-CLS	6.70E + 02	5.02E + 02	6.07E + 02	1.98E + 03
	ALPSO-MLS	7.01E + 02	3.26E + 02	5.12E + 02	1.08E + 04
	PSO	4.19E + 02	**1.08E – 02**	8.63E + 01	1.78E + 04

4 The Aircraft Maintenance Technician Scheduling Model

Aircraft maintenance technician scheduling (AMTS) is an NP-hard problem with the consideration of aircraft dimension, maintenance technician dimension, maintenance task dimension, and maintenance shift dimension.

4.1 Objective

For most aircraft maintenance companies, the aim of AMTS is to minimize the total cost, which can be divided into manpower cost, delivery delay cost, and overwork cost, as shown in formula (6),

$$\min\left\{\sum_{m\in M}\sum_{t\in T}\sum_{s\in S} amts\cdot\theta m+\sum_{t\in T}\max\left\{\sum_{m\in M}\sum_{s\in S} amts\cdot\tau mts-dt,0\right\}\right.$$
$$\left.+\sum_{m\in M} om\cdot\max\left\{\sum_{t\in T}\sum_{s\in S} amts\cdot\tau mts-mwm,0\right\}\right\} \tag{6}$$

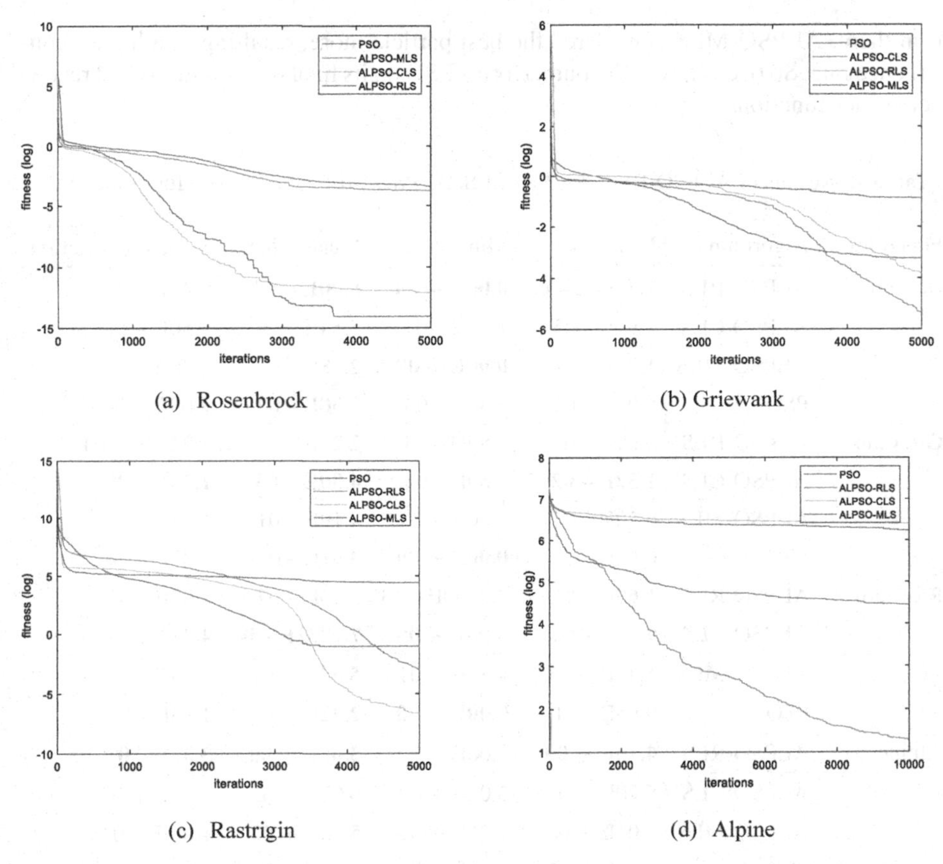

(a) Rosenbrock (b) Griewank

(c) Rastrigin (d) Alpine

Fig. 5. Convergence curves of ALPSO-RLS, ALPSO-CLS, ALPSO-MLS, and PSO.

where M, T, and S represent the total number of maintenance technician, the total number of shifts, and the total number of tasks. The other meanings of variables are listed in Table 5.

Table 5. Meanings of variables.

Variables	Meaning of variables
θm	Working cost of technician m
τmts	Working time of technician m to finish task s in shift t
dt	Delivery time of shift t
mwm	Maximum working hours of technician m

(*continued*)

Table 5. *(continued)*

Variables	Meaning of variables
Ps	Pre-tasks of task *s*
wsm	Working state of technician *m*
lm	License level of technician *m*, $l_m \in \{1, 2, 3\}$
Rls	License level *l* can meet the requirements of task *s*
amts	If technician *m* is assigned to task *s* in shift *t*
tst	If task *s* in shift *t* can be performed
om	If technician *m* needs to work overtime
cs	If task *s* is finished

4.2 Constraints

$$tst \le \frac{1}{|Ps|} \cdot \sum_{s' \in Ps} cs', \forall s \in S, \forall t \in T \tag{7}$$

$$\sum_{t \in T} \sum_{s \in S} amts \cdot \tau mts \le mwm, \forall m \in M \tag{8}$$

$$\sum_{m \in M} amts = 1, \forall t \in T, \forall s \in S \tag{9}$$

$$amts \le lmRls, \forall t \in T, \forall m \in M, \forall s \in S \tag{10}$$

$$amts \le wsm, \forall t \in T, \forall m \in M, \forall s \in S \tag{11}$$

Constraints (7) ensure that aircraft maintenance tasks are completed in sequence. Constraints (8)-(10) mean that only the free technicians can be scheduled and one task can only be assigned to one technician with qualified license level. Constraints (11) ensure that maintenance personnel have enough rest time.

4.3 Experiments and Results

Suppose there is one aircraft parking in the maintenance hanger and waiting for maintenance service. The numbers of technicians with the three license levels are all set to 5, whose unit costs are 0.5, 1.0, 1.5, respectively. Besides, there are five maintenance shifts and the numbers of tasks in each shift are set to 3, 4, 3, 2, 2, respectively. The required license levels of each task are 2, 1, 3, 1, 2, 1, 3, 1, 1, 1, 2, 3, 3, 1. Thereafter, ALPSO and PSO algorithms are used to solve the AMTS model. The results are shown in Fig. 6 and Table 6.

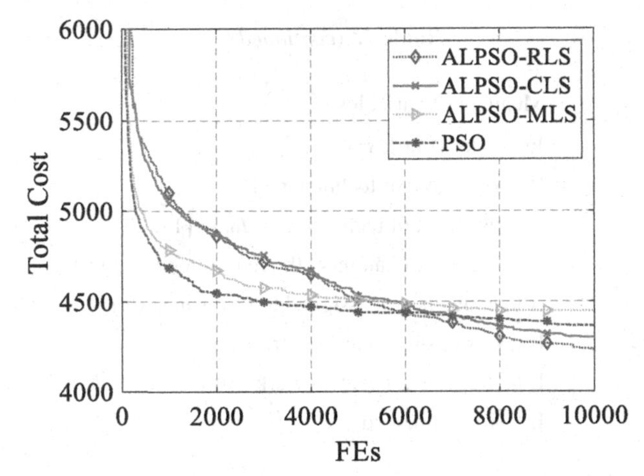

Fig. 6. Convergence curves of ALPSO-RLS, ALPSO-CLS, ALPSO-MLS, and PSO on the total cost of AMTS.

As we can see from Fig. 6, both ALPSO-RLS and ALPSO-CLS can minimize the total cost better than PSO. Besides, the phenomenon of premature convergence has been significantly alleviated in ALPSO-RLS and ALPSO-CLS. ALPSO-RLS gets the smallest total cost in the end and saved down the largest amount of total cost. From Table 6, we can see that ALPSO-RLS gets the best results.

Table 6. Total cost of AMTS based on ALPSO-RLS, ALPSO-CLS, ALPSO-MLS, and PSO.

Algorithms	Max	Mean	Min	Standard deviation	Cost saved
ALPSO-RLS	**4714.882**	**4235.196**	**3805.186**	**221.2569**	**2914.814**
ALPSO-CLS	4735.366	4303.026	3849.577	251.4595	2150.423
ALPSO-MLS	5062.041	4448.487	3904.479	292.2962	2095.521
PSO	5000.688	4366.24	3825.754	271.9244	2174.246

5 Conclusions

This paper introduces the alternative learning particle swarm optimization, which slows down the trend of premature convergence and effectively alleviate falling into local optimization. The results of three ALPSO algorithms on four benchmark functions are better than the standard PSO in most cases. When applying ALPSO to tackle aircraft maintenance technician scheduling problem, ALPSO-RLS can obtain the optimal total cost.

In the future, ALPSO algorithm can be expended to deal with multi-objective optimization models. In addition, other improvement methods can be taken into account, such as multi-swarm strategy and algorithm hybridization.

Acknowledgement. This study is supported by Natural Science Foundation of Guangdong (2022A1515012077).

References

1. Eberhart, R.C., Kennedy, J.: Particle swarm optimization. In: Proceedings of ICNN 1995-International Conference on Neural Networks, pp.1942–19448, IEEE, Perth, Australia (1995)
2. Engelbrecht, A.P.: Particle swarm optimization: global best or local best? In: 2013 BRICS Congress on Computational Intelligence and 11th Brazilian Congress on Computational Intelligence, pp. 124–135. IEEE, Ipojuca, Brazil (2013)
3. Mendes, R., Kennedy, J., Neves, J.: The fully informed particle swarm: simpler, maybe better. IEEE Trans. Evol. Comput. **8**(3), 204–210 (2004)
4. Lin, A., Sun, W., Yu, H., Wu, G., Tang, H.: Global genetic learning particle swarm optimization with diversity enhanced by ring topology. Swarm Evol. Comput. **44**, 571–583 (2019)
5. Chen, Y., Li, L., Peng, H., Xiao, J., Wu, Q.T.: Dynamic multi-swarm differential learning particle swarm optimizer. Swarm Evol. Comput. **39**, 209–221 (2018)
6. Lim, W.H., Isa, N.A.M.: Particle swarm optimization with increasing topology connectivity. Eng. Appl. Artif. Intell. **27**, 80–102 (2014)
7. Niu, B., Zhu, Y., He, X., et al.: MCPSO: a multi-swarm cooperative particle swarm optimizer. Appl. Math. Comput. **185**(2), 1050–1062 (2007)
8. Liang, J.J., Suganthan, P.N.: Dynamic multi-swarm particle swarm optimizer with local search. In: The 2005 IEEE Congress on Evolutionary Computation, pp. 522–528, IEEE, Pasadena, CA, USA(2005)
9. Lin, A., Sun, W., Yu, H., Wu, G., Tang, H.: Adaptive comprehensive learning particle swarm optimization with cooperative archive. Appl. Soft Comput. **77**, 533–546 (2019)
10. Cheng, R., Jin, Y.: A social learning particle swarm optimization algorithm for scalable optimization. Inf. Sci. **291**, 43–60 (2015)
11. Qin, Q., Cheng, S., Zhang, Q., Li, L., Shi, Y.: Particle swarm optimization with interswarm interactive learning strategy. IEEE Trans. Cybern. **46**(10), 2238–2251 (2016)
12. Niu, B., Li, L., Chu, X.: Novel multi-swarm cooperative particle swarm optimization. Comput. Eng. Appl. **45**(3), 28–34 (2009)
13. Tan, Y., Shi, Y. (eds.): ICSI 2021. LNCS, vol. 12689. Springer, Cham (2021). https://doi.org/10.1007/978-3-030-78743-1

A Surrogate-Assisted Ensemble Particle Swarm Optimizer for Feature Selection Problems

Jiang Zhi(✉), Zhang Yong, Song Xian-fang, and He Chunlin

China University of Mining and Technology, Xuzhou 221116, China
19826084236@163.com

Abstract. For feature selection problems on high-dimensional data, this paper proposes a surrogate-assisted ensemble particle swarm feature selection algorithm, by combining the global search ability of evolutionary algorithm with the fast search ability of filter method. A space partition method based on K-nearest neighbors is proposed to select represent samples as surrogate. The proposed ensemble algorithm is applied to several datasets. Experimental results show that the proposed algorithm can obtain feature subsets with higher classification accuracy in less computing time.

Keywords: Particle swarm optimization · Feature selection · Surrogate model

1 Introduction

The purpose of feature selection is to select a group of key features from all the features of the data set, so as to optimize specified indicators while reducing the learning cost [1]. Traditional feature selection algorithms are commonly divided into three categories: filter, wrapper and embedded methods [2]. The filter method is fast in the calculation cost, but difficult to eliminate redundant features, the wrapper method can get a high classification accuracy, but it is computationally expensive relatively.

In recent years, more and more scholars began to pay attention to evolutionary feature selection algorithms. In order to solve feature selection problems in large-scale data, Xu et al. [3] proposed an evolutionary algorithm based on repeated analysis of decision space and objective space. Wang et al. [4] proposed a new differential evolution algorithm, by using a new adaptive mechanism and a weighted model to improve the robustness of algorithm. Wang et al. [5] proposed a feature selection algorithm based on artificial bee colony algorithm and representative sample mechanism. Song et al. [6] proposed a particle swarm optimization algorithm based on feature and class tag. However, since classifiers are needed to continuously evaluate the classification accuracy of individuals (i.e. feature subsets), these algorithms have the problem of "expensive computational cost".

Instead of a single model, ensemble learning can combine multiple models with different characteristics to solve the same problem. Hence it can get better result in general [7]. Chen et al. [8] designed two ensemble methods by combining three different types

Y. Tan et al. (Eds.): ICSI 2022, LNCS 13344, pp. 160–166, 2022.
https://doi.org/10.1007/978-3-031-09677-8_14

of benchmark feature selection methods. Das et al. [9] proposed a feature selection algorithm based on ensemble parallel processing dual-objective genetic algorithm. However, in the face of high-dimensional data, ensemble feature selection algorithms tend to fall into local optimization because of the lack of effective global search strategy.

In view of this, this paper proposed a surrogate-assisted ensemble particle swarm optimization algorithm. By combining the global search capability of evolutionary algorithm, the fast search capability of filter algorithm and the low cost of surrogate-assisted strategy, the proposed algorithm can significantly improve the overall performance of feature selection. Experimental results show that the proposed algorithm can obtain feature subsets with higher classification accuracy in less computing time.

This paper is organized as follows. Section 2 focuses on the related work. Section 3 details the proposed algorithm. Sections 4 verifies the effectiveness of the proposed algorithm. Section 5 concludes the paper.

2 Related Work

2.1 Traditional Feature Selection Method

Traditional feature selection algorithms are commonly divided into three categories [8]: filter, wrapper and embedded methods. The filter method completes feature selection by scoring and ranking the importance of features [10]. The wrapper method uses learning algorithm to evaluate the performance of each feature subset in feature selection process [11]. In terms of the classification accuracy of feature subset, wrapper method is superior to filter method, but it consumes high computational cost. The embedded method combines the feature selection process with the classifier learning process [12], it uses classifiers to deduce the importance of features, but its classification performance depends heavily on the classifiers.

The feature selection method based on evolutionary algorithm is called evolutionary feature selection. Because the global search strategy can find the optimal or suboptimal solution of the problem, the method has gradually become a hot research technology to solve the problem of feature selection in recent years. Typical algorithms such as genetic algorithm [13], ant colony algorithm [14] and artificial bee colony algorithm [15] are applied to feature selection. As a kind of heuristic search technology based on population, particle swarm optimization (PSO) has the advantages of simple concept, convenient implementation and fast convergence speed, it has been widely used in FS problems [16, 17]. However, as mentioned above, with the rapid increase in the number of feature dimensions and instances, the existing methods still suffer from the problem of high computational cost.

2.2 Ensemble Feature Selection

Ensemble feature selection uses multiple same or different base selectors to process data sets simultaneously and ensemble the feature selection results of these base selectors. Existing ensemble feature selection methods can be divided into two categories: homogeneous and heterogeneous [18]. In the homogeneous approach, the same feature selection method is used, but with different training data subsets. For the heterogeneous approach, a number of different feature selection methods are applied, but over the same training data [19]. Tsai et al. [20] analyzed serial and parallel ensemble feature selection methods. Barbara et al. [21] analyzed the influence of different base feature selection methods on ensemble framework of high dimensional genomic data. Jiang et al. [22] proposed a hybrid ensemble feature selection algorithm based on particle swarm optimization.

Compared with traditional feature selection algorithms, the ensemble feature selection algorithm performs better when dealing with high-dimensional datasets containing a small number of samples [21]. However, as mentioned above, the existing methods still have the disadvantages of high computational cost when facing large-scale data sets. A few scholars have started to study the hybrid strategy of different ensemble feature selection algorithms. Chiew et al. [23] proposed a two-stage hybrid ensemble feature selection framework for Phishing detection system based on machine learning. Tu et al. [24] proposed a multi-strategy integrated gray wolf optimization algorithm. These papers still face the disadvantage of local convergence due to the lack of effective global search mechanism.

3 The Proposed Algorithm

This section introduces the proposed surrogate-assisted ensemble feature selection algorithm based on PSO (MEPSO) in detail. The proposed algorithm includes two stages: the ensemble feature selection stage, the surrogate-assisted evolutionary optimization stage.

3.1 The Ensemble Feature Selection Stage

The purpose of the ensemble feature selection stage is to remove irrelevant or weakly correlated features quickly. In this stage, heterogeneous ensemble based on filter methods is used to obtain a set of better feature subset. Three classical filter methods, namely Information Gain, Relief and mRMR, are selected to obtain three feature ranking lists. Then, the median reduction method is used to aggregate these feature ranking results, and a reduced feature subset is obtained.

3.2 The Surrogate-Assisted Evolutionary Optimization Stage

In this stage, an evolutionary algorithm with good global capability is used to remove redundant features. Firstly, some representative samples are selected from the original sample set to construct a surrogate model, and then on the basis of the surrogate model, PSO is used to remove the redundant features, and finally the feature subset is obtained.

In order to build surrogate model more reasonable, the majority and minority classes in data are separated. In order to make the selected samples representative, boundary samples and central samples are selected to construct the representative sample set of the majority class. A method based on k-nearest neighbor (K-RS) is designed to divide sample space. The method determines whether K nearest neighbors of a sample belong to the same class as the sample by the distance between them. If so, the sample is a center sample. Otherwise, the sample is a class boundary sample. Figure 1 shows a simple example. As shown in the figure, one of the K neighbors of sample point X_1 belongs to other classes, so this point belongs to the class boundary sample. The K neighbors of sample X_2 belong to the same class, so this point is a class center sample.

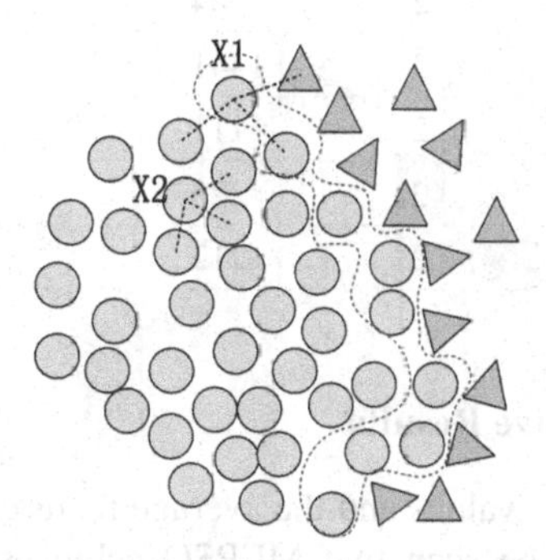

Fig. 1. The class boundaries region when K = 3. Here the samples in red dotted box belong to class boundary region of majority class

This paper uses the bare-bones PSO based feature selection algorithm (BBPSO-FS) in [25] to generate a candidate feature subset. Compared with other PSO-based FS algorithms, BBPSO-FS does not need to set key parameters including inertia weights and learning factors.

The binary code is used to represent a particle. If a feature is selected, its corresponding element is "1"; otherwise, it is 0. KNN classifier is used to evaluate the particle fitness as follows:

$$Fit(X_i) = \frac{\#correctly\ classified\ samples}{\#\ total\ samples} \tag{1}$$

Noted that the representative samples selected above are used to construct a surrogate to replace the whole original sample set when evaluating a particle's fitness.

Moreover, a local search strategy is also used to modify the optimal feature subset obtained by PSO. Firstly, the unselected features are ranked using the three filter methods in the stage 1, and three new feature ranking results are obtained. Secondly, according to the feature grade values from low to high, the features are successively put into the optimal feature subset obtained by PSO until the performance of the subset is no longer improved.

4 Experiments

In order to test the performance of the proposed algorithm, 5 datasets from UCI are selected. Table 1 shows the basic information of these datasets. To verify the performance of the proposed MEPSO, this section compares it with two FS algorithms, i.e., SAPSO-FS [26] and HPSO-SSM [27]. In all the three algorithms. the maximum iteration times is set to be100, the swarm size is set to be 45. Classification accuracy is used to evaluate the performance of the algorithm. Using 10-fold cross-validation method to evaluate feature subset, in order to ensure the fairness of comparison.

Table 1. Information of the datasets

Datasets	#samples	#features	#classes	# samples of majority class	#samples of minority class
LSVT	126	311	2	84	42
Semeion	1593	267	2	1441	152
Sonar	208	61	2	111	97
Ionosphere	351	35	2	225	126
SPECTHeart	267	45	2	212	55

4.1 Analysis of Comparative Results

Table 2 shows the average *Ac* values and the average feature subset sizes (***d****) obtained by three algorithms. It can be seen that MEPSO achieves the highest classification accuracy *Ac* on all the five datasets, and is significantly better than these comparison algorithms. More intuitively, by calculating the difference of classification accuracy between MEPSO and the two comparison algorithms in each data set, it can be seen that the classification accuracy of MEPSO is about 14.3% higher than other algorithms. For the 3 out of 5 datasets, MEPSO obtains the smallest average feature subset sizes (***d****). Furthermore, Table 3 shows the running time of MEPSO and the two comparison algorithms. We can see that the proposed MEPSO algorithm achieves the shortest running time on all datasets.

Table 2. Average *Ac* and *d**values obtained by MEPSO and two comparison algorithms

Datasets	Average *Ac*			Average *d**		
	MEPSO	HPSO-SSM	SAPSO	MEPSO	HPSO-SSM	SAPSO
LSVT	93.1	66.5	75.0	**34**	70	261
Semeion	96.9	57.7	92.4	160	50	98
Sonar	91.8	68.2	84.8	36	**4**	49
Ionosphere	100.0	80.3	85.8	**6**	10	11
SPECTHeart	99.3	70.2	77.2	**13**	33	35

Table 3. Average running time of MEPSO and the two comparison algorithms (mints)

Datasets	MEPSO	HPSO-SSM	SAPSO
LSVT	**0.1**	1.1	2.7
Semeion	**1.4**	1.7	5.4
Sonar	**0.1**	2.3	2.4
Ionosphere	**0.1**	1.1	2.7
SPECTHeart	**0.2**	1.1	2.4

5 Conclusions

By combining the global search ability of evolutionary algorithm and the fast search ability of filter ensemble algorithm, this paper proposed an ensemble particle swarm feature selection algorithm (MEPSO). Comparing with two typical evolutionary feature selection algorithms (SAPSO and HPSO-SSM), the results show that the proposed algorithm can significantly improve the classification accuracy without increasing the running time. Ensemble feature selection under the protection of privacy will be the focus of future research.

References

1. Abdel-Basset, M., El-Shahat, D., El-Henawy, I., Albuquerque, V.H.C., Mirjalili, S.: A new fusion of grey wolf optimizer algorithm with a two-phase mutation for feature selection. Expert Syst. Appl. **139**, 112824 (2020)
2. Li, Y., Li, T., Liu, H.: Recent advances in feature selection and its applications. Knowl. Inf. Syst. **53**(3), 551–577 (2017). https://doi.org/10.1007/s10115-017-1059-8
3. Xu, H., Xue, B., Zhang, M.: A duplication analysis-based evolutionary algorithm for biobjective feature selection. IEEE Trans. Evol. Comput. **25**(2), 205–218 (2021)
4. Wang, X.B., Wang, Y.H., Wong, K.C., Li, X.T.: A self-adaptive weighted differential evolution approach for large-scale feature selection. Knowl.-Based Syst. **235**, 107633 (2022)
5. Wang, X.H., Zhang, Y., Sun, X.Y., Wang, Y.L., Du, C.H.: Multi-objective feature selection based on artificial bee colony: an acceleration approach with variable sample size. Appl. Soft Comput. **88**, 106041 (2020)
6. Song, X.F., Zhang, Y., Gong, D.W., Sun, X.Y.: Feature selection using bare-bones particle swarm optimization with mutual information. Pattern Recogn. **112**, 107804 (2021)
7. Khurshid, F., Zhu, Y., Xu, Z., Ahmad, M.: Enactment of ensemble learning for review spam detection on selected features. Int. J. Comput. Intell. Syst. **12**(1), 387–394 (2019)
8. Chen, C.W., Tsai, Y.H., Chang, F.R., Lin, W.C.: Ensemble feature selection in medical datasets: combining filter, wrapper, and embedded feature selection results. Expert Syst. **37**(5), e12553 (2020)
9. Das, A.K., Das, S., Ghosh, A.: Ensemble feature selection using bi-objective genetic algorithm. Knowl.-Based Syst. **123**, 116–127 (2017)
10. Tallón-Ballesteros, A.J., Riquelme, J.C., Ruiz, R.: Filter-based feature selection in the context of evolutionary neural networks in supervised machine learning. Pattern Anal. Appl. **23**(1), 467–491 (2019). https://doi.org/10.1007/s10044-019-00798-z
11. Gokalp, O., Tasci, E., Ugur, A.: A novel wrapper feature selection algorithm based on iterated greedy metaheuristic for sentiment classification. Expert Syst. Appl. **146**, 113176 (2020)
12. Khanji, C., Lalonde, L., Bareil, C., Lussier, M.T., Perreault, S., Schnitzer, M.E.: Lasso regression for the prediction of intermediate outcomes related to cardiovascular disease prevention using the transit quality indicators. Med Care **57**(1), 63–72 (2019)
13. Sikora, R., Piramuthu, S.: Framework for efficient feature selection in genetic algorithm based data mining. Eur. J. Oper. Res. **180**(2), 723–737 (2007)
14. AlFarraj, O., AlZubi, A., Tolba, A.: Optimized feature selection algorithm based on fireflies with gravitational ant colony algorithm for big data predictive analytics. Neural Comput. Appl. **31**(5), 1391–1403 (2018). https://doi.org/10.1007/s00521-018-3612-0
15. Cura, T.: A rapidly converging artificial bee colony algorithm for portfolio optimization. Knowl.-Based Syst. **233**, 107505 (2020)

16. Song, X.F., Zhang, Y., Guo, Y.N., Sun, X.Y., Wang, Y.L.: Variable-size cooperative coevolutionary particle swarm optimization for feature selection on high-dimensional data. IEEE Trans. Evol. Comput. **24**(5), 882–895 (2020)
17. Song, X.F., Zhang, Y., Gong, D.W., Gao, X.Z.: A fast hybrid feature selection based on correlation-guided clustering and particle swarm optimization for high-dimensional data. IEEE Trans. Cybern. (2021). 109/TCYB.2021.3061152
18. Seijo-Pardo, B., Porto-Díaz, I., Bolón-Canedo, V., Alonso-Betanzos, A.: Ensemble feature selection: homogeneous and heterogeneous approaches. Knowl.-Based Syst. **118**, 124–139 (2017)
19. Pereira, T., Ferreira, F.L., Cardoso, S.: Ensemble feature selection for stable biomarker identification and cancer classification from microarray expression data. Comput. Biol. Med. **142**, 105208 (2021)
20. Tsai, C.F., Sung, Y.T.: Ensemble feature selection in high dimension, low sample size datasets: parallel and serial combination approaches. Knowl.-Based Syst. **203**, 106097 (2020)
21. Pes, B., Dessi, N., Angioni, M.: Exploiting the ensemble paradigm for stable feature selection: a case study on high-dimensional genomic data. Inform. Fusion **35**, 132–147 (2017)
22. Jiang, Z., Zhang, Y., Wang, J.: A multi-surrogate-assisted dual-layer ensemble feature selection algorithm. Appl. Soft Comput. **110**, 107625 (2021)
23. Chiew, K.L., Tan, C.L., Wong, K., Yong, K.S.C., Tiong, W.K.: A new hybrid ensemble feature selection framework for machine learning-based phishing detection system. Inf. Sci. **484**, 153–166 (2019)
24. Tu, Q., Chen, X.C., Liu, X.C.: Multi-strategy ensemble grey wolf optimizer and its application to feature selection. Appl. Soft Comput. **76**, 16–30 (2019)
25. Zhang, Y., Gong, D.-W., Sun, X.-Y., Geng, N.: Adaptive bare-bones particle swarm optimization algorithm and its convergence analysis. Soft. Comput. **18**(7), 1337–1352 (2013). https://doi.org/10.1007/s00500-013-1147-y
26. Xue, Y., Xue, B., Zhang, M.J.: Self-adaptive particle swarm optimization for large-scale feature selection in classification. ACM Trans. Knowl. Discov. Data **13**(5), 1–27 (2019)
27. Chen, K., Zhou, F.Y., Yuan, X.F.: Hybrid particle swarm optimization with spiral-shaped mechanism for feature selection. Expert Syst. Appl. **128**, 140–156 (2019)

Generational Exclusion Particle Swarm Optimization with Central Aggregation Mechanism

Chenhan Wu[1], Zijian Zhu[1], Zhong Tian[1(✉)], and Chen Zou[2]

[1] Shenzhen College of International Education, Shenzhen 518048, Guangdong, China
siennazhong@outlook.com

[2] College of Management, Shenzhen University, Shenzhen 518060, China

Abstract. This paper proposes Generational Exclusion Particle Swarm Optimization with Central Aggregation Mechanism (GEPSO-CAM) to improve the global searching ability of the standard particle swarm optimization (SPSO). First, in the generational exclusion strategy, particles will not only approach the optimal particle, but also stay away from the worst particle with a certain probability when updating positions. Then, in the central aggregation mechanism, particles learn from both their historical optimal positions and the central positions in the swarm. Finally, the proposed algorithm is compared with SPSO and bacterial foraging optimization (BFO) algorithms based on four benchmark functions. The experimental results show that the GEPSO-CAM can effectively alleviate falling into local optimal solutions and improve the accuracy of optimal solutions.

Keywords: Generational exclusion strategy · Central aggregation · Particle swarm optimization

1 Introduction

In 1995, the standard particle swarm optimization (SPSO) algorithm was initially proposed by Eberhart et al. [1], as one of the most popular intelligent algorithms, motivated by the searching and social behavior of the swarm particles in a particular searching space. The velocity of each particle is updated by comparing the searching experience of itself and that of the whole swarm.

A large variety of research has been done in the past decades to find new methods to improve the convergence of SPSO algorithm, such as incorporating new learning strategies [2–4], combining of social learning strategies [5–7] and so on. However, during the experiments, it was found that the SPSO can hardly avoid falling into local optimal solution, especially when dealing with complex multimodal functions. Therefore, based on the previous work [2, 8–10], we present a new algorithm to balance the exploration search and exploitation search using a central-directed strategy and generational exclusion. When a particle updates its position, it will learn from the central particle and the best particle of the swarm. In addition, it will stay away from the worst individual in the entire swarm generationally.

Y. Tan et al. (Eds.): ICSI 2022, LNCS 13344, pp. 167–175, 2022.
https://doi.org/10.1007/978-3-031-09677-8_15

The rest of this paper is organized as follows. Section 2 gives a brief overview of the standard particle swarm optimization algorithm. Section 3 introduces the generational exclusion strategy and central aggregation mechanism in detail. Section 4 offers the experimental results, and comparison with other algorithms. Section 5 gives the overall conclusions and future opportunities.

2 Standard Particle Swarm Optimization Algorithm

In SPSO algorithm, the motion of swarm particles has two uppermost elements, including velocity and the particle's current position. The position equals to the sum of the initial position with the particle velocity multiplied by the time interval Δt, assuming that each iteration is uniform. All particles aim to find the optimal location in the test area by attractive operations. The time interval of the algorithm is [1, K], and the particle swarm optimization algorithm gives the candidate solution in the form of position, and it is reevaluated in each iteration. The iterative search process of each particle shows two kinds of behaviors, e.g., cognitive behavior (to record the optimal position of the particle and the tendency to return to that position) and social behavior (to observe the rest of the colony and tend to move towards the best position). The update formulas for velocity and position are as follows.

The formula relies on the upgradation of the position (x_i) and velocity (v_i) of the i th particle, shown in the formulas below:

$$v_i = v_i + c_1 \times r_1 \times (Pbest_i - x_i) + c_2 \times r_2 \times (Gbest_i - x_i) \tag{1}$$

$$x_i = x_i + v_i \times \Delta t \tag{2}$$

where v_i and x_i represent the position and velocity, c_1 and c_2 are two acceleration coefficients, r_1 and r_2 are random numbers in the range of [0.0, 1.0]. $Pbest_i$ is the

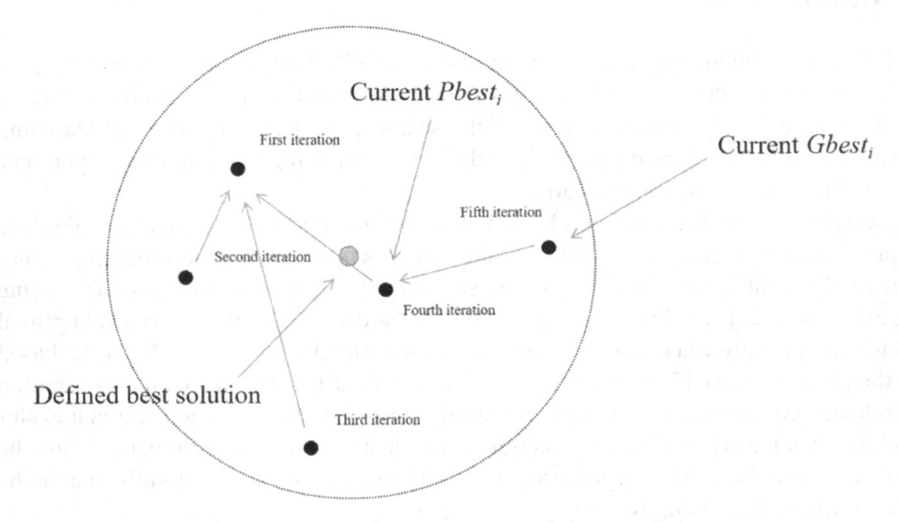

Fig. 1. Learning from the *Pbest* and G*best* particles.

historical optimal position of particle i and $Gbest_i$ is the current global optimal position (see Fig. 1).

3 Generational Exclusion Particle Swarm Optimization with Central Aggregation Mechanism

With the increase of dimension size, optimization process becomes increasingly complicated. More searching space is thus needed, which might lead to an inferior convergence performance and fall into local optimal solutions, especially in multimodal functions. In addition, the situation that minority of particles may not follow the choices of most particles should also be taken into consideration. Therefore, the algorithm should be more realistic and follow the evolution rules of the nature to improve the efficiency for seeking best solution. The choices of minority particles in a swarm should be included because these choices sometimes lead to a better result and avoid the whole swarm falling into a local optimal solution. If this situation can be considered, the convergence performance might be better. Thus, we propose the generational exclusion particle swarm optimization with central aggregation mechanism (GEPSO-CAM) in order to add the choices from minority groups by adding an exclusion process in a certain time of iterations.

3.1 Generational Exclusion Particle Swarm Optimization (GEPSO)

Naturally, when the relative population of an area is overpopulated or the population pressure increases to a certain degree, a part of the citizens will leave their current cities or countries to find a more superior and suitable one, which is called the population

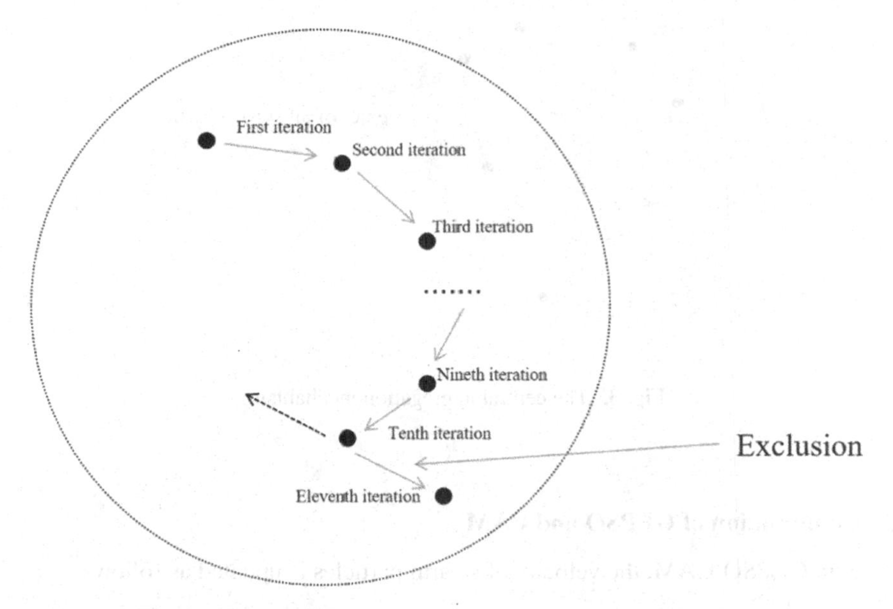

Fig. 2. The generational exclusion particle swarm optimization.

diffusion. Therefore, following the natural rule, the choices of migrating living areas are included in the generational exclusion particle swarm optimization with central aggregation mechanism (GEPSO-CAM) by adding exclusion movement per certain number of an iteration. This is one part of the factors in GEPSO-CAM that determines the motion of particles. For every 10 iterations (based on previous experiments 10 iterations is used in the later experimental study), an exclusion process will occur between particle (see Fig. 2). Therefore, when it comes to the 11th iteration, it will change the current trend of motion and move towards a opposite direction as the implement of generational exclusion strategy.

3.2 Central Aggregation Mechanism (CAM)

Throughout the development of human beings, from the primitive times to the current era, people are moving increasingly closer to each other from separated families with 3–4 people living in caves to concentrated community with 300–400 people living in mansions. Due to the contribution of economic gravity, people are gradually concentrating in order to seek a more suitable region for development. Inspired by this phenomenon, it can be concluded that gathering towards the middle of various individuals might actually leads to a better consequence. Thus, the central position value (the middle of individuals) of the whole swarm group is included in GEPSO-CAM. Therefore, another part of the factors determines the final velocity in GEPSO-CAM is created by the average solution of a particle swarm (see Fig. 3). All particles are moving towards the center position of all swarm particles.

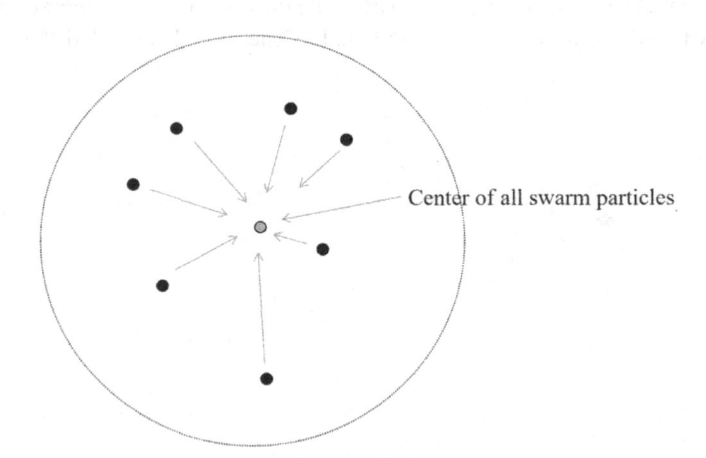

Fig. 3. The central aggregation mechanism.

3.3 Combination of GEPSO and CAM

Hence, in GEPSO-CAM, the velocity of swarm particles is updated as follows:

$$v_i = v_i + c_1 \times r_1 \times (Pbest_i - x_i) + c_2 \times r_2 \times \left(Averagex_i - x_i\right) \quad (3)$$

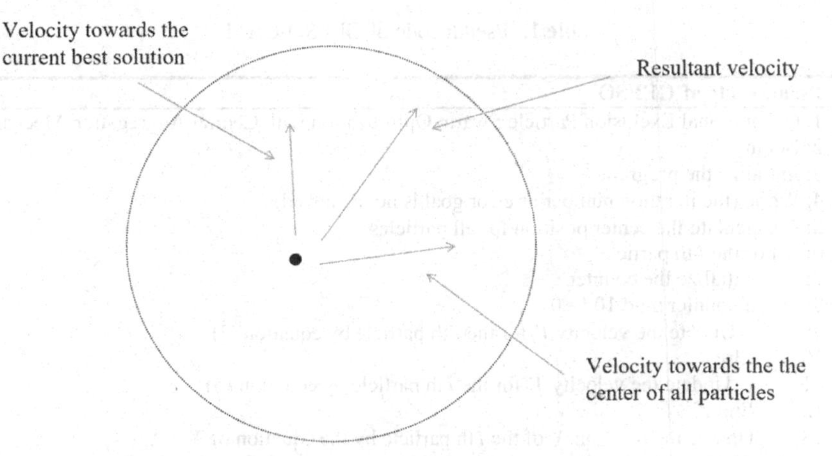

Fig. 4. A figure for the resultant velocity of GEPSO-CAM.

$$x_i = x_i + v_i \times \Delta t \tag{4}$$

$$v_i = -0.5 \times [v_i + c_2 \times r \times (Gworst_i - x_i)] if\ Iter\ mod\ 10 = 0 \tag{5}$$

(3) and (4) are due to the Central Aggregation Mechanism, (5) is due to the Generational Exclusion Particle Swarm Optimization. Average x_i is defined as the mean position of all swarm particles. $Gworst_i$ is the global worst particle in each iteration. *Iter* means the total number of iterations. The velocity of each particle depends on two elements: the velocity towards the center of all particles and the velocity towards the current best particles (see Fig. 4). The pseudocode of GEPSO-CAM algorithm is shown in Table 1.

Table 1. Pseudocode of GEPSO-CAM.

Pseudocode of GEPSO
1: Generational Exclusion Particle Swarm Optimization with Central Aggregation Mechanism
2: Begin
3: Initialize the program
4: While (the iteration number or error goal is not achieved)
5: Calculate the center position for all particles
6: For the i th particle
7: Initialize the counter
8: If counter mod 10 != 0
9: Update the velocity V_i for the i th particle by equation (3)
10: Else
11: Update the velocity V_i for the i th particle by equation (5)
12: End if
13: Update the position X_i of the i th particle by the equation of $X_i = X_i + V_i$
14: End for
15: End while
16: End

4 Experiment and Result

4.1 Parameter Setting

SPSO and BFO algorithms [10] are chosen to test the performance of the GEPSO-CAM on five benchmark functions, as shown in the Table 2. The related parameters are illustrated in Table 3 as well.

Experiments are carried out under the same number of swarm size which is 160, each benchmark run for 20 times, the inertia weight gradually decreases from 0.9 to 0.4, the acceleration coefficient of these function is 2.0, iteration is set to 5000, and the variable dimension is 10.

Table 2. Search range for four benchmark functions.

Functions	Mathematical formulas	Search ranges
Apline	$f(x) = \sum_{i=1}^{d} \lvert x_i sin(x_i) + 0.1x_i \rvert$	[0, 10]
Griewank	$f_4(x) = 1 + \frac{1}{4000}\sum_{i=1}^{n} x_i^2 - \prod_{i=1}^{n} cos\left(\frac{x_i}{\sqrt{i}}\right)$	[–600, 600]
Rastrigin	$f(x) = 10d + \sum_{i=1}^{d}\left[x_i^2 - 10cos(2\pi x_i)\right]$	[–5.12, 5.12]
Rosenbrock	$f_3(x) = \sum_{i=1}^{n-1}\left((x_i - 1)^2 + 100\left(x_{i+1} - X_i^2\right)^2\right)$	[–2.048, 2.048]

Table 3. Parameters of three algorithms.

No.	Algorithms	Parameters
1	GEPSO-CAM	*SwarmSize* = 160, Run for 20 times, $c_1 = c_2 = 2$, *inertia weight* = 0.9 to 0.4, *iteration* = 5000–1, *Dimension* = 10
2	SPSO	*SwarmSize* = 160, Run for 20 times, $c_1 = c_2 = 2$, *inertia weight* = 0.9 to 0.4, *iteration* = 5000–1, *Dimension* = 10
3	BFO	*SwarmSize* = 160, Run for 20 times, $c_1 = c_2 = 2$, *inertia weight* = 0.9 to 0.4, *iteration* = 5000–1, *Dimension* = 10

4.2 Experimental Results and Analysis

Figure 5 and Table 4 present the results of four benchmark function and the convergence characteristic related to the value of max metric, min metric, mean metric and standard deviation.

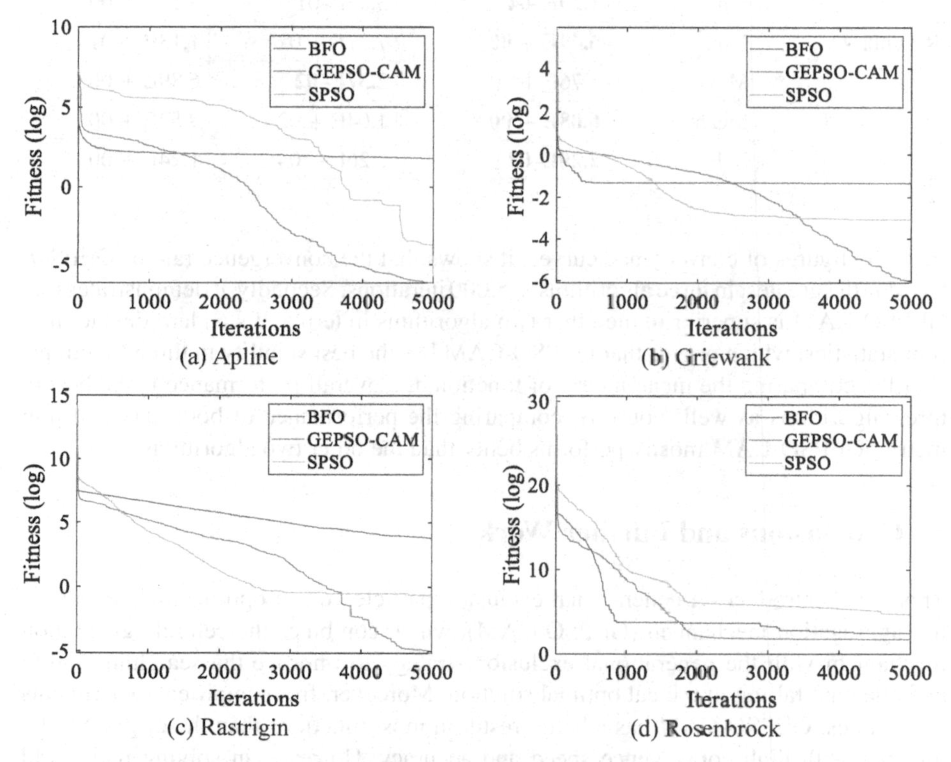

Fig. 5. Convergence curves of GEPSO-CAM, SPSO, and BFO.

Comparing the four benchmark functions in Fig. 5, including Apline, Griewank, Rastrigin, and Rosenbrock, we can conclude four features of GEPSO-CAM. Firstly,

Table 4. Comparison among GEPSO-CAM, SPSO, and BFO on four functions.

Functions	Metrics	GEPSO-CAM	SPSO	BFO
Apline	Max	**6.90E–03**	4.53E + 00	5.21E + 00
	Min	**4.80E–41**	9.89E–12	2.93E + 00
	Mean	**1.70E–03**	2.27E–01	4.26E + 00
	Std	**6.55E–06**	1.03E + 00	5.90E–01
Griewank	Max	**5.10E–03**	8.60E–02	2.58E–01
	Min	0.00E + 00	**7.40E-03**	1.50E–01
	Mean	**1.50E–03**	3.31E–02	2.21E–01
	Std	**2.36E–06**	4.46E–04	3.17E–02
Rastrigin	Max	**3.76E–02**	1.99E + 00	2.46E + 01
	Min	**0.00E + 00**	**0.00E + 00**	1.38E + 01
	Mean	**5.80E–03**	5.47E–01	2.00E + 01
	Std	**1.20E–04**	3.62E–01	3.17E + 00
Rosenbrock	Max	**6.49E + 00**	7.37E + 02	1.13E + 01
	Min	5.76E + 00	**2.37E–02**	6.89E + 00
	Mean	**6.08E + 00**	1.04E + 02	9.52E + 00
	Std	**2.20E–02**	3.26E + 04	1.24E + 00

from the figures of convergence curves, it shows that the convergence rate of GEPSO-CAM is the greatest in three algorithms in 5,000 iterations. Secondly, it demonstrates that GEPSO-CAM is superior to the other two algorithms in terms of standard deviation in four statistics, which proves that GEPSO-CAM has the best stability in three functions. Thirdly, comparing the mean metric of function, the overall performance is the best in three algorithms as well. Fourthly, comparing the performance of both max and min metric, GEPSO-CAM mostly performs better than the other two algorithms.

5 Conclusions and Further Work

This paper introduces a generational exclusion particle swarm optimization with central aggregation mechanism (GEPSO-CAM), which combines the central aggregation mechanism with the generational exclusion strategy to improve the searching ability and alleviate falling into local optimal solution. Moreover, from empirical experiments and studies, GEPSO-CAM has a better result in most functions than the original PSO, together with high convergence speed and accuracy. However, in solving real-world problems, the stability of GEPSO-CAM may not be steady enough, which requires further study in the future. Therefore, in the future GEPSO-CAM will be considered to solve the real-world problems.

Acknowledgement. This study is supported by Natural Science Foundation of Guangdong (2022A1515012077).

References

1. Kaveh, A.: Particle swarm optimization. In: Advances in Metaheuristic Algorithms for Optimal Design of Structures, pp. 9–40. Springer, Cham (2014). https://doi.org/10.1007/978-3-319-05549-7_2
2. Liang, J.J., Qin, A.K., Suganthan, P.N., Baskar, S.: Comprehensive learning particle swarm optimizer for global optimization of multimodal functions. IEEE Trans. Evol. Comput. **10**(3), 281–295 (2006)
3. Ren, Z., Zhang, A., Wen, C., Feng, Z.: A scatter learning particle swarm optimization algorithm for multimodal problems. IEEE Trans. Cybern. **44**(7), 1127–1140 (2014)
4. Bonyadi, M.R., Michalewicz, Z.: Analysis of stability, local convergence, and transformation sensitivity of a variant of the particle swarm optimization algorithm. IEEE Trans. Evol. Comput. **20**(3), 370–385 (2016)
5. Cheng, R.: A social learning particle swarm optimization algorithm for scalable optimization. Inf. Sci. **291**, 43–60 (2015)
6. Chen, W.N., et al.: Particle swarm optimization with an aging leader and challengers. IEEE Trans. Evol. Comput. **17**(2), 241–258 (2013)
7. Zhang, H., Liang, Y., Zhang, W., Xu, N., Guo, Z., Wu, G.: Improved PSO-based method for leak detection and localization in liquid pipelines. IEEE Trans. Industr. Inform. **14**(7), 3143–3154 (2018)
8. Niu, B., Zhu, Y., He, X.X., Wu, H.: MCPSO: a multi-swarm cooperative particle swarm optimizer. Appl. Math. Comput. **185**(2), 1050–1062 (2007)
9. Niu, B., Li, L.: An Improved MCPSO with center communication. In: Proceedings of 2008 International Conference on Computational Intelligence and Security, pp. 57–61 (2008)
10. Niu, B., Wang, J.W., Wang, H.: Bacterial-inspired algorithms for solving constrained optimization problems. Neurocomputing **148**, 54–62 (2015)

Solving the Unit Commitment Problem with Improving Binary Particle Swarm Optimization

Jianhua Liu[1,2](✉), Zihang Wang[1,2], Yuxiang Chen[1,2], and Jian Zhu[1,2]

[1] School of Information Science and Engineering, Fujian University of Technology, Fuzhou 350108, China
jhliu@fjnu.edu.cn

[2] Fujian Provincial Key Laboratory of Big Data Mining and Applications, Fujian University of Technology, Fuzhou 350108, China

Abstract. Unit commitment is a traditional mixed-integer non-convex problem and an optimization task in power system scheduling. The traditional methods of solving the Unit commitment problem have some problems, such as slow solving speed, low accuracy and complex calculation. Therefore, intelligent algorithms have been applied to solve the unit combination problem with continues and discrete feature, such as Particle Swarm Optimization, Genetic Algorithm. In order to improve the solution quality of Unit commitment, this paper proposes the adaptive binary Particle Swarm Optimization with V-shaped transfer function to solve the unit commitment problem, and adopts the policy of the segmented solution. By comparison with some classical algorithm in the same unit model, the experimental results show that solving the UC problem by using improved algorithm with segmented solution has higher stability and lower total energy consumption.

Keywords: Binary Particle Swarm Optimization · Segmented solution · Unit commitment problem

1 Introduction

The Unit Commitment (UC) Problem is a hot problem in the power systems, which reasonably arrange the on/off state and the load distribution of the generator unit during a dispatch period in order to make the total operating cost of the unit system reach the minimum under some constraints condition [1]. There are many methods to be proposed to obtain the solution of UC problem. Traditional methods include Lagrangian Relaxation (LR) [2], Dynamic Programming (DP) [3], Mixed Integer Programming (MIP) [4] etc. Although the classical methods have the advantage of fast solution for small-scale systems, it is difficult to obtain high-quality optimal solutions for large-scale systems.

With the development of intelligent computation, more and more intelligent algorithms are used to solve the UC problem [5–7]. The Genetic Algorithm (GA) is used to optimize the fuel cost of unit for UC [5], the result shows the saving of fuel cost. Sun

Y. Tan et al. (Eds.): ICSI 2022, LNCS 13344, pp. 176–189, 2022.
https://doi.org/10.1007/978-3-031-09677-8_16

et al. [6] proposed the intelligent Water Droplet Algorithm modeling mechanism (IWD) for the UC Problem, the results show that the intelligent Water Droplet Algorithm has strong global search ability in UC Problem. Kumar et al. [7] provides a computational methodology based on Monarch Butterfly Optimization (MBO) to solve UC Problem, and the comparative analysis shows the effectiveness in terms of operating costs and execution time in relation to other techniques. The Particle Swarm Optimization (PSO), as a popular swarm intelligent algorithm, has been used to optimize the power system problem of UC because it's simply and easy to operate. PSO is extended as the Binary Particle Swarm Optimization (BPSO) in order to be applied to binary problem. The on/off state of the generator units of UC Problem is binary problem, so the BPSO is used to solve the UC Problem. In the recent ten years, more and more researcher used PSO and BPSO to solve UC Problems. Song et al. [8] proposed an improved Particle Swarm Culture Algorithm, and by selecting the particle global optimal position through the individual evolution and parameter adjustment to solve the unit commitment. Zhai, et al. [9] adopted the dual Particle Swarm Optimization algorithm and propose a dimensionality reduction idea to solve the UC Problem, which converted from optimizing the entire dispatching cycle to doing each dispatching time respectively and orderly. Ismail et al. [10] considered BPSO algorithms and Dynamic Programming (DP) to solve the UC Problem, the results show that BPSO algorithm satisfies all the constraints of the UC Problem and minimizes the total operation cost. Liu et al. [11] used a new strategy to generate particles on the basis of discrete Particle Swarm Optimization, and introduced the concept of optimization window and heuristic rules, the simulation results clearly show that the proposed method is effective. However, the BPSO algorithm used above is still based on the original BPSO algorithm and its performance is insufficient. This is because the traditional BPSO used the S-shaped transfer function, which leads to a stronger global search ability and a weaker local search ability in the later stage, so that it is hard to obtain the optimal solution [12].

This paper proposes a V-shaped transfer function meet the requirements of BPSO (VBPSO) to overcome the strong global search ability and enhance the local search ability in the later stage of the algorithm. Meanwhile, this paper also proposes an adaptive Binary Particle Swarm Optimization (ABPSO) to enhance particle diversity and the global search ability at early stage and local search ability at last stage. Finally, the combining BPSO are applied to the unit commitment problem, and the experimental result shows the effectiveness of the proposed method compared with the methods listed in the reference.

The rest of this paper are organized as follows: Sect. 2 introduces some related work on BPSO and UC Problem. The proposed VBPSO and ABPSO are presented in Sect. 3. In Sect. 4, the methods of handling constraints are described. The experimental results are presented in Sect. 5 and Sect. 6 concludes the paper.

2 Related Work

2.1 Binary Particle Swarm Optimization

Particle Swarm Optimization (PSO) algorithm is a popular intelligent algorithm for solving optimization problems. The basic PSO is designed to solve continuous optimization problems. However, the UC Problem is a discrete problem, it is necessary to transform the continuous search space to the binary one. A binary version of PSO, Binary PSO, was first introduced in 1997 [13]. In BPSO, the velocity updating equation is similar to that of PSO as the Eq. (1).

$$v_{id} = w * v_{id} + c_1 * rand() * (p_{id} - x_{id}) + c_2 * rand() * (p_{gd} - x_{id}) \tag{1}$$

The transfer function is employed to convert velocity to the value in interval [0, 1] in BPSO, and the transfer function is shown as Eq. (2).

$$s(v_{id}) = \frac{1}{1 + e^{-v_{id}}} \tag{2}$$

where v_{id} is the next velocity of the ith particle in the dth dimension, $|v_{id}| < v_{max}$. and v_{max} is set to a constant. Equation (2) is the sigmoid function which generates a value in range [0, 1] and the particle's position at the next generation in the binary search space, x_{id}, is updated in terms of the Eq. (3).

$$x_{id} = \begin{cases} 1 & if\ rand() \le s(v_{id}) \\ 0 & otherwirse \end{cases} \tag{3}$$

where rand() is a random number in U ~ (0,1).

2.2 Mathematical Model of the UC Problem

Unit commitment (UC) aims to minimize the economic cost under condition of some physical constraints in the power system [14]. In general, the UC Problem regards the economic cost of the unit within 1h as an independent time unit without considering the interaction of each unit time.

The total cost of the UC Problem can be found by summing the operational cost for each unit over the time with the start-up cost, which is denoted as Eq. (4).

$$F = \sum_{i=1}^{N} \sum_{t=1}^{T} \left[F_i(P_{it}) + S_{it}(t)\left(1 - u_{i,t-1}\right)\right] u_{it} \tag{4}$$

where F represents the total economic cost, which is composed of fuel cost and start-up cost. The start-up cost is the cost of the unit when started up. $F_t(P_{it})$ represents a function of the fuel cost of the unit i in operation. P_{it} represents the generating capacity of unit i in hour t. $F_t(P_{it})$ is calculated by Eq. (5).

$$F_i(P_{it}) = a_i P_{it}^2 + b_i P_{it} + c_i \tag{5}$$

where a_i, b_i, c_i represents fuel cost coefficients of the unit i, respectively.

The $S_{it}(t)$ in Eq. (4) represents the start-up cost of the unit i in the hour t. And u_{it} is the start-up of the unit i, which is 0 or 1 at the hour t. The start-up cost is an indispensable part of the economic cost of the unit, which is expressed as Eq. (6).

$$S_{it}(t) = S_{0,i} + S_{1,i}(1 - e^{-\frac{T_{it}^{off}}{\tau_i}}) \tag{6}$$

where $S_{0,i}$, $S_{1,i}$, τ_i are the start-up cost parameters of the unit i, T_{it}^{off} is the continuous time that the unit i is in OFF state at the hour t.

UC Problem contains some constraints in real world, so it is solved on the premise of satisfying the following constraints.

(1) The system power balance constraint, which is shown as Eq. (7).

$$\sum_{i=1}^{N} P_{it}u_{it} = D_t \tag{7}$$

where D_t is the total power demand of the system in the hour t, and the restricted power in the hour t must be equal to the value of D_t,that is, the sum of the power of all units should be equal to the total system power demand D_t.

(2) Spinning reserve constraint, which is shown as Eq. (8).

$$\sum_{i=1}^{N} P_{i,\max}u_{it} \geq D_t + R_t \tag{8}$$

where $P_{i,max}$ indicates the power maximum limit of unit i when it starts, R_t represents the spinning reserve value in hour t. In practice, in order to cope with the unexpected extra load, it is necessary to preserve the spinning power, which is important to keep the power balance.

(3) Minimum continuous on/off times constraints, which is defined as Eq. (9) and Eq. (10).

$$T_{\text{it}}^{on} \geq M_i^{on} \tag{9}$$

$$T_{\text{it}}^{off} \geq M_i^{off} \tag{10}$$

where T_{it}^{on} and T_{it}^{off} is the continuous operating time and continuous shutdown time of the unit i in hour t. M_i^{on} and M_i^{off} is the minimum continuous operating and minimum continuous shutdown time required by the unit i.

(4) Ramp rate constraint, which is shown as Eq. (11),

$$\left|P_{\text{it}} - P_{i,t-1}\right| \leq p_i^{\max v} \tag{11}$$

where $p_i^{\max v}$ is the maximum ramp rate of the unit i, that is, the power of the start-up unit in adjacent periods should be increased or decreased between the given maximum and minimum value.

(5) The maximum and minimum constraint of unit output, which is shown as Eq. (12),

$$P_{i,\min} \leq P_{it} \leq P_{i,\max} \tag{12}$$

where $P_{i,min}$ and $P_{i,max}$ represents the maximum and minimum limit of the power of the unit i respectively. The power supplied from the generation unit must be lie in the range between the maximum and minimum power.

3 The Improving Binary Particle Swarm Optimization

In the section, two methods are adopted to improve Binary PSO. One policy is to change the transfer function of BPSO, the other is the change equation of position updating of BSPO.

(1) the policy of the V-shaped transfer function
The original BPSO has S-shaped transfer function of sigmoid function, which makes global search ability of BPSO too strong, and hard to converge to the global optimal particles [15]. The V-shaped transfer function in BPSO is proposed to improve the BPSO, which function curve is a V-shaped and symmetrical over the *Y* axis. Here, a new V-shaped transfer function is put forward as shown in Eq. (13).

$$S(v_{id}) = \frac{(v_{id})^2}{1 + (v_{id})^2} \tag{13}$$

Comparing with the traditional BPSO, the VBPSO algorithm with the V-shaped transfer function can enhance the later local search ability of the algorithm and improve the performance of the original BPSO algorithm.

(2) The policy of adaptive mutation
According to the influence of particle mutation on the algorithm's search ability, a particle position updating method with adaptive change of mutation probability is proposed. The position of the particle is updated with Eq. (14) that is different with the Eq. (3). In Eq. (14), the binary bit changes when *rand*() < *R,* and the *R* is changed over iteration in Eq. (15), So the Binary PSO with Eq. (14) and Eq. (15) has the adaptive feature. The policy can enhance the global search ability at early stage and local search ability at last stage.

$$x_{id} = \begin{cases} \sim x_{id} & \textit{if } \text{rand}() \le R \\ x_{id} & \textit{otherwise} \end{cases} \tag{14}$$

$$R = (1 - \frac{t}{T})^2 \tag{15}$$

4 Constraint Handling

In order to prevent the damage of units, the solution of the UC Problem must be meet some constraints. In addition, for the users' demand of the power generation, how to handle these constraints is very important. The methods of handling constraints are presented in this section.

4.1 Sufficient Condition for a Feasible Solution

First, it must be decided whether the on/off state of the unit is a feasible solution after the particle of PSO updating by considering the spinning reserve constraint at Eq. (8). Otherwise, if the number of start-up units were too large so that the power allocation of units cannot meet the maximum and minimum limit in Eq. (12) [17], a new deciding conditions about feasible solution is added to solve the situation, So, the new deciding conditions can be shown as Eq. (16),

$$\sum_{i=1}^{N} u_{it} P_{i,\min} \leq \eta D_t. \tag{16}$$

where η is set to random values in the intervals (0, 1), the left of Eq. (16) is the sum of minimum power of the unit. The total demand power multiplied by the decimal will become smaller in right formula. Under this condition, only when it be greater than the minimum power, it can allocate the power. The purpose of adding the judgment conditions is to keep the power of the operating unit not to be allocated when the minimum power of the start-up unit at a certain time is lower than the total demand power of the current hour t. The power allocation for the operating unit can be realized if the minimum power of the operating unit is lower than the reduced total demand power.

4.2 Power Balance Constraints

The sum of the generating power of units in a dispatch time must be equal to the electrical load in the current time which satisfies the constraint Eq. (7). First, for the determined on/off state unit, it should find out the last unit, and then allocates the load to all units. Second, subtracting the sum of the last unit's load from the total demand load. If the result is greater than the maximum constraint value of the last unit, the power of units except the last unit should be increased randomly and appropriately. The constraints of the maximum and minimum limit constraints of unit output Eq. (12) should be met to make the power reach the balance constraint.

4.3 Ramp Rate Constraint

In the process of power allocation, the total operating coal consumption of the unit can be calculated, and the total coal consumption value can be used as the evaluation function to obtain the optimal power allocation of the units by using the continuous PSO algorithm. The Ramp rate constraint Eq. (11) can be used in solving the problem of unit power allocation and coal consumption, the penalty function is added in the unit coal consumption, the objective function is shown in Eq. (17),

$$\min F(P_{it}, U_{it}) = \sum_{i=1}^{N} \sum_{t=1}^{T} [F_i(P_{it}) + S_{it}(1 - U_{i,t-1})]U_{it} + \sum_{i=1}^{N} \sum_{t=1}^{T} \alpha[\min(0, (p_i^{\max v} - |P_{i,t-1} - P_{it}|))]^2 \tag{17}$$

where α is penalty coefficient which takes 1200. The ramp rate constraint is satisfied when the penalty coefficient is equal to 0.

4.4 Start-Up Priority of the Unit

To get better unit on/off state and less consumption cost, 20% of the units can be started up all the time during the dispatching period, because each unit has its characteristics and the power consumption are different [18]. Based on these characteristics, the unit starting priority λ_i can be determined by the unit operating cost when full load operation. The lower unit operating cost, the greater the start-up priority of the unit, and the top 20% unit with higher priority will be selected to keep running. For the test units used in this paper, the top 20% with higher start-up priority are units 9 and 10, so units 9 and 10 were selected to keep on during the whole dispatch period.

4.5 Segment Solution

The general method to solve the UC Problem is regard all dispatch periods as a whole [19]. Especially in handling continuous on/off time constraints, if the on/off state of the unit in 24 periods is adjusted based on different unit constraints after updating, it has certain blindness. the size of data processed is large, so it is difficult to find the best on/off state of the unit. Because the whole dispatch period is composed of 24 scheduling periods, it can convert the whole dispatch period into 24 scheduling periods (Divide Time, DT), reduce the solution dimension to obtain a more reasonable unit on/off state and the total energy consumption of unit operation.

The optimization of the whole scheduling cycle is transformed into the sequential and separate optimization of each scheduling time, so the optimization process of each scheduling time is similar. First of all, for the first period, it refers to the determined optimal unit state at the previous scheduling time to judge whether the unit has met the minimum continuous start-stop time constraint. If the constraint were met, updating the on/off states of the unit by using the improved algorithm VBPSO or ABPSO. If it reaches a certain number of iteration and still does not meet the judgment conditions of the feasible solution, the maximum iteration must be set larger. If the iteration reaches the maximum value of the iteration, the units which have met the minimum stop time constraint and has higher starting priority will be on in turn until the judgment conditions of the feasible solution are met.

5 Results and Discussion

5.1 Test System and Parameter Settings

This paper uses classical examples to verify the proposed solution of the UC Problem. The BPSO algorithm with two policies (ABPSO) is used as the binary algorithm and the continuous particle swarm algorithm uses the standard PSO. The parameters are set as recommended in the corresponding reference papers, population size is 20, the maximum iteration is 200, the learning factors are both 2 and the dispatch period is one day, which is divided into 24 time periods. In test system, the unit parameters, 24-h load data and other parameters was adopted in [20], as shown in the Table 1 and 2. In order to verity the performance of the algorithm, this paper conducts tests on these algorithms, each algorithm was run 30 times on the selected 10-unit benchmark problem.

Table 1. Unit data

Unit	a	b	c	P_{imin}	P_{imax}	$S_{0,i}$	$S_{1,i}$	τ	P_i^{maxv}	M_i^{on}	M_i^{off}
1	0.0051	2.2034	15	15	60	0	85	3	0.3	2	2
2	0.00396	1.9101	25	20	80	0	101	3	0.4	2	2
3	0.00393	1.8518	40	30	100	0	114	3	0.5	2	2
4	0.00382	1.6966	32	25	120	0	94	4	0.6	3	3
5	0.00212	1.8015	29	50	150	0	113	4	0.75	3	3
6	0.00261	1.5354	72	75	280	0	176	6	1.4	5	5
7	0.00289	1.2643	49	120	320	0	187	8	1.6	5	5
8	0.00148	1.213	82	125	445	0	227	10	2.225	8	8
9	0.00127	1.1954	105	250	520	0	267	12	2.6	8	8
10	0.00135	1.1285	100	250	550	0	280	12	2.75	8	8

Table 2. Total system load and system spinning reserve capacity in t period

t	1	2	3	4	5	6	7	8	9	10	11	12
Dt	200	198	194	190	184	187	182	170	151	141	132	120
Rt	140	139	136	133	129	131	127	119	106	99	92	88
t	13	14	15	16	17	18	19	20	21	22	23	24
Dt	120	1160	1140	1160	1260	1380	1560	1700	1820	1900	1950	1990
Rt	84	81	80	81	88	97	109	119	127	133	138	139

5.2 Solution Results and Analysis

Firstly, the non-segmentation BPSO (BPSOUT), segmentation PSO (MDPSODT) [20], segmentation ABPSO and VBPSO were used to compare in solving the energy consumption problem of the unit. As shown in the Table 3, the result of using MDPSODT based on segment solution is significantly lower than the non-segmentation BPSOUT algorithm, which can verify the effectiveness of segment solution. In addition, the final result of the two improved algorithms is lower than that of the MDPSODT, which shows the effectiveness of the two improved methods.

Table 3. Results of different algorithms

Algorithm	BPSOUT	MDPSODT	VBPSO	ABPSO
Total Energy. cons	82499.2	79665.8	79043.1	79030.8

Table 4 shows the comparison of the variances of the original BPSO algorithm without segmentation and the two improved methods. It can be seen from Table 4 that the variances of the two improved methods is smaller which indicated the two methods have low volatility and stability in solving UC Problems. In order to reflect the above analysis more intuitively, Fig. 1, 2 and Fig. 3 reveals the results and average values of 30 experiments. From the Fig. 1, 2 and Fig. 3, it can be clearly seen that the result of the non-segmented BPSOUT algorithm has the largest volatility in the average value, and the algorithm is unstable. The solutions obtained by the improved methods have the smallest volatility near the average value and higher stability.

Table 4. Variances of different algorithms

Algorithm	BPSOUT	VMBPSO	ABPSO
Variances	81.836	3.915	3.743

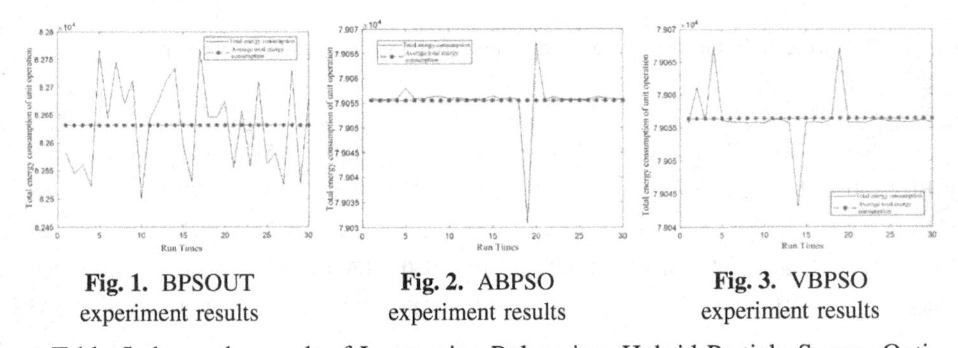

Fig. 1. BPSOUT experiment results

Fig. 2. ABPSO experiment results

Fig. 3. VBPSO experiment results

Table 5 shows the result of Lagrangian Relaxation, Hybrid Particle Swarm Optimization (HPSO) and Genetic Algorithm (GA) which are the classical algorithms from the UC Problem in reference [20–22]. The parameters are set according to the above references, the inertia weight w is set to 0.7298, the learning factor c takes value 1.49618 in HPSO. In GA, the population size is 91, the iteration is 60. It can be seen from the Table 5 that compared with other classical algorithms, the two algorithms proposed in this paper have the lowest total energy consumption. The result shows that the proposed algorithms have obvious advantages compared with other algorithms. Figure 4 shows the difference in the total cost by these algorithms. It is not difficult to see from Fig. 4 that the two proposed algorithms have obvious effects on cost reduction compared with other classical algorithms.

Table 5. Comparison of solution results of different algorithms

Algorithm	LRM	HPSO	GA	VBPSO	ABPSO
Cost	81245.5	81118.3	79807.0	79043.054	79030.807

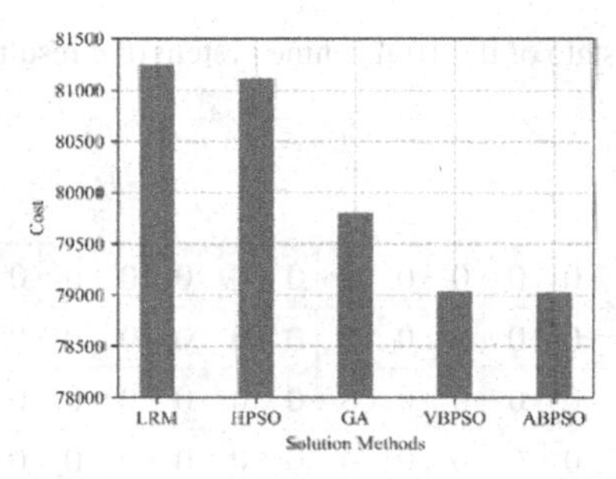

Fig. 4. Total cost produced by different methods

Table 6. The optimal unit state of the 10-machine system (the result of the BPSO algorithm)

Unit	Dispatch period																							
	1–24 h																							
1	0	0	1	1	0	0	0	1	1	1	1	1	0	0	0	1	1	1	1	1	1	1	1	0
2	1	1	0	0	0	0	0	0	1	1	0	0	0	0	1	1	0	0	0	0	0	0	0	0
3	0	0	0	0	0	1	1	0	0	0	0	0	0	1	1	1	0	0	1	1	0	0	1	1
4	0	0	0	1	1	1	1	1	1	0	0	0	0	1	1	1	0	0	0	1	1	1	1	0
5	0	0	0	1	1	1	0	0	0	0	1	1	1	1	1	0	0	0	0	1	1	1	1	1
6	1	1	1	0	0	0	0	0	0	0	0	1	1	1	1	1	1	1	1	1	1	1	1	1
7	1	1	1	1	1	1	1	1	1	1	1	1	0	0	0	0	0	1	1	1	1	1	1	1
8	1	1	1	1	1	1	1	1	1	0	0	0	0	0	0	0	0	1	1	1	1	1	1	1
9	1	1	1	1	1	1	1	1	1	1	1	1	1	1	1	1	1	1	1	1	1	1	1	1
10	1	1	1	1	1	1	1	1	1	1	1	1	1	1	1	1	1	1	1	1	1	1	1	1

Table 7. The optimal unit state of the 10-machine system (the result of the ABPSO algorithm)

Unit	Dispatch period																							
	1–24 h																							
1	0	0	0	0	0	0	0	0	0	0	0	0	0	0	0	0	0	0	0	0	0	0	0	1
2	0	0	0	0	0	0	0	0	0	0	0	0	0	0	0	0	0	0	0	0	0	0	0	0
3	0	0	0	0	0	0	0	0	0	0	0	0	0	0	0	0	0	0	0	0	0	0	0	0
4	1	1	1	1	0	0	0	0	0	0	0	0	0	0	0	0	0	0	0	0	0	1	1	1
5	1	1	1	1	1	1	1	1	1	0	0	0	0	0	0	0	0	0	0	1	1	1	1	1
6	1	1	1	1	1	1	1	0	0	0	0	0	0	0	0	0	0	0	0	0	0	0	0	0
7	1	1	1	1	1	1	1	1	1	1	1	1	1	1	1	1	1	1	1	1	1	1	1	1
8	1	1	1	1	1	1	1	1	1	1	1	1	1	1	1	1	1	1	1	1	1	1	1	1
9	1	1	1	1	1	1	1	1	1	1	1	1	1	1	1	1	1	1	1	1	1	1	1	1
10	1	1	1	1	1	1	1	1	1	1	1	1	1	1	1	1	1	1	1	1	1	1	1	1

Table 8. The optimal unit state of the 10-machine system (the result of the VBPSO algorithm)

Unit	Dispatch period																							
	1–24 h																							
1	0	0	0	0	0	0	0	0	0	0	0	0	0	0	0	0	0	0	0	0	0	0	0	1
2	0	0	0	0	0	0	0	0	0	0	0	0	0	0	0	0	0	0	0	0	0	0	0	0
3	0	0	0	0	0	0	0	0	0	0	0	0	0	0	0	0	0	0	0	0	0	0	0	0
4	1	1	1	1	0	0	0	0	0	0	0	0	0	0	0	0	0	0	1	1	1	1	1	1
5	1	1	1	1	1	1	1	1	1	0	0	0	0	0	0	0	0	0	0	0	1	1	1	1
6	1	1	1	1	1	1	1	0	0	0	0	0	0	0	0	0	0	0	0	0	0	0	0	0
7	1	1	1	1	1	1	1	1	1	1	1	1	1	1	1	1	1	1	1	1	1	1	1	1
8	1	1	1	1	1	1	1	1	1	1	1	1	1	1	1	1	1	1	1	1	1	1	1	1
9	1	1	1	1	1	1	1	1	1	1	1	1	1	1	1	1	1	1	1	1	1	1	1	1
10	1	1	1	1	1	1	1	1	1	1	1	1	1	1	1	1	1	1	1	1	1	1	1	1

Table 9. The result of BPSO

t	Cons	t	Cons
1	4303.218	13	2636.816
2	4200.061	14	2603.123
3	4158.786	15	2506.633
4	4124.212	16	2531.588
5	3824.805	17	2670.621
6	4013.982	18	3003.720
7	3803.030	19	3395.254
8	3568.024	20	3678.195
9	3175.776	21	3809.123
10	2846.038	22	3981.682
11	2813.601	23	4189.270
12	2455.258	24	4206.377

Finally, Tables 6, 7 and 8 shows the obtained optimal unit states, Table 9, 10 and 11 shows the consumption cost of each period. It can be seen that the two groups of units using the improved algorithm have little difference on the on/off state. But compared with the conventional BPSO algorithm, the number of start-up units is less, and the utilization rate of the units is higher. In totally, the two improved algorithms using the segment solution have obtained good results in solving the traditional UC Problem.

Table 10. The result of ABPSO

t	Cons	t	Cons
1	4303.218	13	2636.816
2	4200.061	14	2603.123
3	4158.786	15	2506.633
4	4124.212	16	2531.588
5	3824.805	17	2670.621
6	4013.982	18	3003.720
7	3803.030	19	3395.254
8	3568.024	20	3678.195
9	3175.776	21	3809.123
10	2846.038	22	3981.682
11	2813.601	23	4189.270
12	2455.258	24	4206.377

Table 11. The result of VBPSO

t	Cons	t	Cons
1	4303.218	13	2636.816
2	4200.061	14	2603.123
3	4158.786	15	2506.633
4	4124.212	16	2531.588
5	3824.805	17	2670.621
6	4013.982	18	3003.720
7	3803.030	19	3395.254
8	3568.024	20	3678.195
9	3175.776	21	3809.123
10	2846.038	22	3981.682
11	2813.601	23	4189.270
12	2455.258	24	4206.377

6 Summary

Aiming at the blindness of the whole solution in adjusting the on/off state of the units, resulting in low solution quality and high total energy consumption for unit operation, this paper adopts the segmented solution, and applies two improved algorithms to solve the UC Problem. The final experimental results indicate that two improved methods proposed in this paper obtain higher stability, lesser number of start-up units and lower total energy consumption compared with the conventional BPSO.

References

1. Abduladheem, I.A., Nasser, H.A.: Solving the unit commitment problem in large systems using hybrid PSO algorithms. IOP Conf. Ser. Mater. Sci. Eng. **1105**(1) (2021)
2. Guido, P., de Souza Mauricio, C.: Formulations and a Lagrangian relaxation approach for the prize collecting traveling salesman problem. Int. Trans. Oper. Res. **29**(2), 729–759 (2021)
3. Patra, S., Goswami, S.K., Goswami, B.: Fuzzy and simulated annealing based dynamic programming for the unit commitment problem. Expert Syst. Appl. **36**(3p1), 5081–5086 (2009)
4. Amani, A., Alizadeh, H.: Solving hydropower unit commitment problem using a novel sequential mixed integer linear programming approach. Water Resour. Manag. (2021, prepublish)
5. Aniket, A., Kirti, P.: Optimization of unit commitment problem using genetic algorithm. Int. J. Syst. Dyn. Appl. (IJSDA) **10**(3), 21–37 (2021)
6. Sun, Y., Wu, Y., Liang, L., et al.: Generation scheduling of thermal power units based on intelligent water droplet algorithm. Power Energy **40**(02), 120–125 (2019)
7. Vineet, K., Ram, N.: Monarch butterfly optimization-based computational methodology for unit commitment problem. Electr. Power Componen. Syst. **48**(19–20), 2181–2194 (2021)
8. Xiao, S., Ye, L., Jiajun, L., et al.: Multi-system joint scheduling based on the improved particle swarm culture algorithm. Power Syst. Clean Energy **32**(06), 77–84 (2016)
9. Zhai, J., Ren, J., Li, Z., Zhou, M.: Dual particle swarm optimization base on dimensionality reduction for unit commitment problem. J. North China Electr. Power Univ. (Nat. Sci. Ed.) **43**(01), 32–38 (2016)
10. Ismail, A.A., Hussain, A.N.: Unit commitment problem solution using binary PSO algorithm. In: 2019 2nd International Conference on Engineering Technology and Its Applications (IICETA) (2019)
11. Liu, Y., Hou, Z., Jiang, C.: Unit commitment via an enhanced binary Particle Swarm Optimization algorithm. Autom. Electr. Power Syst. **30**(04), 35–39 (2006)
12. Liu, J., Yang, R., Sun, S.: The analysis of binary Particle Swarm Optimization. J. Nanjing Univ. (Nat. Sci.) **47**(05), 504–514 (2011)
13. Kennedy, J., Eberhart, R.C.: A discrete binary version of the particle swarm algorithm. In: 1997 IEEE International Conference on Systems, Man, and Cybernetics. Computational Cybernetics and Simulation. IEEE (1997)
14. Abdi, H.: Profit-based unit commitment problem: a review of models, methods, challenges, and future directions. Renewable Sustain. Energy Rev. (2020)
15. Jiang, L., Liu, J., Zhang, D., Bu, G.: Application analysis of V-shaped transfer function in binary particle swarm optimization. Comput. Appl. Softw. **38**(04), 263–270 (2021)
16. Jiang, L., Liu, J., Zhang, D., Bu, G.: An adaptive mutation binary Particle Swarm Optimization algorithm. J. Fujian Univ. Technol. **18**(03), 273–279 (2020)
17. Qiu, H., Gu, W., Liu, P., et al.: Application of two-stage robust optimization theory in power system scheduling under uncertainties: a review and perspective. Energy **251**, 123942 (2022)
18. Senjyu, T., Shimabukuro, K., Uezato, K., et al.: A fast technique for unit commitment problem by extended priority list. IEEE Trans. Power Syst. **18**(2), 882–888 (2003)
19. Li, Z., Tan, W., Qin, J.: An improved dual particle swarm optimization algorithm for unit commitment problem. Proc. CSEE **32**(25), 189–195 (2012)

20. Han, X., Liu, Z.: Optimal unit commitment considering unit's ramp-rate limits. Power Syst. Technol. **18**(06), 11–16 (1994)
21. Hu, J., Guo, C., Guo, C.: A hybrid particle swarm optimization method for unit commitment problem. Proc. CSEE **24**(4), 24–28 (2004)
22. Cai, C., Cai, Y.: Optimization of unit commitment by genetic algorithm. Power Syst. Technol. **21**(1), 44–47, 51 (1997)

An Improved Particle Swarm Optimization Algorithm for Irregular Flight Recovery Problem

Tianwei Zhou[1,2], Pengcheng He[1,2], Churong Zhang[1,2], Yichen Lai[1,2], Huifen Zhong[1,2,3], and Xusheng Wu[2,4](✉)

[1] College of Management, Shenzhen University, Shenzhen 518060, China
[2] Great Bay Area International Institute for Innovation, Shenzhen University, Shenzhen 518060, China
1257451869@qq.com
[3] Faculty of Business and Administration, University of Macau, Macau 999078, China
[4] Shenzhen Health Development Research and Data Management Center, Shenzhen 518060, China

Abstract. As with the rapid development of air transportation and potential uncertainties caused by abnormal weather and other emergencies, such as Covid-19, irregular flights may occur. Under this situation, how to reduce the negative impact on airlines, especially how to rearrange the crew for each aircraft, becomes an important problem. To solve this problem, firstly, we established the model by minimizing the cost of crew recovery with time-space constraints. Secondly, in view of the fact that crew recovery belongs to an NP-hard problem, we proposed an improved particle swarm optimization (PSO) with mutation and crossover mechanisms to avoid prematurity and local optima. Thirdly, we designed an encoding scheme based on the characteristics of the problem. Finally, to verify the effectiveness of the improved PSO, the variant and the original PSO are used for comparison. And the experimental results show that the performance of the improved PSO algorithm is significantly better than the comparison algorithms in the irregular flight recovery problem covered in this paper.

Keywords: Crew recovery · Irregular flight · Particle swarm algorithm · Cross-over mechanism · Mutation mechanism

1 Introduction

In post Covid-19 era in China, any inevitable imported cases may lead to regional quarantine and circuit breaker mechanisms for airlines. Moreover, abnormal weather may also introduce abnormal situations. It's vital for airlines to quickly schedule their irregular flights in face of flight delays or cancellations.

According to the *Normal Statistical Method of Civil Aviation Flight* released by the Civil Aviation Administration of China [1], normal flights refer to those depart 10 min or shorter after scheduled departure time without sliding back, veering or preparing for

Y. Tan et al. (Eds.): ICSI 2022, LNCS 13344, pp. 190–200, 2022.
https://doi.org/10.1007/978-3-031-09677-8_17

landing, or arriving within 10 min before scheduled arrival time. And irregular flights refer to those who do not obey the above conditions. When irregular flight occurs, how to quickly recover the flight plan becomes an urgent problem. The irregular flight recovery problem could be separated into several parts, including the route, flight, aircraft, crew and passenger recoveries [2]. In this paper, we mainly focus on the crew recovery problem. The crew recovery serves as a connecting link between the preceding and the following. A good crew recovery plan allows for the perfect implementation of the recovered route and maximizes the convenience of subsequent passenger recovery [3]. If there are problems with crew assignments, the flight route needs to be re-routed. It can cause huge losses to the airline, while also reducing passenger satisfaction with the airline and affecting its reputation. Therefore, airlines need a comprehensive crew recovery system to deal with the negative impact of irregular flights.

In recent years, several scholars have studied this problem in numerous perspectives. Doi et al. [4] and Quesnel et al. [5] separately considered fair working time and crews' preferences. Antunes et al. [6] focused on the robustness of primary schedules. Sun et al. [7] considered the impact of flying time on the irregular flight. Zhou et al. [8] developed an ant colony system for multiple objectives taking fairness and satisfaction into account. However, those papers indeed take many humanized objectives into consideration, and the measurement of cost and consistency of flights can be further polished. To solve this problem, in our model, we introduced variable costs to lower the complexity of calculation and reflect time-space constraints to embody the continuity of the flight task list.

As the crew recovery problem holds NP-hard characteristics, normal mathematics methods are difficult to get a satisfying solution, especially for the large-scale case. However, heuristic algorithms are outstanding for their large searching scales and fast calculating speed, which exactly fits our requirements. Among the heuristic algorithms, particle swarm optimization is remarkable for its easier implementation and fewer adjustment parameters. Xia et al. [9] proposed triple archives particle swarm optimization to obtain higher solution accuracy and faster convergence speed. Xu et al. [10] and Kiran [11] separately introduced dimensional learning strategy and distribution-based update rule to the primary PSO. Ibrahim et al. [12] combined the slap swarm algorithm with PSO to solve the feature selection problem. Zhang et al. [13] introduced a dynamic neighborhood-based learning strategy and competition mechanism to improve PSO's performance in solving multi-objective problems. Overall, the improved PSO algorithms have superior performance and they have successfully applied to industrial engineering problems. However, few papers use PSO to solve the crew recovery problem. There is a large research space to solve the crew recovery problem based on a new PSO algorithm. Thus, we purpose an improved PSO combining crossover and mutation mechanisms to solve the crew recovery problem in this paper.

The main contributions of this paper include three aspects. Firstly, crossover and variation mechanisms are introduced to address the problems of traditional PSO and improve the performance of the algorithm. Secondly, to further verify the effectiveness of the new algorithm, the variation mechanism is also led separately for comparison with the improved PSO that introduces both crossover and variation mechanisms. Thirdly, a new coding scheme is established based on the characteristic of the problem and interfaced with the new algorithm.

The remainder of this paper is listed as follows. Section 2 states the model of the crew recovery problem. Section 3 describes the improved particle swarm optimization. Section 4 explains the encoding scheme. Section 5 presents the simulation results against comparative algorithms. Section 6 concludes the paper and points out the future directions (Table 1).

Table 1. Definition of symbols

Symbols	Meaning of symbols
F	Flight set
C	Crew set
T	Crew task set
A	Airport set
n	Flight subscripts
m	Crew task subscripts
s	Crew superscripts
n_1, n_2	Subscripts of two continuous flights in crew task
b_{nm}	Parameter of crew task m containing flight n
c_n	The cost of canceling flight n
v_m^s	The variable cost brought by crew s when executing task m
t_m^s, t_s	The total time of crew s's executing task m and the total flight time of crew s
$t_{n_1}^m, t_{n_2}^m$	The prior flight's arrival time and posterior flight's departure time included in two continuous flights of crew task m
$a_{n_1}^m, a_{n_2}^m$	The prior flight's arrival airport and posterior flight's departure airport included in two continuous flights of crew task m
x_m^s	Whether crew s executes crew task m
y_n	Whether flight n is canceled

2 Model of Crew Recovery Problem

This section introduces the model of the crew recovery problem. The objective function is described below.

$$\min z = \sum_{s \in C} \sum_{m \in T} v_m^s x_m^s + \sum_{n \in F} c_n y_n. \tag{1}$$

This optimization objective function demands the lowest cost of crew recovery, including (i) the variable cost of two crew tasks; (ii) the cost of canceling one flight.

Moreover, constraints (2)–(8) are listed below.

$$s.t. \sum_{s \in C} \sum_{m \in T} b_{nm} x_m^s + y_n = 1, \tag{2}$$

$$\sum_{m \in T} x_m^s \leq 1, \forall s \in C, \tag{3}$$

$$t_m^s < t_s, \forall s \in C, \forall m \in T, \tag{4}$$

$$t_{n_1}^m < t_{n_2}^m, \forall m \in T, \forall n_1 \in F, \forall n_2 \in F, \tag{5}$$

$$a_{n_1}^m = a_{n_2}^m, \forall m \in T, \forall n_1 \in F, \forall n_2 \in F, \tag{6}$$

$$x_m^s \in \{0, 1\}, \forall s \in C, \forall m \in T, \tag{7}$$

$$y_n \in \{0, 1\}, \forall n \in F. \tag{8}$$

Constraint (2) ensures each flight can only be executed by one crew or be canceled. Constraint (3) restricts that each crew can only execute at most one crew task list. Constraint (4) imposes that the total flight time of a crew executing one crew task list must be shorter than the crew flight time regulated by airlines. Constraint (5) restricts that, in the same crew task list, the former flight's arrival time must be earlier than the latter one's departure time. Constraint (6) prescribes, in the same crew task list, that the former flight's arrival airport must be the same as the latter one's departure airport. Constraints (7)–(8) are the range of decision variables.

3 Improved Particle Swarm Optimization

This section presents the proposed improved particle swarm optimization by introducing crossover and mutation mechanisms to the primary particle swarm optimization.

3.1 Primary Particle Swarm Optimization

In the primary particle swarm optimization (PSO), we firstly initialize particles' scale, dimensions, velocity and location according to the optimization problem. Then calculate the solution in their present conditions, picking their own best solution as the personal best and the best solution of the swarm as the global best. After that, update particles' velocity and location according to certain formulas. And calculate the new solution with the new velocity and location. Finally, iterate the above processes until the termination criteria is met and output the best solution and its location.

3.2 PSO with Crossover and Mutation Mechanisms

Due to the shortcomings of prematurity and local optima, we improve the algorithm by introducing crossover and mutation to the primary PSO.

Crossover Mechanism. Crossover is a method that generates a new individual by recombining certain parts of its parent individuals. The operation is to randomly pick two individuals from the swarm, select the crossover location and choose whether to crossover according to the crossover rate p_c, which is between 0.25 and 1.

Mutation Mechanism. Mutation is an operation that changes the value in a certain dimension of the individual with a relatively small probability. The detailed operation is to generate a random number between 0 and 1. If the number is smaller than the mutation probability, which is between 0.001 and 0.01, in this iteration each dimension of this particle will randomly mutate within constraint.

We name the improved PSO as Mutation Crossover Particle Swarm Optimization (MCPSO). Moreover, we will introduce an algorithm with only a mutation mechanism as a comparing algorithm (Mutation Particle Swarm Optimization, MPSO). Compared with the primary PSO, MCPSO adds mutation and crossover operations after the update of velocity and location. And, due to the introduction of the crossover mechanism, neighboring particles can learn from each other through a crossover in each dimension. This enhances the region learning ability of the particles and facilitates the algorithm to escape from the local optimum. Figure 1 shows the process of the improved PSO and Fig. 2 shows the process of solving the crew recovery problem with MCPSO.

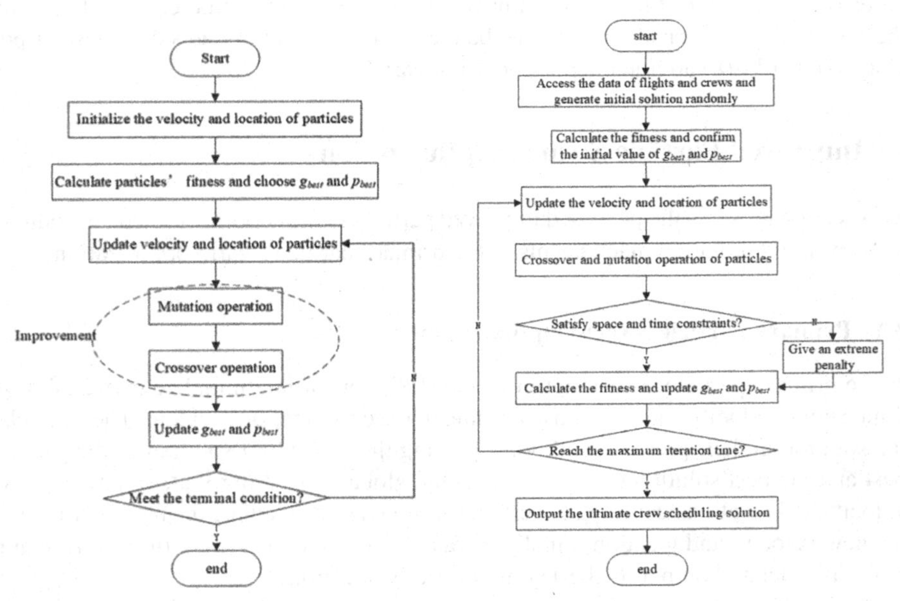

Fig. 1. Flow chart of MCPSO

Fig. 2. Flow chart of MCPSO solving crew recovery problem

4 Encoding Scheme

According to the time-space network diagram and multi-commodity flow model, the crew recovery problem can be converted into reallocating crews for each flight. For each flight, they can only choose one crew, meanwhile satisfying related constraints. Combining the characteristics of PSO, the solution to the crew recovery problem can be treated as the sequence number of executable crews selected by each flight. Each particle in the swarm represents a feasible solution, each dimension of the particle represents the flight and the value is the crew's number picked by the flight.

Based on this coding idea, in the swarm, $x_i = (x_{i1}, x_{i2}, x_{i3}......x_{in})$ is the location of ith particle, among which the dimension n should be equal to the total number of flights, x_{in} represents the value of nth dimension in i^{th} particle. Corresponding to the crew recovery problem, x_{in} refers to the crew number selected by n^{th} flight in i^{th} flight schedule. Figure 3 illustrates the encoding method for the crew recovery problem.

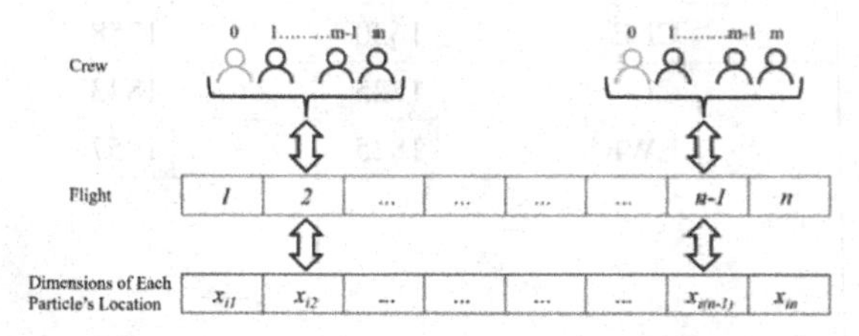

Fig. 3. Encoding method of crew recovery problem

5 Experiments and Results

5.1 Parameter Settings

This paper used data in Table 2 and Table 3 [14] to verify the performance of MCPSO in solving the crew recovery problem. Table 2 is the primary flight schedule list. Table 3 is the primary crew schedule list, including 6 crews and a backup crew.

Table 2. Primary flight schedule list

Flight	Departure airport	Arrival airport	Departure time	Arrival time	Flight time
1481	BOS	CLE	730	930	158
1519	BOS	GSO	1015	1210	155
1687	CLE	BOS	740	940	156
789	CLE	EWR	1100	1225	119
1867	CLE	GSO	1335	1450	113

(continued)

Table 2. *(continued)*

Flight	Departure airport	Arrival airport	Departure time	Arrival time	Flight time
1609	CLE	GSO	1650	1805	112
1568	CLE	GSO	2150	2305	110
1601	EWR	GSO	700	843	117
1779	EWR	GSO	830	1015	121
1690	EWR	CLE	955	1134	124
1531	EWR	GSO	1155	1330	130
1431	EWR	GSO	1300	1440	136
1626	GSO	EWR	1220	1353	129
1670	GSO	CLE	1240	1355	124
1678	GSO	CLE	1545	1700	108
1591	GSO	CLE	1630	1758	121
1720	GSO	CLE	1725	1843	116
1698	GSO	EWR	1825	1957	130

Table 3. Primary crew schedule list

Crew	Flight number	Departure airport	Arrival airport	Departure time	Arrival time	Flight time
E1	1601	EWR	GSO	700	843	117
	1626	GSO	EWR	1220	1353	129
E2	1779	EWR	GSO	830	1015	121
	1670	GSO	CLE	1240	1355	124
	1609	CLE	GSO	1650	1805	112
E3	1690	EWR	CLE	955	1134	124
	1867	CLE	GSO	1335	1450	113
	1678	GSO	CLE	1545	1700	108
E4	1531	EWR	GSO	1155	1330	130
	1720	GSO	CLE	1725	1843	116
	1568	CLE	GSO	2150	2305	110
V1	1687	CLE	BOS	740	940	156
	1519	BOS	GSO	1015	1210	155

(continued)

Table 3. (*continued*)

Crew	Flight number	Departure airport	Arrival airport	Departure time	Arrival time	Flight time
	1698	GSO	EWR	1825	1957	130
V2	1481	BOS	CLE	730	930	158
	789	CLE	EWR	1100	1225	119
	1431	EWR	GSO	1300	1440	136
	1591	GSO	CLE	1630	1758	121

Table 4. Parameter list

Symbol	Meaning	Value	Symbol	Meaning	Value
I	Maximum iteration time	5000	v_{max}	Upper limit of particle's velocity	10
D	Dimension of particle	17	v_{min}	Lower limit of particle's velocity	– 10
N	Number of particles	20	c_1, c_2	Self-learning and social learning rate	1.5
w_{max}	Upper limit of inertia weight	0.9	p_c(MCPSO)	Probability of crossover	0.75
w_{min}	Lower limit of inertia weight	0.4	p_m(MPSO)	Probability of mutation	0.05

Table 4 is the parameter settings of the algorithms. Note that, the parameter settings are based on the PSO original papers. And the canceling cost is 100 thousand yuan per time, the crew switching cost is 20 thousand yuan per time and the backup crew using cost is 30 thousand yuan per time. In the experiment, flight 1867 is canceled due to weather reason. It is preferred that the recovery crew's schedule holds the lowest cost and the minimum change compared against the original schedule.

5.2 Experimental Results

We calculated the data 10 times with PSO, MPSO, MCPSO and two variants of MCPSO respectively, and compared their results. Note that, to validate the sensitivity of MCPSO to the crossover rate, the crossover rates of the two variants are set to 0.6 and 0.9, respectively, while the rest of the parameters are the same as MCPSO settings. Table 5 shows the number of times each strategy is used. Figure 4 is the cost of three algorithms. Figure 6 presents the average convergence of the three algorithms with multiple runs. Figure 5 and Fig. 7 display the comparison between MCPSO and the two variants, where the crossover rate is 0.6 for MCPSO_1 and 0.9 for MCPSO_2. Based on the results, it can be seen that the original PSO has a poor optimal-seeking ability, and its final solution

Table 5. Optimal crew recovery strategy usage table using different algorithms

Algorithm \ Strategy / Amount	Cancel Flights	Use Backup Crew	Switch Crew Task
PSO	0	0	15
MPSO	0	0	2
MCPSO	0	1	0

has the highest crew recovery cost and a large number of exchange crew tasks so such a recovery scheme is not satisfactory. The performances of the MPSO and MPCSO with additional variation mechanisms have been significantly improved. As shown in Fig. 6, MCPSO and MPSO have an approximate average convergence capacity. However, by comparing the optimal result between the two algorithms in Fig. 4, we can find that the strategy found by MCPSO is better than MPSO. The reason is that the additional crossover mechanism in MCPSO can increase the diversity of solutions, thus sometimes helping the algorithm to find better solutions beyond the local optimum. Furthermore, as shown in Fig. 5 and Fig. 7, the optimal solution of MCPSO is better than the two variants. This illustrates that a moderate crossover rate can improves the diversity of solutions.

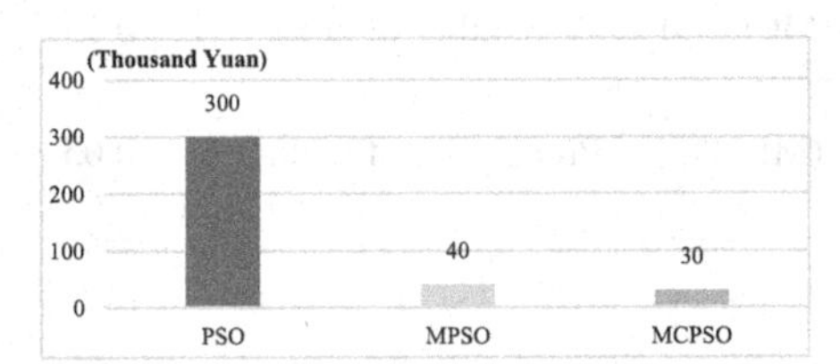

Fig. 4. Optimal crew recovery cost comparison among different algorithms

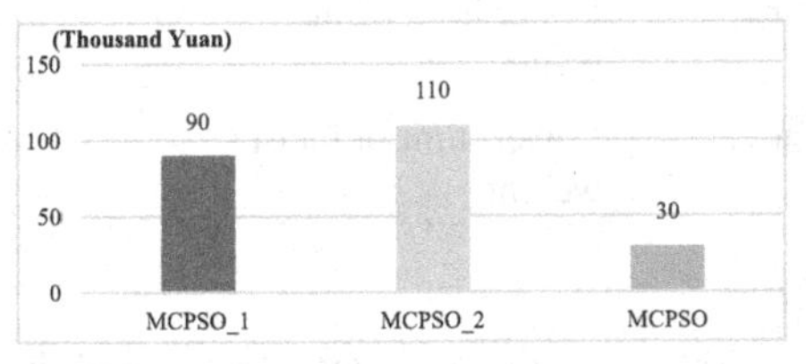

Fig. 5. Optimal crew recovery cost comparison among different *pc*

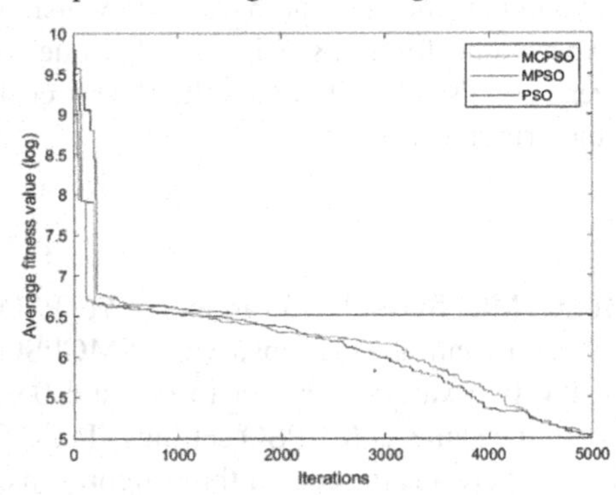

Fig. 6. Average crew recovery cost comparison among different algorithms

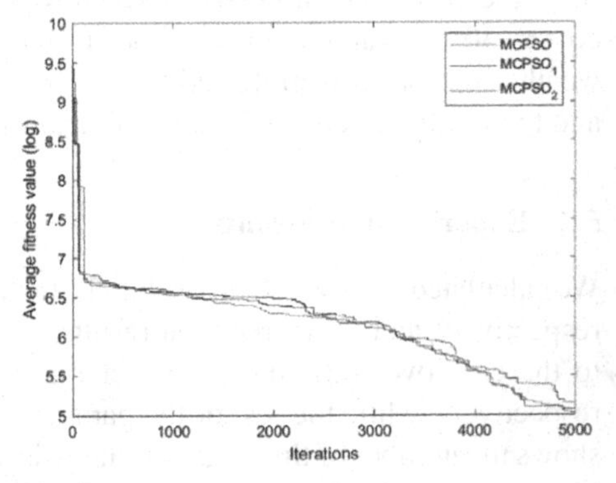

Fig. 7. Average crew recovery cost comparison among different p_C

6 Conclusions and Future Directions

In this paper, we studied the irregular flight crew recovery problem, established a crew recovery model and proposed MCPSO and corresponding coding strategy to solve it. By analyzing the improved PSO with comparison algorithms in actual flight instances, it can be concluded that the improved PSO can effectively solve the irregular flight crew recovery problem and finally output a new crew schedule with the lowest recovery cost. In the future, we will utilize MCPSO to solve other irregular flight recovery problems, such as aircraft recovery, passenger recovery, and so on. Moreover, multi-objective optimization is another direction to further explore the considered problem.

Acknowledgment. The study was supported in part by the Natural Science Foundation of China Grant No. 62103286, No. 71971143, No. 62001302, in part by Social Science Youth Foundation of Ministry of Education of China under Grant 21YJC630181, in part by Guangdong Basic and Applied Basic Research Foundation under Grant 2021A1515011348, 2019A1515111205, 2019A1515110401, 2020A1515010752, in part by Natural Science Foundation of Guangdong Province under Grant 2020A1515010749, 2020A1515010752, in part by Key Research Foundation of Higher Education of Guangdong Provincial Education Bureau under Grant 2019KZDXM030, in part by Natural Science Foundation of Shenzhen under Grant JCYJ20190808145011259, in part by Shenzhen Science and Technology Program under Grant RCBS20200714114920379, in part by Guangdong Province Innovation Team under Grant 2021WCXTD002.

References

1. Gao, Z., Zhang, S.: Statistics of civil aviation flight normality data. China Civil Aviat. 04, 48–50 (2008)
2. Hassan, L.K., Santos, B.F., Vink, J.: Airline disruption management: a literature review and practical challenges. Comput. Oper. Res. **127**, 105137 (2021)
3. Wen, X., Sun, X., Sun, Y., et al.: Airline crew scheduling: models and algorithms. Transp. Res. Part Logistics Transp. Rev. **149**, 102304 (2021)
4. Doi, T., Nishi, T., Voß, S.: Two-level decomposition-based matheuristic for airline crew rostering problems with fair working time. Europ. J. Oper. Res. **267**(2), 428–438 (2018)
5. Quesnel, F., Desaulniers, G., Soumis, F.: Improving air crew rostering by considering crew preferences in the crew pairing problem. Transp. Sci. **54**(1), 97–114 (2020)
6. Antunes, D., Vaze, V., Antunes, A.P.: A robust pairing model for airline crew scheduling. Transp. Sci. **53**(6), 1751–1771 (2019)
7. Sun, X., Chung, S.H., Ma, H.L.: Operational risk in airline crew scheduling: do features of flight delays matter? Decis. Sci. **51**(6), 1455–1489 (2020)
8. Zhou, S.Z., Zhan, Z.H., Chen, Z.G., et al.: A multi-objective ant colony system algorithm for airline crew rostering problem with fairness and satisfaction. IEEE Trans. Intell. Transp. Syst. **22**(11), 6784–6798 (2020)
9. Xia, X., Gui, L., Yu, F., et al.: Triple archives particle swarm optimization. IEEE Trans. Cybern. **50**(12), 4862–4875 (2019)
10. Xu, G., Cui, Q., Shi, X., et al.: Particle swarm optimization based on dimensional learning strategy. Swarm Evol. Comput. **45**, 33–51 (2019)
11. Kiran, M.S.: Particle swarm optimization with a new update mechanism. Appl. Soft Comput. **60**, 670–678 (2017)

12. Ibrahim, R.A., Ewees, A.A., Oliva, D., Abd Elaziz, M., Lu, S.: Improved salp swarm algorithm based on particle swarm optimization for feature selection. J. Ambient. Intell. Humaniz. Comput. **10**(8), 3155–3169 (2018). https://doi.org/10.1007/s12652-018-1031-9
13. Zhang, X.W., Liu, H., Tu, L.P.: A modified particle swarm optimization for multimodal multi-objective optimization. Eng. Appl. Artif. Intell. **95**, 103905 (2020)
14. Wei, G., Yu, G., Song, M.: Optimization model and algorithm for crew management during airline irregular operations. J. Comb. Optim. **1**(3), 305–321 (1997)

Improved Aircraft Maintenance Technician Scheduling with Task Splitting Strategy Based on Particle Swarm Optimization

Bowen Xue[1], Haiyun Qiu[1], Ben Niu[1,2](✉), and Xiaohui Yan[3]

[1] College of Management, Shenzhen University, Shenzhen 518060, China
drniuben@gmail.com

[2] Greater Bay Area International Institute for Innovation, Shenzhen University, Shenzhen 518060, China

[3] School of Mechanical Engineering, Dongguan University of Technology, Dongguan 523808, China

Abstract. Working overtime is commonly existed in aircraft maintenance, which may aggravate the fatigue of maintenance technicians and is not conducive to ensuring the accuracy and efficiency of the maintenance work. To tackle this problem, this paper improves the basic aircraft maintenance technician scheduling (AMTS) model while controlling the maintenance time and costs, and proposes an aircraft maintenance technician scheduling model with task splitting strategy (AMTS-TSS). In the AMTS-TSS model, the task splitting strategy is designed especially for those complex and time-consuming tasks that require being completed across maintenance shifts. Then, with the aim of reducing the maintenance time, the technician assignment approach based on the work efficiency is discussed when splitting tasks. Finally, particle swarm optimization (PSO) is applied to test the performance of the two models. The experimental results show that compared with the AMTS model, the AMTS-TSS model can not only effectively reduce the total maintenance cost, but also shorten the average maintenance time while preventing the overtime work.

Keywords: Aircraft maintenance scheduling · Task splitting strategy · Technician exchange approach · Particle swarm optimization

1 Introduction

Aircraft maintenance not only provides security for airline operations and ensures the continuous airworthiness of aircrafts, but also promises available aircrafts on time for airline flight plans every day [1]. According to the International Air Transport Association (IATA), the maintenance cost can account for 9–10% of the total cost of airlines [2]. Therefore, while ensuring the aircraft maintenance quality, many airlines aim to strengthen the maintenance cost control, reduce the maintenance cost and pursue economic benefits.

Y. Tan et al. (Eds.): ICSI 2022, LNCS 13344, pp. 201–213, 2022.
https://doi.org/10.1007/978-3-031-09677-8_18

Aircraft maintenance is a work with high pressure and difficult working environment. Aircraft maintenance technicians are often faced with the problem of overtime work due to the complex and time-consuming maintenance tasks [3]. In this situation, it is difficult for the technicians to complete maintenance tasks with quality and quantity under fatigue conditions, which brings additional overtime costs for airlines. In the early research on aircraft maintenance technician scheduling, authors often take the aircraft maintenance team as a whole to allocate work shifts [4, 5], and take the minimization of maintenance workforce and total cost as the optimization objectives [6, 7]. In recent years, some studies gradually began to pay attention to the workload allocation of maintenance technicians, and refined the aircraft maintenance scheduling to the technician allocation of each maintenance task in each maintenance shift [3, 8]. Chen et al. sequenced the maintenance task of multiple aircrafts, and then assigned technicians for each task. Following Chen et al., a more complex aircraft maintenance technician scheduling model was proposed by Niu et al., which can realize the distributed task-technician assignment. However, these studies only minimize the penalty cost of overtime work from the perspective of improving the problem-solving algorithms, and do not effectively solve the problem of overtime work from the perspective of model improvement.

Therefore, aiming at minimizing the total cost, this paper proposes a task splitting strategy and a technician exchange approach especially for the maintenance tasks leading to consecutive working shifts. Also, in order to balance the workload of technicians holding different license levels, the fairness of workload distribution is also optimized as a part of the total maintenance cost. Besides, due to the fact that aircraft maintenance technician scheduling problem corresponds to dimensions of aircraft, shift, maintenance task, maintenance technician and license level, it is an NP-hard problem, which is suitable to be solved by heuristic algorithms. As one of the most representative heuristic algorithms, particle swarm optimization (PSO) [9] has the advantages of high precision and fast running speed, and is widely used in maintenance scheduling problems [10, 11]. Hence, this paper uses PSO to verify the effectiveness of task splitting strategy in aircraft maintenance technician scheduling problem.

The reminder of this paper is organized as follows. Section 2 gives the basic aircraft maintenance technician scheduling model. Section 3 introduces the task splitting strategy and the technician assignment approach in detail. The introduction of PSO and the encoding scheme designed for it are illustrated in Sect. 4. In Sect. 5, experiments and results are shown to test the effectiveness of the AMTS-TSS model in controlling maintenance cost and time. Finally, Sect. 6 concludes this paper.

2 Basic Aircraft Maintenance Technician Scheduling Model

The purpose of aircraft maintenance technician scheduling is to assign technicians to each task of multiple aircrafts to be maintained. In Sect. 2.1, the process of basic aircraft maintenance technician scheduling is described in detail. Then, the objective function and constraints of this model are given in Sect. 2.2.

2.1 Problem Description

Each aircraft to be repaired will be equipped with a group of maintenance technicians with different license levels. The maintenance tasks on each aircraft are arranged in order. One maintenance task may require different levels of technicians to complete together. Technicians holding higher license levels can replace those with lower license levels to complete their work. When assigning tasks to technicians, the requirements of maintenance tasks for license levels and the number of technicians at each level should be met. Additionally, each working day is divided into three maintenance shifts and each shift lasts for eight hours. In particular, since different combination of technicians require different time to complete the same task, the maintenance shift of a task cannot be predetermined. Therefore, the current maintenance task is located through its position in the task sequence. The related variables and definitions are given in Table 1.

Table 1. Definition of variables.

Variables	Meaning of variables
A	Number of aircrafts
M_i	Number of maintenance technicians of aircraft i
T	Number of maintenance shifts
S_i	Number of maintenance tasks of aircraft i
λ_{im}	Unit cost of technician m of aircraft i
avg_i	Average working time of all technicians of aircraft i
NC_{im}	Times that technician m works overtime of aircraft i
PS_{is}	Previous tasks of task s of aircraft i
ws_{im}	Working state of technician m of aircraft i
Aq_{is}	Task s of aircraft i can be assigned to technicians with license q
wt_{ism}	Work time of technician m to participate in task s of aircraft i
rt_{ism}	Time required for technician m to complete task s of aircraft i
rm_{is}	Number of technicians required by task s of aircraft i
nm	Normal working hours of technician m
st_{is}	Start time of task s of aircraft i
lt_{ist}	Left time of task s of aircraft i in shift t
H	Length of a shift
mt_{is}	Maintenance time of task s of aircraft i
mt_{is}^0, mt_{is}^1	Maintenance time of task s of aircraft i with no technician exchange approach and with technician exchange approach, respectively
AT_i	Arrival time of aircraft i

(continued)

Table 1. *(continued)*

Variables	Meaning of variables
CT	Current time
MT_i	Maintenance time of aircraft i
q_m	Whether technician m holding license level q, $q \in \{1, 2, 3\}$
in_i	Whether aircraft i has entered the maintenance hangar
out_i	Whether aircraft i has left the maintenance hangar
k_{mis}	Whether technician m participates in task s of aircraft i
t_{is}	Whether task s of aircraft i can be implemented
f_{is}	Whether task s of aircraft i is complicated

2.2 Objective and Constrains

Objective. In this paper, the optimization objective is to minimize the total maintenance costs, including workforce cost, overtime cost and workload distribution fairness cost, as shown in formula (1). The workforce cost is the sum of labor cost of all technicians of all aircrafts participating in all tasks. wt_{ism} represents the working time of technician m to participate in task s of aircraft i and λ_{im} is the unit cost of technician m. Overtime cost is expressed by the number of times that technicians work overtime, where NC_{im} is the number of technician m working overtime. The workload distribution fairness cost is represented by the standard deviation of the workload of all technicians. avg_i is the average working time of technicians of aircraft i. p_1, p_2, and p_3 are the penalty factors.

$$\min \sum_{i \in A} \left\{ p_1 \cdot \sum_{s \in S_i} \sum_{m \in M_i} wt_{ism} \cdot \lambda_{im} + p_2 \cdot \sum_{m \in M_i} NC_{im} + p_3 \cdot \frac{1}{|M_i|} \sqrt{\sum_{s \in S_i} \left[\sum_{m \in M_i} kmis \cdot (wt_{ism} - avg_i) \right]^2} \right\}. \tag{1}$$

Constraints. On the one hand, only qualified maintenance tasks can be performed. Constraints (2) and (3) define the aircrafts parking in the maintenance hangar. Constraints (4)–(5) indicate that only the maintenance tasks on the aircraft parking in the maintenance hangar can be performed. Then, constraint (6) means that a maintenance task can be performed only after its previous tasks have been completed.

$$in_i = \begin{cases} 0, & if\ AT_i > CT \\ 1, & else \end{cases}, \forall i \in A. \tag{2}$$

$$out_i = \begin{cases} 0, & if\ \sum_{s \in S_i} fis = S_i \\ 1, & else \end{cases}, \forall i \in A. \tag{3}$$

$$t_{is} \le in_i, \forall i \in A, \forall s \in S_i. \tag{4}$$

$$t_{is} \leq out_i, \forall i \in A, \forall s \in S_i. \tag{5}$$

$$tis \leq \frac{1}{|PSis|} \cdot \sum_{s' \in PS_{is}} fis', \forall i \in A, \forall s \in S_i. \tag{6}$$

On the other hand, the qualifications and the requirements of maintenance tasks should be considered when assigning maintenance technicians. Technicians in available state are supposed to be assigned to those tasks that can be performed, as shown in constraint (7) and (8). Besides, constraints (9) and (10) mean that the assignment of technicians needs to meet the requirements of maintenance tasks on license level and number of workforce. Constraint (11) indicates that the total workload of technicians cannot exceed the specified workload limit. Constraints (12) and (13) define the maintenance time of each aircraft.

$$km_{is} \leq t_{is}, \forall i \in A, \forall s \in S_i. \tag{7}$$

$$km_{is} \leq ws_{im}, \forall i \in A, \forall m \in M_i, \forall s \in S_i. \tag{8}$$

$$km_{is} \leq q_m \cdot A_{qis}, \forall i \in A, \forall m \in M_i, \forall s \in S_i. \tag{9}$$

$$\sum_{m \in M_i} k_{mis} = rm_{is}, \forall i \in A, \forall s \in S_i. \tag{10}$$

$$\sum_{s \in S_i} k_{mis} \cdot wt_{ism} \leq nm, \forall i \in A, \forall m \in M_i. \tag{11}$$

$$mt_{is} = rt_{ism}, \forall i \in A, \forall m \in M_i, \forall s \in S_i. \tag{12}$$

$$MT_i = \sum_{s \in S_i} mt_{is}, \forall i \in A. \tag{13}$$

3 Aircraft Maintenance Technician Scheduling Model with Task Splitting Strategy

Since it is uncertain which shift a maintenance task belongs to, it often occurs that the maintenance task needs to be completed across shifts. At this time, if the maintenance technicians are not replaced by another group, they will work overtime. Therefore, this section proposes the task splitting strategy especially for the tasks across shifts on the basis of the AMTS model, and forms the AMTS-TSS model. Particularly, compared with the AMTS model, AMTS-TSS model only adds the improvement of operation process. In addition to the calculation of the maintenance time mt_{is}, the objective function and other constraints are the same as AMTS model. The variables and definitions in the AMTS-TSS model is also shown in Table 1.

In the aircraft maintenance technician scheduling process, the left time for task s in shift t before assigning technicians to it is lt_{ist}, which is calculated by formula (14). Then, when the required time rt_{ism_1} for technicians m_1 to complete task s is longer than lt_{ist}, as shown in formula (15), another group of technicians m_2 meeting the same requirements as m_1 will also be selected, and the required time for m_2 to complete task s is rt_{ism_2}. Particularly, if rt_{ism_2} is less than lt_{ist}, then task s will not be split and will be totally completed by technicians m_2. Otherwise, the task splitting strategy will be executed.

$$lt_{ist} = H \cdot t - st_{is}. \quad (14)$$

$$rt_{ism_1} > lt_{ist}, \forall i \in A, \forall m_1 \in M_i, \forall s \in S_i, \forall t \in T. \quad (15)$$

With the task splitting strategy, task s is split into two subtasks s_1 and s_2. It can be seen that the work efficiency of technicians m_1 and m_2 in completing task s is not the same. Therefore, in order to avoid overtime work as well as shorten the maintenance time, the following parts discuss how to assign technicians m_1 and m_2 to subtasks s_1 and s_2.

There are two approaches to assign m_1 and m_2 to subtasks s_1 and s_2 after splitting. The first is assigning technicians m_1 to subtask s_1 and technicians m_2 to subtask s_2, called as no technician exchange approach. The second is assigning technicians m_2 to subtask s_1 and technicians m_1 to subtask s_2, called as technician exchange approach. The choice of these two approaches depends on the total maintenance time of task s, i.e., the sum of the time of the two subtasks.

3.1 No Technician Exchange Approach

In this section, the work time for technicians m_1 as well as the maintenance time of subtask s_1 is equal to lt_{ist}. Thus, the work progress of subtask s_1 is lt_{ist}/rt_{ism_1} and the left work progress of subtask s_2 is $\left(1 - lt_{ist}/rt_{ism_1}\right)$. Therefore, the work time for technicians m_2 to complete subtask s_2 is $\left(1 - lt_{ist}/rt_{ism_1}\right) \cdot rt_{ism_2}$. Then the total maintenance time mt_{is}^0 of task s with no technician exchange approach is shown in formula (16).

$$mt_{is}^0 = lt_{ist} + \left(1 - lt_{ist}/rt_{ism_1}\right) \cdot rt_{ism_2}. \quad (16)$$

3.2 Technician Exchange Approach

With the technician exchange approach, the work time for technician m_2 to complete subtask s_1 is equal to lt_{ist}, so the work progress of subtask s_1 is lt_{ist}/rt_{ism_2}. Then the left work progress for technician m_1 to complete subtask s_2 is $\left(1 - lt_{ist}/rt_{ism_2}\right)$ and the work time lasts for $\left(1 - lt_{ist}/rt_{ism_2}\right) \cdot rt_{ism_1}$. Hence, the total maintenance time mt_{is}^1 of task s with technician exchange approach is shown in formula (17).

$$mt_{is}^1 = lt_{ist} + \left(1 - lt_{ist}/rt_{ism_2}\right) \cdot rt_{ism_1}. \quad (17)$$

3.3 Technician Assignment for Split Subtasks

By comparing mt_{is}^0 and mt_{is}^1, it can be decided whether using technician exchange approach for the subtasks s_1 and s_2 with the aim of minimizing the maintenance time of task s. The calculation results are summarized as follows. The total maintenance time of task s can be defined in formula (18).

- If $0 < lt_{ist} \leq \frac{rt_{ism_1} \cdot rt_{ism_2}}{rt_{ism_1} + rt_{ism_2}}$, then $mt_{is}^0 \leq mt_{is}^1$, and the no technician exchange approach is applied, the total maintenance time of task s is $mt_{is} = mt_{is}^0$.
- If $\frac{rt_{ism_1} \cdot rt_{ism_2}}{rt_{ism_1} + rt_{ism_2}} < lt_{ist} \leq H$, then $mt_{is}^0 > mt_{is}^1$, and the technician exchange approach is applied, the total maintenance time of task s is $mt_{is} = mt_{is}^1$.

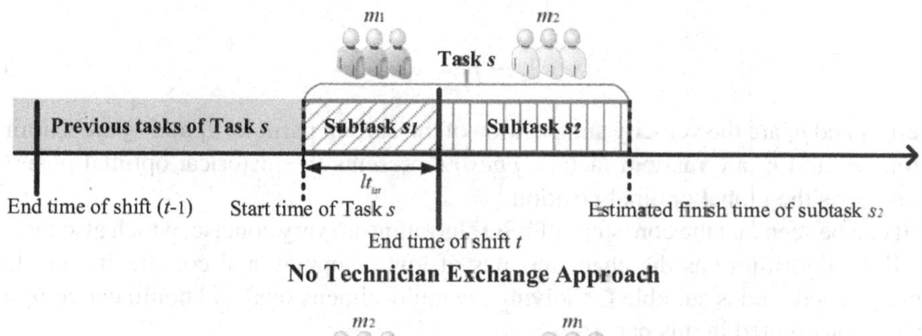

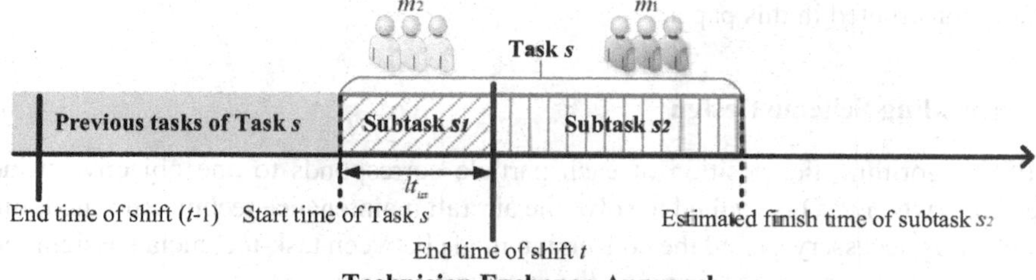

Fig. 1. Task splitting strategy and technician exchange approaches.

$$mt_{is} = \begin{cases} lt_{ist} + \left(1 - lt_{ist}/rt_{ism_1}\right) \cdot rt_{ism_2}, & \text{if } 0 < lt_{ist} \leq \frac{rt_{ism_1} \cdot rt_{ism_2}}{rt_{ism_1} + rt_{ism_2}} \\ lt_{ist} + \left(1 - lt_{ist}/rt_{ism_2}\right) \cdot rt_{ism_1}, & \text{if } \frac{rt_{ism_1} \cdot rt_{ism_2}}{rt_{ism_1} + rt_{ism_2}} < lt_{ist} \leq H \end{cases}. \tag{18}$$

Finally, Fig. 1 demonstrates the task splitting strategy and two technician exchange approaches. The solid red line indicates the end time of shift t, i.e., $H \cdot t$. When the start time of task s is earlier than $H \cdot t$ and the estimated finish time is later than $H \cdot t$, task s needs to be split into subtasks s_1 and s_2. With the no technician exchange approach, technicians m_1 are assigned to subtask s_1 and technicians m_2 are assigned to subtask s_2. With the technician exchange approach, technicians m_1 are assigned to subtask s_2 and technicians m_2 are assigned to subtask s_1.

4 Particle Swarm Optimization and Encoding Scheme

This section briefly introduces the particle swarm optimization (PSO) [9] algorithm, and designs an encoding scheme that can connect both the aircraft maintenance technician scheduling model and the solution algorithm.

4.1 Particle Swarm Optimization

PSO algorithm is a biological heuristic algorithm based on the foraging behavior of birds [9]. Each particle in the algorithm approaches a better position by learning from its historical optimal position and the global optimal position in the whole swarm. The velocity and position of each particle are updated as follows.

$$v_i = v_i + c_1 \cdot r_1 \cdot (pbest_i - p_i) + c_2 \cdot r_2 \cdot (gbest - p_i). \tag{19}$$

$$p_i = p_i + v_i. \tag{20}$$

where v_i and p_i are the velocity and position of the current particle, c_1 and c_2 are learning factors, r_1 and r_2 are random factors. $pbest_i$ represents the historical optimal position and $gbest$ is the global optimal position.

It can be seen that the core steps of PSO algorithm are very concise, which also means that PSO algorithm has the characteristics of low computational complexity and fast running speed, and is suitable for solving the multi-dimensional and nonlinear complex models constructed in this paper.

4.2 Encoding Scheme Design

In PSO algorithm, the position of each particle corresponds to one objective value. Therefore, when PSO is applied to solve the aircraft maintenance technician scheduling models, it is necessary to find the conversion mode between task-technician assignment scheme and the particle position, i.e., the encoding scheme.

The encoding scheme of aircraft maintenance technician scheduling model is illustrated in Table 2. "0" represents the technicians who do not meet the license level requirements of maintenance tasks. Bold "1" indicates the technicians chosen for each maintenance task. For instance, technicians *M7* and *M10* are assigned to Task 4.

Table 2. Encoding scheme of aircraft maintenance technician scheduling models.

Technicians	Task 1	Task 2	Task 3	Task 4	Task 5	Task 6	Task 7
M1	0.27	0.63	0.85	0.57	0.26	0.53	1
M2	0.95	0.62	1	0.27	0.60	1	1
M3	1	0.76	1	0.67	0.27	0.35	1
M4	0.77	0.79	1	0.83	0.80	1	0.53
M5	0.88	1	1	0.06	0.91	1	0.48

(continued)

Table 2. (*continued*)

Technicians	Task 1	Task 2	Task 3	Task 4	Task 5	Task 6	Task 7
M6	1	1	0	0.20	1	0.42	0.10
M7	0.30	1	0	1	0.34	0.02	0.31
M8	0.06	0.88	0	0.50	0.13	0.72	1
M9	0.86	0.18	0	1	1	0.27	1
M10	0.60	0.28	0	1	0	1	1
M11	1	0.95	0	0.15	0	0.13	0.63

5 Experiments and Results Analysis

This section introduces the simulation setup, gives the results of solving the two models using PSO algorithm, and compares the results of the AMTS-TSS model with the basic AMTS model.

5.1 Simulation Setup

The data used in the experiments is improved on the basis of the data obtained from the investigation of an aircraft maintenance enterprise. Suppose that there are seven aircrafts waiting to be repaired. The arrival time, number of maintenance technicians at each level, number of tasks and the requirements for each task of each aircraft are listed in Table 3. For example, aircraft *A1* arrives at 4:00 with 13 technicians, the number of technicians holding license level 1, 2, 3 are 5, 6, 2, respectively. There are 6 maintenance tasks for *A1*. The first task of *A1* requires one technician with a level of at least 1 and two technicians with a level of at least 2.

Table 3. Simulation data of aircrafts to be repaired.

A	AT_i	M_i	S_i	rm_{is}
A1	4:00	13 (5, 6, 2)	6	(1 2 0; 2 0 0; 0 2 0; 0 0 1; 1 1 0; 0 2 0)
A2	5:15	14 (3, 5, 6)	9	(1 2 0; 2 0 0; 0 2 0; 0 1 1; 1 1 0; 0 0 2; 1 1 0; 0 2 0; 1 2 0)
A3	8:00	8 (3, 3, 2)	4	(1 1 0; 2 0 0; 0 2 0; 2 0 0)
A4	9:36	9 (4, 3, 2)	4	(1 0 1; 2 0 0; 3 0 0; 1 1 0)
A5	10:00	11 (2, 4, 5)	7	(1 2 0; 2 1 0; 0 0 2; 3 0 0; 0 2 0; 1 0 1; 1 2 0)
A6	12:00	9 (4, 3, 2)	3	(2 0 0; 1 1 0; 1 0 0)
A7	16:00	11 (3, 5, 3)	8	(0 2 0; 1 1 0; 1 2 0; 1 0 1; 3 0 0; 1 1 0; 1 0 1; 1 2 0)

The average unit costs of technicians with the three license levels are 0.5, 1 and 1.5. The time they require to complete each maintenance task follows a random distribution between 0.5 and 6.5 h, and the standard deviation is 1.427 h. The values of the penalty factors p_1, p_2 and p_3 are 5, 1, and 1, respectively. During a working day, the end time of each shift is 8:00, 16:00 and 24:00.

Then, PSO algorithm is used to solve the AMTS and AMTS-TSS models. In PSO, the fitness function evaluations are set to 2000, the learning factors c_1 and c_2 are 2, and the inertia weight $w = 0.9$. The experiment is carried out 10 times.

5.2 Results and Analysis

The final optimal solutions of the maintenance time and cost of each aircraft are given in Table 4, where the "*Gap*" represents the improvement of the AMTS-TSS model in reducing maintenance time and controlling maintenance cost compared with the AMTS model. It can be seen that with the task splitting strategy, the maintenance time and cost of most aircrafts are reduced compared with the basic model. Besides, in the AMTS-TSS model, the average maintenance time and total maintenance cost have been effectively optimized. The optimization ratios of average maintenance time and total maintenance cost are similar, which are 13.41% and 4.71% respectively. Particularly, although the maintenance cost of several aircrafts in the AMTS-TSS model is higher than that in the AMTS model, it has less impact on the overall optimization utility.

Table 4. Optimal solution of maintenance time and cost.

A	Maintenance time			Cost per aircraft		
	AMTS model	AMTS-TSS model	Gap	AMTS model	AMTS-TSS model	Gap
A1	17.90	10.75	−39.97%	172.82	115.46	−33.19%
A2	25.67	23.36	−8.99%	366.59	330.39	−9.87%
A3	11.60	10.60	−8.61%	131.92	123.03	−6.74%
A4	9.22	7.94	−13.88%	108.39	90.24	−16.75%
A5	19.62	18.29	−6.78%	326.80	300.06	−8.18%
A6	7.85	7.33	−6.68%	81.02	107.93	33.22%
A7	25.67	23.36	−8.99%	298.28	323.89	8.58%
	Average maintenance time		*Gap*	*Total cost*		*Gap*
	16.79	14.52	−13.41%	1485.84	1391.00	−4.71%

Then, in order to further discuss the role of the task splitting strategy, taking the aircraft *A7* in Table 4 as an example, this paper makes statistics on the scheduling and splitting of maintenance tasks on this aircraft in the two models. The results are shown in Table 5 and Table 6, including the start time and status of each task and "(+1)" indicates the second working day. By comparing Table 5 and Table 6, it can be found that in the AMTS model, Task 3 and Task 5 require the same group of technicians to work overtime, because the start time of Task 4 and Task 6 are later than the end time of their previous shifts. In the AMTS-TSS model, it can be seen that Task 3 and Task 6 that originally need to be completed overtime are split and assigned to two groups of technicians to complete, while Task 8 avoids overtime work by being assigned to another group of maintenance technicians.

Table 5. Optimal scheduling scheme of aircraft *A7* in the AMTS model.

Tasks	Task start time	Task status
Task 1	16:00	Normal work
Task 2	19:30	Normal work
Task 3	23:00	**Overtime work**
Task 4	(+1) 1:42	Normal work
Task 5	(+1) 6:18	**Overtime work**
Task 6	(+1) 9:18	Normal work
Task 7	(+1) 11:54	Normal work
Task 8	(+1) 13:36	Normal work

Table 6. Optimal scheduling scheme of aircraft *A7* in the AMTS-TSS model.

Tasks	Task start time	Task status
Task 1	16:00	Normal work
Task 2	18:00	Normal work
Task 3	21:39 (+1) 0:00	Task splitting strategy and no technician exchange approach are applied. Technicians *M5, M7, M10* are assigned to subtask s_1 and technicians *M3, M4, M6* are assigned to subtask s_2 Normal work
Task 4	(+1) 0:21	Normal work
Task 5	(+1) 3:45	Normal work
Task 6	(+1) 6:57 (+1) 8:00	Task splitting strategy and no technician exchange approach are applied. Technicians *M1* and *M4* are assigned to subtask s_1 and technicians *M2* and *M6* are assigned to subtask s_2 Normal work
Task 7	(+1) 9:25	Normal work
Task 8	(+1) 11:07	First, when technicians *M1, M6, M10* are chosen to finish Task 8, it needs to be split. However, when technicians *M2, M3, M7* are chosen, Task 8 can be completed before the end of the current shift. Finally, Task 8 is assigned to technicians *M2, M3, M7* without task splitting strategy Normal work

6 Conclusion

In order to solve the problem of aircraft maintenance technicians working overtime, this paper improves the basic aircraft maintenance technician scheduling model, introduces the task splitting strategy, and matches the technicians and maintenance tasks according to the work efficiency of different technicians for the purpose of saving maintenance time. In order to verify the effectiveness of the task splitting strategy, PSO algorithm is applied to test the basic model and improved model. The comparative experimental results show that compared with the basic model, the improved model can effectively shorten the average maintenance time of all aircraft and reduce the total maintenance cost while avoiding working overtime.

In the future, more research can be carried out to extend the current study. First, more discussion about task splitting conditions can be considered, such as adding a time threshold to flexibly adjust the splitting strategy. In addition, other intelligent optimization methods can be tried to solve the aircraft maintenance technician scheduling problems, such as bee colony algorithm, bacterial foraging optimization algorithm, and ant colony algorithm.

Acknowledgement. The study is supported by The National Natural Science Foundation of China (Nos. 71971143, 61703102), Major Project of Natural Science Foundation of China (No. 71790615), Integrated Project of Natural Science Foundation of China (No. 91846301), Social Science Youth Foundation of Ministry of Education of China (Nos. 21YJC630052, 21YJC630181), Key Research Foundation of Higher Education of Guangdong Provincial Education Bureau (No.2019KZDXM030), Natural Science Foundation of Guangdong Province (Nos. 2020A1515010749, 2020A1515010752), Guangdong Basic and Applied Basic Research Foundation (No. 2019A1515110401) , and Natural Science Foundation of Shenzhen City (No. JCYJ20190808145011259), Guangdong Province Innovation Team (No. 2021WCXTD002).

References

1. Qin, Y., Zhang, J.H., Chan, F.T., Chung, S.H., Niu, B., Qu, T.: A two-stage optimization approach for aircraft hangar maintenance planning and staff assignment problems under MRO outsourcing mode. Comput. Ind. Eng. **146**, 106607 (2020)
2. IATA's Maintenance Cost Task Force, Airline Maintenance Cost Executive Commentary Edition 2019. https://www.iata.org/contentassets/bf8ca67c8bcd4358b3d004b0d6d0916f/mctg-fy2018-report-public.pdf. Accessed 11 Sept 2020
3. Niu, B., Xue, B., Zhou, T., Kustudic, M.: Aviation maintenance technician scheduling with personnel satisfaction based on interactive multi-swarm bacterial foraging optimization. Int. J. Intell. Syst. **37**(1), 723–747 (2022)
4. Yang, T.H., Yan, S., Chen, H.H.: An airline maintenance manpower planning model with flexible strategies. J. Air Transp. Manag. **9**(4), 233–239 (2003)
5. Yan, S., Yang, T.H., Chen, H.H.: Airline short-term maintenance manpower supply planning. Transp. Res. Part Policy Pract. **38**(9–10), 615–642 (2004)
6. De Bruecker, P., Van den Bergh, J., Beliën, J., Demeulemeester, E.: A model enhancement heuristic for building robust aircraft maintenance personnel rosters with stochastic constraints. Eur. J. Oper. Res. **246**(2), 661–673 (2015)

7. Beliën, J., Demeulemeester, E., De Bruecker, P., Van den Bergh, J., Cardoen, B.: Integrated staffing and scheduling for an aircraft line maintenance problem. Comput. Oper. Res. **40**(4), 1023–1033 (2013)
8. Chen, G., He, W., Leung, L.C., Lan, T., Han, Y.: Assigning licenced technicians to maintenance tasks at aircraft maintenance base: a bi-objective approach and a Chinese airline application. Int. J. Prod. Res. **55**(19), 5550–5563 (2017)
9. Eberhart, R., Kennedy, J.: Particle swarm optimization. In: Proceedings of ICNN'95-International Conference on Neural Networks. vol. 4, pp. 1942–1948. IEEE (1995)
10. Pereira, C.M., Lapa, C.M., Mol, A.C., Da Luz, A.F.: A particle swarm optimization (PSO) approach for non-periodic preventive maintenance scheduling programming. Prog. Nucl. Energy **52**(8), 710–714 (2010)
11. Lin, D., Jin, B., Chang, D.: A PSO approach for the integrated maintenance model. Reliab. Eng. Syst. Saf. **193**, 106625 (2020)

Ant Colony Optimization

Mobile Robot Path Planning Based on Angle Guided Ant Colony Algorithm

Yongsheng Li[1], Yinjuan Huang[1](✉), Shibin Xuan[1], Xi Li[1], and Yuang Wu[2]

[1] School of Artificial Intelligence, Guangxi University for Nationalities, Nanning 530006, Guangxi, China
270549548@qq.com

[2] College of Electronic Information, Guangxi University for Nationalities, Nanning 530006, Guangxi, China

Abstract. Ant colony algorithm is easy to fall into local optimum and its convergent speed is slow when solving mobile robot path planning. Therefore, an ant colony algorithm based on angle guided is proposed in this paper to solve the problems. In the choice of nodes, integrate the angle factor into the heuristic information of the ant colony algorithm to guide the ants' search direction and improve the search efficiency. The pheromone differential updating is carried out for different quality paths and the pheromone chaotic disturbance updating mechanism is introduced, then the algorithm can make full use of the better path information and maintain a better global search ability. According to simulations, its global search is strong and it can range out of local optimum and it is fast convergence to the global optimum. The improved algorithm is feasible and effective.

Keywords: Ant colony algorithm · Mobile robot · Path planning · Angle guided · Chaos

1 Introduction

Mobile robot path planning is one of the most basic and critical issues in the field of mobile robot research [1–3]. Its purpose is to find a path with the shortest distance between the start point and the end point under the condition of known robot environment information and the path does not pass through any obstacles [4]. The method of solving the path planning problem of mobile robots can be divided into two categories: traditional algorithms and intelligent algorithms. Traditional algorithms include artificial potential field method, fuzzy logic algorithm, viewable method, free space method, etc. [5, 6]. Since the path planning problem was proposed in the 1970s, these traditional algorithms have played an important role in the field of robot path planning and have achieved many research results. However, with the continuous expansion of mobile robot application fields, such as practical applications in marine science, industrial field and military operations, these traditional path planning optimization methods will have certain defects in dealing with these complex environments. For example, the artificial potential field method is easy to fall into a local minimum, and there is a problem of unreachable goals.

Y. Tan et al. (Eds.): ICSI 2022, LNCS 13344, pp. 217–229, 2022.
https://doi.org/10.1007/978-3-031-09677-8_19

The visualization method is very inefficient and cannot meet the real-time requirements of path planning. Fuzzy control algorithm is difficult to establish fuzzy rule base in complex and changeable environment and lacks intelligent obstacle avoidance strategy for dynamic obstacles [7]. In recent years, with the rise of artificial intelligence, more and more intelligent algorithms have been proposed and applied to the path planning optimization of mobile robots to overcome the limitations of traditional path planning algorithms. One of the important characteristics of these intelligent algorithms is that their operation mechanism is very similar to the biological group behavior or ecological mechanism in nature, and the efficiency of these intelligent algorithms is usually higher than that of traditional algorithms. The typical ones are genetic algorithm, ant colony algorithm, particle swarm optimization algorithm, artificial neural network algorithm, firefly algorithm, artificial bee colony algorithm, invasive weed algorithm and so on. Khaled Akka et al. [8] proposed an improved ant colony algorithm to solve the robot path planning problem, using the stimulus probability to help ants select the next node, and using new pheromone update rules and dynamic adjustment of evaporation rate to accelerate the convergence speed and expand the search space. Long s et al. [9] proposed an improved ant colony algorithm, which realized the efficient search ability of mobile robot in complex map path planning, and established the grid environment model. Faridi et al. [10] proposed a multi-objective dynamic path planning method for multi robot based on improved artificial bee colony algorithm. This method improves the artificial bee colony algorithm and applies it to the neighborhood search path planner and the algorithm avoids falling into local optimum by adding appropriate parameters into the objective function. Kang Yuxiang et al. [11] proposed an improved particle swarm optimization algorithm for robot path planning. According to the principle that variables in gradient descent method change along the negative gradient direction, an improved particle velocity update model is proposed. In order to improve the search efficiency and accuracy of particles, the adaptive particle position update coefficient is added.

Ant colony algorithm (ACO) is a heuristic random search algorithm, which is derived from the simulation of natural ant colony searching for the shortest path from nest to food source [12–14]. Ant colony algorithm finds the optimal path through the positive feedback generated by pheromone accumulation, which has the characteristics of robustness, positive feedback and distributed computing. Moreover, it is easy to program and implement, does not involve complex mathematical operations, and has no high requirements on computer software and hardware. The robot path planning problem can be simulated as a group behavior of ants looking for the optimal path for food. Therefore, many scholars apply ant colony algorithm to solve various robot path planning problems and have achieved some results. However, the basic ant colony algorithm generally has shortcomings such as too long search time, premature maturity, and stagnation when solving robot path planning problems. In view of this, this paper proposes an improved ant colony algorithm, which integrates the angle factor into the heuristic information of the ant colony algorithm, guides the ants' search direction. The pheromone is updated differently, and at the same time, chaotic disturbance is added to the path that may fall into the local optimum to make the solution jump out of the local extreme value interval. Simulation experiments show that the algorithm can effectively solve the robot path planning problem.

2 Environment Modeling

To solve the robot path planning problem, environment modeling is often needed. Grid method is a classical method of environment modeling [15], which cuts the working environment of mobile robot into grids, and these grids are of the same size and connected with each other. Each grid corresponds to the corresponding position information, with black representing obstacle information and white representing barrier free. Then the raster map is encoded. In this paper, we assume that the working environment of the robot is a complex static two-dimensional space. In the global path planning, the working environment is n × n grid environment. The black grid uses 1 to represent obstacles and the white grid uses 0 to represent the free feasible area. Obstacles less than one grid are still treated as a grid. The grid numbers are 1, 2, 3,…, N from top to bottom and from left to right. Taking the lower left corner of the grid as the coordinate origin, the horizontal axis from left to right is the positive direction of X axis, and the vertical axis is from bottom to top as the positive direction of Y axis. The length of each grid is taken as the unit length, and each grid is marked as a node. The environment model established is shown in Fig. 1 (n = 20).

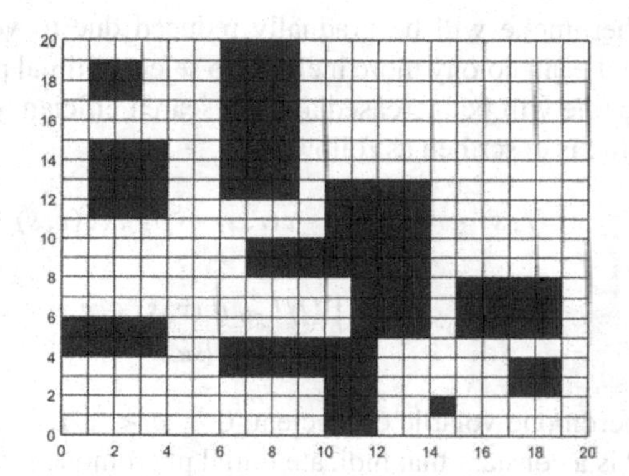

Fig. 1. Grid model diagram

3 Basic Ant Colony System

The ant colony optimization algorithm (ACO) is proposed in the early 1990s. In 1996, ant colony system is proposed by Dorigo and Gmabardella [12–14], the performance of the ACO is effectively improved and they made three improvements as follows:

(1) A new selection strategy that combination of deterministic selection and random selection is adapted, which both can utilize the advantage of prior knowledge and can tendentiously explore. For an ant at node r to move to the next city s, the state transition rule is given by the following formula.

$$s_k = \begin{cases} \arg \underset{u \in allowed}{max} \{[\tau(r,u)]^{\alpha}[\eta(r,u)]^{\beta}, & q \leq q_0 \\ S \; q > q_0 \end{cases} \tag{1}$$

$$p_{ij}^k(t) = \begin{cases} \frac{[\tau_{ij}(t)]^\alpha \cdot [\tau_{ij}(t)]^\beta}{\sum_{s \in allowed_k}([\tau_{ij}(t)]^\alpha \cdot [\tau_{ij}(t)]^\beta)} \cdot j \in allowed_k \\ 0 \qquad\qquad otherwise \end{cases} \tag{2}$$

where s_k is the next node of ant k, q is the random number draw from [0,1], q_0 is a parameter ($0 \leq q_0 \leq 1$). S is a random variable selected by the probability distribution given in Eq. (2). $allowed_k$ is a node set that these node can be selected for ant k in the next time, α is the pheromone heuristic factor, which reflects the effect of pheromone by ant accumulates when ants move to the other nodes. β is a heuristic factor, had reflects the degree of the heuristic information is focused when the ants select path. $\tau_{i,j}(t)$ is the pheromone of $path(i, j)$ at t time, η_{ij} is the visibility of $path(i, j)$, which is corresponding with the inverse of the distance from node i to node j:

$$\eta_{ij} = 1/d_{ij} \tag{3}$$

(2) Only the global optimal ant path performs global updating rule. After each iteration, the pheromone is enhanced only occur on the path walked by the best ant. For other pathway, the pheromone will be gradually reduced due to volatile mechanism, which can make the ant colony more inclined to select optimal path. Consequently, the convergence rate will be increased and the search efficiency will be enhanced. Global update rule is described as follows:

$$\tau(r, s) \leftarrow (1 - \rho) \cdot \tau(r, s) + \rho \cdot \Delta\tau(r, s) \tag{4}$$

$$\Delta\tau(r, s) = \begin{cases} Q/L_{gb} \; if \, (r, s) \in g \\ 0 \qquad else \end{cases} \tag{5}$$

here ρ is the pheromone volatile coefficient, $0 < \rho < 1$, L_{gb} is the current global optimal path. Q is a constant that indicate initial pheromone intensity between two nodes.

(3) Using the local update rule. The pheromone will be local updated when these ants build a path, which make the pheromone release by ants to reduce when they pass the path. The local update rule is used to decrease influence on other ants and make them search other edges. Therefore, the ants can avoid that they prematurely converge to a same solution.

The local update rule is represented by Eq. (6).

$$\tau(r, s) \leftarrow (1 - \rho) \cdot \tau(r, s) + \rho \cdot \Delta\tau(r, s) \tag{6}$$

$$\Delta\tau(r, s) = (nL_{nn})^{-1} \tag{7}$$

where n is the number of nodes. L_{nn} is a path length generated heuristically by the recent neighborhood.

4 Application of Angle Guided Ant Colony Algorithm(AGACO) in Mobile Robot Path Planning

4.1 Node Selection Strategy with Angle Guidance

In the path planning of mobile robot, the relative position of the starting point and the end point of the path is known. The starting point and the end point are connected by a directed line segment (The direction is from the starting point to the end point.), and the directed line segment is taken as the standard line. It is not difficult to see that the smaller the angle between the path line and the standard line, the more likely it is to be a part of the optimal path. Therefore, the spatial geometric relationship of the angle between the path line and the standard line can be introduced into the algorithm as heuristic information to guide the search direction of ants. As shown in Fig. 2, let S and T be the starting point and end point of path planning, P1, P2, P3, P4 are the path points on the environment map respectively. Assuming that the robot starts from S and passes through P1, the current node is P1, and the next optional node is P2, P3 and P4. The included angles between the standard line and the optional line are θ1, θ2 and θ3. It can be considered that the path segment with smaller angle is more likely to be a part of the global optimal path, and the probability of ant selecting the path is greater. Therefore, the angle factor can be incorporated into the heuristic information of ant colony algorithm to guide the search direction of ants and improve the search efficiency.

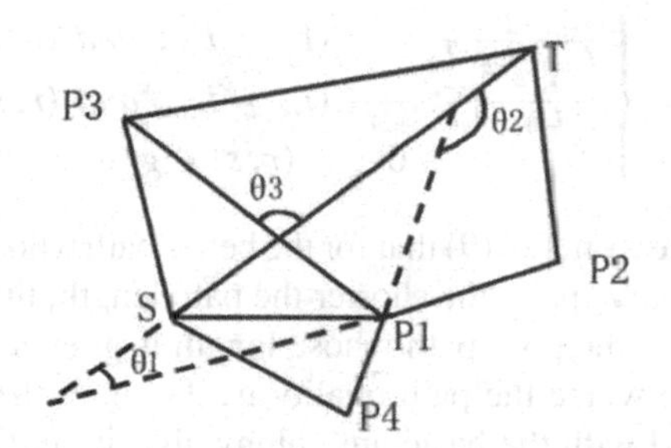

Fig. 2. Schematic diagram of included angle

Based on the above idea, angle factor θ can be introduced when ants select the next node.The calculation formula (3) of heuristic information in node selection formula is modified as follows:

$$\eta_{ij} = 1/\left(\theta_{ij} + 1/n\sum_{j(i)}\theta_{ij}\right) \tag{8}$$

in formula (8), J (i) is the set composed of all optional nodes with paths connected to the current node, n is the number of elements in the set, and θ_{ij} is the angle formed between the optional line and the standard line. It is not difficult to see from formula (8) that the smaller the angle θ_{ij} is, the greater the value of η_{ij} is, and the greater the probability of the optional node being selected. Since the value gap of the included angle information between each optional line and the standard line may be very large, this will lead to a large difference in the amount of heuristic information of each path segment, which makes the ant's selection probability of each path segment too large. Coupled with the positive feedback effect of pheromone sowing, the algorithm is likely to quickly converge to the local extremum. In order to alleviate this problem, $1/n\sum_{j(i)}\theta_{ij}$ is added to formula (8), it reduces the difference of heuristic information of each optional path segment, and increases the probability of path segment selection with small amount of heuristic information, so that the global search ability of the algorithm is strengthened and premature convergence to local extremum is avoided.

4.2 Differential Pheromone Updating Strategy

The improved pheromone updating rules are classified according to the length of the path searched by ants in a cycle, and the pheromone increment on each path is dynamically adjusted according to the path information. After all ants complete a complete path construction, find out the optimal path length lib and the worst path length lworth of this iteration, and calculate the average path length lave searched by all ants. Appropriate pheromone enhancement is applied to the better path whose path length is less than lave, and the pheromone weakening is carried out for the poor path whose path length is larger than Lave. The pheromone increment calculation formula (5) in the global pheromone update formula is modified as follows:

$$\Delta\tau r,s=\begin{cases}\frac{L_{ave}-L_k}{L_{ave}-L_{ib}}\cdot\frac{1}{L_k} & (L_k<L_{ave}\ and\ (r,s)\in \mathrm{g})\\ -\frac{L_k-L_{ave}}{L_{ave}-L_{ib}}\cdot\frac{1}{L_{worst}} & (L_k\geq L_{ave}\ and\ (r,s)\in \mathrm{g})\\ 0 & (r,s)\notin g\end{cases} \tag{9}$$

It is not difficult to see from Eq. (9) that for the better path whose length is less than the average value of the iterative path, the shorter the path length, the larger the pheromone increment of the path; for the poor path whose length is greater than the average value of the iterative path, the worse the path quality is, the more the pheromone reduction of the path is. Compared with the basic ant colony algorithm, the improved algorithm improves the attractiveness of the better path to the ant and reduces the interference of the poor path on the ant path selection through the pheromone differential update of the different quality paths, so that the algorithm can make full use of the information of the better path and maintain a better global search ability.

4.3 Random Perturbation is Applied to the Path Which May Fall into Local Optimum

When solving the path planning problem of mobile robot, ants start from the same source node to find destination nodes. Thus it is easier to fall into local optimum. If the optimal solution of the algorithm is not improved within the set C iterations, it is reasonable to suspect that the algorithm falls into local optimum. In order to solve this problem, we add the chaos disturbance quantity when adjust global pheromone in this paper, so that the solution can jump out of local optimum interval.

Chaos exists widely in the nature [16–19], and it has many features such as "randomness", "ergodicity" and "regularity" etc. It seems chaotic but has a delicate internal structure, and is extremely sensitive to the initial condition. It also can repeatedly traverse all the states in a certain range according to its own rule, so we can use these properties of chaotic motion to optimize the search. Logistic mapping is a typical chaotic system [20, 21], and its iterative formula is defined as follows:

$$z_{i+1} = \mu \cdot z_i \cdot (1 - z_i), i = 0, 1 \cdots, \mu \in (2, 4] \tag{10}$$

where μ is the control parameter, and when $\mu = 4, 0 \leq z_0 \leq 1$, Logistic is completely in a chaotic state.

If the value of objective function does not change optimal in the given C iteration, the algorithm may fall into local optimum, and global pheromone update formula (4) of the algorithm is adjusted as follows:

$$\tau_{ij}(t+1) = (1-\rho)\tau_{ij}(t) + \Delta\tau_{ij} + \xi z_{ij} \tag{11}$$

where, z_{ij} is the chaotic perturbation variable in [0, 1], by Eq. (10) and ξ is disturbance factor.

5 Algorithm Steps Description

The algorithm steps are as follows:

Step 1: According to the known static environment information, the environment model of robot path planning is established by grid method, and the grid serial number, starting point and ending point are set.

Step 2: Initialize the pheromone intensity of all the paths in the environment model. Set various parameters of the algorithm and the number m of the ants, ***DiedaiNum*** is the the maximum number of loop. Initial the number of loop $k = 0$.

Step 3: Set the current loop number $k = k+1$ and set the increment of the pheromone of various paths $\Delta\tau_{ij} = 0, t = 0$, m ants are deployed in the source node. Tabu table are generated for every ant, and source node is deployed in the tabu table.

Step 4: $t = t + 1$, for each ant l do not finish searching, according to the Eq. (1), select the next node j from the current node i(Among them, the calculation formula of heuristic information is the improved formula (8), if node j does not exist, then note the ant has finished searching, else ant l is deployed in the tabu table, and if node j is target node, then note the ant has finished searching,else continue, use the Eq. (4) and (5) to update the local pheromone.

Step 5: repeat Step 4 until the m ants all have finished searching, record all the qualified paths and the optimum paths until the current loop, and record all the pheromone and fitness function value of all the qualified paths. Find out the optimal path length lib and the worst path length lworth of this iteration, and calculate the average path length lave searched by all ants.

Step 6: If in the consecutive C iterations, the optimal path obtained by the algorithm is not significantly improved, then updating the pheromone of optimum paths according to the Eq. (11), otherwise updating the pheromone of optimum paths according to the Eq. (4) (Among them, the pheromone increment formula in the global pheromone updating formula is the improved formula (9)). If $\boldsymbol{k} < \boldsymbol{DiedaiNum}$, then go to step3, else go to step 7.

Step 7: output the optimum path $\boldsymbol{Path_{best}}$, end.

The flow chart of the algorithm is shown in Fig. 3.

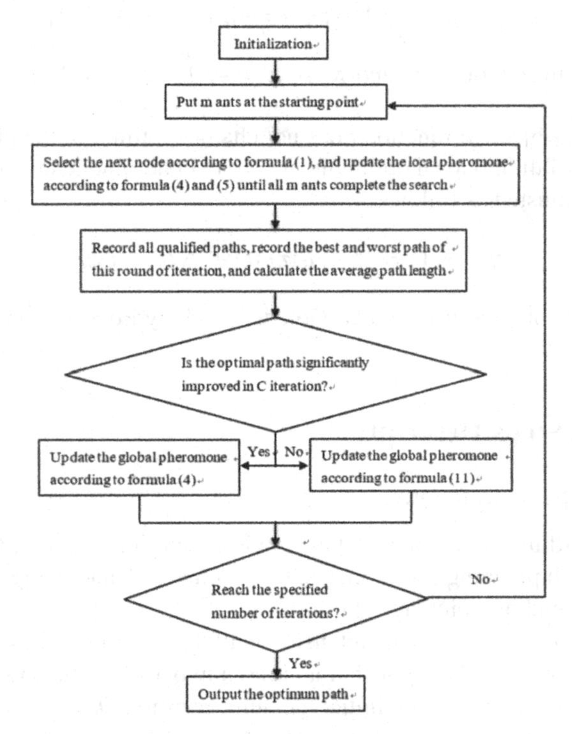

Fig. 3. AGACO flow chart

6 Simulation

6.1 Comparison Between Angle Guided Ant Colony Algorithm and Basic Ant Colony Algorithm (ACO) Guidance

The proposed AGACO is implemented in MATLAB, In order to validate the validity of AGACO, we selected an example to experimentize. Based on the grid method, the environment map of 20 × 20 as shown in Fig. 1 is established. The starting point of the robot is the upper left corner of the grid model, and the target point is the lower right corner. In the figure, black represents obstacle information and white represents obstacle free. Then the raster map is encoded.

Given there are service request of routing, the source node is the upper left corner, the target node is the lower right corner, the parameters is specified as: $\alpha = 1$, $\beta = 2$, $\rho_0 = 0.2$, $\rho = 0.3$, $Q = 2$, $\mu = 3.8$, z_0 take the random number between (0,1) and the chaotic disturbance coefficient $\xi = 1.06$, assign an initial value is 1 to the pheromone of each paths, the number of the ants deployed in the source node is m = 80, the number of iterations is 100.

Table 1 shows the results of 10 times path planning of angle guided ant colony algorithm and basic ant colony algorithm in the environment of Fig. 1. Among them, l_1 is the length of the optimal path found in each iteration of the angle guided ant colony algorithm, l_2 is the length of the optimal path found in each iteration of the basic ant colony algorithm. It can be seen from Table 1 that the angle guided ant colony algorithm has found the optimal solution 34.3848 four times, while the optimal solution found by the basic ant colony algorithm is 36.3848, which is not the optimal solution in this environment. Figure 4 is the optimal path found by ACO and AGACO. Table 1 also shows the time of 100 iterations of the two algorithms. It can be seen that AGACO converges faster, AGACO converges in the 36th generation, while ACO algorithm converges in the 47th generation, and does not converge to the optimal value. Considering the quality of time and solution, the performance of AGACO algorithm is better.

Table 1. Comparison of shortest path length between AGACO and ACO

Serial number	l_1	Time/s	l_2	Time/s
1	34.9706	46.7674	36.9706	61.4479
2	34.9706	44.0747	37.5563	56.1768
3	**34.3848**	45.4479	36.9706	57.2539
4	**34.3848**	46.6238	**36.3848**	60.6255
5	35.7990	47.4885	**36.3848**	**58.4885**
6	35.5563	44.2356	37.5563	57.3682
7	**34.3848**	**45.0628**	36.9706	61.5296
8	35.5563	48.6592	37.2132	55.3863
9	**34.3848**	47.5611	**36.3848**	60.2356
10	34.9706	46.2195	37.7990	61.8793
Average value	34.93626	46.2141	37.0191	59.0392

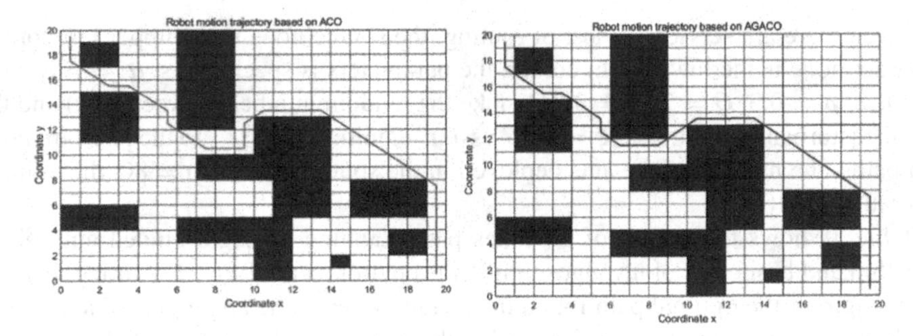

Fig. 4. Robot motion trajectory diagram of ACO and AGACO

Figure 5 is the convergence graph (the coordinate in the graph representatives the route length, and the abscissa representatives the number of iterations) of the two algorithms in the environment shown in Fig. 1. It can be seen from the figure that AGACO converges to the optimal solution of 34.3848 in the 36th generation, while ACO converges to the optimal solution of 36.3848 in the 47th generation. The simulation results show that the algorithm has strong local search capabilities, can jump out of the local optimal, and quickly converge to the global optimal solution. The algorithm is feasible and effective.

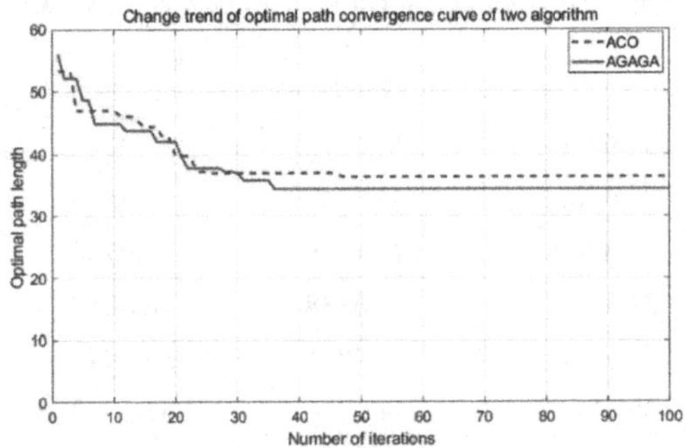

Fig. 5. The optimal path convergence curve of two algorithm

6.2 Comparison Between Angle Guided Ant Colony Algorithm and Particle Swarm Optimization (PSO)

This paper also compares with particle swarm optimization algorithm. Table 2 shows the results of 10 times path planning of angle guided ant colony algorithm and particle swarm optimization algorithm in the environment of Fig. 6. Among them, l_5 is the length of the optimal path found by AGACO every iteration, l_6 is the length of the optimal path found by PSO every iteration. It can be seen from Table 2 that AGACO can find the optimal solution of 28.6274 every time, while PSO finds the optimal solution of 29.2136, which is not the optimal solution in this environment. Figure 6 is the optimal path found by PSO and AGACO. Table 2 also shows the time of 100 iterations of the two algorithms.

It can be seen from Table 2: compared with PSO, agaco has shorter optimization time, shorter optimal path, and higher success rate, which further shows the effectiveness of the improved algorithm.

Table 2. Comparison of shortest path length between AGACO and PSO

Serial number	l_5	Time/s	l_6	Time/s
1	**28.6274**	20.5020	30.0416	22.4259
2	**28.6274**	20.3190	30.0416	22.5928
3	**28.6274**	20.2850	30.6274	23.8637
4	**28.6274**	20.1810	**29.2132**	22.9493
5	**28.6274**	**20.0730**	**29.2132**	**22.3000**
6	**28.6274**	20.1080	30.6274	21.9214
7	**28.6274**	20.4870	31.4558	22.1361
8	**28.6274**	20.4880	30.0416	21.3667
9	**28.6274**	20.6890	29.4558	21.2479
10	**28.6274**	20.3599	30.6274	23.6118
Average value	**28.6274**	20.3492	30.1345	22.4416

Figure 7 is the convergence graph (the coordinate in the graph representatives the route length, and the abscissa representatives the number of iterations) of the two algorithms in the environment shown in Fig. 6. It can be seen from the figure that AGACO converges to the optimal solution of 28.6274 in the 19st generation, while PSO finds the optimal solution of 29.2132, and converges in the 24th generation. AGACO has less convergence algebra, more stability and higher search efficiency, the effectiveness of the algorithm is further verified.

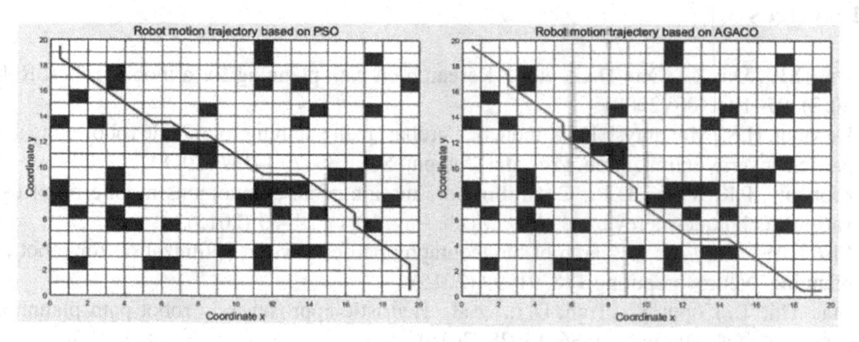

Fig. 6. Robot motion trajectory diagram of PSO and AGACO

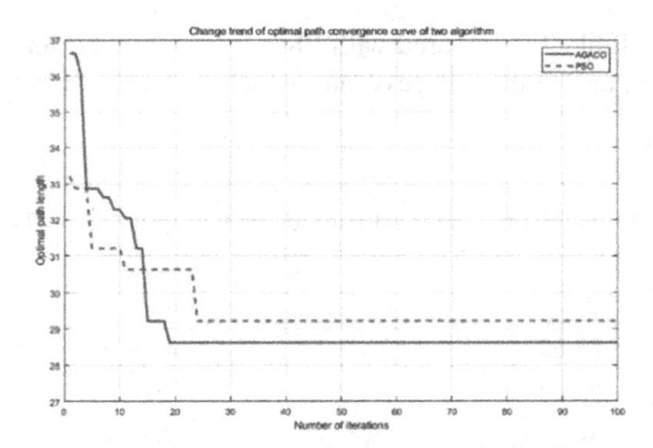

Fig. 7. The optimal path convergence curve of AGACO and PSO

7 Conclusion

In view of the contradiction between premature, stagnation and accelerated convergence of ant colony algorithm, combined with the characteristics of mobile robot path planning problem, the ant colony algorithm is improved. The angle relationship is introduced into the heuristic information of ant colony algorithm, and the node selection strategy of the algorithm is optimized. The pheromone update strategy is differentiated according to the degree of the path found, and the chaotic disturbance update mechanism of the pheromone is introduced, so that the algorithm can jump out of the local extreme value interval. The simulation results show that the algorithm in this paper is feasible and effective. How to improve the algorithm, and apply it to the more complex mobile robot path planning problem. It is worth intensive study in the future.

Acknowledgement. This work was supported by National Natural Science Foundation of China (No. 61866003).

References

1. Qu, D.K., Du, Z.J., Xu, D.G., et al.: Research on path planning for a mobile robot. Robot. **30**(2), 97–101,106 (2008)
2. Dewang, H.S., Mohanty, P.K., Kundu, S.: A robust path planning for mobile robot using smart particle swarm optimization. Procedia Comput. Sci. **133**, 290–297 (2018)
3. Mohanty, P.K., Parhi, D.R.: Controlling the motion of an autonomous mobile robot using various techniques: a review. J. Adv. Mech. Eng. **10**(1), 24–39 (2013)
4. Mo, H.W., Xu, L.F.: Research of biogeography particle swarm optimization for robot path planning. Neurocomputing **148**, 91–99 (2015)
5. Mac Thi, T., Copot, C., Tran, D.T., et al.: Heuristic approaches in robot path planning: a survey. Robot. Auton. Syst. **86**, 13–28 (2016)
6. Zhao, X., Wang, Z., Huang, C.K., et al.: Mobile robot path planning based on an improved A* algorithm. Robot **40**(6), 903–910 (2018)
7. Yu, Z.Z., Li, Q., Fan, Q.G.: Survey on application of bioinspired intelligent algorithms in path planning optimization of mobile robots. Appl. Res. Comput. **26**(11), 3210–3219 (2019)

8. Akka, K., Khaber, F.: Mobile robot path planning using an improved ant colony op-timization. Int. J. Adv. Robot. Syst. **15**(3), 1729881418774673 (2018).https://doi.org/10.1177/1729881418774673(2018)
9. Long, S., Gong, D., Dai, X., et al.: Mobile robot path planning based on ant colony algorithm with A* heuristic method. Front. Neurorobot. **26**, 13–15 (2019)
10. Faridi, A.Q., Sharma, S., Shukla, A., Tiwari, R., Dhar, J.: Multi-robot multi-target dynamic path planning using artificial bee colony and evolutionary programming in unknown environment. Intel. Serv. Robot. **11**(2), 171–186 (2018). https://doi.org/10.1007/s11370-017-0244-7
11. Kang, Y.X., Jiang, C.Y., Qin, Y.H., Ye, C.L.: Robot path planning and experiment with an improved PSO algorithm. Robot **42**(1), 71–78 (2020)
12. Colorni, A., Dorigo, M., Maniezzo, V., et al.: Distributed optimization by ant colonies. In: Varela, F., Bourgine, P. (eds.) Proceedings of the ECAL 1991, European Conference of Artificial Life, pp. 134–144. Elsevier, Paris (1991)
13. Dorigo M.: Optimization,learning and natural algorithms. Ph. D. Thesis, Department of Electronics, Politecnico di Milano. Italy (1992)
14. Dorigo, M., Maniezzo, V., Colorni, A.: Ant system: optimization by a colony of cooperating agents. IEEE Trans. Syst. Man Cybern. -Part **26**(1), 29–41 (1996)
15. Liu, J., Feng, S., Ren, J.H.: Directed D* algorithm for dynamic path planning of mobile robots. J. Zhejiang Univ. (Eng. Sci.) **54**(2), 291–300 (2019)
16. Huang, Y.S., Huang, H.: Chaos With Applications. Wuhan University Press, Wuhan (2005)
17. Shen, X.B., Zheng, K.F., Li, D.: New chaos-particle swarm optimization algorithm. J. Commun. **33**(1), 25–30, 37 (2012)
18. Tang, W., Li, D.P., Chen, X.Y.: ChenChaos theory and research on its applications. Autom. Electric Power Syst. **7** (37), 67–70 (2000)
19. Campbell, D.K.: Nonlinear science--from paradigms to practicalities. Adv. Mech. **19**(3), 376–392 (1989)
20. Zhu, Z.X.: Chaos in nonlinear dynamics. Adv. Mech. **14**(2), 376–392 (1989)
21. Liu, Z.: Nonlinear Dynamics and Chaos Elements. Northeast Normal University Press, Changchun (1994)

A Novel Intelligent Ant Colony System Based on Blockchain

Wei Wu[1], Haipeng Peng[1], Lixiang Li[1(✉)], H. Eugene Stanley[2], Licheng Wang[1], and Jürgen Kurths[3]

[1] State Key Laboratory of Networking and Switching Technology, Beijing University of Posts and Telecommunications, Beijng, China
{wuwei,penghaipeng,lixiang,lichengwang}@bupt.edu.cn

[2] Boston University, Boston, USA
hes@bu.edu

[3] Potsdam Institute for Climate Impact Research, Potsdam, Germany
Juergen.Kurths@pik-potsdam.de

Abstract. Swarm intelligence occurs when the collective behavior of low-level individuals and their local interactions form an overall pattern of uniform function. Incorporating swarm intelligence allows us to disregard global models when we explore collective cooperation systems that lack any central control. Blockchain is a key technology in the functioning of Bitcoin and combines network and cryptographic algorithms. A group of agents agrees on a particular status and records the protocol without controlling it. Blockchain and other distributed systems, such as ant colony systems, allow the building of "ants" that are more secure, flexible, and successful. We use the principle of blockchain technology and carry out ant colony research to solve three urgent problems. We use new security protocols, system implementations, and business models to generate ant swarm system scenarios. Finally we combine these two technologies to solve the problems of limitation and reduced future potential. Our work opens the door to new business models and approaches that allow ant colony technologies to be applied to a wide range of market applications.

Keywords: Blockchain · Ant colony system · Security · Service model

1 Introduction

In 2008, a white paper authored by Satoshi Nakamoto, "Bitcoin: A Peer-to-Peer Electronic Cash System" introduced Blockchain, which can be viewed as a global ledger of transactions recorded distributively by a network of agents. Because creating a block is time consuming, it prevents attackers from altering a blockchain [1]. It uses SHA256 cryptographic technique to output an unpredictable numeric value that encapsulates all transactions into a digital fingerprint. Every discrepancy in the input data containing transaction orders, quantities, and receivers produces different output data and generates different digital fingerprints. Since

Y. Tan et al. (Eds.): ICSI 2022, LNCS 13344, pp. 230–246, 2022.
https://doi.org/10.1007/978-3-031-09677-8_20

that introduction in 2008, much research has explored the benefit of blockchain technology when applied to such fields as intellectual property [2,3] and real estate [4–6]. Two of the most promising applications of blockchain technology are Bitcongress and colored coins [7].

New block transactions are valid and it will not affect the validity of previous transactions. Through double expenditures, a network agent ("miner") adds a new block to the end of the blockchain. Blockchain technology uses peer-to-peer networks and cryptographic algorithms to enable a group of agents to agree, securely record the action, and verify the agreement without control permissions. Because of its dispersion, robustness, and fault tolerance, blockchain technology can be applied to transportation, logistics and warehouse automation systems, and even the combination system of cloud computing and emergency networking.

Natural systems and bio-inspired models display swarms that are adaptable to different environments and tasks [8]. Ant swarms [9] are robust to failure and scalability because their coordination is simple and distributed. Thus their global behavior is not a given but is generated by local interactions among the ants. Because of this, research on ant colonies (as shown in Fig. 1) [10] has become increasingly popular [11].

There are many similarities between ant colonies and blockchain.

(1) Both are decentralized.
(2) The operation of both is distributed.
(3) The search mechanism of an ant colony is similar to the competition mechanism of blockchain.
(4) The pheromone mechanism of an ant colony is similar to the storage authentication mechanism of blockchain.

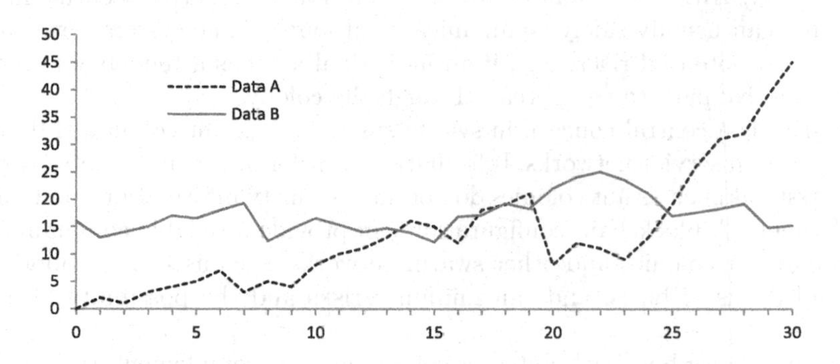

Fig. 1. The ant colony system.

We can thus integrate ant colonies and blockchain. The combination of blockchain technology and ant colony behavior provides solutions to many networking problems [12], and we next examine how blockchain can solve ant clustering problems. We will use transactions encapsulated in blocks to represent ants, which we treat as nodes in a network.

This work could open the door not only to new technical approaches, but also to new business models that make swarm ants technology suitable for innumerable market applications. By applying blockchain technology to ant colony systems and identifying ants as nodes and their shared information as blocks, we have the following discoveries.

(1) The study of ant colonies has produced new security models that increase data confidentiality and validation, and that are applicable to secure real-world applications.
(2) New ways of implementing and executing distributed decision making for collaborative tasks is found. Using a transaction ledger, ant agents vote and reach agreement. This increases ant behavior flexibility without increasing swarm complexity.
(3) Blockchain technology provides an infrastructure that ensures the ant colony system to follow its laws and maintain safety protocols during the integration process. This may have human applications and suggest new business models for group operation.

Thus using blockchain technology to study ant swarms opens the door to new technical approaches to real-world problems and suggests many financial market and business practice applications.

2 Ant Colony Block with Security

A central concern in the operation of an ant colony [13] is security. Previous research has found that each ant must be able to recognize and trust other ants in the colony. Although traditional ant colonies lack a central security mechanism, ants can usually safely communicate without it. The system nevertheless is subject to potential risks, e.g., if an individual ant has a tendency to attack, this destructive pattern can spread through the colony.

Security is a central concern in swarm systems, e.g., ant colonies [14], and a core concern in service networks. Information transfer occurs in all networks, but swarm systems such as ant colonies do not have security information communication channels. A blockchain configuration can provide a reliable communication channel for ant colonies and other swarm networks, such as swarm robotic systems and flocks of birds, and can minimize risks and the possibility of attack [15].

Figure 2 shows how each ant in a colony has two complementary keys. The public key provides core information to each ant via a blockchain communication channel [16] that encrypts the information and functions as a type of username. The private key validates an individual ant's identity, which is inaccessible to other ants, decrypts data, and functions as a type of password. Through these keys an ant can transmit three types of information, (i) their task, the tasks of other ants, and communications with other ants. This communication channel allows both point-to-point interactions and system broadcasts.

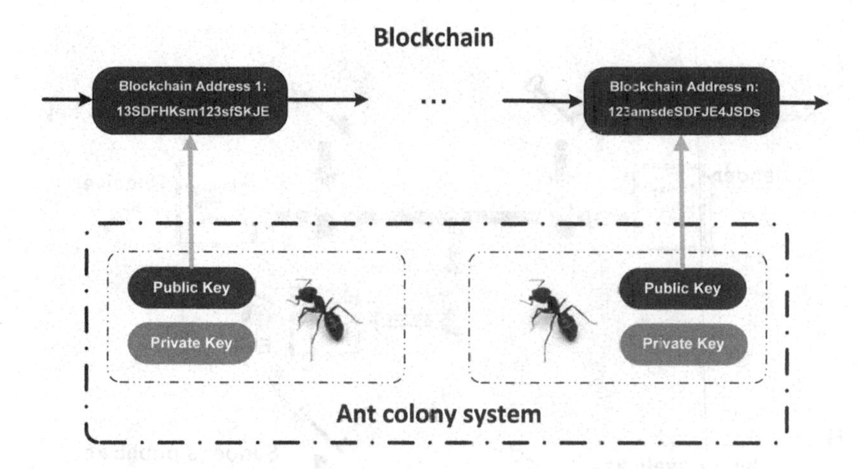

Fig. 2. Ants realize secure communication among each other through their own public key and private key.

Figure 3A shows the secure transmission of data in an ant colony system [17]. We use the sender's public key to encrypt the information to ensure that only the designated receiver has the access to it. Ants share their public keys with others. And any ant in the colony can send information to the intended receivers who have the access to their own private keys to decrypt the message. Because the private key is only known to the receiver, other ants do not have access to the message, and third-party ants cannot receive messages from that particular communication channel.

Along the transmitting of information, a sender can also prove its identity to all ants in colony. Figure 3B shows the framework of public key signature in which the sender uses its private key to sign a message, and subsequently the validity of the message as well as the signature are verified by all other ants in the swarm using the sender's public key.

As shown in Fig. 4, the content of a message encapsulated in a blockchain communication channel can only be read by the intended ant receiver, and thus the cryptographic primitive of signcryption can resist third-party attacks. We use this ant colony block configuration to build an ant colony service system, which can also be applied to other systems, such as swarm robotics and flocks of birds. For example, giving each robot in a system a pair of keys allows them to communicate with each other more securely.

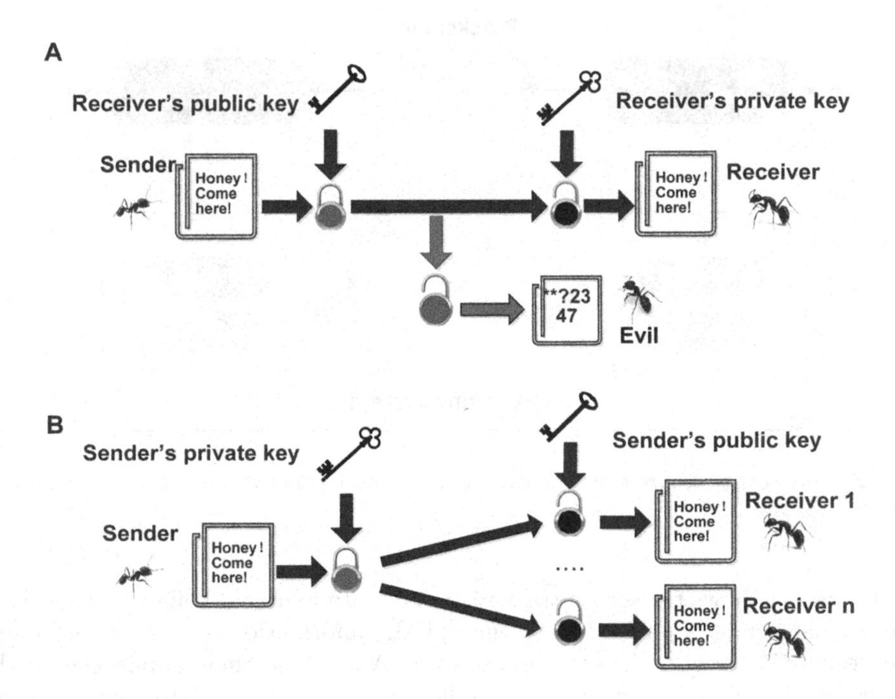

Fig. 3. Two different cryptographic systems based on public and private keys. A) Public key encryption makes it impossible for eavesdropping ants to read real content. B) Private key signature system not only confirms the integrity of the transmitted information, but also confirms the identity of the sending ant.

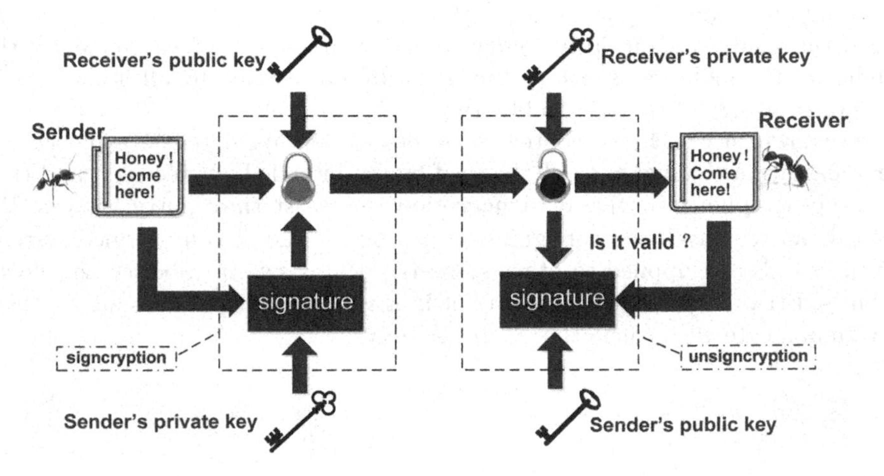

Fig. 4. Public signcryption system not only realizes the confidentiality of the transferred contents, but also enables the receiver to verify whether the contents come from the designate ants.

3 New System Models Based on Blockchain and Ant Colony

We can use this ant colony system to solve many real-world problems. For example, we can improve both individual initiative and colony cooperation [18].

3.1 Collaborative Service Model (CSM)

Because service collaboration is essential in an ant colony [19], we propose a collaborative service model (CSM). In the life of an ant colony, there are many service patterns For instance, whenever the queen wants food, she secretes a pheromone that enables workers to locate her and bring food to her. This pattern is also found in machine networks, e.g., multiple addresses are needed in a cluster collaboration. In ant colony systems, we also need to know how ants receive requests and how a worker ant is chosen to undertake certain tasks.

The CSM helps us simplify the multiple addresses involved in collaborative swarm missions. The part of the transaction is a request for help that is sent to the ant colony's service system. Then a worker ant is designated to respond to the queen's invocation of the transaction. This action unlocks such information as the location of the queen, or the token included in the transaction address to complete the action. Both collaborative planning and autonomous behavior can occur in the ant colony service system.

The use of blockchain technology provides additional benefits to the ant colony using the CSM approach. Because all protocols and all related transactions are stored in the blockchain, the mechanism of automatically synchronizing of blockchain enables other ants to access to all protocols and information previously created and stored in the blockchain, and thus the training for newly entering ants in the colony is saved.

Traditionally, information pheromone plays an important role in the communication process of ants. Ants realize mutual information exchanging via information pheromones exchanging, and thus achieve group intelligence cooperation. There is a set of glands in the ants. They use different chemicals (pheromones) to convey more than 20 kinds of meanings, such as: "There is food in front", "I need help", "I am hungry" and so on. This information can only be normally identified and authenticated by the ants within the same colony. For simple tasks, such as when the queen is hungry, it will excrete the pheromone expression message "I am hungry." Then, other ants will help it to eat after interpreting the information. This means a completion of the cooperation. That is, the queen uses her private key to sign the pheromone, and other ants use the queen's public key to identify the pheromone. This kind of process indicates the completion of a simple cooperation.

In the completion of complex tasks, such as nesting, the self-organization cooperation of ant colony plays a powerful role. After an ant secretes pheromone to send out the message "I need help," other nearby ants in the colony identify the information, and secrete pheromones to reinforce the information again, further expressing the message "I need help". As the number of ants increases, more

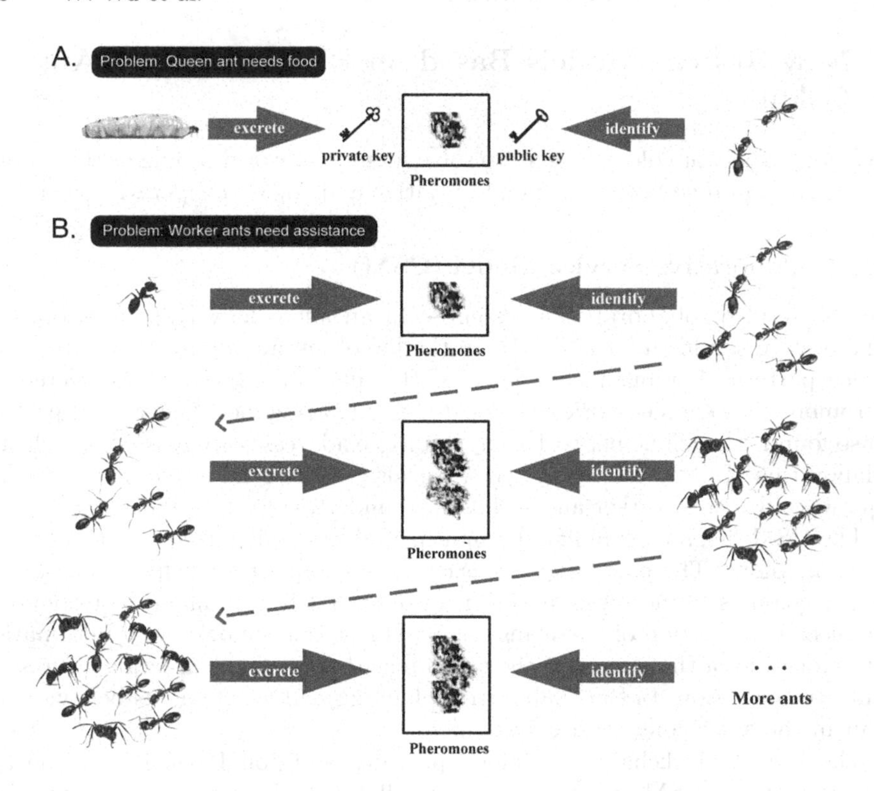

Fig. 5. Potential capabilities of threshold-signature addresses in swarm collaborative missions. An ant queen has the necessity to avoid obstacle, hungry, etc. A part of the signature of the transactions that convey queen's requests for help, is created and distributed to the service system of ant colony. Then, some worker ants are signed in response the invocation of the transaction. And they will unlock related information such as queen's location, or even the token that is included in the threshold-signature address to complete the action. Collaborative planning and more autonomous behavior can occur in this ant colony. For example, positive feedback mechanism of pheromone makes ant colonies to complete self-organization cooperation.

pheromones will attract more ants to cooperate. This positive feedback mechanism (see Fig. 5) of pheromones makes ant colony to complete self-organization cooperation.

Because of the limited scope of the effects of pheromones, the pheromone-based cooperation model can be considered as a cooperation mode based on local interaction, just like the alliance chain cooperation mode in terms of blockchain technology. The introduction of public chain into ant colony intelligence not only enhances the security of ant colony information transmission, but also introduces a new information dissemination mode - bulletin board mode - a global information dissemination mode for ant colonies. In bulletin board mode, the blockchain can be regarded as a bulletin board with cryptographic protection, an in-erasable and growing bulletin board that allows all ants with keys to share existing information on the blockchain, instead of local information only. The bulletin board propagation mode of information and a large amount of shared information will further promote new modes of intelligent cooperation of ant colonies, such as network-wide task division mode and bidding model.

Applying blockchain technology to ant colonies can also allow a more advanced collaboration model by using threshold-signature ants [20]. Threshold-signature requires a valid signature towards some message is in fact come from multiple valid signature slides towards the same message and each slide is produced by a valid private key. Complex collaboration tasks are specifically designed for heterogeneous robot colonies that are easily formalized, published, and performed.

3.2 Decision-as-a-Service (DaaS) Model

The distributed decision-making algorithm is essential in the development of a swarm intelligence system. For example, it can use ant swarms to connect to a distributed sensor [21] that enables agents to access information from multiple viewpoints and improve data quality. Ants in the colony must globally agree on goals in order to form shapes and avoid obstacles. Thus a distributed decision making protocol is needed to ensure consistent overall results [22].

Although collective decision making strategies have been used in such ant applications as election, dynamic task allocation and obstacle avoidance, we still do not know how to deploy a large number of distributed decision-making agents. The trade-off between speed and accuracy must be taken into account before the process of collective decision-making can be deployed. We thus must know how to make ant decision-making in the distributed system more autonomous, flexible, and responsive to new challenges. Using blockchain ensures that all participants in a decentralized network will have the same worldview. For example, blockchain allows a distributed voting system to be built that enables swarms to reach agreements.

The Daas model allows blockchain technology to be used to enable ant swarm decision making. One swarm member perceives an object of interest, creates two addresses that represent possible options, and registers the options in the blockchain. The swarm then votes on which option to choose. When a cluster member needs an agreement, the swarm issues special transactions and creates the ant colony address associated with each option open for selection. After the information is written in a block, other cluster members can vote, for example, to move a token to the address corresponding to the selected option. This kind of distributed decision making protocols can be quickly and securely acquired and can be auditable, and where every ant can monitor the voting process (as shown in Fig. 6A). Obviously, the introduction of the public blockchain into the ant colony intelligence will have a profound impact on collective decision making, making it possible to make a wider range of voting decisions, especially for large-scale inter-colonies decision-making with competition.

In the collective decision making of ant colony, positive feedback caused by pheromone plays a magnifying effect, such that slight changes in colony preferences (for example, referendum preferences), changes in behavioral characteristics (such as chemotaxis), or changes in individual knowledge. All these have an important impact on collective decision making. This kind of positive feedback is very important for recruiting teammates (such as recruiting teammates to prey and establish the best path). But the intelligent decision is not only to recruit teammates, but also to adjust the number of recruits based on available resources. If only the amplification effect is only possible, it can only lead to recruitment. More and more ants make the decision-making process uncontrollable and make it impossible to decide the best number of ants. Negative feedback, such as depletion of food or overcrowding of food sites or near saturation of the nest storage room, counteracts or "weakens" the amplification effect.

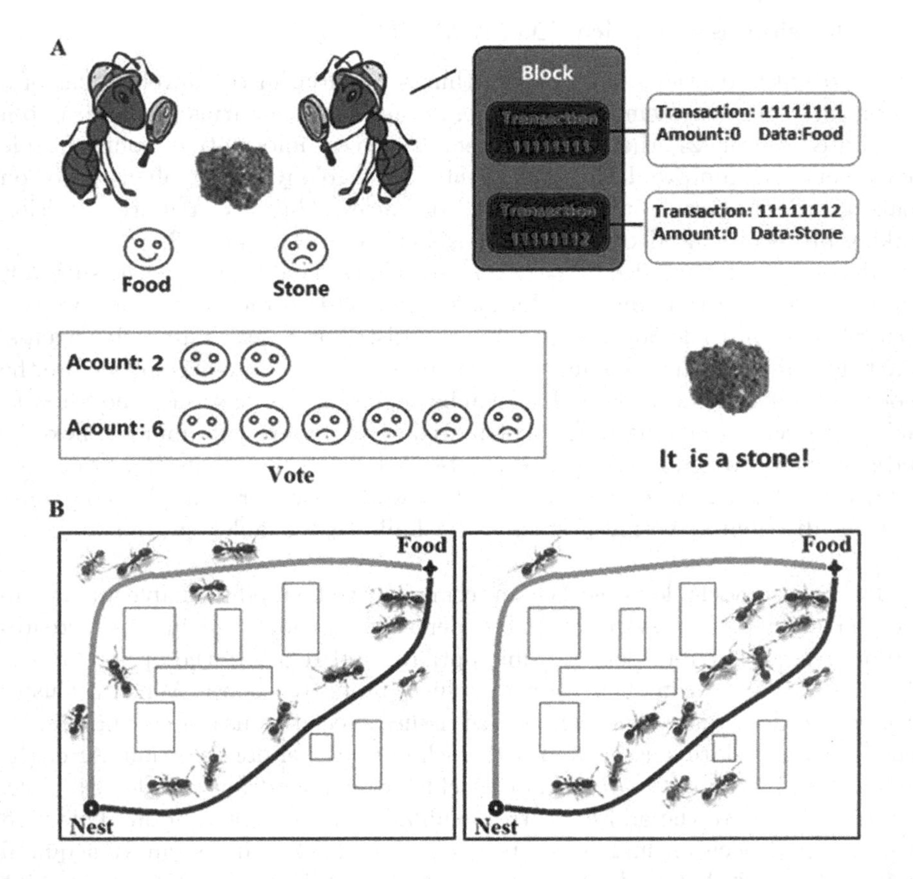

Fig. 6. Two models of decision making in ant swarms. A) Ants directly cast their votes through blockchains and each ant can vote only once. B) Ant indirectly cast their votes via excreting pheromones: The more pheromones on the path, the greater the probability that ants will choose this path, where each ant is allowed to vote multiple times. Finally, the candidate, represented by the optimal path, is selected and the ant colony realizes an intelligent decision.

The weakening mechanism of negative feedback provides a stabilizing force and plays a regulatory role by adjusting the number of foraging eaters based on factors such as the number and size of the food, the size of the storage room, and so on. Obviously, the subtle interaction between this positive feedback and negative feedback will lead to the emergence of the best model and intelligent collective decision making (as shown in Fig. 6B). It can be expected that the positive feedback and negative feedback mechanism of the ant colony combined with the blockchain will have significant effects and will lead to large-scale intelligent decision making across populations.

3.3 Sensing-as-a-Service (SaaS) Model

In our application of blockchain to topics [23] that reach beyond currency, we do not forget that the primary application of blockchain technology is economic, and that what we have discovered in our ant colony study can be applied back to the economy. Thus the findings when applying blockchain technology to ant colonies are also applicable to industrial and economic research [24].

One of the applications that the ant colony behavior is adopted to economics uses the data exchange process that occurs between an ant and a data requester. Service is a new business metric applied to the Internet of Things. The SaaS model creates multi-sided markets that produce sensor data, i.e., customers pay for data provided by sensors [25]. This model is analogous to the control area of the intelligent sensor network of a city that requires users to develop more flexible and adaptive control techniques.

We combine swarm ants and blockchain technology and develop two SaaS models: Customer customization model and sensor recommendation model (as shown in Fig. 7). By the former, individual ants are registered as a group that can be found by requester, and the sequesters customized services to sensing ants according to different tasks. By the latter, individual sequesters are registered as a group that can be found by sensing ants, and the sensing ants recommend data, active sensed by themselves, to potential customers. In more complex scenarios, ant swarms can use the distributed collaboration described in subsection CSM. The requester obtains the access to this data through ant transactions, and the perceived services become available to the swarm and individual ants.

Blockchain technology plays an essential role for competitive companies, which allows an appropriate competitive framework for group interaction by taking into account transaction sequence and timestamp considerations. For example, multiple different tasks and multiple working ants coming from competitive swarms can be matched via a blockchain-based bidding mechanism.

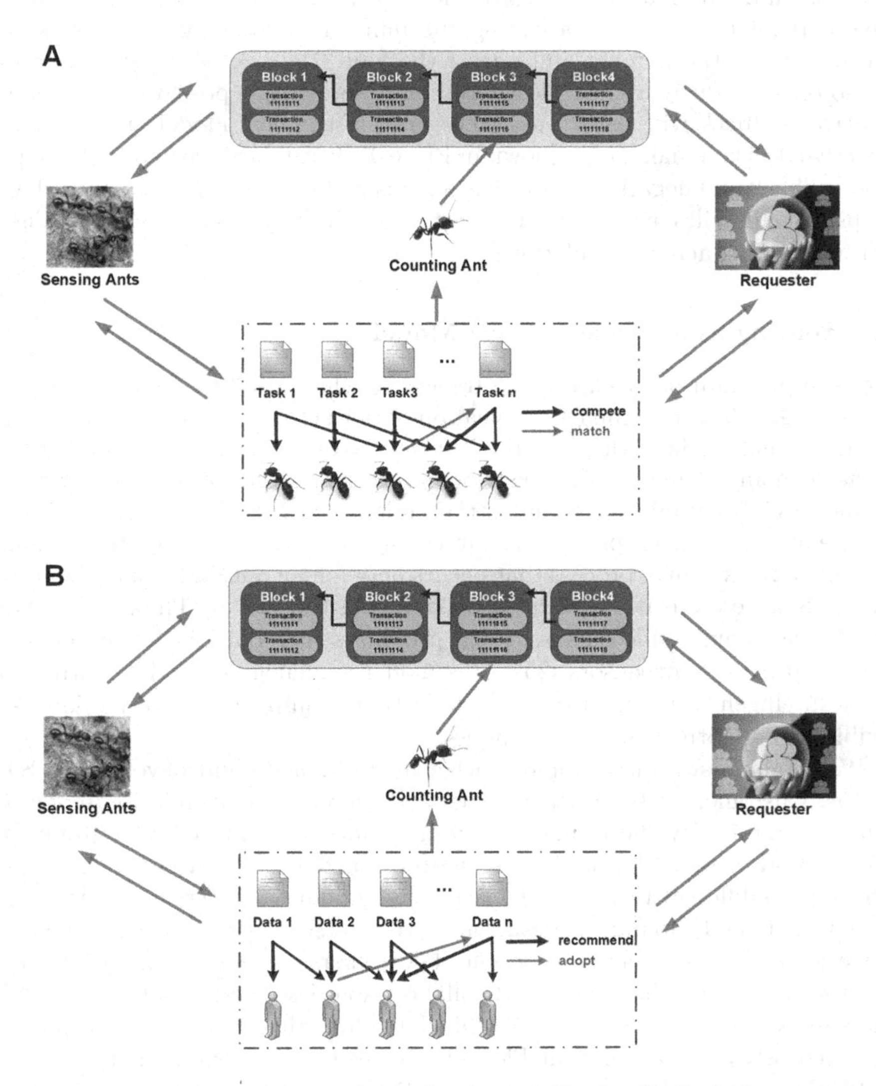

Fig. 7. Outlines of the working model in developing effective SaaS platforms via combining swarm ants and blockchain technology. The requester ant asks for the full list of these ants and their perceived services, multiple tasks and multiple worker ants are matched via a blockchain-based bidding mechanism. And all the information is recorded into blockchain via some counting ants whom are selected via certain consensus mechanism, say wins out in solving some difficult problem via ant swarm intelligent algorithms. This model can be further divided into two sub-models: customer customization model and sensor recommendation model. The former focuses on passive sensing in the sense that the sensing task is invoked by customers, while the latter emphasizes active sensing even without sensing requests from customers.

4 Discussion

Blockchain technology enables a group of agents, via combing a peer-to-peer communication network and cryptographic algorithms, to reach agreements on particular problems and securely record these agreements distributively without a centralized control mechanism. Although the technology is still in its infancy, it has been able to extend functionality beyond its original application and, in conjunction with others, create a national model for emerging technologies. The latest developments in the field have increased the research focus on ant colony systems. Research on ant colony systems helps us understand agent autonomy, decentralized control, and the emergence of collectives.

Blockchain technology and swarm intelligence have a lot common when applied to certain fields. Although swarm intelligence has been applied to optimization field, and blockchain technology is widely used in finance, they still have shortcomings. In response we suggest two improvements related to, (i) proof of work, and (ii) system implementation.

4.1 Proof of Work

The process of calculation attempts made by miners to seek acceptable answers is called proof of work (POW) [26]. For instance, C is a target value set by the blockchain system, and we use *nonce* as a POW counter, i.e., the initial value of *nonce* is 0, and the value *nonce* is increased by 1 after each calculation. When the miner obtains a calculation result that is less than C, the process of calculation is considered to be done and POW is calculation amounts in order to get such answer. The secure hash algorithm (SHA) is commonly used to determine Bitcoin's proof of work. For example, we add an integer value *nonce* to the end of the string "Hello, world!" and use SHA256 until there is "0000" at the beginning of the string. We increase the *nonce* value to achieve a new string by SHA256. Using this rule, we carry out 4251 calculations to find this Hash value.

SHA256 is a standard Bitcoin method and has many advantages, but it has few applications to swarm intelligence, including that found in ant colonies. Thus we propose a new POW such that SHA256 can be applied to environments other than Bitcoin. We use ant colony optimization (ACO) in a new POW to identify optimal paths [9]. Its mathematical form is

$$\tau_{ij}(t+n) = \rho \times \tau_{ij}(t) + \Delta\tau_{ij}, \tag{1}$$

$$\Delta\tau_{ij} = \sum_{k=1}^{m} \Delta\tau_{ij}^{k}, \tag{2}$$

and

$$p_{ij}^{k}(t) = \begin{cases} \frac{(\tau_{ij}(t))^{\alpha}(\eta_{ij})^{\beta}}{\sum_{k \in allowed_k} (\tau_{ik}(t))^{\alpha}(\eta_{ik})^{\beta}} & \text{if} :j \in allowed_k \\ 0 & \text{else} \end{cases}, \tag{3}$$

where $\tau_{ij}(t)$ is the pheromone intensity on path (i, j) at time t, $\rho\,(0 \leq \rho \leq 1)$ is a constant that represents the redundancy after trace pheromone volatilization, $1-\rho$ is the pheromone volatilization between time $(t,t+1)$, $\Delta\tau_{ij}^{k}$ is the pheromone number that ant k leaves under unit path length in time interval $(t, t+n)$, $\eta_{ij} = \frac{1}{d_{ij}}$ is the visibility of path (i, j), α and β are the parameters of controlling the path and the visibility, and $allowed_k = \{N - tabu_k\}$, where $tabu_k(s)$ is element s in the tabu table that the ants access to city i in a recent trip. The parameters in the ant colony algorithm have a great impact on the performance of the algorithm [27]. Specifically, the value of α indicates that the importance degree of information related to each node. That is to say, the larger the value of α becomes, the more likely an ant is to choose the route it has passed before. The value of β indicates that the importance degree of heuristic information. Namely, the larger the value of β becomes, the more likely an ant is to choose the route of a nearby city. And the value of ρ reflects the strength of the influence between individuals in an ant colony, which plays a key role in determining the convergence rate of the ant colony algorithm.

In the ant colony algorithm, ants are placed in different cities, and each side has initial pheromone intensity values $\tau_{ij}(0)$. The first element of an tabu table for each ant is set at the beginning city. The ant then moves from city i to city j, using the transition probability function between two cities to select the city. The tabu table is filled after all ants have finished a cycle. We then calculate the total length L_k that each ant k has traveled and update τ_{ij} using Eq. (1). We also save the shortest path found by the ants and empty all tabu tables. We repeat this process until the cyclic counter reaches a maximum $NC_{\max}$ or all ants are following the same route.

We thus provide a new POW method based on ACO. In the ant colony optimization service system, a new block is created every 10 min, and we adjust the difficulty of the optimization algorithm to ensure the creation time of a block. A unique block identifier is the optimal path found by ACO, and we use this optimal path instead of the hash to identify the new block. Each block thus has an optimal path to the previous block in sequence.

4.2 A Semi-centralized Ant Colony System Based on Blockchain

Combining blockchain technology and swarm control techniques allows us to solve problems that go beyond security and distributed decision-making [28]. Little research has focused on a combination of different ant colony behaviors, but blockchain technology could allow multiple blockchains to be hierarchically connected ("tacks") to enable ant colony agents to alter group behaviors using their own blockchain and such parameters as diversity and permission values.

Open source projects such as multi-chain combinations and fixed sidechain algorithms could provide a way of creating multiple blockchains, which connects ledgers that run in parallel. We here use a typical blockchain configuration to examine the distribution diversity of network agents in the ant colony. Here the control behind the transaction decision is part of the blockchain, and it is decentralized and distributed.

Figure 8 shows how a semi-concentrated blockchain ant colony system works [29]. In addition to providing competitive strategy, we also introduce a mechanism by which leaders can direct the fulfilling of tasks. Usually a leader is a mature ant or the ant queen. Since individual ants have a weak storage and computing power, most blocks are stored in the nest, and complex operations, e.g., a query search for blocks, are also carried out in the nest. This greatly improves the ability of the cluster to function.

Cluster members can also create a parallel fixed chain by moving a portion of their assets to another chain. The different parameters of this sidechain can be optimized using parameters in the main chain. Figure 8 shows how a centralized

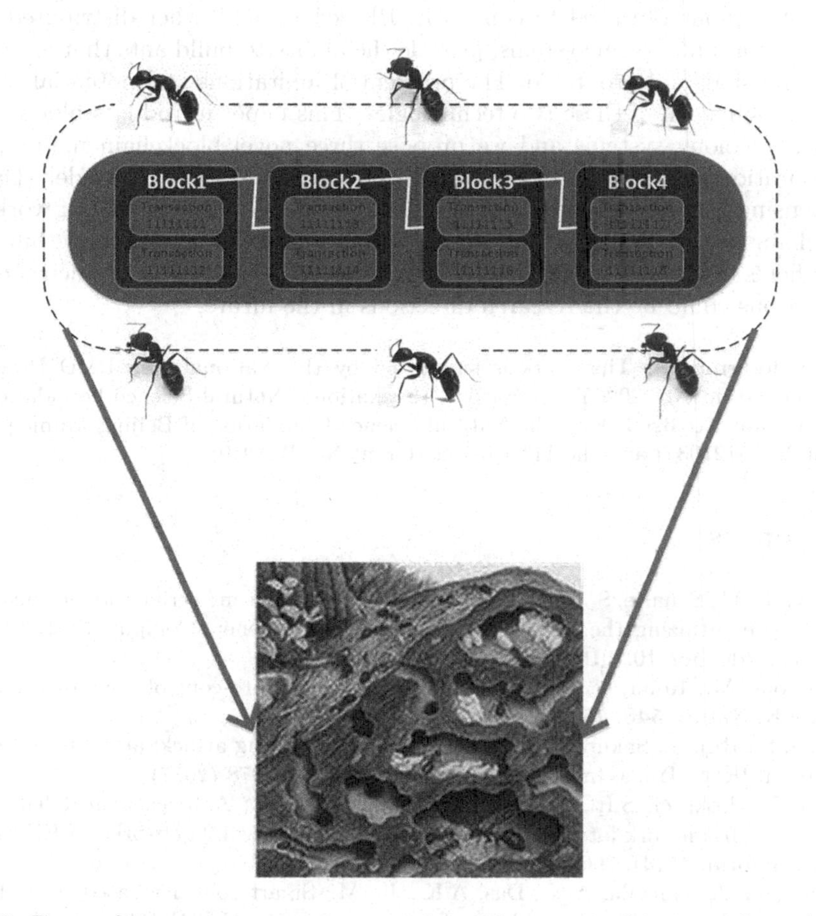

Fig. 8. By sending a transaction to a special address, several agents of an already established blockchain might create different branches of the blockchain ledgers. The mining diversity parameter is changed to produce a single miner configuration. This configuration emphasizes a centralized approach in which only the miner can take control of the block creation process, and thus transforming the blockchain into a leader-follower control scheme. If necessary, these branches can achieve consistent again via certain consensus mechanism.

mining scheme can be transformed from a decentralized mining scheme. Thus a single ant's control can generate different ant behaviors without increasing the complexity of the ant controller.

5 Summarization

The characteristic of swarm intelligence is that the collective behavior of low-level individuals and their environment local interaction which forms the overall pattern of uniform function. The blockchain is an emerging technology in bitcoin field. A group of agents can agree on a particular status and record the protocol without the need to control it. Blockchain and other distributed systems, such as ant swarm systems, provide the ability to build ants that are more secure, flexible, and profitable. The problem of limitations and potential future is the combination of these two technologies. This paper introduces blockchain-based ant colony systems and we propose three novel blockchain models, i.e. Collaboration-as-a-Service model (CaaS), Decision-as-a-Service model (DaaS) and Sensing-as-a-Service model (SaaS), and ant-colony-based proof of work for blockchains, which could be applied not only in financial transactions, but also in the fields of wireless sensor networks, internet of things, etc. And such specific applications could be the research directions in the future.

Acknowledgments. This work is supported by the National Key R&D Program of China (Grant No. 2020YFB1805403), the National Natural Science Foundation of China (Grant No. 62032002), the Natural Science Foundation of Beijing Municipality (Grant No. M21034) and the 111 Project (Grant No. B21049).

References

1. Prybila, C., Schulte, S., Hochreiner, C., Weber, I.: Runtime verification for business processes utilizing the Bitcoin blockchain. Future Gener. Comput. Syst. (2017). https://doi.org/10.1016/j.future.2017.08.024
2. Andoni, M., Robu, V., Flynn, D.: Blockchains: crypto-control your own energy supply. Nature **548**, 158–158 (2017)
3. Bag, S., Ruj, S., Sakurai, K.: Bitcoin block withholding attack: analysis and mitigation. IEEE Trans. Inf. Forensics Secur. **12**, 1967–1978 (2017)
4. Qu, Y., Pokhrel, S.R., Garg, S., Gao, L., Xiang, Y.: A blockchained federated learning framework for cognitive computing in industry 4.0 networks. IEEE Trans. Ind. Inform. **17**(4), 2964–2973 (2021)
5. Vangala, A., Sutrala, A.K., Das, A.K., Jo, M.: Smart contract-based blockchain-envisioned authentication scheme for smart farming. IEEE Internet Things J. **8**(13), 10792–10806 (2021)
6. Li, Y., Cao, B., Liang, L., Mao, D., Zhang, L.: Block access control in wireless blockchain network: design, modeling and analysis. IEEE Trans. Veh. Technol. **70**(9), 9258–9272 (2021)
7. Vranken, H.: Sustainability of bitcoin and blockchains. Curr. Opin. Environ. Sustain. **28**, 1–9 (2017)

8. Kennedy, J.: Review of Engelbrecht's fundamentals of computational swarm intelligence. Genet. Program Evolvable Mach. **8**, 107–109 (2007)
9. Carabaza, S.P., Besada-Portas, E., Lopez-Orozco, J.A., de la Cruz, J.M.: Ant colony optimization for multi-UAV minimum time search in uncertain domains. Appl. Soft Comput. (2017). https://doi.org/10.1016/j.asoc.2017.09.009
10. Zhang, W., Gong, X., Han, G., Zhao, Y.: An improved ant colony algorithm for path planning in one scenic area with many spots. IEEE Access **5**, 13260–13269 (2017)
11. Boubertakh, H.: Knowledge-based ant colony optimization method to design fuzzy proportional integral derivative controllers. J. Comput. Syst. Sci. Int. **56**(4), 681–700 (2017). https://doi.org/10.1134/S1064230717040050
12. Korb, O.: Efficient ant colony optimization algorithms for structure- and ligand-based drug design. Chem. Cent. J. **3**, O10 (2009)
13. Zuo, L., Shu, L., Dong, S., Zhu, C., Hara, T.: A multi-objective optimization scheduling method based on the ant colony algorithm in cloud computing. IEEE Access **3**, 2687–2699 (2015)
14. Zhou, J., et al.: A multi-objective multi-population ant colony optimization for economic emission dispatch considering power system security. Appl. Math. Model. **45**, 684–704 (2017)
15. Aste, T., Tasca, P., Di Matteo, T.: Blockchain technologies: the foreseeable impact on society and industry. Computer **50**, 18–28 (2017)
16. Li, X., Jiang, P., Chen, T., Luo, X., Wen, Q.: A Survey on the security of blockchain systems. Future Gener. Comput. Syst. (2017). https://doi.org/10.1016/j.future.2017.08.020
17. Kamali, M.Z.M., Kumaresan, N., Ratnavel, K.: Solving differential equations with ant colony programming. Appl. Math. Model. **3910**, 3150–3163 (2015)
18. Hsin, H.K., Chang, E.J., Lin, C.A., Wu, A.Y.: Ant colony optimization-based fault-aware routing in mesh-based network-on-chip systems. IEEE Trans. Comput.-Aided Des. Integr. Circuits Syst. **33**, 1693–1705 (2014)
19. Gaifang, D., Xueliang, F., Honghui, L., Pengfei, X.: Cooperative ant colony-genetic algorithm based on spark. Comput. Electr. Eng. **60**, 66–75 (2017)
20. Yue, W., Ma, W., Miao, Q., Wang, S.: Multimodal continuous ant colony optimization for multisensor remote sensing image registration with local search. Swarm Evol. Comput. (2017). https://doi.org/10.1016/j.swevo.2017.07.004
21. Arnay, R., Fumero, F., Sigut, J.: Ant colony optimization-based method for optic cup segmentation in retinal images. Appl. Soft Comput. **52**, 409–417 (2017)
22. Naeem, M., Pareek, U., Lee, D.C.: Swarm intelligence for sensor selection problems. IEEE Sens. J. **128**, 2577–2585 (2012)
23. Yang, Z., Sun, J., Zhang, Y., Wang, Y.: Understanding SaaS adoption from the perspective of organizational. Comput. Hum. Behav. **45**, 254–264 (2015)
24. Byk, J., Del-Claro, K.: Ant-plant interaction in the Neotropical savanna. Popul. Ecol. **53**, 327–332 (2011)
25. Sharif, S., Watson, P., Taheri, J., Nepal, S., Zomaya, A.Y.: Privacy-aware scheduling SaaS in high performance computing environments. IEEE Trans. Parallel Distrib. Syst. **28**, 1176–1188 (2017)
26. Shi, N.: A new proof-of-work mechanism for bitcoin. Financ. Innov. **2**(1), 1–8 (2016). https://doi.org/10.1186/s40854-016-0045-6
27. Jangra, R., Kait, R.: Analysis and comparison among ant system; ant colony system and max-min ant system with different parameters setting. In: 2017 3rd International Conference on Computational Intelligence & Communication Technology (CICT), pp. 1–4 (2017)

28. Samanta, C.K., Padhy, S.K., Panigrahi, S.P., Panigrahi, B.K.: Hybrid swarm intelligence methods for energy management in hybrid electric vehicles. IET Electr. Syst. Transp. **3**, 22–29 (2013)
29. Filho, J.C.M., de Souza, R.N., Abrao, T.: Ant colony input parameters optimization for multiuser detection in DS/CDMA systems. IEEE Latin Am. Trans. **12**, 1355–1364 (2014)

Implementation of the Test Data Generation Algorithm Based on the Ant Colony Optimization Pheromone Model

Serdyukov Konstantin(✉) and Tatyana Avdeenko

Novosibirsk State Technical University, 630073 Novosibirsk, Russia
{serdyukov.2011,avdeenko}@corp.nstu.ru

Abstract. In present paper we investigate an approach to intelligent support of the software white box testing process based on evolutionary paradigm. As a part of this approach, we solve the urgent problem of generating the optimal set of test data that provides maximum statement coverage of the code when it is used in the testing process. Earlier approaches that have been explored have shown the need to adjust the value of k for different programs, since its value has a significant impact on the quality of the fitted test data. To eliminate this problem, we propose to use the pheromone model, which is used in Ant Colony Optimizations in order to shift the focus of data generation to unexplored paths.

Keywords: Genetic algorithm · Test data generation · Fitness function

1 Introduction

The classic software engineering lifecycle includes such stages as reliability requirements engineering, design resulting in the software architecture, programming (coding), testing, debugging and maintenance. Software testing, defined as a process of the testing program evaluation, aimed at verifying actual behavior of the program code and its expected behavior on a special set of tests (the so-called test cases). Testing is one of the most expensive and labour-consuming stage and can take up to 40–60% of the total software development time [1].

Test data generation is a complex and time-consuming process which needs a lot of effort and budget. Therefore, automation of this process, at least partially, is an urgent research problem, the solution of which could improve the efficiency of the software testing. One of the goals of the automatic test data generation is to create such a multitude of test data that would ensure a sufficient level of quality of the final product by checking most of the various code paths, i.e. would provide maximum code coverage to satisfy some criteria (for example, statement or branch coverage).

There are different approaches to solving the problem of automating test data generation. For example, there are approaches based on a constraint-based algorithm [2], constraint logic programming and symbolic execution [3] and constraint handling rules

Y. Tan et al. (Eds.): ICSI 2022, LNCS 13344, pp. 247–258, 2022.
https://doi.org/10.1007/978-3-031-09677-8_21

[4]. To automate the software testing process heuristic approach for the Data-flow diagram could be used. Studies of automation methods using this kind of diagrams were carried out in papers [5–8]. Some of the researchers suggest using hybrid approaches. For example, in [9] was proposed to combine Random Strategy, Dynamic Symbolic Execution, and Search Based Strategy. The paper [10] proposes a hybrid approach based on Memetic Algorithm for generating test data. The work [11] compares different methods for generating test data, including genetic algorithms, random search, and other heuristic methods.

To improve the process of test data generation, some researchers suggest to use UML diagrams in collaboration with various methods. Genetic algorithm can be used to find the critical path in the program [12, 13] or to select test data to many parallel paths [14]. Also, the code could be displayed in the form of Classification-Tree Method [15, 16]. The developed ADDICT prototype (AutomateD test Data generation using the Integrated Classification-Tree methodology) was studied in the paper [17].

As follows from the above, many researchers focus on evolutionary approaches to solving this problem, in particular, on the genetic algorithm and its hybrid modifications. However, it should be noted that traditionally genetic algorithm is used to find the most fitted chromosome, which is a set of test data that ensures passage along the most complex (long) path in the Control Flow Graph [18, 19]. Many data sets that provide maximum code coverage can be found by repeating this procedure multiple times with preliminary zeroing of the code operation weights corresponding to the chromosomes found earlier [20]. So, the fitness function of the genetic algorithm has a simple form, but the process of finding all the data sets is quite long and non-optimal. To increase speed of the test data generation was proposed to include additional term in fitness function [21]. Additional term is used to increase diversity of the algorithm and the convergence speed. It was noted, that the relation coefficient between two term is must be found manually for every testing code. To increase the universality, this coefficient should have less impact on the final result, so we propose to modify the algorithm with the pheromone model adopted from Ant Colony Optimization. Our research confirmed greater diversity of the test data in form of better coverage rate for any value of the coefficient.

The paper is organized as follows. Section 1 gives introduction to the problem and literature review. Section 2 discusses theoretical issues of the research and describe the method of test data generation. In Sect. 3 we present the results of the conducted research. Section 4 provides the conclusions.

2 Theoretical Background

2.1 Genetic Algorithm for Test Data Generation

Genetic Algorithm (GA) borrows its idea and terminology from the biological world. In such a way, it uses different representations for potential solutions referred to as chromosomes, genetic operators such as crossover and mutation used to generate new child solutions, and also selection and evaluation mechanisms derived from the nature.

In accordance with the terminology of GA, we define a population of individuals consisting of m chromosomes $\{x_1, x_2, \ldots, x_m\}$, where each chromosome $x_i =$

$[var_1^i, var_2^i, \ldots, var_N^i]$ corresponding to one set of test data consists of N genes (values of N input variables).

The main GA cycle for generating test data includes the following steps (steps 2–6 are repeated iteratively until the specified coverage value or number of generations is reached):

- Initialization. The initial population is formed randomly, taking into account the bounds of the input variables. The population size m is chosen based on the size of the program under test (rather, the number of possible paths).
- Fitness function calculation. Each chromosome of the population is estimated by a fitness function.
- Selection. The best 20% of chromosomes are transfer to the next generation unchanged; the remaining 80% of the next generation will be obtained by crossover.
- Crossover. Half of the chromosomes of the next generation are formed by crossover 20% of the best chromosomes of the previous generation with each other. The remaining chromosomes will be obtained by random crossover between all the chromosomes of the previous generation. Crossover occurs by choosing a random constant $\beta_i \in [0, 1]$ for each $l = \overline{1, N}$ and sequential blending, where the l-th offspring gene is a linear combination of the corresponding genes of the parental chromosomes:

$$var_l^{offspring} = \beta_l * var_l^{mother} + (1 - \beta_l) * var_l^{father}, \; l = \overline{(1, N)} \tag{1}$$

- Mutation. With a given mutation probability, each gene can change its value to random within the given bounds. The main purpose of mutation is to achieve greater diversity.
- Formation of test data sets in the form of a pool of elite chromosomes. In each generation, chromosomes of the population are selected into the pool of elite chromosomes. Only those chromosomes that provide additional code coverage in comparison with the previous coverage include into the pool.

2.2 Formulation of the Fitness Function for Maximum Statement Coverage

In this section, we will formulate the fitness function of the genetic algorithm in such a way that to maximize the coverage of code statements by both individual test cases and the whole test cases population.

The first step of white-box testing is to translate the source code into a Control Flow Graph (CFG) as oriented graph $CFG = (V, R, v0, vE)$, where V is set of graph nodes, R is set of edges, $v0$ и vE are input and output nodes, respectively, $v0 \in V$, $vE \in V$.The CFG makes it easier to specify in detail the control elements that must be covered, so we can define $v_j \in V$ as a separate node of CFG, in other words, one or more statements of the code. Different initial data of the program lead to traversing along different paths of the CFG, ensuring the execution of only quite specific (not all) statements of the program. Let us denote $g(x_i)$ a vector that is an indicator of the coverage of the graph nodes by a path initiated by a specific set of the test case $x_i - g(x_i) = (g_1(x_i), g_2(x_i), \ldots, g_n(x_i))$, where

$$g_j(x_i) = \begin{cases} 1, & \text{if path initiated by test case } x_i \text{ traverses through the node } v_j; \\ 0, & \text{otherwise} \end{cases}$$

If we define vector $\{w_1, w_2, \ldots, w_n\}$ as weights assigned to the test program statements, then the fitness function for the individual chromosome x_i can be formulated as follows

$$F_1(x_i) = \sum_{j=1}^{n} w_j g_j(x_i). \quad (2)$$

To ensure a greater diversity of the population, it is necessary to introduce into the fitness function a term that gives preference to chromosomes that provide the greatest possible distance from each other all paths that are generated by test cases of the population's chromosomes.

The developed fitness function is based on the idea given in paper [22]. We correct some inconsistencies in the formulas and propose more balanced relation of terms in the final formula of the fitness function.

In order to calculate the j-th similarity coefficient $sim_j(x_{i_1}, x_{i_2})$ of two chromosomes x_{i_1} and x_{i_2} we compare if the node v_j of the CFG is covered or uncovered by both paths initiated by these two test cases

$$sim_j = (x_{i_1}, x_{i_2}) = \overline{g_J(x_{l_1}) \oplus g_J(x_{l_2})}, \; j = \overline{1, n}. \quad (3)$$

The more matching bits are there between the two paths, the greater is the similarity value between the chromosomes. The following formula takes into account weights of corresponding CFG nodes

$$sim(x_{i_1}, x_{i_2}) = \sum_{j=1}^{n} w_j \cdot sim_j(x_{i_1}, x_{i_2}). \quad (4)$$

The value of similarity between the chromosome x_i and the rest of the chromosomes in the population is calculated as

$$f_{sim}(x_i) = \frac{1}{(m-1)} \sum_{\substack{s=1 \\ s \neq i}}^{m} sim(x_s, x_i) \quad (5)$$

Now we can determine the maximum value of path similarity in the whole population

$$\overline{f_{sim}} = \max_{i=\overline{1,m}} f_{sim}(x_i) \quad (6)$$

So, we can formulate the term of fitness function responsible for the diversity of paths in a population. It is

$$F_2(x_i) = \overline{f_{sim}} - f_{sim}(x_i). \quad (7)$$

Thus, the fitness function for the chromosome x_i is calculated by the formula

$$F(x_i) = F_1(x_i) + k \cdot F_2(x_i), \quad (8)$$

where $F_1(x_i)$ and $F_2(x_i)$ are defined by formulas (1) and (6). The first term $F_1(x_i)$ determines the complexity of the path initialized by the chromosome x_i, and the second term $F_2(x_i)$ determines the remoteness of this path from other paths in the population. The constant k determines relation between the two terms and is chosen dynamically.

The influence of the value of the parameter k on the fitness function is quite significant [21]. As past studies have shown, at zero $k = 0$, the coverage is minimal, reaching its maximum value at $k = 10$, after which it begins to decline. Obviously, choosing the right k can significantly affect the final results. However, this value of k obtained in studies is the best only within the tested SUT. For programs with different code size and branching, this value may not be the best. Therefore, in order to increase the universality of the algorithm, it is necessary to reduce the influence of k, so that choice of its value have less impact on the process of test data generation.

The fitness function is formed from two components – $F_1(x_i)$ and $F_2(x_i)$, as shown in the formula (7). The type of dependence $F_2(x_i)$, for determining the distance of paths from each other, has been well studied, and the necessary modifications have been proposed and studied for it to increase the diversity of the population.

At the same time, the $F_1(x_i)$ component has a rather simple form and has not yet been sufficiently developed, so it has the potential for further improvement. Accordingly, for the subsequent modification of the proposed method for generating test data, we can concentrate on possible changes in $F_1(x_i)$, that is, complicate the method for determining and calculating the weights of operations.

Of particular interest, in this case, are other evolutionary algorithms, since they are based on the same paradigm as in GA. For example, PSO is one of the Swarm Intelligence (SI) algorithms, but not its only representative. Its other representatives are the Ant Colony Optimization (ACO), the Artificial Bee Colony Algorithm (ABC), the Cuckoo Search (CS) and many other algorithms based on the collective interaction of various elements or agents.

The Ant Colony Optimization (ACO) [22] is one of the algorithms that allows to solve problems of finding a route search on graphs. It is based on the simulation of the behavior of an ant colony. Ants, passing along certain paths, leave behind a trail of pheromones. The better solution was found, the more pheromones will be on one way or another. In the next generation, ants already form their paths based on the number of pheromones - the more pheromones on a certain path, the more ants will be directed to this path and continue to explore it. In this way, the colony gradually explores the entire solution space, gradually cover better and better paths.

Directly using the ant colony algorithm as-is is not possible, since the output to certain paths is initialized by different data sets, and the only way to change the path is to manipulate the initial data. However, the idea of using "pheromones" to prioritize pathfinding may have a positive effect on providing greater population diversity.

To implement this idea, it is proposed to modify the component $F_1(x_i)$ of the fitness function. The new function additionally introduces the Ph_i parameter to dynamically change the value of the operation weights. We have identified two main approaches that can be used to simulate the effect of pheromones:

1. Direct approach. All operations initially have a weight factor of 0. Gradually, the weights of operations increase as they cover certain paths. With each generation, the weight of covered statements will increase.

2. Inverted approach. All operations initially have a weight factor of 100, that is, the maximum weight. With each new generation, the weight of covered statements will decrease, shifting the direction of generation to uncovered paths.

As a result of the implementation of the modification, the function $F_1(x_i)$ will have the following form

$$F_1(x_i) = \sum\nolimits_{j=1}^{n} w_j g_j(x_i) Ph_j, \tag{9}$$

where Ph_j – the weight factor of a particular operation.

Since the ultimate goal of introducing this modification is to increase the diversity of the population, the complete exclusion or inclusion of operations is not the best method of implementation. If only certain paths are included, then the genetic algorithm with each new generation will generate data only for these paths, excluding those for which the data was not generated, even if they are potentially more difficult. At the same time, if paths are excluded, then for significantly branched test codes, the data may also not be selected, since the algorithm may not have enough time to reach them.

The solution to this problem is the alternation of inclusion and exclusion upon reaching a certain value Ph_j. If the covered statement has reached the value $Ph_j = 1$, it begins decreasing. If, on the contrary, the coefficient Ph_j has decreased to 0, then it will start increasing. Thus, operations can gradually either increase the value of $F_1(x_i)$ for certain test sets, or decrease it, thereby ensuring sufficient population diversity without focusing only on certain paths.

The only difference between the approaches is the generation of the initial population – whether the initial sets will be formed completely randomly at $Ph_j = 0$, followed by a search for complex paths, or sets for more complex paths will be generated initially, after which their successive exclusion will begin.

3 Research

To investigate the GA work with the proposed fitness function (7) and modification of the $F_1(x_i)$ (8), a software-under-test (SUT) was developed containing many conditional statements defining a sufficient number of different paths of the program code.

To begin with, it is worth considering methods for changing the value of Ph_j. As mentioned above, in order to exclude the concentration of test data generation only on certain paths, it was decided to alternate the inclusion and exclusion of operations depending on their value Ph_j Then the main question is to determine the rate of growth and decrease of this value.

We are proposing two main groups of methods for switching inclusion and exclusion, on which the rate of change of Ph_j depends. The first group of methods involves covering a certain number of generations, after which a switch occurs. We have identified the following boundaries:

- Achievement of half generations (Half method). After the statement has been covered in half of the generations, a switch occurs. That is, at best, switching will occur once for frequently covered operations. For rarely covered, switching may not occur.

- Achievement of a quarter of generations (Quarter method). Switching will occur if the operation was covered by at least a quarter of all generations. The statement can switch up to 3 times.
- Achievement of the tenth generations (Tenth method). When covering at least 1/10 of all generations, the operation will switch. This method implies a fast change in the value of Ph_j, when the operation can switch up to 9 times.

The other group contains only one method, which depends on the number of statement coverages in one generation. The more test sets cover the certain statement, the greater the change in the value of Ph_j will be. That is, for example, if the operation was covered by 90 out of 100 chromosomes in one generation, then for the next generation the weight of this operation will change to $\Delta Ph_j = \frac{90}{100} = 0,9$. Thus, the more datasets that cover one or another operation, the more often the switching will occur. Figure 1 shows comparison of the different direct approaches, the method without modifications, and the hybrid method (Count-) based on the inverted approach.

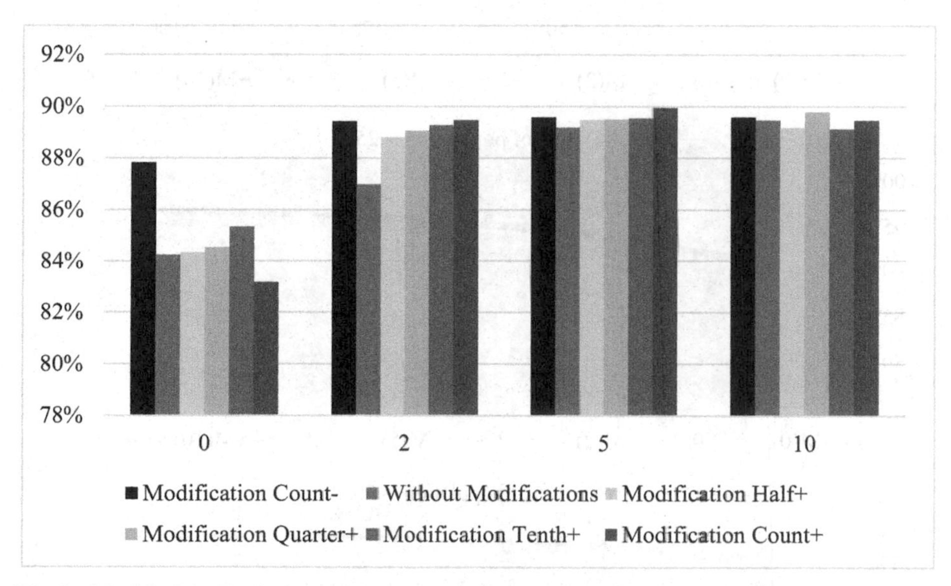

Fig. 1. Modified methods for different k. Population size is 50, number of generations is 25. (Color figure online)

Red color shows coverage without modification. Shades of blue are different switching methods based on the direct approach. In general, these methods show slightly better coverage. The inverted methods, which are not shown in the figure, show slightly better coverage than the direct methods, but in general, not so much higher as to unambiguously state the effectiveness of the modification.

Of greatest interest in the figure is the black column, which shows the coverage of the hybrid method based on inverted approach depending on the coverage (Count-). Instead of switching inclusion and exclusion, the weight of the operation is reduced depending on the number of sets that covered it in one generation. At the same time, if the statement was not covered at all, the value of Ph_j for it increased to 1. Thus, the algorithm more often tries to generate data for those operations that have not yet been covered. At the same time, often covered operations cease to play a significant role in the generating data sets. This method showed the greatest efficiency in the formation of test sets, so it will be used for other studies.

On the Fig. 2 presented the results with and without modification for different k with various number of generations and population size.

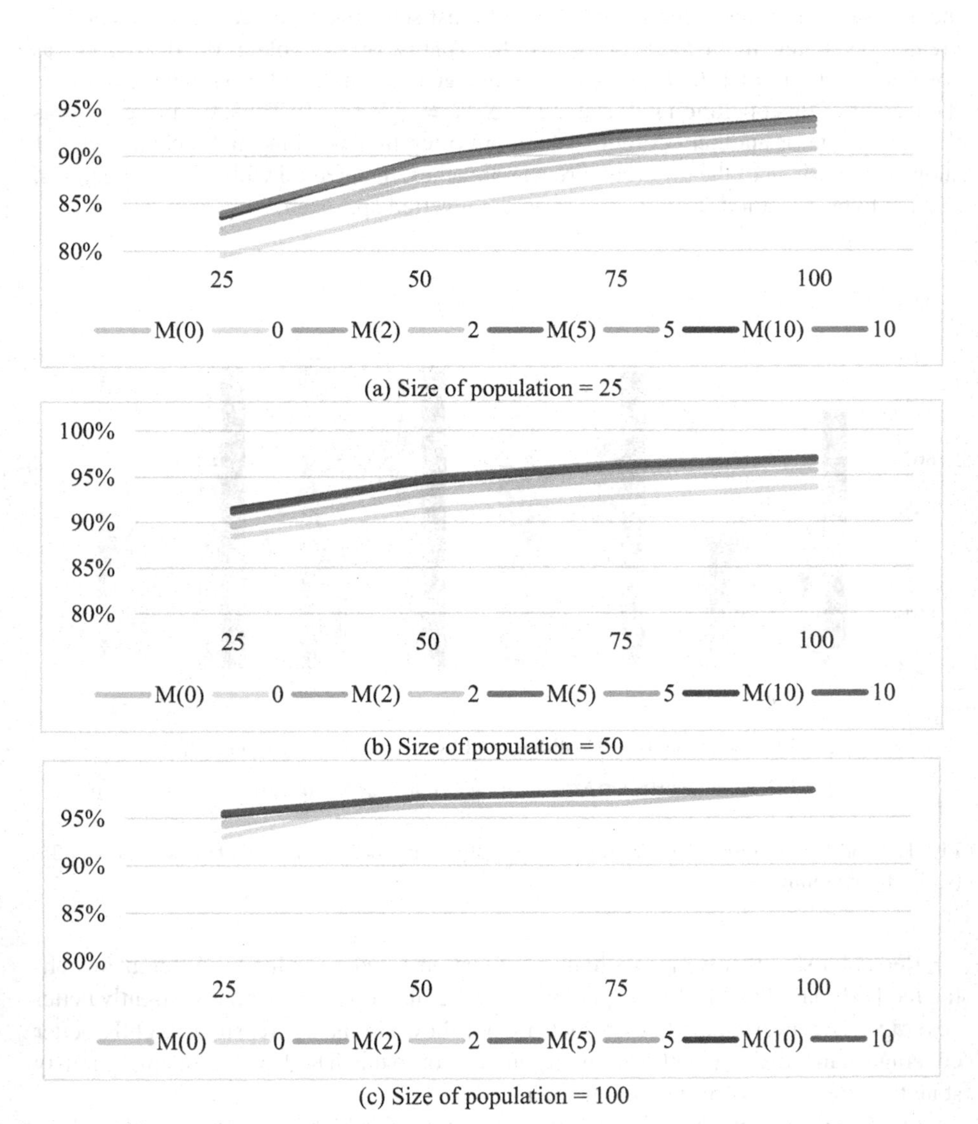

Fig. 2. Comparison of coverage with different population size and generation number

It is clearly seen in the figures that the modification allows to significantly increase the coverage even without using the optimal value $k = 10$ for this SUT. Even with k = 0, that is, without using the second parameter of the fitness function, the coverage achieved is on average higher. The use of any value of k with modification makes it possible to achieve the maximum possible coverage for the given parameters of the genetic algorithm.

A more accurate visualization is shown in Fig. 3, which presents a coverage with a population size of 25 chromosomes at 50 generations.

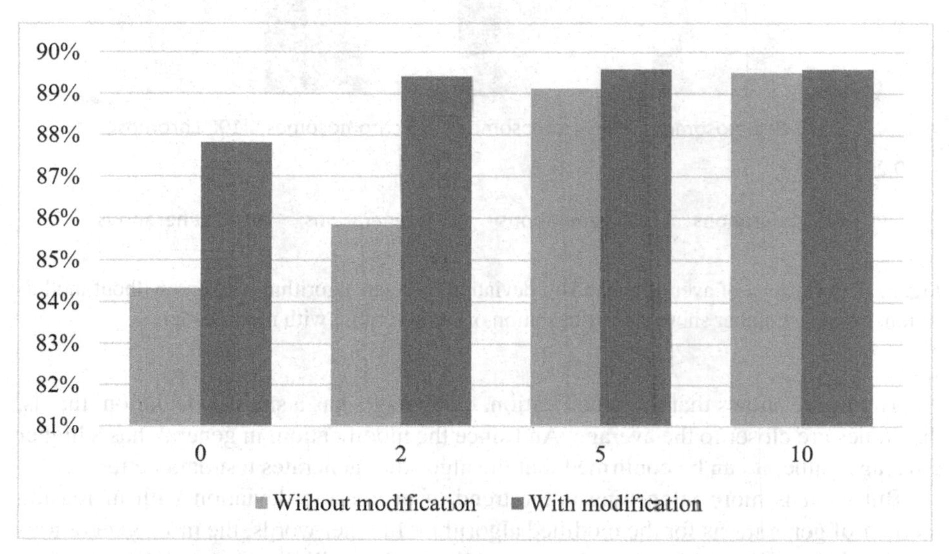

Fig. 3. Code coverage with different *k* with and without modification. Population size is 25, number of generations is 50

Obviously, the modification shows better coverage. More importantly, the maximum coverage for these parameters of the genetic algorithm is achieved using any value of k, that is, the main goal of introducing the modification is achieved – k ceases to play a significant role in achieving maximum coverage, which allows the algorithm to be used without the need to search for the optimal k for the SUT.

Interesting results were also obtained from the study of the standard deviation of the generated values. Figure 4 shows the difference in *k*-averaged standard deviation between algorithms with and without modification.

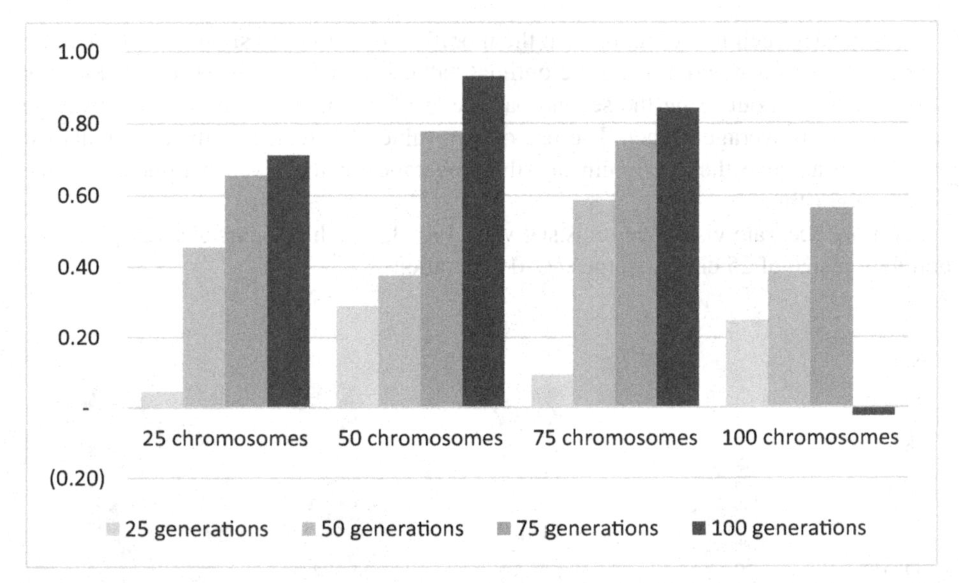

Fig. 4. Discrepancy of averaged standard deviation between algorithm with and without modification. Positive number shown lesser deviation of the algorithm with modification

The figure shows that the modification, on average, has a smaller deviation, that is, the values are closer to the average. And since the modification, in general, has a higher coverage value, it can be confirmed that the algorithm generates test data better.

But what is more interesting is the trend of decreasing deviation with increasing number of generations for the modified algorithm. In other words, the more generations the algorithm will conduct, the more efficiently the data will be generated. When examining the coverage as such, this is not noticeable, since the coverage is generally quite high, but the study of the deviation made it possible to determine the additional benefits of the modification.

4 Conclusion

The article presents a study of modifying the previously developed method for generating test data sets to eliminate the significant influence of the change in the fitness function parameter k, which is responsible for the relation between the parameters of path complexity and population diversity. The results showed that the modification with a hybrid approach to the determination of Ph_i showed noticeably better coverage values. But more importantly, the maximum possible coverage with certain parameters of the genetic algorithm was achieved using absolutely any k, which eliminates the need to search for it for each tested program or develop additional methods for its dynamic determination.

References

1. Kumar, M., Chaudhary, J.: Reviewing automatic test data generation. Int. J. Eng. Sci. Comput. **7**(5), 11432–11435 (2017)
2. Richard, A.D., Jefferson, A.O.: Constraint-based automatic test data generation. IEEE Trans. Softw. Eng. **17**(9), 900–910 (1991)
3. Meudec, C.: ATGen: automatic test data generation using constraint logic programming and symbolic execution. Softw. Test. Verification Reliabil. **11**(2), 81–96 (2001)
4. Gerlich, R.: Automatic test data generation and model checking with CHR. In: 11th Workshop on Constraint Handling Rules (2014)
5. Girgis, M.R.: Automatic test data generation for data flow testing using a genetic algorithm. J. Univ. Comput. Sci. **11**(6), 898–915 (2005)
6. Weyuker, E.J.: The complexity of data flow criteria for test data selection. Inf. Process. Lett. **19**(2), 103–109 (1984)
7. Khamis, A., Bahgat, R., Abdelaziz, R.: Automatic test data generation using data flow information. Dogus Univ. J. **2**, 140–153 (2011)
8. Singla, S., Kumar, D., Rai, M., Singla, P.: A hybrid PSO approach to automate test data generation for data flow coverage with dominance concepts. J. Adv. Sci. Technol. **37**, 15–26 (2011)
9. Liu, Z., Chen, Z., Fang, C., Shi, Q.: Hybrid test data generation. state key laboratory for novel software technology. In: ICSE Companion 2014 Companion Proceedings of the 36th International Conference on Software Engineering, pp. 630–631 (2014)
10. Harman, M., McMinn, P.A.: Theoretical and empirical study of search-based testing: local, global, and hybrid search. IEEE Trans. Softw. Eng. **36**(2), 226–247 (2010)
11. Maragathavalli, P., Anusha, M., Geethamalini, P., Priyadharsini, S.: Automatic test-data generation for modified condition. decision coverage using genetic algorithm. Int. J. Eng. Sci. Technol. **3**(2), 1311–1318 (2011)
12. Doungsaard, C., Dahal, K., Hossain, A.G., Suwannasart, T.: An automatic test data generation from UML state diagram using genetic algorithm. IEEE Computer Society Press, 47–52 (2007)
13. Sabharwal, S., Sibal, R., Sharma, C.: Applying genetic algorithm for prioritization of test case scenarios derived from UML diagrams. IJCSI Int. J. Comput. Sci. **8**(3(2)), (2011)
14. Doungsaard, C., Dahal, K., Hossain, A., Suwannasart, T.: GA-based automatic test data generation for UML state diagrams with parallel paths. In: Yan, X.T., Jiang, C., Eynard, B. (eds.) Advanced Design and Manufacture to Gain a Competitive Edge. Springer, London, pp. 147–156 (2008). https://doi.org/10.1007/978-1-84800-241-8_16
15. Grochtmann, M., Grimm, K.: Classification trees for partition testing. Softw.Test. Verification Reliab. **3**(2), 63–82 (1993)
16. Chen, T.Y., Poon, P.L., Tse, T.H.: An integrated classification-tree methodology for test case generation. Int. J. Software Eng. Knowl. Eng. **10**(6), 647–679 (2000)
17. Ramamoorthy, C.V., Lee, R., Lee, K.W. (eds.): SERA 2003. LNCS, vol. 3026. Springer, Heidelberg (2004). https://doi.org/10.1007/b97161
18. Serdyukov, K., Avdeenko, T.: Investigation of the genetic algorithm possibilities for retrieving relevant cases from big data in the decision support systems. In: CEUR Workshop Proceedings, vol.1903, pp. 36–41 (2017)
19. Praveen, R.S., Tai-hoon, K.: Application of genetic algorithm in software testing. Int. J. Softw. Eng. Appl. **3**(4), 87–96 (2009)
20. Serdyukov, K., Avdeenko, T.: Researching of methods for assessing the complexity of program code when generating input test data. In: CEUR Workshop Proceedings, vol. 2667, pp. 299–304 (2020)

21. Avdeenko, T., Serdyukov, K.: Genetic algorithm fitness function formulation for test data generation with maximum statement coverage. In: Tan, Y., Shi, Y. (eds.) ICSI 2021. LNCS, vol. 12689, pp. 379–389. Springer, Cham (2021). https://doi.org/10.1007/978-3-030-78743-1_34
22. Zhu, E., Yao, C., Ma, Z., Liu, F.: Study of an improved genetic algorithm for multiple paths automatic software test case generation. In: Tan, Y., Takagi, H., Shi, Y. (eds.) ICSI 2017. LNCS, vol. 10385, pp. 402–408. Springer, Cham (2017). https://doi.org/10.1007/978-3-319-61824-1_44
23. Dorigo, M., Birattari, M., Stutzle, T.: Ant colony optimization. IEEE Comput. Intell. Mag. **1**(4), 28–39 (2006)

Genetic Algorithm and Evolutionary Computation

Reasoning About Order Crossover in Genetic Algorithms

M. Saqib Nawaz[1], Saleha Noor[2], and Philippe Fournier-Viger[1](✉)

[1] School of Computer Science and Software Engineering, Shenzhen University, Shenzhen, China
{msaqibnawaz,philfv}@szu.edu.cn

[2] School of Information Science and Engineering, East China University of Science and Technology, Shanghai, China

Abstract. The formal modeling and verification of algorithms is a challenging task, but it is a necessary requirement for the proof of correctness. Evolutionary computation and theorem proving approach of formal methods are two different domains in theoretical computer science. Using Prototype Verification System (PVS), this paper presents a method of formal specification, reasoning and verification for order crossover operator in Genetic Algorithms (GAs) and their rudimentary properties. Order crossover operator is first formally specified in PVS specification language. Some other operators used in the definitions of order crossover are also specified. PVS theorem prover is then used to prove some properties of order crossover and operators.

Keywords: Order crossover · Genetic algorithms · PVS · Specification · Verification

1 Introduction

There are many complex problems in real life where the goal is to find best solution(s) in very large search space(s). In such problems, optimization algorithms [1] are commonly used to find approximate solutions. Genetic algorithms (GAs) [2] are search algorithms based on the biological evolution and Darwinian theory principles. In theoretical computer science, GA are well-known for solving various types of optimization problems. Besides optimization problems, GAs are now used in many other fields and systems that include bioinformatics, control engineering, scheduling applications, artificial intelligence, robotics and safety critical systems. As evolution/optimization-based algorithms are now widely used in many systems, so their formal analysis and reasoning is an important and interesting research problem.

In this work, we focus on formal analysis and verification of order crossover operator in GA using theorem proving approach of formal verification methods. In interactive theorem proving, the systems that need to be analyzed are first modeled using an appropriate mathematical logic. Important system properties are then proved using theorem provers [3, 4].

Y. Tan et al. (Eds.): ICSI 2022, LNCS 13344, pp. 261–271, 2022.
https://doi.org/10.1007/978-3-031-09677-8_22

More specifically, we use the Prototype Verification System (PVS) [5], which is an interactive theorem prover. Specification language of PVS is built on higher-order logic (HOL) and PVS prover is based on *sequent calculus* where each proof goal is a *sequent* consisting of formulas called *antecedents* (hypothesis) and *consequents* (conclusions. Using a proof system, one of the initial efforts in the formalization of the GAs foundations can be found in [6]. Single-point crossover (SPC) and multi-point crossover (MPC) are formally specified and verified in PVS [7]. Taking the work done in [6, 7], the main activities in this paper involve:

1. Using the PVS specification language for providing a formal knowledge by writing the specification of order crossover operator in GA, and
2. Using the PVS theorem prover for the reasoning in the modeled knowledge for order crossover and proving its properties.

The rest of the article is organized as follows: Sect. 2 discusses the related work. In Sect. 3, formal specification of order crossover operator is carried out in PVS. Section 4 provides the verification details for order crossover. Finally, in Sect. 5 conclusions are drawn.

2 Related Work

This work on formal analysis of crossover operators in GA is not the first one. Aguado et al. [8] used the Coq [9] ITP (interactive theorem prover) to formally verify crossover operator (generalized version) of genetic algorithms. Work done in [8] was divided in three parts. In first part, they defined multiple crossover operation. Given definition of multiple crossover operator generalizes the definition given in [6]. Multiple crossover that was defined in first part does not depend on crossing points order. This was proved in second part. In third part, an effective definition of crossover operator was presented. Furthermore, Aguado et al. [10] formally specified and verified position based and order-based crossover in Coq. They used lists for the representation of chromosomes (strings).

Nawaz et al. [7] used PVS for the formal specification and verification of two crossover (SPC and MPC) operators. Zhang et al. [11] used the HOL4 proof assistant [12] for the formalization of crossover operators. Moreover, [13] provided the formal semantics for crossover and mutation operators of GA by representing operators by designs in Unifying Theories of Programming (UTP) [14] semantic framework.

3 Order Crossover Modeling

The crossover operator in GA is used to recombines two selected strings to get new better strings. The two selected strings are called parent strings and the resulting strings obtained after the crossover operation are known as off-spring strings. Crossover operators in GA can lead the population to converge on one of the best solutions. In GA literature, many crossover operators exist but most adopted are SPC and MPC operators

[15]. For strings having small length, SPC operator is suitable whereas MPC is more appropriate for long strings [16].

In SPC operation, single crossing point is randomly selected in both of the parent strings. Both strings are then split into sub-strings with crossing point and these sub-strings are combined to form new off-spring strings. Let two parent strings are:

$$x = x_1,\ x_2, \ldots\ldots, x_n$$

$$y = y_1,\ y_2, \ldots\ldots, y_n$$

Let position j $(1 \le j \le n)$ in both parent strings is selected as crossing point. New off-springs that will replace these two parent strings will be:

$$x' = x_1, \ldots, x_j,\ y_{j+1}, \ldots, y_n$$
$$y' = y_1, \ldots, y_j,\ x_{j+1}, \ldots, x_n$$

De Jong [17] in his work generalized the SPC operator and considered the MPC operator. In the later operator, two crossover points are chosen in the strings and sub-strings between the points are exchanged to form new strings. Let x and y (already defined above) be the two parent strings with length n and j, and k be the two crossover points with $k > j$, then:

$$x' = x_1, \ldots, x_j, y_{j+1}, \ldots, y_k,\ x_{k+1}, \ldots, x_n$$
$$y' = y_1, \ldots, y_j, x_{j+1}, \ldots, x_k,\ y_{k+1}, \ldots, y_n$$

The formal modeling and verification of SPC and MPC can be found in [7].

The most widely used encoding technique for the representation of chromosomes in GA is the use of strings $x = x_1, x_2, \ldots, x_n$ of binary values ($x_i \in (0, 1)$). However, in some application such as Graph Coloring Problem, Vertex Coloring Problem and Traveling Salesman Problem, the use of permutation produces better result than standard binary encoding [18]. Order crossover operator of GA, invented by Davis [18], was specifically developed for permutation encoding representation. It is important to remark that for the representation of chromosomes in order crossover, permutations of (possibly) repeated elements are considered. For convenience, permutation between two strings say x and y is denoted by $x \approx y$.

Order crossover is explained with a simple example. Assume that population P is a non-empty finite set and x, y are the two finite sequences (strings) of elements of P. Let two parent strings x and y are:

$$x = c\ b\ a\ f\ e\ d\ g\ h\ i$$
$$y = g\ h\ i\ f\ a\ c\ b\ d\ e$$

Two natural numbers are randomly chosen as crossing points (for example j = 3 and $k = 6$). Substring x' obtained from x contains elements of x between j and k. So $x' = f$ e d. In y, elements of x' are removed, obtaining $y' = g\ h\ i\ a\ c\ b$. Finally, the substring x' is inserted in y' to get the child string:

$$Child\ String = g\ h\ i\ f\ e\ d\ a\ c\ b$$

Therefore, order crossover operation generates a string which has between *j* and *k* the elements from string *x* and the elements from the string *y* in the other positions. Note that order crossover depends on parent strings *x, y* and crossing points *j, k*. Order crossover can be defined as:

Definition 1. Let *x, y* be the strings of finite sequences and $x, y \in P$. Let $j, k \in \mathbb{N}$ (natural number). Order crossover ($\odot$) operator is defined as:

$$(x, y)(j, k) = \left\{ \left(\left(y, \left(x\Updownarrow_{(j,k)}\right)\neg\right)\uparrow_j\right)o\left(x\Updownarrow_{(j,k)}\right)o\left(\left(y, \left(x\Updownarrow_{(j,k)}\right)\neg\right)\downarrow_k\right) \right\}$$

Here **o**[1] represents the concatenation and ↑, ↓, ⇕ and ¬ are four operators called *Head, Tai, Cut* and *delete_seq,* respectively. Before specification of four operators, we declared in PVS the type of *P* as a not-empty finite set of sequences, *p* and *q* are two finite sequences of type *P* and crossover points *j, k* are of type natural number.

```
P = Setof[seq]
p, q = VAR Finseq[P]
j, k = VAR NAT
```

Here, *seq* is the unspecified type of sequences. Like PVS syntax, we have used square brackets (brackets []) for type constructors and type parameters. We have used some already pre-defined types of PVS prelude like *below[nat]* and *finseq. Finseq[P]* is record (type) of two fields, *length* and *seq. Length* represents the total length of the finite sequence and *seq* represents a finite sequence which is defined as a function *below[length] → P.*

```
TYPE = [# length: nat, seq:
                    ARRAY[below[length] → P]#]
```

Finite sequences in PVS always begin from 0 upto *n – 1* where *n* represents the length of the sequence. Formal specification of ↑, ↓, ⇕ and ¬ in PVS is as follows.

Head (↑) Specification in PVS

We define *Head* operator in PVS with the help of a total function. This function takes a number (crossover point) and a sequence of finite length as input and returns a finite sequence as output with following type definition:

```
Head:[Nat × Finseq → Finseq],
Head(j, p):Finseq[P] = IF j < length(p) THEN(#length:=
           j, seq:= λ(i:below[j]): p`seq(i)#) ELSE IF
           j = 0 THEN empty_seq ELSE p
           ENDIF ENDIF
```

[1] We have used the symbol **o** for concatenation according to PVS syntax.

Tail (↓) Specification in PVS
Tail operator is also specified as a total function that takes a natural number and a finite sequence and returns a sequence of finite length according to given type definition:

Tail:[Nat × Finseq → Finseq],

```
Tail(j, p): Finseq[P] = IF j < length(p) THEN
        (#length:= p`length - j, seq:= (λ(i:
         below[p`length - j]): p`seq(j+i))#)
        ELSE IF j=0 THEN p ELSE empty_seq
        ENDIF ENDIF
```

Cut (⇕) Specification in PVS
In PVS, we have specified *Cut* as a total function that takes two numbers (crossover points) and a finite sequence as input and returns a finite sequence as output with given type definition:

Cut:[Nat × Nat × Finseq → Finseq],

```
Cut(j, k, p): Finseq[P] = IF k,j < length(p) ∧
    j<k THEN (#length:= k - j, seq:= (λ(i:
    below[k - j]): p`seq(j + i))#)
   ELSE IF j > length(p) > k THEN Head(k, p)
   ELSE IF k > length(p) > j THEN Head(j, p)
   ELSE empty_seq ENDIF ENDIF ENDIF
```

For the delete_seq (¬), two functions are declared that are named *first* and *remaining*. *First* function selects the first element from the selected string *x*, if *x* is a non-empty sequence. *Remaining* function returns a string in which first element from *x* is excluded.

```
first(x: (ne_seq?)): P = x`seq(0)

remaining(x:(ne_seq?)):{d:finseq[P]|d`length <
    x`length} = LET len = x`length - 1 IN (#
    length:=  len,  seq := (λ(i:below(len)):
                                x`seq(i+1))#)
```

Whereas, *ne_seq?* is a *Boolean* predicate that returns *true* if the length of the string is greater than 1. Two more total functions *find* and *remove* are also declared. *Find* function returns location of a specific element in a sequence, whereas *remove* function removes the element returned by the *find* function.

```
find(a:P, x, (m: upto(x`length))):RECURSIVE
                    upto(x`length) =
     IF m = x`length THEN m
     ELSIF x`seq(m) = a THEN m
     ELSE find(a, x, m + 1) ENDIF
     MEASURE x`length - m

   remove(x, (i: below(x`length))): finseq[P]  =
     LET newlen = x`length - 1
     IN x WITH [`length := newlen, seq :=
                          (λ(j:below(newlen)):
     IF j<I THEN x`seq(j) ELSE x`seq(j+1) ENDIF)]
```

Recursive function *find* locates the position of the first occurrence of element *a* from a sequence *x* (if *a* appears in *x*) and *remove* function eliminates this element from the *x*.

Delete_seq (¬) Specification in PVS

The total Function *delete_seq(y, x)* ($\neg_{y,x}$) is declared that takes two parent sequences *x* and *y* and delete elements of *y* from sequence *x* with following type definition:

delete_seq: [Finseq × Finseq → Finseq],

```
delete_seq(y, x): RECURSIVE Finseq[P] =
 IF ne_seq?(y)
 THEN LET a = first(y), i = find(a, x, 0)
 IN IF i < x`length THEN
   delete_seq(remaining(y), remove(x, i))
 ELSE delete_seq(remaining(y), x) ENDIF
 ELSE empty_seq ENDIF MEASURE y`length
```

Order Crossover Specification

In PVS, order crossover operator is specified as a total function that takes two strings (finite sequences) and two natural numbers as input and returns a string as output.

```
ordercrossover(x, y, j, k):finseq[P] =
 LET s = cut(j, k, x), d = delete_seq(s, y)
 IN IF ne_seq?(d) THEN head(j,d) o s o
             tail(k,d) ELSE p ENDIF
```

4 Verification

PVS is based on HOL, but proving HOL properties is not fully automatic due to the undecidability in HOL. Thus, human assistance is required in the process of proof searching and optimization [19]. The proof development process in PVS is interactive in nature and it follows the *sequent-style proof representation.* A user first provides the property (in the form of a lemma or theorem) that is called a proof goal. User then applies proof commands, inference rules and decision procedures to solve the proof goal. The action resulting from a proof command, inference rule or decision procedure is referred to as a proof step (*PPS*) here. A *PPS* may either prove the goal or generates another sequent or divides the main goal into sub-goals. The proof development process for a theorem or lemma is completed when the sequent or all the subgoals are proved.

All theorems in this work are proved with PVS theorem prover by mostly using following commands and decision procedures: *skosimp, expand, lift-if, prop, grind, assert, apply-extensionality: hide t, typepred!, induct id* and *ground. Induct* command is used to prove the properties involving recursive functions. Next, we proved some properties for the four operators, followed by the order crossover properties verification.

4.1 Head, Cut, Tail and Delete_seq Operators Properties

Some of the properties of *Head, Cut* and *Tail* operators that are proved using PVS are included in the following lemmas.

Lemma 1. *Assume that parent string* $p = ()$*, then:*

(1) $p{\uparrow_j} = ()$,
(2) $p{\downarrow_j} = ()$ *and.*
(3) $p{\Updownarrow_{(j,\,k)}} = ()$.

```
c_e_p: LEMMA Head(j, empty_seq) = empty_seq
                               ∧
              LEMMA Tail(j, empty_seq) = empty_seq
                               ∧
           LEMMA Cut(j, k, empty_seq) = empty_seq
```

In lemma 1, it is proved that if parent string is empty then the *Head, Cut* and *Tail* operators will return empty strings.

Lemma 2. *Assume that crossover points j, k* = *0, it holds that:*

(1) $p{\uparrow_0} = ()$,
(2) $p{\downarrow_0} = p$ *and.*
(3) $p{\Updownarrow_{(0,\,0)}} = ()$.

```
co_j,k_0: LEMMA length(p) > 0 ⊃ Head(0, p) = empty_seq
                                  ∧
             LEMMA length(p) > 0 ⊃ Tail(0, p) = p
                                  ∧
          LEMMA length(p) > 0 ⊃ Cut(0,0,p) = empty_seq
```

In other words, if crossover points *j*, $k = 0$ then *Head* and *Cut* will output empty strings while *Tail* operation will generate same parent string *p*.

Lemma 3. *Let* $p \in P$ *and crossover point* $j \in \mathbb{N}$. *It can be proved that* $(p\uparrow_j)$ **o** $(p\downarrow_j) = p$.

```
s_con: LEMMA Head(j, p) o Tail(j, p) = p
```

Head and *Tail* operators divide the parent string into two sub-strings. If these two operators are applied on the same parent string and generated sub-strings are concatenated, then they should be equal to the parent string. Hand written proof of this lemma is described below that mainly follows the proof done by PVS but many specifics are omitted.

Proof. Let total length of the string *p* is *n*. We have three cases for crossover point *j*.

Case 1 is when $j < n$. In this case from the specifications of *Head* and *Tail* operators, $p\uparrow_j = p_1, p_2,\ldots, p_j$ and $p\downarrow_j = p_{j+1}, p_{j+2},\ldots, p_n$. By concatenating these two, we get $p_1, p_2,\ldots, p_j$ **o** $p_{j+1}, p_{j+2},\ldots, p_n = p_1, p_2,\ldots, p_{n-1}, p_n =$ **p**.

Case 2 is when $j = 0$. In this case, from specifications of *Head* and *Tail* operators, $p\uparrow_j = ()$ and $p\downarrow_j = p$. By concatenating these two, we get $()$ **o** $p =$ **p**.

Case 3 is where $j > n$. In this case, $p\uparrow_j = p$ and $p\downarrow_j = ()$. By concatenating these two, we get p **o** $() =$ **p**.

This completes the proof.

Similarly, a parent string is divided into three sub-strings by *Head*, *Cut* and *Tail* operators, provided with two crossover points *j* and *k*. So, concatenation of these three operators should result in same parent string.

```
m_con: LEMMA Head(j, p) o Cut(j, k, p) o Tail(k, p) = p
```

Some other properties of ↑, ⇕ and ↓ that are proved in PVS are included in the following proposition.

Proposition 1. *If* $p, q \in P$ *and* $j, k \in \mathbb{N}$.

- *if* $p\uparrow_j = q$, *then* $p = q$
- $length(p\downarrow_j) = length(p) - j$
- *if* $length(p) = length(q)$, *then* $length(p\uparrow_j) = length(q\uparrow_j)$
- *if* $length(p) = length(q)$, *then* $length(p\downarrow_j) = length(q\downarrow_j)$
- *if* $length(p) = length(q)$, *then* $length(p\Updownarrow_{(j,k)}) = length(q\Updownarrow_{(j,k)})$

Lemma 4. *Let P be a Gene-Set and* $x, y \in P$. *It holds that* $x \approx y$ *iff* $x \subseteq y$ *and* $y \subseteq x$.

Lemma 5. *Let P be a Gene-Set, x, y be individuals of P. It is proved that:*

$length(\neg_{y,x}) = length(x) - length(y)$, *if* $length(x) > length(y)$.

Proof. This lemma is proved with the help of induction. After induction, the definition of *delete_seq* function is expanded. The proof is completed by applying PVS commands and inference rules on expanded definition.

The following two lemmas are for the *delete_seq* properties.

Lemma 6. *Let P be a Gene-Set and x, y ∈ P, then:*

1) If $x \subseteq y$ *and if* $\neg_{y,x} = ()$*, then* $x \approx y$ *and.*
2. If $x \approx y$*, then* $\neg_{y,x} = ()$*.*

Here, *()* represents an empty sequence. Lemma 6 ensures that if elements in both strings *x, y* are same than $\neg_{y,x}$ returns an empty sequence.

Lemma 7. *Let x, y ∈ P satisfying* $x \subseteq y$*. Then* $(\neg_{y,x})\ o\ y \approx x$*.*

4.2 Order Crossover Properties

Order crossover properties are included with the following theorems.

Theorem 1. *Let x, y are two finite sequences such that* $x \approx y$ *and crossing points j, k* $\in \mathbb{N}$*, then:*

$$\odot(x,\ y)_{(j,k)} \approx x.$$

Proof. This theorem is proved by taking result 2 of Lemma 6 that is if $x \approx y$, then $\neg_{y,x} = ()$ and for two crossover points *j* and *k*, $x{\uparrow_j}\ o\ x{\downarrow_j} = x$ and $x{\uparrow_j}\ o\ x{\Updownarrow_{(j,k)}}\ o\ x{\downarrow_k} = x$.

Theorem 2. *Let P be a Gene-Set, x, y be individuals of P and j, k are the crossover points. If j, k = 0, then:*
$\odot(x,\ y)_{(0,0)} \approx x.$

Proof. `∀(x, y: Finseq[P]): ordercrossover(x, y, 0, 0)= x.`
{After *skolemizing* and *flattening*}.

```
ordercrossover(x`, y`, 0, 0) = x`
```

{After expanding the definition of *ordercrossover*, above consequent becomes:}

```
IF ne_seq?(delete_seq(cut(0, 0, x`), y`))
THEN head(0, delete_seq(cut(0, 0, x`), y`)) o
cut(0, 0, x`) o tail(0, delete_seq(cut(0, 0,
  x`), y`)) ELSE x ENDIF = x`
```

Propositional simplification is applied on the sequent obtained after expanding the definitions of *delete_seq, head, cut* and *tail* functions. Proof of this theorem is completed by applying repeated *skolemization* and *instantiation.*

In theorem 2, it is proved that when both crossover points are 0, then there is no point for crossing in the strings and the order crossover will return the same parent string x.

The type system of PVS is not algorithmically decidable and theorem proving may be needed for establishing the type-consistency in PVS specifications [5]. We type checked the specifications for proof obligations also known as type correctness conditions (TCC's). Some generated TCCs were discharged automatically by the PVS prover, while some were interactively proved and some were removed by modifying the specifications. These TCC's also helped in finding errors in specifications.

5 Conclusion

Starting from work done by [6, 7], we formally specified and verified the definition order crossover operator of GA in PVS. Similarly, *Head, Cut, Tail* and *delete_seq* operators that are used in order crossover have been specified and their properties are proved. For order crossover, the permutations of a finite sequence of elements for chromosomes representations are used which can be applied on a wide range of applications. Theorem proving in general is a tough job and it requires technical expertise and good knowledge about theorem provers and the modeling and reasoning process. As shown in this work, the formal proof of a goal in PVS mainly depends on the specifications along with different combinations of proof commands, inference rules and decision procedures. This is also the case for other ITPs. It is important to point out here that all the proofs in this work can also be done with any other HOL-based theorem prover like Coq [10] and HOL4 [12] but these provers lack proof trees.

Generally, it will be hard for the readers to understand definitions and theorems specifications in PVS syntax without good knowledge on PVS. For that purpose, efforts are made to identify and justify all the specification in notations that is close to both PVS syntax and general mathematical notation. Moreover, the reasoning process was explained as simple as possible. For future research, one interesting idea is to use pattern mining and evolutionary computations techniques for proof learning and searching in PVS. Some recent works in this regard can be found in [20, 21].

References

1. Kochenderfer, M.J., Wheeler, T.A.: Algorithms for Optimization. MIT Press, Cambridge (2019)
2. Holland, H.H.: Adaptation in Natural and Artificial Systems. University of Michigan Press, Ann Arbor (1975)
3. Hasan, O., Tahar, S.: Formal verification methods. In: Encyclopedia of Information Science and Technology, 3rd edn., pp 7162–7170. IGI Global (2015)
4. Nawaz, M.S., Malik, M., Li, Y., Sun, M., Lali, M.I.: A survey on theorem provers in formal methods, CoRR, abs/1902.03028 (2019)
5. Owre, S., Shankar, N., Rushby, J.M., Stringer-Calvert, D.W.J.: PVS version 2.4, system guide, prover guide, PVS language reference (2001)

6. Uchibori, A., Endou, N.: Basic properties of genetic algorithms. J. Formal. Math. **8**, 151–160 (1999)
7. Nawaz, M.S., Lali, M.I., Pasha, M.A.: Formal verification of crossover operator in genetic algorithms using Prototype Verification System (PVS). In: Proceedings of International Conference on Emerging Technologies, pp. 1–6 (2013)
8. Aguado, F., Doncel, J.L., Molinelli, J.M., Perez, G., Vidal, C.: Genetic algorithms in Coq: generalization and formalization of the crossover operator. J. Formal. Reason. **1**, 25–37 (2008)
9. Bertot, Y., Casteran, P.: Interactive Theorem Proving and Program Development: Coq'Art: The Calculus of Inductive Construction. Springer, Heidelberg (2003). https://doi.org/10.1007/978-3-662-07964-5
10. Aguado, F., Doncel, J.L., Molinelli, J.M., Pérez, G., Vidal, C., Vieites, A.: Certified genetic algorithms: crossover operators for permutations. In: Moreno Díaz, R., Pichler, F., Quesada Arencibia, A. (eds.) EUROCAST 2007. LNCS, vol. 4739, pp. 282–289. Springer, Heidelberg (2007). https://doi.org/10.1007/978-3-540-75867-9_36
11. Zhang, J., Kang, M., Li, X., Liu, G.Y.: Bio-inspired genetic algorithms with formalized crossover operators for robotic applications. Front. Neurorobot. **11**, 56 (2017)
12. Slind, K., Norrish, M.: A brief overview of HOL4. In: Proceedings of Theorem Proving in Higher-Order Logic, pp. 28–32 (2008)
13. Nawaz, M.S., Sun, M.: A formal design model for genetic algorithms operators and its encoding in PVS. In: Proceedings of International Conference on Big Data and Internet of Things, pp 186–190 (2018)
14. Hoare, C.A.R., He, J.: Unifying Theories of Programming. Prentice Hall International, Englewood Cliffs (1998)
15. Noor, S., Lali, M.I., Nawaz, M.S.: Solving job shop scheduling problem with genetic algorithms. Sci. Int. **27**, 3367–3371 (2015)
16. Mitchell, M.: An Introduction to Genetic Algorithms (Complex Adaptive Systems). A Bradford Book, England (1998)
17. De Jong, A.: An analysis of the behavior of a class of genetic adaptive systems. Ph.D thesis, Ann Arbor, MI, USA (1975)
18. Davis, L.: Handbook of Genetic Algorithms. Van Nostrand Reinhold, New York (1991)
19. Hong,W., Nawaz, M.S., Zhang, X., Li, Y., Sun, M.: Using Coq for formal modeling and verification of timed connectors. In: Proceedings of Software Engineering and Formal Methods Workshops, pp. 558–573 (2018)
20. Nawaz, M.S., Sun, M., Fouriner-Viger, P.: Proof searching in PVS using simulated annealing. In: Proceedings of International Conference on Swarm Intelligence, pp. 253–262 (2021)
21. Nawaz, M.S., Sun, M., Fouriner-Viger, P.: Proof guidance in PVS with sequential pattern mining. In: Proceedings of International Conference on Fundamentals of Software Engineering, pp. 45–60 (2019)

Copy and Paste: A Multi-offspring Genetic Algorithm Crossover Operator for the Traveling Salesman Problem

Chia E. Tungom[1,2], Ben Niu[1], Tongtong Xing[3], Ji Yuan[2], and Hong Wang[1](✉)

[1] College of Management, Shenzhen University, Shenzhen 518060, China
ms.hongwang@szu.edu.cn
[2] Onewo Space-Tech Service Co., Ltd., Shenzhen 518049, China
yuanj36@vanke.com
[3] Greater Bay Area International Institute for Innovation, Shenzhen University, Shenzhen 518049, China

Abstract. The Multi-Offspring Genetic Algorithm (MOGA) is a GA variant that was proposed specifically for the Traveling Salesman Problem (TSP). Like the Base GA (BGA), the first genetic algorithm designed to solve the TSP, MOGA has a crossover operator which is non trivial to implement. This paper proposes a copy and paste crossover operator for the multi-offsprings genetic algorithm which is easy to implement and also effective in generating a family of diverse offsprings. The algorithm named Copy and Paste Multi-offspring Genetic Algorithm (CP-MOGA) from the crossover operator design. Crossover and mutation in CP-MOGA is designed to cater for exploration and exploitation by carefully choosing a gene insertion section and mutation point such that two parents produce two predominantly exploratory, two predominantly exploitative and two moderately exploratory and exploitative offsprings, thereby balancing the exploration exploitation trade-off. Simulation results on twelve instances of the Traveling Salesman Problem show that the proposed algorithm outperforms MOGA and BGA in most cases.

Keywords: Genetic algorithm · Crossover · Traveling Salesman Problem

1 Introduction

The Traveling Salesman Problem (TSP) is an important test bed for combinatorial optimization algorithms and a model for many real world applications. It is an NP-hard problem [1–3,5] with the goal to find the shortest path to visit all cities [4]. Solving the TSP is of great importance because of its many real world applications ranging from Scheduling and routing [8,9], vehicle path planning

Y. Tan et al. (Eds.): ICSI 2022, LNCS 13344, pp. 272–281, 2022.
https://doi.org/10.1007/978-3-031-09677-8_23

[10–12,18], integrated circuit design [13], poly-genetic trees [14], machine learning [15], green logistics [13], efficient fuel management [16] and a host of others. In the amazon last mile challenge [17], one of the best performing algorithms for routing in package delivery was a variant of GA [19].

GA is one of the most successful intelligent meta-heuristics, however, most GA variants suffer from scalability and exploration in high dimension which has prompted most of the recent advances in GA to tackle these problems. MOGA [6] was developed based on the idea of mathematical ecological theory and biological evolution to tackle the above mentioned shortcomings. In MOGA, the number of offsprings from a single generation is relatively higher than those from earlier variants of GA and thus increases competitiveness for survival of the fittest. MOGA helps tackle the problems of premature convergence, local search, exploration and faster convergence to high quality solutions faced by GA. Building from MOGA, a GA called multi-inversion-based-genetic algorithm was developed for path planning of unmanned surface vehicles [7]. However MOGA's crossover operator is fairly complex and a higher ratio of similar offsprings are generated as a result of the increased number of offsprings.

To mitigate the shortcomings of MOGA, this study proposes a copy and paste crossover technique for MOGA which is less complex, trivial to implement and boosts the robustness and convergence speed of MOGA while preserving the schemata of existing solutions. Solutions are evolved by copying a section of a chromosome from one parent and pasting in another parent forming a new offspring and therefore it is named Copy and Paste Multi-offspring Genetic Algorithm (CP-MOGA). We focus on simplicity in our design and the contribution of this study is as follows:

1. We propose a relatively simple and effective crossover operator for generating a diverse set offsprings from a pair of parents.
2. Our design technique results in a faster and more stable MOGA that utilizes it's larger number of offsprings for exploration and exploitation.

The rest of this paper is organized as follows: Sect. 2 discusses related GAs, Sect. 3 introduces the proposed algorithm, Sect. 4 shows simulation results and finally Sect. 5 concludes the study.

2 Related Genetic Algorithms

In this section we discuss BGA [20] and MOGA [6] both of which are similar in their evolutionary process are the bases of our proposed method. MOGA is slightly different in that it produces four offsprings as opposed to two by the BGA. Note that the crossover and mutation operators discussed here are not for the original GA but for the variant designed for the TSP.

BGA was the first GA variant for the TSP in which solutions to a given problem undergo evolution by crossover and mutation. The algorithm uses uniform crossover to generate n offspring from n parents (every two parents give birth to two children). The evolutionary procedure for solutions is outlined below.

In the initialization phase, a population of solutions is randomly generated based on the given problem. Solutions are then evolved by crossover where two parents A and B are selected by roulette wheel and their genes exchanged by uniform crossover technique to generate two offsprings. After crossover an offspring undergoes mutation by introducing a small change to the offspring with a given probability. The mutation operator for BGA is same as that discussed in Sect. 3 where the offsprings selected for mutation are split into three sections, then the middle section is reversed and rejoined. We don't go into the details of BGA but refer the reader to [20].

The multi-offspring genetic algorithm builds from GA with the idea that parents can have more than two off-springs. In MOGA each parent generates four off-springs after crossover. Although, $2n$ offsprings are generated in each generation for MOGA, only the best n offsprings including elitist members (best k solutions in a population) survive to the next generation. By preserving elite parents and having many children with an additional crossover method, MOGA increases competitiveness among offsprings and thereby outperforms BGA. The evolutionary process of MOGA is quite similar to that of BGA with the main difference being in the crossover method used to generate the extra two offspring.

3 Copy and Paste Multi-offspring Genetic Algorithm

Our design aims for simplicity and robustness. The proposed algorithm mitigates the shortcomings of MOGA by simplifying it's crossover operations for easy implementation and enhancing it's search exploration ability, speed of convergence and robustness.

Before describing the solution construction and evolutionary procedure for CP-MOGA, We begin by defining important parameters for GA and formulating the objective function for the TSP:

Parameters: In every GA we need to define the mutation probability P_m, the initial population size n and the number of cities m.

Objective function: This is the goal we aim to achieve which is to find the shortest path to visit all cities with the constraint that we visit every city once. The distance is our measure and is computed between two cities from route coordinates if not given directly. The distance between any two cities i and j can be calculated using Eq. (1). After visiting all cities the total distance is the solution to the TSP as shown in Eq. (2). We outline formulation of the objective as follows

Given two cities i and j with coordinates (x_i, y_i) and (x_j, y_j), the distance between them is given by

$$D_{i,j} = \sqrt{(x_i - x_j)^2 + (y_i - y_j)^2} \tag{1}$$

As an example, to compute the objective function for a 10 city problem with a solution represented by $ABCDEFGHIJ$. This Means a traveling salesman goes from A to B, B to C, ... up to I to J and J back to A ($A > B > C >$

$D > E > F > G > H > I > J > A$) the total distance of the path is given by $D_{AB} + D_{BC} + D_{CD} + D_{DE} + D_{EF} + D_{FG} + D_{GH} + D_{HI} + D_{IJ} + D_{JA}$. For m cities, the distance covered or path length is represented mathematically by the Eq. 2.

$$D_{1-m} = D_{m,i} + \sum_{i=1}^{m-1} D_{i,i+1} \tag{2}$$

where D_{1-m} is the total cost or distance to travel from city 1 to city m and back to city 1.

To find the shortest path, we propose CP-MOGA with an evolutionary procedure outlined in the next section.

3.1 Initialization

For generalization and illustration purposes, a solutions is represented as an array of non-repeating integers where each integer represents a city. To initialize solutions, randomly generate a set of n non-repeating integer arrays each of size m. For instance if we have 8 cities and a population size of 3, we will have three different solution like $ParentA$ and $ParentB$ in Fig 1.

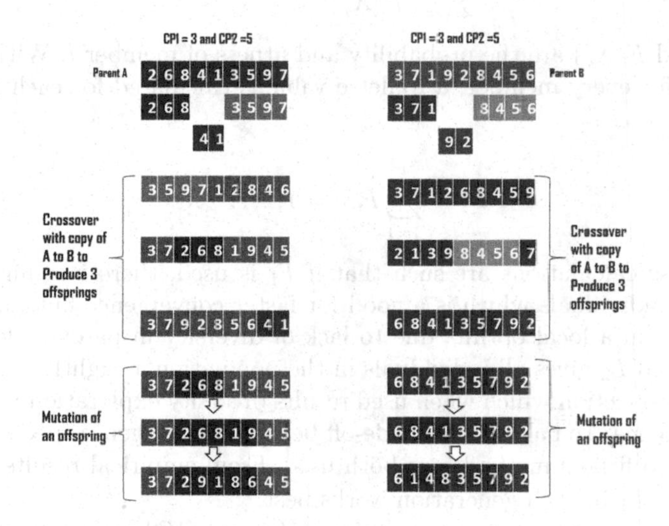

Fig. 1. Crossover and mutation for offspring generation in CP-MOGA

3.2 Crossover Operation

Crossover involves the exchange of gene sections e.g. [1,4] from ParentA to ParentB in Fig. 1 to form the offspring [1–9]. Two parents are needed for crossover (in some cases more than two parents are used). A roulette wheel [6] method is used to select a pair of parents for crossover. A roulette wheel probability based on the fitness or solution quality is computed and use for selection. In this study, two methods are used to compute the fitness. The fitness of a member in both of these methods is proportional to its position in the sorted population (top position high probability). In the first approach, the probability of each member is computed by Eq. (3)

$$F_1(X_i^{'}) = \beta(1-\beta)^{i-1}, i = 1, 2, \dots, n \tag{3}$$

where $X_i^{'}$ is the ith individual in the population of n sorted members. $\beta \in [0, 1]$ is a fitness control parameter often chosen to be between 0.01 and 0.3.

The second fitness function formulation is as shown in Eq. (4)

$$F_2(X_i^{'}) = \frac{n-i-1}{n}, i = 1, 2, \dots, n \tag{4}$$

After computing the fitness of individuals, the probability for an individual to be selected for crossover is also needed and is computed with Eq. (5) using the fitness values

$$P_i = \frac{F(X_i^{'})}{\sum_{i=1}^{n} F(X_i^{'})}, i = 1, 2, \dots, n \tag{5}$$

where P_i and $F(X_i^{'})$ are the probability and fitness of member i. With a known probability for every member, a roulette value is calculated for each individual using Eq. 6.

$$PP_i = \sum_{j=1}^{i} P_i, i = 1, 2, \dots, n \tag{6}$$

the fitness calculations are such that if F_1 is used, there is a high bias for selecting fit individuals which is a good for faster convergence but might result in stagnation in a local optima due to lack of diversity in parent selection. On the other hand F_2 gives all individuals in the population a slightly equal chance for crossover selection which when used results diversity exploration but leads to slow convergence. To balance the trade-off between convergence speed and diversity, these two fitness methods are both used. From empirical results randomly picking F_2 or F_1 in each generation works best.

In the crossover phase, two gene points $CP1$ and $CP2$ are randomly selected while ensuring both points are not same. The minimum distance between the two points can be set is the difference in positions between two points which is $|CP1 - CP2|$, in our case we set to 2. Each parent has three gene sections as can be seen in Fig. 1 which would be used for crossover. For instance ParentA denoted as [1–9] is clipped at gene points 3 and 5 producing [1,2,4,6,8], and [3,5,7,9] individual genes to be used for crossover. same is the case with ParentB

For ease of explanation the section between CP1 and CP2 will be referred to as the middle section, from ParentA this is [1,4]. The gene after the middle section we refer to as the last section [3,5,7,9] and that before as the first section [2,6,8] as shown in Fig. 1. To illustrate the crossover procedure for offspring generation proposed in this study, a 9-city problem is used. Let us say the two individuals selected for crossover are ParentA and ParentB with their chromosomes or path as shown in Fig. 1.

We start by selecting two points CP1 and CP2 using a uniform distribution $U(1, m-1)$ where m is the size of the problem or number of cities. After choosing CP1 and CP2, the crossover procedure involving the two parents is as follows:

a. One gene sections of ParentA is randomly chosen
b. Elements (cities) of the chosen gene section from a are removed from ParentB
c. The chosen gene section from a is placed in the first section of ParentB (Exploratory offsprings).
d. repeat a and b but this time the chosen gene section is placed in the last section of ParentB (Exploitative offsprings).
e. repeat a and b, then a random point is chosen between the first and last city genes of ParentB obtained in b and the chosen gene section from a placed in it's place (Exploratory and Exploitative offsprings).
f. The same procedure from step a to e but this time with the chosen section from ParentB. This procedure generates 6 off-springs from two parents.

The procedure is illustrated in Fig. 1 and we can observe it's fairly easy to comprehend and implement in practice.

3.3 Mutation Operation

After crossover, mutation is used to make minor variations to some of the offsprings generated. Every offspring is mutated with a given mutation probability Pm. So for a total of $3n$ offsprings in the population, about $Pm * 3n$ of them undergo mutation. For a chosen mutatable off-spring, mutation is carried out as follows:

a. Two mutation points are chosen the same way it's done for crossover
b. The middle section is flipped or reversed

After mutation, the new population is made up of the offsprings and q elitists' members. The pseudo code of the newly proposed algorithm is outlined in Algorithm 1. We the pick a number of offspring $U(n, n+10)$ plus q elites to the next generation. The number $U(n, n+10)$ proved to result in faster convergence in higher dimension from empirically results and this is because more diverse offspring are included in the offspring pool.

4 Simulation Experiments

To verify the performance of the proposed algorithm, we run simulations on twelve instances of TSP taken from the TSPLIB library and compare results with MOGA and GA. We choose these two algorithms because MOGA built from BGA and our proposed algorithm advances MOGA. Note that in the presented results Figures and Tables CP-MOGA is written in short as CPGA in order for it to fit.

The test problems are grouped into four classes based on their difficulty (number of cities or tours) as seen in Table 1, 2, 3 and 4. As can be noticed, the problem dimension is the integer number it's name.

Table 1. Results for Group 1

Problem	burma14			bays29			att48		
	BGA	MOGA	CPGA	BGA	MOGA	CPGA	BGA	MOGA	CPGA
Mean	30.885109	30.891714	30.878504	9154.974872	9201.661804	9115.258268	34632.91129	34874.43995	34348.41482
Worse	31.208766	31.208766	30.878504	9541.735863	9586.436729	9396.474986	35516.73784	36955.82849	35336.64984
Best	30.878504	30.878504	30.878504	9074.148048	9074.148048	9074.148048	33600.56146	33523.70851	33523.70851
Percentage mean error	0.00046237	0.00064718	0	0.100900881	0.11352624	0.066876969	0.47034421	0.634515085	0.412513196
Average time	0.581897	0.771466	0.521452	1.13729	1.369196	1.197614	3.243276	3.084515	3.212694
Average number of iteration	49.54	39.6	18.08	176.94	101.56	92.44	298.58	227.98	243.02

Table 2. Results for Group 2

Problem	eil51			berlin52			gr96		
	BGA	MOGA	CPGA	BGA	MOGA	CPGA	BGA	MOGA	CPGA
Mean	447.460878	451.449726	441.822101	8033.954236	8046.445949	7898.289124	545.487584	551.7375	537.80619
Worse	469.357385	475.366105	451.738011	8501.189585	8618.048791	8322.099894	582.87376	590.35064	558.917249
Best	433.732389	433.832676	428.981647	7544.365902	7598.442341	7544.365902	527.661487	516.484663	521.794053
Percentage mean error	0.0708224	0.09125721	0.05122501	0.235724736	0.200797684	0.212895879	0.13415381	0.19275797	0.10192889
Average time	2.679047	2.364255	2.562738	2.936803	3.666442	4.298606	12.169715	16.790744	15.735424
Average number of iteration	363.02	242.58	298.3	334.94	259.44	279.58	824.76	727.84	744.04

Table 3. Results for Group 3

Problem	krob100			u159			ch150		
	BGA	MOGA	CPGA	BGA	MOGA	CPGA	BGA	MOGA	CPGA
Mean	23888.48519	23955.85733	23375.74848	54245.43587	51892.42215	49849.72113	7893.345588	7846.609787	7480.838388
Worse	25135.8973	25016.67681	24421.49431	57773.00632	55904.57708	52558.53654	8393.746121	8287.17289	7829.220716
Best	22934.66166	22973.86279	22710.56018	50947.77786	47515.64792	47451.98712	7539.195445	7466.068508	7193.629826
Percentage mean error	0.523916745	0.457455586	0.431830332	1.673775513	1.933515641	1.08524162	0.214324007	0.191285928	0.145726143
Average time	10.765955	16.833239	14.43211	18.53412	33.267559	33.932948	17.112645	32.352859	33.106483
Average number of iteration	823.56	765.2	723.2	1000	1000	1000	1000	1000	1000

Table 4. Results for Group 4

Problem	att532			pr1002			pcb3038		
	BGA	MOGA	CPGA	BGA	MOGA	CPGA	BGA	MOGA	CPGA
Mean	319153.683	280486.3579	253281.8309	2.00E+06	1.75E+06	1.58E+06	2.75E+06	2.46E+06	2.34E+06
Worse	330378.2853	297752.2516	262105.8877	2.01E+06	1.79E+06	1.62E+06	2.78E+06	2.49E+06	2.35E+06
Best	307877.2744	272749.4699	245884.0903	1.98E+06	1.71E+06	1.55E+06	2.72E+06	2.42E+06	2.32E+06
Percentage mean error	7.726524021	9.199652098	5.248380021	9.74190557	27.0083722	24.00189996	25.01461661	24.79822852	11.58194729
Average time	115.748354	209.312204	200.086791	379.222364	628.966004	627.628679	2769.984054	5097.698321	6337.731706
Average number of iteration	1000	1000	1000	1000	1000	1000	1000	1000	1000

From the experiments, we observe CP-MOGA outperforms both BGA and MOGA on all TSP instances as shown in Tables 1, 2, 3 and 4. It converges faster than BGA and MOGA on majority of the problems and is more robust with the lowest error rate on all problems. The running time is comparable to that of MOGA and BGA on some instances, which is due to the number of offsprings generated in one generation.

For Group 1 problems, CP-MOGA outperforms its counterparts on all fronts, consistently finding the best solution with a very small error margin. The number of iterations needed to find the optimal solution is shorter and thus reducing the run time. In Group 2 the performance is overall comparable to that of group 1 with a slight increase in error rate. The increase is as a result of an increase in problem dimension. In this group, the algorithm does not find the optimal solution but comes very close. Group 3 problems have a significantly higher size and the optimal solution is ultimately difficult to find but the proposed algorithm performs best. As the problem size increases above 500 in Group 4 instances, all algorithms struggle to find an optimal solution. In this case, the algorithms are not efficient enough. This problem can easily be tackled by including a local search method in the algorithm.

In summary, the proposed CP-MOGA is efficient in exploration and exploitation and converges faster and to a better solution than compared to BGA and MOGA on all 12 instances of the TSP. The algorithm is also more robust in its solutions with reduced error rate and smaller number of generations taken to find better solutions.

5 Conclusion

In this study, a CP-MOGA is presented with a novel copy and paste crossover operator. The proposed crossover operator results in the generation of six offsprings from two parents. The operator enhances diversity and exploration thereby increasing competitiveness for survival of the fittest among offsprings in a generation. The algorithm is tested on twelve instances of TSP showing better performance in terms of error rate, solution quality, and convergence speed. The Strategy introduced in this study is highly efficient in problems of dimensions less than a hundred and like most optimization algorithms requires further improvement for instanced with high dimension though it still performs better than MOGA and BGA. In higher Dimensional problems, the algorithm requires a longer running time and there is a significant increase in error rate along with premature convergence. In future studies this problems will be handled by introducing a local search method or other efficient repair methods.

Acknowledgment. This work is partially supported by The National Natural Science Foundation of China (Grants Nos. 71901152, 71971143), Guangdong innovation team project "intelligent management and cross innovation" (2021WCXTD002), Scientific Research Team Project of Shenzhen Institute of Information Technology (SZIIT2019KJ022), and Guangdong Basic and Applied Basic Research Foundation (Project No. 2019A1515011392).

References

1. Graey, M.R., Johnson, D.S.: Computers and Intractability: A Guide to the Theory of NP-Completeness. Freeman W.H, San Francisco (1979)
2. Papadimitriou, C.H., Stegilitz, K.: Combinatorial Optimization: Algorithms and Complexity. Prentice Hall of India Private Limited, India (1997)
3. NP-hardness. https://en.wikipedia.org/wiki/NP-hardness. Accessed 27 Jan 2022
4. Philip, A., Taofiki, A.A., Kehinde, O.: A genetic algorithm for solving traveling salesman problem. Int. J. Adv. Comput. Sci. Appl. **2**(1), 26–29 (2011)
5. Lawler, E.L., Lenstra, J.K., Rinnooy Kan, A.H.G., Shmoys, D.B.: The Traveling Sales-man Problem: A Guided Tour of Combinatorial Optimization. Wiley, Chichester (1985)
6. Wang, J., Ersoy, O.K., He, M., Wang, F.: Multi-ospring genetic algorithm and its application to the traveling salesman problem. Appl. Soft Comput. **43**(1), 415–423 (2016)
7. Xin, J., Zhong, J., Yang, F., Cui, Y., Sheng, J.: An improved genetic algorithm for path-planning of unmanned surface vehicle. Sensors (Basel) **19**(11), 2640 (2019)
8. Ha, Q.M., Deville, Y., Pham, Q.D., Hà, M.H.: A hybrid genetic algorithm for the traveling salesman problem with drone. J. Heuristics **26**(2), 219–247 (2019). https://doi.org/10.1007/s10732-019-09431-y
9. Liu, F., Zeng, G.: Study of genetic algorithm with reinforcement learning to solve the TSP. Expert Syst. Appl. **36**(3), 6995–7001 (2009)
10. Kivelevitch, E., Cohen, K., Kumar, M.: A market-based solution to the multiple traveling salesmen problem. J. Intell. Robot. Syst. **72**, 21–40 (2013)

11. Murray, C.C., Chu, A.G.: The flying sidekick traveling salesman problem: optimization of drone-assisted parcel delivery. Transp. Res. Part C Emerg. Technol. **54**, 86–109 (2015)
12. Agatz, N., Bouman, P., Schmidt, M.: Optimization approaches for the traveling salesman problem with drone. Transp. Sci. **52**, 965–981 (2018)
13. Kirkpatrick, S., Gelatt, C.D., Vechi, M.P.: Optimization by simulated annealing. Science **220**(4598), 671–680 (1983). New series
14. Korostensky, C., Gonnet, G.H.: Using traveling salesman problem algorithms for evolutionary tree construction. Bioinformatics **16**(7), 619–627 (2000)
15. Roberti, R., Wen, M.: The electric traveling salesman problem with time windows. Transp. Res. Part E Logist. Transp. Rev. **89**(7), 32–52 (2016)
16. Ruiz, E., Albareda-Sambola, M., Fern'andez, E., Resende, M.G.C.: A biased random-key genetic algorithm for the capacitated minimum spanning tree problem. Comput. Oper. Res. **57**, 95–108 (2015)
17. Winkenbach, M., Parks, S., Noszek, J.: Technical Proceedings of the Amazon Last Mile Routing Research Challenge. https://dspace.mit.edu/handle/1721.1/131235. Accessed 02 Sep 2021
18. Park, H., Son, D., Koo, B., Jeong, B.: Waiting strategy for the vehicle routing problem with simultaneous pickup and delivery using genetic algorithm. Expert Syst. Appl. **165**, 113959 (2021)
19. Liu, S.: A powerful genetic algorithm for traveling salesman problem. Preprint arXiv:1402.4699 (2014)
20. Seniw, D.: A genetic algorithm for the traveling salesman problem. M.Sc. thesis, University of North Carolina, at Charlotte (1996). http://www.heatonresearch.com/articales/65/page1.html

Genetic Algorithm for Bayesian Knowledge Tracing: A Practical Application

Shuai Sun[1,2], Xuegang Hu[1,2](✉), Chenyang Bu[1,2](✉), Fei Liu[1,2], Yuhong Zhang[1,2], and Wenjian Luo[3]

[1] Key Laboratory of Knowledge Engineering with Big Data (the Ministry of Education of China), Hefei University of Technology, Hefei 230009, China
{sun_shuai,feiliu}@mail.hfut.edu.cn
{jsjxhuxg,chenyangbu,zhangyuhong}@hfut.edu.cn
[2] School of Computer Science and Information Engineering, Hefei University of Technology, Hefei 230009, China
[3] School of Computer Science and Technology, Harbin Institute of Technology, Shenzhen 518055, China
luowenjian@hit.edu.cn

Abstract. Online intelligent tutoring systems have developed rapidly in recent years. Analyzing educational data to help students personalize learning has become a research hotspot. Knowledge Tracing (KT) aims to assess students' changing cognitive states of skills by analyzing their performance on answers. As a representative KT model, Bayesian Knowledge Tracing (BKT) has good interpretability due to the use of the Hidden Markov Model. However, BKT needs to model students' performance on different skills separately. If BKT simultaneously traces the cognitive states of students' multiple skills, its time complexity increases exponentially with the number of skills. Therefore, we introduce a genetic algorithm to solve this problem and propose a Multi-skills BKT. This approach allows the BKT model to handle multiple skills simultaneously. Experiments on real datasets show that the model has a significant improvement in prediction performance over the BKT.

Keywords: Bayesian knowledge tracing · Multiple knowledge skills · Genetic algorithm

1 Introduction

Various E-learning systems are emerging today, such as large open online courses and intelligent tutoring systems [21]. The booming development of online education has generated a large amount of educational data. Analyzing educational data to help personalize learning or teaching has become a research hotspot [1].

Supported by the National Natural Science Foundation of China (under grants 61806065, 62120106008, 62076085, and 61976077), and the Fundamental Research Funds for the Central Universities (under grant JZ2020HGQA0186).

Y. Tan et al. (Eds.): ICSI 2022, LNCS 13344, pp. 282–293, 2022.
https://doi.org/10.1007/978-3-031-09677-8_24

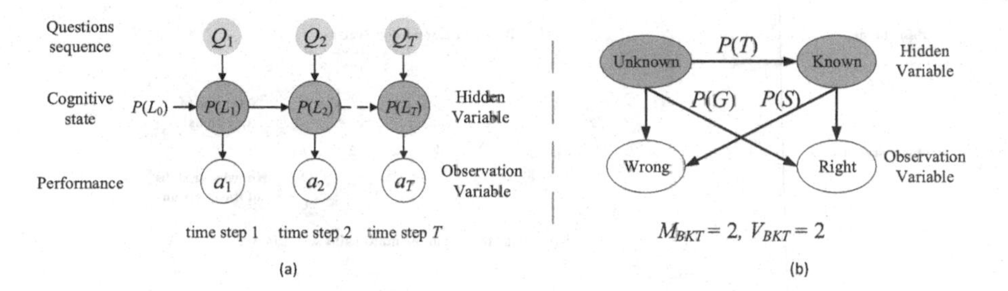

Fig. 1. Schematic diagram of the BKT model. (a) The framework of BKT. (b) The relationship between hidden states ($P(T)$) and the relationship between hidden states and observation states ($P(G)$, $P(S)$) [4].

Knowledge Tracing (KT) aims to trace the changes of students' cognitive states by analyzing the dynamic data generated by students in the process of answering questions [10,12]. KT can trace students' cognitive states and predict students' performance in future responses, which has important research significance and application prospects [13,18]. Existing KT models can be roughly divided into two categories, i.e., probability-graph-model based methods and deep-learning-based methods [10]. The representative model of the former is BKT proposed by Corbett et al. [4]. BKT uses Hidden Markov Models (HMM) [2] to model the KT task. The Deep Knowledge Tracing (DKT) model proposed by Piech et al. [20] in 2015 applies deep learning methods to KT task. Currently BKT has better interpretability compared to DKT.

The schematic diagram of the BKT is shown in Fig. 1, which uses HMM to model students' behavior [4]. A student's cognitive state at time t, i.e., $P(L_t)$, is considered to be only related to its cognitive state at time $t-1$. Here, the concept "cognitive state" refers to the degree of mastery of knowledge skills, which is a hidden variable. Only two cognitive states, i.e., known or unknown, are considered in BKT. And students' performance ("right" or "wrong") in answering exercises at time t, i.e., a_t, is a observable variable and is considered to be only related to their cognitive state [4]. The EM algorithm [23] is used to optimize the BKT parameters, including $P(L_0)$, $P(T)$, $P(S)$ and $P(G)$. And the complexity of Expectation-Maximization algorithm (EM) increases linearly with the number of hidden states [9].

However, the number of hidden states is exponentially related to the number of knowledge skills. Therefore, the existing BKT models optimized by the EM algorithm can only assume that each exercise only examines one knowledge skill. The reason is that if each exercise involves multiple knowledge skills, the time complexity will increase exponentially with the number of knowledge skills, resulting in unacceptable solution times. In fact, in a series of questions answered by students, the knowledge skills examined by each question may be different, and even one question may examine several knowledge skills. We call it "multi-skills knowledge tracing problem".

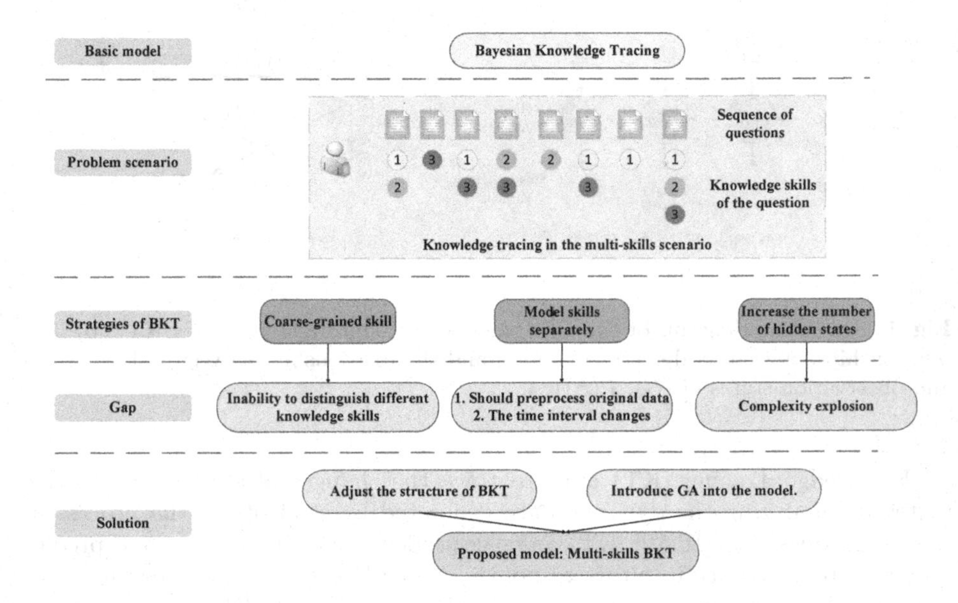

Fig. 2. The motivation of this study.

Although BKT can use the following strategies to solve problems with multiple knowledge skills, each strategy has its own shortcomings.

- Strategy 1 (Treat multiple knowledge skills as a whole): Regard all knowledge skills involved in an exercise as a new coarse-grained knowledge skill. We call this strategy "coarse-grained skill", which cannot distinguish the differences between different knowledge skills.
- Strategy 2 (Model skills separately): Use BKT to model each knowledge skill. The original data needs to be regrouped by knowledge skills. The time interval after processing is different from the original data.
- Strategy 3 (Increase the number of hidden states): If BKT wants to trace students' cognitive states of K knowledge skills at the same time, the number of hidden states (M_{BKT}) will be 2^K. Then, the BKT's time complexity is exponentially correlated with the number of knowledge skills K.

To solve the above problems, we propose multi-skills BKT model. We adjust the structure of BKT first. Instead of directly saving the state of students' overall cognition, the current cognition is indirectly calculated using the students' cognition state of each knowledge skill. Then, to address the problem of exponential explosion when tracing multiple skills simultaneously, the genetic algorithm (GA) [8] is introduced to optimize the model. The motivation of this study is summarized in Fig. 2.

Our contributions are as follows:

- We take the advantage of GA to solve the exponential explosion problem in the optimization process when BKT facing the multi-skills scenarios. There-

fore, The model broadens the practical application potential of the BKT model.
- Experiments show that the models have achieved better prediction performance in the multi-skills scenarios.

The structure of this paper is distributed as follows. The second section introduces the related work, mainly including the notations, knowledge tracing tasks, and the details of the traditional BKT model. In Sect. 3, the multi-skills BKT model is proposed. In Sect. 4, the experiments are discussed. Finally, the conclusion is presented in Sect. 5.

2 Related Work

2.1 Notations

For clarity, notations are listed in Table 1.

Table 1. Notations

Notations	Descriptions
Q_t	Question answered at time step t;
a_t	The student's performance at time step t;
y_t	The prediction score at time step t;
$P(L_t)$	The cognitive state of student at time step t;
$P(T), P(S), P(G)$	Transition, slipping and guessing probabilities, respectively;
M_{BKT}	The number of hidden states (i.e., unknown, know) in the BKT, the value is 2;
V_{BKT}	The number of observable states (i.e., wrong, right) in the BKT, the value is 2;
KC_j	The jth knowledge skill;
K	The number of knowledge skills;
C_j^t	The student's cognitive state of the jth knowledge skill at time step t;
W_j	The weight of the jth knowledge skill in the prediction score;
$P_e(L_t)$	The cognitive state at time step t after reassessing based on student's performance at time step t;

2.2 Knowledge Tracing

KT aims to obtain the students' cognitive state by analyzing students' performances. By knowledge tracing, domain experts can predict the performances of students in future learning behaviors, and provide intelligent recommendations for teachers and students.

Many models have been designed to solve KT tasks, mainly including Bayesian-method based models and deep-learning based models [10]. BKT is a representative of KT based on the Bayesian method, which will be described in detail in Sect. 2.3. With the development of online education platforms, the KT model and its variants have received extensive attention. Pardos et al. [19] introduced the item difficulty into the KT model, and proposed KT-IDEM. Considering the hierarchical structure and relationship between different skills, Käser et al. [11] used DBN to model students. Liu et al. [12] proposed the FBKT and T2FBKT models, introducing fuzzy theory into the BKT to handle continuous scoring scenarios. DKT proposed by Piech et al. [20] is a representative of KT based on the deep learning method. Nagatani et al. [16] improved the DKT model, considering the forgetting behavior of students to model and proposed the DKT with forgetting. A series of interpretable methods inspired by cognitive psychometrics are introduced to knowledge tracing, and attentive knowledge tracing (AKT) is generated [7]. Vie et al. [22] used factorization machines (FMs) to solve KT tasks. Yang et al. [24] proposed a graph-based KT model named GIKT to solve the multi-skills problem and data sparsity.

2.3 Bayesian Knowledge Tracing

BKT [4] uses the hidden Markov model (HMM) to model the students' learning data. In the BKT, whether a student can answer the question correctly mainly depends on the student's cognitive state. For a question, students may not master the knowledge skill but guess the correct answer (called "guessing"), or they may master the knowledge skill but get the wrong answer because of careless or other factors (called "slipping"). In summary, the relationship between hidden states ($P(T)$) and the relationship between hidden states and observation states ($P(G), P(S)$) are shown in Fig. 1 and described as follows.

At time step t, the formulas for calculating the student's performance are as follows [4]:

$$P(y_t = right|L_t) = P(L_t)(1 - P(S)) + (1 - P(L_t))P(G) \tag{1}$$

$$P(y_t = wrong|L_t) = P(L_t)P(S) + (1 - P(L_t))(1 - P(G)) \tag{2}$$

KT estimates the students' current cognitive states based on the known performance of the students. When the answer is right or the answer is wrong, the updated formulas correspond to Eq. (3) or Eq. (4), respectively [4]:

$$P_e(L_t) = \frac{P(L_t)(1 - P(S))}{P(L_t)(1 - P(S)) + (1 - P(L_t))P(G)} \tag{3}$$

$$P_e(L_t) = \frac{P(L_t)P(S)}{P(L_t)P(S) + (1 - P(L_t))(1 - P(G))} \tag{4}$$

The assumption of the BKT model is that after students complete the exercises, their cognitive state can change from unknown to known; and that once a student has mastered a certain knowledge component, they will not forget it [16]. Therefore, the transfer formula of the student's cognitive state at the next moment is Eq. (5) [4]:

$$P(L_{t+1}) = P_e(L_t) + (1 - P_e(L_t))P(T) \tag{5}$$

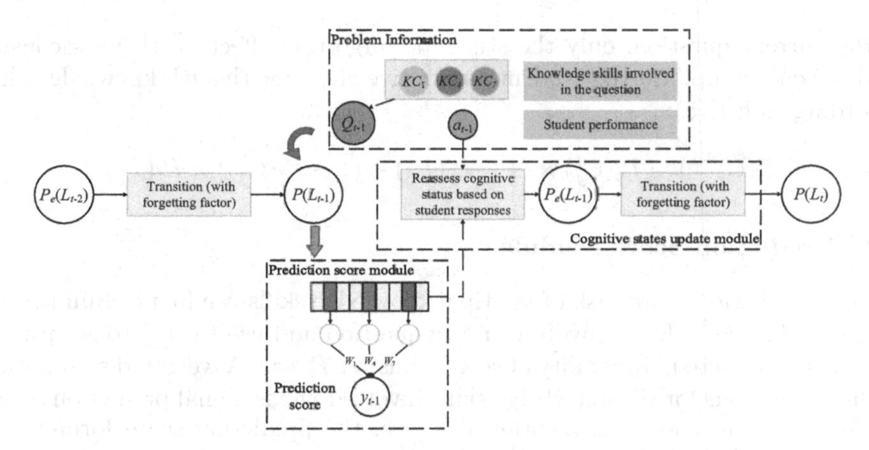

Fig. 3. Multi-skills BKT. The KC nodes in the "problem information" box represents the knowledge skills involved in the current question Q. The boxes marked with the same color in the "Predicted Score Module" box represent the students' mastery of the corresponding knowledge skills. $P(L_{t-1})$ denotes the cognitive state at time step $t-1$, which is inferred from the time step $t-2$. And $P_e(L_{t-1})$ is the cognitive state after reassessment based on the student's performance at time step $t-1$. (Color figure online)

3 Proposed Model

The overall architecture is shown in Fig. 3. To deal with the multi-skills KT tasks, our model introduces the "cognitive states update module" and "predictive score module" based on the BKT. To solve the "exponential explosion problem" of the parameter optimization algorithm caused by multiple skills, the model introduces a genetic algorithm.

3.1 Cognitive States Update Module

There are two cognitive states for each knowledge skill, named "known" and "unknown". At time step t, the overall cognitive state of the student can be expressed as $P(L_t) = P(C_1^t, C_2^t, ..., C_K^t)$. From time step t to time step $t+1$, the process of student's cognitive state transfer can be described as Eq. (6):

$$P(L_{t+1}) = P(L_t) \times P(L_{t+1}|L_t) = P(L_t) \times P(C_1^{t+1}, ..., C_K^{t+1}|C_1^t, ..., C_K^t). \quad (6)$$

Under the assumption that each knowledge component is independent of each other, the above formula can be reduced to Eq. (7):

$$P(L_{t+1}) = P(L_t) \times \prod_{i=1}^{K} P(C_i^{t+1}|C_i^t). \quad (7)$$

In the process of answering questions, students' cognitive states will change over time. We should evaluate the student's cognitive state according to his/her performance at each time step. As for the knowledge skills that are not examined in the current question, only the students' forgetting effect on them is considered. Then, we update the student's cognitive state for the jth knowledge skill according to Eq. (8):

$$P(C_j^{t+1}) = P_e(C_j^t) \times (1 - P_j(F)) + (1 - P_e(C_j^t)) \times P_j(T). \tag{8}$$

3.2 Prediction Score Module

For the prediction score task of multiple knowledge skills, we first obtain the set of knowledge skills KC_{in} involved in the question and use Eq. (1) to separately obtain the predicted probability of correct answer $Preds$. A weighted summation of the predictions for the knowledge skills involved yields a final prediction score.

According to the ideas mentioned above, the prediction score formula for multiple knowledge skills is as follows:

$$P(y_t = right) = \sum_{j \in KC_{in}} W_j \left[P(C_j^t) \times (1 - P_j(S)) + (1 - P(C_j^t) \times P_j(G)\right], \tag{9}$$

where $\sum W_j = 1$ and $j \in KC_{in}$.

For the weight of each knowledge skill W_j, we have two strategies:

- Strategy 1 (Average weights): It is considered that each knowledge skill has the same weight, that is $W_j = \frac{1}{NUM(KC_{in})}, j \in KC_{in}$;
- Strategy 2 (Evolution weights): W_j is generated by genetic algorithm optimization.

3.3 Parameter Optimization

BKT will face the problem of "exponential explosion" when processing the multi-skills knowledge tracing task (due to the EM algorithm). Therefore, our model introduces a genetic algorithm to optimize model parameters.

As shown in Fig. 4, the chromosome coding in our model includes $P(L_0)$, $P(T)$, $P(S)$, $P(G)$, $P(F)$ and $W(Optional)$ for each knowledge skill. These parameters are all in the range of $[0, 1]$, so we choose the real number encoding method. Each individual in the population is represented using a vector of length $5 * K$ or $6 * K$ (with parameter W), where K is the number of knowledge skills.

To evaluate the quality of individuals in the population, we use the AUC of the prediction results on the data set as the fitness value of the individual. For the evolution of the population, we use the "tournament selection" strategy to obtain the breeding population. According to the characteristics of the task, we choose the "simulated binary crossover" [5] and the "bounded polynomial mutation" [6] as the crossover and mutation operators, respectively, and then obtain the next generation population.

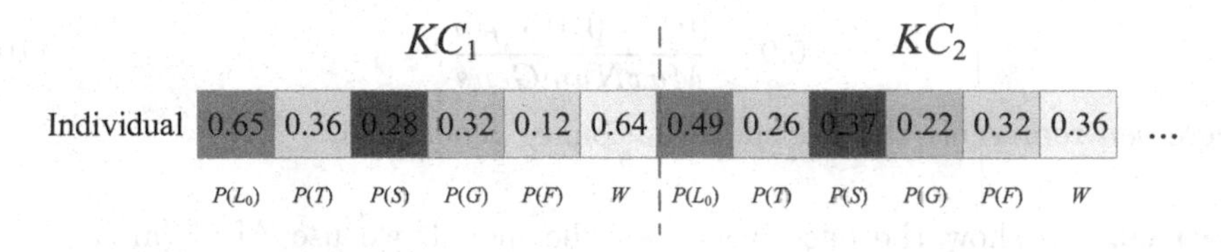

Fig. 4. The chromosome coding in our model. *KC* in the figure represents the parameters of each knowledge skill, and *individual* represents the chromosome encoding of the individuals in the population.

4 Experiments

4.1 Setup

Herein, we introduce the datasets, the parameter settings, the metrics, and the compared models.

Datasets. We use two real-world datasets in the experiments. The datasets come from the 2010 KDD Educational Data Mining Challenge. Details are as follows:

- Algebra I 2005–2006, denoted as Algebra05 [1].
- Bridge to Algebra 2006–2007, denoted as Bridge06 [2].

We first cleaned the data, removed the data without knowledge skill information, and removed the data with less than 15 steps in the student's response time step. The information of two datasets after preprocessing is shown in Table 2:

Table 2. Dataset statistics.

Dataset	#students	#skills	#problems	#interaction
Algebra05	545	112	1083	545969
Bridge06	1111	493	17837	1553671

Parameters. The parameters of the genetic algorithm we use are as follows, and other parameters are consistent with the original BKT paper. The population size *PopSize* is set to 100. The individual crossover and mutation rates change dynamically with evolutionary generations according to Eq. (10):

[1] Stamper, J., Niculescu-Mizil, A., Ritter, S., Gordon, G.J., & Koedinger, K.R. (2010). Algebra I 2005–2006. Challenge data set from KDD Cup 2010 Educational Data Mining Challenge. Find it at http://pslcdatashop.web.cmu.edu/KDDCup/downloads.jsp.

[2] J. Stamper, A. Niculescu-Mizil, S. Ritter, G. J. Gordon, and K.R.Koedinger, Bridge to Algebra 2006–2007. Challenge data set from KDD Cup 2010 Educational Data Mining Challenge. Find it at http://pslcdatashop.web.cmu.edu/KDDCup/downloads.jsp.

$$0.9 - \frac{(0.9 - 0.4) * gen}{MaxNumGens}, \tag{10}$$

where gen represents the current evolutionary generation.

Metrics. To show the effectiveness of the model, we use AUC (area under curve) as the metric of both our model and the compared models. AUC is an evaluation index to measure the performance of the binary classification model, which indicates the probability that the positive sample value correctly predicted by the classifier is greater than the negative sample value.

Compared Models. To demonstrate the feasibility of the proposed model, the compared models are listed as follows:

- BKT [4]: The traditional Bayesian knowledge tracing using the EM algorithm to optimize the parameters;
- GA-BKT: Use GA to optimize the parameters of traditional BKT;
- Multi-skills BKT (E-W): Our model with the "Evolution weights" strategy;
- Multi-skills BKT (A-W): Our model with the "Average weights" strategy.

4.2 Comparison Experiments

In this paper, the experiments adopted the method of five-fold cross-experiment verification to reduce the accidental error of the results.

Table 3. Comparison of prediction performance (AUC)

Model	Algebra05	Bridge06
BKT	0.634	0.621
GA-BKT	0.642	0.667
Multi-skills BKT (E-W)	0.680	0.650
Multi-skills BKT (A-W)	0.684	0.656

Table 3 shows that GA-BKT outperforms the traditional BKT. For example, on the dataset Bridge06, the AUC metric of GA-BKT is 0.667, while the AUC metric of BKT is 0.621. This shows that optimizing the parameters of BKT with GA is effective. Moreover, Table 3 and Fig. 5 also show that two variants of the Multi-skills BKT (E-W, A-W) also outperform BKT in performance. Compared with the traditional BKT, the Multi-skills BKT improves the AUC by about 0.05 and 0.03 on these two datasets, respectively. It shows that our proposed method to solve the multi-skill KT task has a certain effect. On the Algebra05 dataset, multi-skills BKT improves the AUC by about 0.04 compared to GA-BKT. But it drops by about 0.01 on the Bridge06 dataset. It may be because there are few exercises involving multiple knowledge skills in this dataset, and these questions involving multiple knowledge skills have a high similarity.

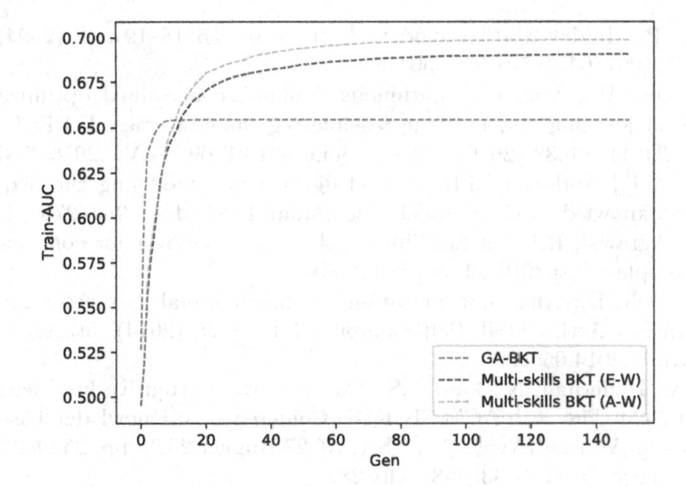

Fig. 5. Convergence plots of the AUC metric for the three models on the Algebra05 training set.

5 Conclusion

Personalized learning and teaching based on analyzing educational data has received extensive attention. As a representative KT model, BKT has good interpretability. Existing BKT models use the EM algorithm to optimize parameters, and the time complexity of the algorithm increases exponentially with the number of knowledge skills. Therefore, most existing BKT models assume that each exercise examines only one knowledge skill. In fact, a single exercise might examine multiple knowledge skills, a situation that is difficult for traditional BKT to handle. Therefore, we adjust the structure of the BKT, and introduce GA into the model to solve the problem that the time complexity of BKT increases exponentially with the number of knowledge skills. Finally, the experiments demonstrate the proposed multi-skills BKT model achieves good performance and the genetic algorithm is beneficial for multiple knowledge skills KT scenarios.

In the future, we will explore more possibilities to combine genetic algorithms with KT to expand the application scenarios of KT. First, we will use more datasets to demonstrate the effectiveness of our method. Then we will try more genetic algorithms, such as the species-based particle swarm optimization [14, 15], the memetic algorithm [17], etc. Fourth, Long-term KT (e.g. cross-semesters KT) can be viewed as a dynamic constrained optimization problem [3] to be studied.

References

1. Bakhshinategh, B., Zaiane, O.R., ElAtia, S., Ipperciel, D.: Educational data mining applications and tasks: a survey of the last 10 years. Educ. Inf. Technol. **23**(1), 537–553 (2017). https://doi.org/10.1007/s10639-017-9616-z

2. Blunsom, P.: Hidden Markov models. Lect. Notes **15**(18–19), 48 (2004). https://doi.org/10.1002/0471650129.dob0318
3. Bu, C., Luo, W., Yue, L.: Continuous dynamic constrained optimization with ensemble of locating and tracking feasible regions strategies. IEEE Trans. Evol. Comput. **21**(1), 14–33 (2016). https://doi.org/10.1109/TEVC.2016.2567644
4. Corbett, A.T., Anderson, J.R.: Knowledge tracing: modeling the acquisition of procedural knowledge. User Model. User-Adap. Inter. **4**(4), 253–278 (1994)
5. Deb, K., Agrawal, R.B., et al.: Simulated binary crossover for continuous search space. Complex Syst. **9**(2), 115–148 (1995)
6. Deb, K., Deb, D.: Analysing mutation schemes for real-parameter genetic algorithms. Int. J. Artif. Intell. Soft Comput. **4**(1), 1–28 (2014). https://doi.org/10.1504/IJAISC.2014.059280
7. Ghosh, A., Heffernan, N., Lan, A.S.: Context-aware attentive knowledge tracing. In: KDD 2020: The 26th ACM SIGKDD Conference on Knowledge Discovery and Data Mining, Virtual Event, CA, USA, 23–27 August 2020, pp. 2330–2339 (2020). https://doi.org/10.1145/3394486.3403282
8. Goldberg, D.E.: The Design of Innovation: Lessons from and for Competent Genetic Algorithms, vol. 1. Springer, New York (2002). https://doi.org/10.1007/978-1-4757-3643-4
9. Gupta, M.R., Chen, Y.: Theory and Use of the EM Algorithm. Now Publishers Inc (2011)
10. Hu, X., Liu, F., Bu, C.: Research advances on knowledge tracing models in educational big data. J. Comput. Res. Dev. **57**(12), 2523–2546 (2020). https://doi.org/10.7544/issn1000-1239.2020.20190767
11. Käser, T., Klingler, S., Schwing, A.G., Gross, M.: Dynamic Bayesian networks for student modeling. IEEE Trans. Learn. Technol. **10**(4), 450–462 (2017). https://doi.org/10.1109/TLT.2017.2689017
12. Liu, F., Hu, X., Bu, C., Yu, K.: Fuzzy Bayesian knowledge tracing. IEEE Trans. Fuzzy Syst. (2021). https://doi.org/10.1109/TFUZZ.2021.3083177
13. Liu, Q., et al.: EKT: exercise-aware knowledge tracing for student performance prediction. IEEE Trans. Knowl. Data Eng. **33**(1), 100–115 (2021). https://doi.org/10.1109/TKDE.2019.2924374
14. Luo, W., Sun, J., Bu, C., Liang, H.: Species-based particle swarm optimizer enhanced by memory for dynamic optimization. Appl. Soft Comput. **47**, 130–140 (2016). https://doi.org/10.1016/j.asoc.2016.05.032
15. Luo, W., Sun, J., Bu, C., Yi, R.: Identifying species for particle swarm optimization under dynamic environments. In: 2018 IEEE Symposium Series on Computational Intelligence (SSCI), pp. 1921–1928. IEEE (2018). https://doi.org/10.1109/SSCI.2018.8628900
16. Nagatani, K., Zhang, Q., Sato, M., Chen, Y.Y., Chen, F., Ohkuma, T.: Augmenting knowledge tracing by considering forgetting behavior. In: The World Wide Web Conference, WWW 2019, San Francisco, CA, USA, 13–17 May 2019, pp. 3101–3107 (2019). https://doi.org/10.1145/3308558.3313565
17. Neri, F., Cotta, C.: Memetic algorithms and memetic computing optimization: a literature review. Swarm Evol. Comput. **2**, 1–14 (2012). https://doi.org/10.1016/j.swevo.2011.11.003
18. Pardos, Z., Bergner, Y., Seaton, D., Pritchard, D.: Adapting Bayesian knowledge tracing to a massive open online course in edX. In: Educational Data Mining 2013. Citeseer (2013)

19. Pardos, Z.A., Heffernan, N.T.: KT-IDEM: introducing item difficulty to the knowledge tracing model. In: Konstan, J.A., Conejo, R., Marzo, J.L., Oliver, N. (eds.) UMAP 2011. LNCS, vol. 6787, pp. 243–254. Springer, Heidelberg (2011). https://doi.org/10.1007/978-3-642-22362-4_21
20. Piech, C., et al.: Deep knowledge tracing. Adv. Neural. Inf. Process. Syst. **28**, 505–513 (2015)
21. Romero, C., Ventura, S.: Educational data mining and learning analytics: an updated survey. Wiley Interdiscip. Rev. Data Min. Knowl. Discov. **10**(3), e1355 (2020). https://doi.org/10.1002/widm.1355
22. Vie, J.J., Kashima, H.: Knowledge tracing machines: factorization machines for knowledge tracing. In: The Thirty-Third AAAI Conference on Artificial Intelligence, AAAI 2019, The Thirty-First Innovative Applications of Artificial Intelligence Conference, IAAI 2019, The Ninth AAAI Symposium on Educational Advances in Artificial Intelligence, EAAI 2019, Honolulu, Hawaii, USA, 27 January–1 February 2019, vol. 33, pp. 750–757 (2019). https://doi.org/10.1609/aaai.v33i01.3301750
23. Wu, X., et al.: Top 10 algorithms in data mining. Knowl. Inf. Syst. **14**(1), 1–37 (2008). https://doi.org/10.1007/s10115-007-0114-2
24. Yang, Y., et al.: GIKT: a graph-based interaction model for knowledge tracing. In: Hutter, F., Kersting, K., Lijffijt, J., Valera, I. (eds.) ECML PKDD 2020. LNCS (LNAI), vol. 12457, pp. 299–315. Springer, Cham (2021). https://doi.org/10.1007/978-3-030-67658-2_18

Reinforced Event-Driven Evolutionary Algorithm Based on Double Deep Q-network

Tianwei Zhou[1], Wenwen Zhang[1,2], Junrui Lu[1,2], Pengcheng He[1], and Keqin Yao[3(✉)]

[1] College of Management, Shenzhen University, Shenzhen 518060, China
[2] Great Bay Area International Institute for Innovation, Shenzhen University, Shenzhen 518060, China
[3] Shenzhen Health Development Research and Data Management Center, Shenzhen 518060, China
szhealth-yao@163.com

Abstract. The real-world optimization task has long been viewed as a noteworthy challenge owing to its enormous search space. To deal with this challenge, evolution algorithm, especially differential evolution algorithm, attracts our attention owing to the excellent robustness. However, for traditional evolution algorithm, how to determine suitable parameters and strategies is a troublesome problem. To deal with the question the reinforced event-driven evolutionary algorithm (REDEA) based on double deep q-network is proposed which embed the double deep q-learning network into differential evolution algorithm with an event-driven controller. To verify the feasibility and superiority of our proposed algorithm, CEC 2013 test suits are utilized and four state-of-arts evolutionary algorithms are involved as the comparisons. The experimental results present that the proposal algorithm obtains comparable capability in most functions.

Keywords: Evolutionary algorithm · Double deep q-network · Event-driven · Differential evolution

1 Introduction

Optimization tasks are of high importance for real life. Evolutionary algorithms (EA) are efficient approaches for optimization, such as genetic algorithm [1], particle swarm optimization algorithm [2], ant colony algorithm [3] bacterial foraging optimization algorithm [4] etc. Some of these methods have been well applied in the industry. Reinforcement Learning(RL) is a kind of machine learning method. RL has been applied in many industries such as electricity [5], wireless communications [6], manufacturing scheduling [7] and other fields. The aim of RL is to train an agent which can perform the right action under different states. Unlike supervised learning or unsupervised learning, the RL directly

Y. Tan et al. (Eds.): ICSI 2022, LNCS 13344, pp. 294–304, 2022.
https://doi.org/10.1007/978-3-031-09677-8_25

interacts with the problem or environment. It learns from the interactions with the environment instead of from training data. For optimization, the RL can compare different solutions and choose one of them based on state and some work has been applied to real world [8,9]. Deep reinforcement learning (DRL) is the combination of RL and deep learning. Since the deep q-network (DQN) was proposed in [10], the DRL has created a lot of buzz around RL. Comparing with RL, with neural network and based on pre-training, DRL perform better in the problem with high dimension.

EA performs excellent in some complex NP-hard optimization tasks. In most circumstances, the parameter setting and strategy selection have significant influence on the efficiency of EA. Once problem is altered slightly, the EA will have to be modified. At the same time, RL or DRL can make decision under different states. Recently, some scholars began to focus on the combination between EA and RL to improve performances of the algorithm. The combined algorithms can be splinted into two categories. The first one trains the agent using online information. [11] used the theory of multi-armed Bandit to learn the selection of strategies online. [12] merged SCGA based on GA and reinforcement learning, then tested it on traveling salesman problem. [13] proposed a multi-objective DE algorithm using reinforcement learning strategies, which uses information entropy to analyze the objective function and determine the optimal probability distribution of the algorithms search strategy set. [14] composed the bacterial foraging optimization algorithm [4] and q-learning [15] to improve the search efficiency and balancing local and global search. These algorithms belong to online learning, which means they get the information no more from the iteration itself. And the second one takes the strategies that train the agent offline first and apply to the question un-trained to validate. [16,17] took the DRL methods. They both trained the agent on training function set offline first and then applied to the testing function set. Comparing with the online method, the offline training methods will make the decision more accurate according to the training phase. But the online strategies can make use of the information directly from the process to solve the problem.

The idea of event-driven comes from networked control area [18]. In event-driven algorithm, the event-driven mechanism will be designed to detect the occurrence of the interested situation. Then, corresponding services will be provided to satisfy specific destination. Based on this idea, the parameter controller is designed to improve the parameter design process. In this paper, in order to overcome the drawbacks of DE and utilize benefits of DRL the reinforced event-driven evolutionary algorithm (REDEA) was proposed. Firstly, to enhance the search ability and decrease the impact of single strategy which means only one mutation strategy will be used in the iteration process, we associate DE with DRL to formulate a strategy selector. Secondly, considering the dimension factor of the problem and the change of fitness value in the iterative process, an event driven parameter controller is designed.

This paper is organized as follows. In Sect. 2, we introduce differential algorithm and double deep q-network. In Sect. 3, the reinforced event-driven evolutionary algorithm (REDEA) proposed. Section 4 shows the experimental results and analysis, and the conclusions and discussion are presented in Sect. 5.

2 Pre Knowledge

2.1 Differential Evolution Algorithm

Differential evolution (DE) algorithm was proposed by [19,20]. It is an evolutionary algorithm with the advantage of fast convergence and little parameters. The typical DE includes four elemental steps - initialization, mutation, crossover and selection, and only the last three steps are repeated in DE iteration process. As a global optimum algorithm, DE begins with a randomly population X.

Mutation. After initialization, DE creates a donor mutant vector $V_i^{(t)}$ corresponding to each population member X. Usually, the mutation strategy of DE can be denoted as DE/M/N. M means which individual in the population will be chosen to manipulate and N is the number of subtract vector used in mutation process. Below are four most frequently referred mutation strategies:

$$\text{DE/rand/1: } V_i^{(t)} = X_{R_1^i}^{(t)} + F(X_{R_2^i}^{(t)} - X_{R_3^i}^{(t)}) \tag{1}$$

$$\text{DE/rand/2: } V_i^{(t)} = X_{R_1^i}^{(t)} + F(X_{R_2^i}^{(t)} - X_{R_3^i}^{(t)}) + F(X_{R_4^i}^{(t)} - X_{R_5^i}^{(t)}) \tag{2}$$

$$\text{DE/best/1: } V_i^{(t)} = X_{best}^{(t)} + F(X_{R_1^i}^{(t)} - X_{R_2^i}^{(t)}) \tag{3}$$

$$\text{DE/current\_to\_pbest/1: } V_i^{(t)} = X_{pbest}^{(t)} + F(X_{R_1^i}^{(t)} - X_{R_2^i}^{(t)}) \tag{4}$$

The $X_{R_1^i}^{(t)}$, $X_{R_2^i}^{(t)}$, $X_{R_3^i}^{(t)}$, $X_{R_4^i}^{(t)}$ and $X_{R_5^i}^{(t)}$ are mutually exclusive integers casually chosen from scope $[1, Np]$ at t_{th} generation. The $X_{best}^{(t)}$ is the individual which has the best fitness value in the whole population of t_{th} generation. The $X_{pbest}^{(t)}$ is the top individual in the population. The scaling factor F is a positive control parameter used to scale different vectors.

Crossover. After mutation, crossover operator will be performed. Binomial crossover is one of the frequently used crossover operators. The scheme can be expressed as

$$u_{i,j}^{(t+1)} = \begin{cases} v_{i,j}^{(t)}, \text{ if } j = K \text{ or } rand_{i,j}[0,1] \leq Cr \\ x_{i,j}^{(t)}, \text{ otherwise} \end{cases} \tag{5}$$

where Cr is the crossover rate, K is a randomly chosen number in $\{1, 2, , d\}$, $v_{i,j}^{(t)}$ and $x_{i,j}^{(t)}$ is respectively j_{th} dimension of $V_i{}^{(t)}$ and $X_i{}^{(t)}$, and $rand_{i,j}[0, 1]$ is a uniform random number in $[0, 1]$.

Selection. A greedy selection strategy will be used:

$$x_i^{(t+1)} = \begin{cases} u_i^{(t+1)} & \text{if } f(u_i^{(t+1)}) < f(x_i^{(t)}) \\ x_i^{(t)} & \text{otherwise} \end{cases} \tag{6}$$

where $f(u_i^{(t+1)})$ and $f(x_i^{(t)})$ represent the fitness value of $u_i^{(t+1)}$ and $x_i^{(t)}$ correspondingly.

2.2 Double Deep Q-network

The deep q-network (DQN) is one of DRL. It was proposed by [10] in 2016. It was further extended by double deep q-networks (DDQN) [21] to solve the problem of overestimating the Q value. The computing method of DQN can be expressed by:

$$y_j = \begin{cases} R_j, & is\_end_j \ is \ true \\ R_j + \gamma \max_{a'} Q'(\phi(S_j'), A_j', w'), & else \end{cases} \tag{7}$$

where y_j and R_j is the target Q value and the reward of j_{th} sample, γ is attenuation factor, Q' is the predicted Q value, $\phi(S_j')$ is eigenvectors of S_j', A_j' is j_{th} action set, w' is parameter of objective q-network, is_end_j is termination condition, and a' is the action selected from A_j' by $\epsilon-$greedy method.

Although two Q networks are used, the calculation of y_j is still obtained by greedy method. For DDQN, it is no longer directly to find the maximum Q value in each action through the target Q network. The movement of the maximum Q value was found first, then this selected action $a^{max}(S_j', w)$ was used to calculate the target Q value in the target network.

$$y_j = R_j + \gamma Q'(\phi(S_j'), \arg\max_{a'} Q(\phi(S_j'), a, w), w') \tag{8}$$

3 Reinforced Event-Driven Evolutionary Algorithm

3.1 Framework of the REDEA

The RL has five main components, that is agent, environment, action, reward and state. The agent chooses action a from actions set A to perform in the environment. Then it will update the policy based on the reward and new state S' that the environment returned from the network.

The aim is to train an agent which can choose a mutation strategy under different states. Obviously, the action of agent is selecting mutation strategy for every individual $x_i^{(t)}$ at each generation. The environment is DE itself. DE will take the mutation strategy from the agent and implement it for every individual, then return the state and reward to the agent.

The Reward R is calculated by equation (9) where $f_{optimum}$ means the optimal value of the test problem.

$$R = \max\left\{\frac{f(x_i) - f(u_i)}{f(u_i) - f_{optimum}}, 0\right\} \tag{9}$$

The State S contains the following features:

- Fitness change: $f(x_i) - f_{bestsofar}$, the change of fitness value between the i_th solution of the population and the best value found up to this step in a single run.
- Stagnation time: numbers of iteration that the $f(x_i)_{bestsofar}$ remain the same.
- Iteration stage: $\frac{iteration}{iteration_{max}}$, the ratio of $iteration$ in the $iteration_{max}$.
- Distance: $\boldsymbol{x}_i - \boldsymbol{x}_j$, the distance of i_{th} solution and a random solution in the population.

The Algorithm 1 describes the process of REDEA.

Algorithm 1 REDEA

Require: Iterative wheel number T, state feature dimension N, action set A, step length α, attenuation factor γ, exploration rate ε, current q-network Q, target q-network parameter update frequency C, the number of training functions U.
  Initialize $DE(F, Np, Cr)$
  Initialize value Q corresponding to state S and actions A randomly
  Initialize value Q' corresponding to state S and actions A randomly
  for t from 1 to T **do**
    Initialize $X_0 = [x_1, x_2, x_{Np}]$ randomly
    repeat
      for i from 1 to Np **do**
        if $rand(0.1) < \varepsilon$ **then**
          randomly choose action a_t from actions set A
        else
          select $a_t = a^{max}(S'_j, w)$
        end if
        Apply a_t to x_t generate v_t
        Do binomial crossover on v_t and x_t to generate u_t
        Evaluate u_t and compare with x_t
      end for
      Calculate and return the Reward R_t and State S' and $\phi(S')$
      if t mod $C = 1$ **then**
        Update $W = W'$
      end if
    until $N <$ MAXNFE or optimum has been reached
  end for
  return W

3.2 Event-Driven Parameter Control

The control parameter scale factor F and crossover rate Cr perform huge impact on the iteration process. Problems have their unique features, thus, the best choice of the two parameters for different questions is not the same. One of the most important features is the dimension of target function. The objective function with higher dimensions require more exploration in the searching space. On the contrary, the one with lower dimensions needs more concentration on exploitation.

To increase the impact of different strategies chosen in the iteration process, we propose the event-driven parameter control. It will determine the parameter according to the specific strategy determined by dimension of questions and the fitness value from iteration process.

Crossover Rate Cr. For Cr, according to the observation of the experiment, when the dimension of objective function is more than 10. The value for every individual in the population will change according to the following scheme

$$Cr_i^{(t+1)} = \begin{cases} Cr_i^{(t)}, & f(x_i^{(t+1)}) \leq f(x_i^{(t)}) \\ randn[0.5, 0.1], & \text{otherwise} \end{cases} \tag{10}$$

Here, $randn[0.5, 0.1]$ is a normal distribution of which the means is 0.5, and the standard deviation is 0.1. For minimization problems, when the fitness value of $(t+1)_{th}$ generation is bigger than the t_{th} generation, the Cr will pick a new value from $randn[0.5, 0.1]$.

Scale Factor F. For rand mutation operators, F will follow this scheme where NFE and MAXNFE respectively represent the number of evaluation and max number of evaluation

$$F_i^{(t+1)} = \begin{cases} 1.2F_p, & \text{NFE} < 0.2\text{MAXNFE} \\ 0.8F_p, & \text{NFE} < 0.6\text{MAXNFE} \\ 0.2F_p, & \text{NFE} < 0.8\text{MAXNFE} \end{cases} \tag{11}$$

For best/pbest mutation operators, F will be updated following this scheme

$$F_i^{(t+1)} = \begin{cases} 0.2F_p, & \text{NFE} < 0.2\text{MAXNFE} \\ 0.8F_p, & \text{NFE} < 0.6\text{MAXNFE} \\ 1.2F_p, & \text{NFE} < 0.8\text{MAXNFE} \end{cases} \tag{12}$$

F_p is the initial scaling factor. The decrease on rand mutation will increase the exploitation efficiency at first. With the increment of iteration, the disturbance is decreased to focus more on exploitation. The strategy for best/pbest mutation operators will perform the opposite effect.

4 Experiments

4.1 Experimental Details

The proposal REDEA algorithm was tested on the CEC 2013 benchmark constrained optimization problems with 10D, 30D, where D represents the dimension of questions. $(f_1 - f_5)$ is uni-modal function. $(f_6 - f_{20})$ is basic multi-modal function and $(f_{21} - f_{28})$ is composition function. 28 testing functions are separated into training function sets which are $(f_1 - f_{20})$ and testing function sets $(f_{21} - f_{28})$. We compared REDEA with typical DE [20] and the following state-of-the-art DE algorithms.

SaDE [22]: The classical adaptive DE method. In SaDE trail vector generation strategies and their associated control parameter values are self-adapted by learning from the past.

JADE [23]: The classical adaptive DE method. JADE proposed the DE/Current-to-pbest/1 mutation strategy firstly.

LSHADE [24]: The TOP3 of the bound constraint competition of CEC 2014 which further extends SHADE [25] with linear population size reduction.

The following parameter are used for REDEA, including population size $N = 100$, initial scale factor $F_p = 0.8$, crossover rate $Cr = 1.0$, discount factor $\gamma = 0.99$, learning rate $\alpha = 0.005$, target q-network parameter update frequency $C = 1000$. Table 1 presents specific parameter settings and operators in compared algorithm.

Table 1. Operators Used and Parameter And Hyper-parameter Settings of The Compared Algorithm

Alg.	Operators		Control parameters	Hyper-parameter
	Mutation	Crossover		
DE	DE/rand/1	bin	$N = 5D$, $F = 0.5, Cr = 0.9$t	NA
SaDE	DE/current-to-pbest/1	bin	$N = 30, 100$ when $D = 10, 30$	$\mu_{Cr} = 0.5, \mu_F = 0.5$, $c = 0.1, p = 0.05$
JADE	DE/current-to-best/2 DE/rand/1	bin	$N = 10D$	$\mu_{Cr} = 0.5, \mu_F = 0.5$
LSHADE	DE/current-to-pbest/1	bin	N=18D	memorysize = 6

The mean and standard deviation of the error values used for analogy were obtained for each function over 30 runs. In the next part, the experimental results are summarized in Table 2 and 3 in which the mean, standard deviation and the results of Wilcoxon rank-sum hypothesis test results are held. The results for MAXNFE are also presented as box plot.

The Mann Whitney Wilcoxon tests are performed to analyze the equality of means between REDEA and the compared algorithms. The results are shown by symbols +, − and ≈ in the tables, which respectively indicate that REDEA performs significantly worse, better or similar to the compared algorithm at a significance level 0.05.

4.2 Results Analysis

Table 2 and Table 3 are the summary of the results when the MAXNFE has been reached. From Table 2, the REDEA performs significantly better than Typical DE, JADE, SaDE and LSHADE on f_1, $(f_{22} - f_{26})$ and f_{28}. On f_{22}, the REDEA performs similar to LSHADE and better than the other three comparison algorithms. From Table 3, when the dimension D raised from 10 to 30, the performance of REDEA lost in some functions, especially for $(f_1 - f_{23})$ and f_{27}. Figure 1 and Fig. 2 show the results more intuitively.

Table 2. The comparison of proposal algorithm on the CEC 2013 test suite for $D = 10$ when MAXNFE has been reached

	REDEA		DE			JADE			SaDE			LAHADE		
	Mean	Std. Dev.	Mean	Std. Dev.	WR	Mean	Std. Dev.	WR	Mean	Std. Dev.	WR	Mean	Std. Dev.	WR
f_{21}	3.95E+02	3.04E+01	4.00E+02	1.16E−13	−	4.00E+02	1.16E−13	−	4.00E+02	1.16E−13	−	4.00E+02	1.16E−13	−
f_{22}	1.74E+03	2.04E+02	2.20E+03	5.87E+02	−	2.69E+03	1.67E+02	−	2.14E+03	6.66E+02	−	9.22E+02	9.96E+02	+
f_{23}	1.79E+03	2.26E+02	1.92E+03	1.83E+02	−	2.09E+03	1.91E+02	−	2.02E+03	8.78E+01	−	2.01E+03	9.16E+01	−
f_{24}	2.10E+02	5.35E+00	2.55E+02	2.25E+01	−	2.97E+02	2.20E+01	−	2.76E+02	2.06E+01	−	2.47E+02	2.82E+01	−
f_{25}	2.11E+02	5.52E+00	2.39E+02	3.72E+01	−	2.78E+02	1.46E+01	−	2.70E+02	2.45E+01	−	2.32E+02	1.83E+01	−
f_{26}	1.77E+02	4.64E+01	3.97E+02	1.67E+01	−	4.00E+02	3.80E−01	−	4.01E+02	3.02E−01	−	2.41E+02	8.09E+01	−
f_{27}	3.92E+02	9.15E+01	4.00E+02	2.79E−14	≈	4.00E+02	1.06E−14	≈	4.00E+02	3.34E−14	≈	4.00E+02	2.79E−14	≈
f_{28}	8.54E+02	3.28E+02	1.02E+03	6.75E+01	−	1.14E+03	5.38E+01	−	1.07E+03	1.08E+02	−	8.54E+02	3.34E+02	−
	+/≈/−		0/1/7			0/1/7			0/1/7			1/1/6		

Table 3. The comparison of proposal algorithm on the CEC 2013 test suite for $D = 30$ when MAXNFE has been reached

	REDEA		DE			JADE			SaDE			LAHADE		
	Mean	Std. Dev.	Mean	Std. Dev.	WR	Mean	Std. Dev.	WR	Mean	Std. Dev.	WR	Mean	Std. Dev.	WR
f_{21}	9.25E+02	2.41E+02	3.53E+02	7.04E+01	+	3.33E+02	6.18E+01	+	3.22E+02	7.17E+01	+	3.10E+02	3.64E+01	+
f_{22}	7.17E+03	3.93E+02	4.35E+03	3.90E+02	+	3.61E+03	1.16E+03	+	1.82E+03	5.32E+02	+	2.48E+03	2.82E+03	+
f_{23}	8.64E+03	4.21E+02	5.62E+03	1.22E+03	+	6.80E+03	4.51E+02	+	6.77E+03	4.08E+02	+	6.57E+03	1.22E+03	+
f_{24}	3.03E+02	7.87E+00	2.13E+02	1.51E+01	+	5.68E+02	1.97E+02	−	5.97E+02	1.73E+02	−	3.06E+02	1.79E+01	−
f_{25}	3.21E+02	4.11E+00	3.41E+02	9.15E+01	−	4.41E+02	3.96E+00	−	4.40E+02	2.77E+00	−	3.80E+02	6.28E+01	−
f_{26}	2.72E+02	4.62E+01	2.38E+02	5.14E+01	+	3.16E+02	3.24E+01	−	3.09E+02	5.40E+01	−	2.89E+02	6.37E+01	≈
f_{27}	1.36E+03	4.85E+01	3.44E+02	8.01E+01	+	5.04E+02	1.02E+02	+	6.08E+02	1.36E+02	+	1.34E+03	4.80E+01	≈
f_{28}	1.52E+03	1.74E+02	2.33E+02	9.59E+01	+	3.61E+03	1.21E+03	−	3.63E+03	1.62E+03	−	3.16E+02	2.48E+02	−
	+/≈/−		7/0/1			4/0/4			4/0/4			3/2/3		

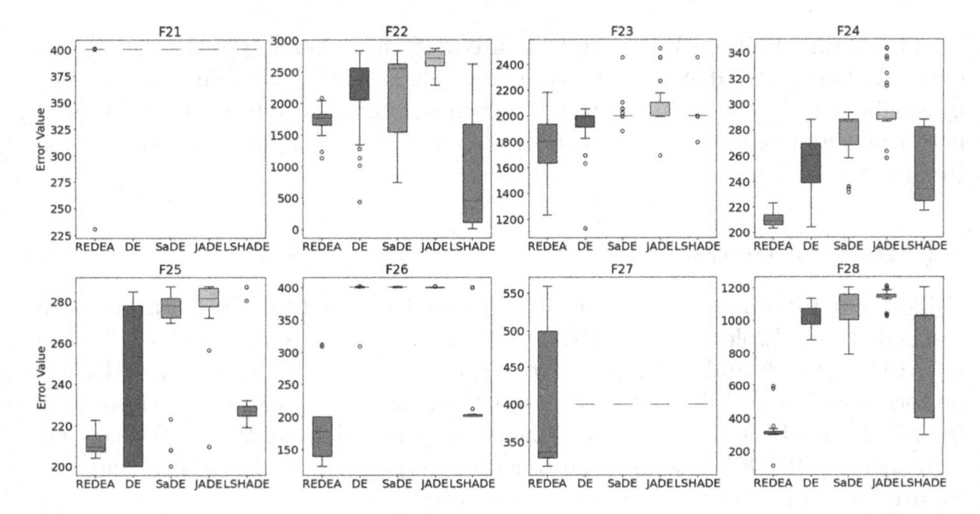

Fig. 1. Function error value obtained by 30 runs on the CEC 2013 test suit for $D = 10$ when MAXNFE has been reached.

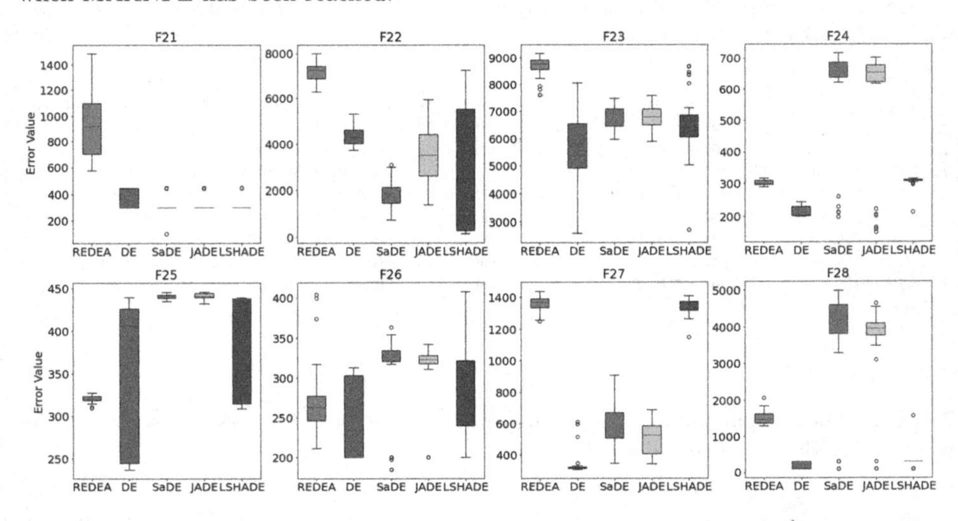

Fig. 2. Function error value obtained by 30 runs on the CEC 2013 test suit for $D = 30$ when MAXNFE has been reached.

5 Conclusion and Discussion

In this work, we proposed the reinforced event-driven evolutionary algorithm based on double deep q-network, which learns from the optimization process of a set of training functions offline. To maximum the effect of different mutation strategies, we also proposed an event-drive parameter control method. Future research will focus on the combination of multi-objective optimization and the deep reinforcement learning. Moreover, analyzing the state feature of specific question to improve the efficiency of training model is still an interesting topic.

Acknowledgement. The study was supported in part by the Natural Science Foundation of China Grant No. 62103286, No. 62001302, No. 71971143, in part by Social Science Youth Foundation of Ministry of Education of China under Grant 21YJC630181, in part by Guangdong Basic and Applied Basic Research Foundation under Grant 2021A1515011348, 2019A1515111205, 2019A1515110401, in part by Natural Science Foundation of Guangdong Province under Grant 2020A1515010749, 2020A 1515010752, in part by Key Research Foundation of Higher Education of Guangdong Provincial Education Bureau under Grant 2019KZDXM030, in part by Natural Science Foundation of Shenzhen under Grant JCYJ20190808145011259, in part by Shenzhen Science and Technology Program under Grant RCBS2020071 4114920379, in part by Guangdong Province Innovation Team under Grant 2021WCXTD002.

References

1. Holland, J.H.: Adaptation in Natural and Artificial Systems: An Introductory Analysis with Applications to Biology, Control, and Artificial Intelligence. MIT press, Cambridge (1992)
2. Kennedy, J., Eberhart, R.: Particle swarm optimization. In: Proceedings of ICNN 1995 - International Conference on Neural Networks, vol. 4, pp. 1942–1948 (1995)
3. Dorigo, M., Birattari, M., Stutzle, T.: Ant colony optimization. IEEE Comput. Intell. Mag. **1**(4), 28–39 (2006)
4. Passino, K.M.: Biomimicry of bacterial foraging for distributed optimization and control. IEEE Control Syst. Mag. **22**(3), 52–67 (2002)
5. Zhang, D., Han, X., Deng, C.: Review on the research and practice of deep learning and reinforcement learning in smart grids. CSEE J. Power Energy Syst. **4**(3), 362–370 (2018)
6. Huang, Y., Xu, C., Zhang, C., Hua, M., Zhang, Z.: An overview of intelligent wireless communications using deep reinforcement learning. J. Commun. Inf. Networks **4**(2), 15–29 (2019)
7. Wang, L., Pan, Z., Wang, J.: A review of reinforcement learning based intelligent optimization for manufacturing scheduling. Complex Syst. Modeling Simul. **1**(4), 257–270 (2021)
8. A. Mirhoseini, et al: Device placement optimization with reinforcement learning. In: International Conference on Machine Learning. PMLR, pp. 2430–2439 (2017)
9. Radaideh, M.I., Shirvan, K.: Rule-based reinforcement learning methodology to inform evolutionary algorithms for constrained optimization of engineering applications. Knowl.-Based Syst. **217**, 106836 (2021)
10. Mnih, V., et al.: Human-level control through deep reinforcement learning. Nature **518**(7540), 529–533 (2015)
11. Gong, W., Fialho, Á., Cai, Z.: Adaptive strategy selection in differential evolution. In: Proceedings of the 12th Annual Conference on Genetic and Evolutionary Computation, pp. 409–416 (2010)
12. Chen, F., Gao, Y., Chen, Z.-Q., Chen, S.-F.: SCGA: controlling genetic algorithms with sarsa (0). In: International Conference on Computational Intelligence for Modelling, Control and Automation and International Conference on Intelligent Agents, Web Technologies and Internet Commerce (CIMCA-IAWTIC 2006), vol. 1, pp. 1177–1183. IEEE (2005)
13. Huang, Y., Li, W., Tian, F., Meng, X.: A fitness landscape ruggedness multiobjective differential evolution algorithm with a reinforcement learning strategy. Appl. Soft Comput. **96**, 106693 (2020)

14. Niu, B., Xue, B.: Q-learning-based adaptive bacterial foraging optimization. In: Chen, X., Yan, H., Yan, Q., Zhang, X. (eds.) ML4CS 2020. LNCS, vol. 12487, pp. 327–337. Springer, Cham (2020). https://doi.org/10.1007/978-3-030-62460-6_29
15. Watkins, C.J., Dayan, P.: Q-learning. Mach. Learn. **8**(3–4), 279–292 (1992)
16. Sharma, M., Komninos, A., López-Ibáñez, M., Kazakov, D.: Deep reinforcement learning based parameter control in differential evolution. In: Proceedings of the Genetic and Evolutionary Computation Conference, pp. 709–717 (2019)
17. Sun, J., Liu, X., Bäck, T., Xu, Z.: Learning adaptive differential evolution algorithm from optimization experiences by policy gradient. IEEE Trans. Evol. Comput. **25**, 666–680 (2021)
18. Zhou, T., Zuo, Z., Wang, Y.: Self-triggered and event-triggered control for linear systems with quantization. IEEE Trans. Syst. Man Cybern. Syst. **50**(9), 3136–3144 (2020)
19. Storn, R.: On the usage of differential evolution for function optimization. In: Proceedings of North American Fuzzy Information Processing, pp. 519–523. IEEE (1996)
20. Storn, R., Price, K.: Differential evolution-a simple and efficient heuristic for global optimization over continuous spaces. J. Global Optim. **11**(4), 341–359 (1997)
21. Van Hasselt, H., Guez, A., Silver, D.: Deep reinforcement learning with double q-learning. In: Proceedings of the AAAI Conference on Artificial Intelligence, vol. 30, no. 1 (2016)
22. Qin, A.K., Huang, V.L., Suganthan, P.N.: Differential evolution algorithm with strategy adaptation for global numerical optimization. IEEE Trans. Evol. Comput. **13**(2), 398–417 (2008)
23. Zhang, J., Sanderson, A.C.: Jade: adaptive differential evolution with optional external archive. IEEE Trans. Evol. Comput. **13**(5), 945–958 (2009)
24. Tanabe, R., Fukunaga, A.S.: Improving the search performance of shade using linear population size reduction. In: IEEE Congress On Evolutionary Computation (CEC). IEEE 2014, pp. 1658–1665 (2014)
25. Tanabe, R., Fukunaga, A.: Evaluating the performance of shade on CEC 2013 benchmark problems. In: 2013 IEEE Congress on Evolutionary Computation, pp. 1952–1959. IEEE (2013)

Offline Data-Driven Evolutionary Optimization Algorithm Using K-Fold Cross

Mengzhen Wang, Yawen Shan, and Fei Xu(✉)

School of Artificial Intelligence and Automation, Huazhong University of Science and Technology, Wuhan 430074, China
fxu@hust.edu.cn

Abstract. In the field of science and engineering, there are many offline data-driven optimization problems, which have no mathematical functions, and cannot use numerical simulations or physical experiments, but can only use the historical data collected in ordinary times to evaluate the quality of candidate solutions during the optimization process. In order to solve offline data-driven optimization problems, offline data-driven evolutionary algorithms use historical data to build surrogate models to simulate the real objective function.

In this paper, an offline data-driven evolutionary optimization algorithm using k-fold cross is proposed. The proposed algorithm uses radial basis function networks as surrogate models and uses the k-fold cross method to build the ensemble surrogate in order to reduce the number of surrogate models in the surrogate and the time cost of the algorithm. To improve the performance of the algorithm, the number of hidden layer neurons and the kernel function in radial basis function network are determined by analyzing the effects of the parameters on the performance of the algorithm. Experimental results on benchmark problems show that the algorithm has good performance and low time cost. Moreover, a similar algorithm uses the parameters of radial basis function network in the proposed algorithm, the performance of the algorithm is improved, which indicates that the parameters have some universality.

Keywords: Offline data-driven optimization · Surrogate model · Evolutionary algorithm · Radial-basis-function network · K-fold cross-validation

1 Introduction

Many real-world optimization problems do not have real objective functions and evaluating the objectives of candidate solutions only based on data, collected from numerical simulations, physical experiments, or usual life. For example, blast furnace optimization problems [1], trauma system design optimization problems [2], and airfoil design optimization problems [3]. Such optimization problems can be called data-driven optimization problems [4].

In order to solve data-driven optimization problems, data-driven evolutionary optimization algorithms (DDEAs) use cheap surrogate models to replace the expensive real fitness functions, which can significantly reduce computational costs [4]. Surrogate

Y. Tan et al. (Eds.): ICSI 2022, LNCS 13344, pp. 305–316, 2022.
https://doi.org/10.1007/978-3-031-09677-8_26

models are built based on historical data. Some regression techniques can be used as surrogate models, such as Kriging models [5], Radial Basis Function Networks (RBFNs) [3], artificial neural networks [6], and polynomial regression [7] models.

Most of the existing algorithms are online DDEAs, which can obtain new data through real fitness evaluations (FEs) to update the surrogate models during the optimization process [3]. Only a small number of algorithms are offline DDEAs. Offline DDEAs cannot update the surrogate models. To improve the quality of the surrogate models, ensemble learning can be used to build surrogate models [8, 9].

In [10], data perturbations are used to generate diverse surrogates models and a selective ensemble method is used to form a final ensemble surrogate to assist evolutionary algorithm (EA). In [3], a large number of surrogate models are built and a small yet diverse subset of surrogate models are adaptively selected as an ensemble surrogate to achieve the best local approximate accuracy. In [11], using three RBFNs as surrogate models, and to overcome the data deficiency, semi-supervised learning is introduced to the offline DDEA process, where tri-training is used to update the surrogate models. The algorithms mentioned above build many surrogate models during the optimization process, which increases the time cost of the algorithms. Moreover, in [3, 10, 11], and [12], all of the surrogate models choose RBFN, but the numbers of the hidden neurons and the kernel functions in RBFN are not identical.

In this paper, an offline data-driven evolutionary optimization algorithm using k-fold cross (DDEA-K) is proposed. DDEA-K chooses RBFN as surrogate models. To reduce the number of surrogate models, refer to k-fold cross-validation [13], an ensemble surrogate is built using the K-fold cross method. To improve the performance of the algorithm, the number of hidden neurons and the kernel function in RBFN are determined by analyzing the influence of the parameters on the performance of the algorithm.

The rest of this paper is organized as follows: Sect. 2 introduces k-fold cross-validation and RBFN; Sect. 3 describes DDEA-K in detail; Sect. 4 chooses five benchmark problems and experimentally verifies the performance of the algorithm; Sect. 5 concludes this paper.

2 Background

2.1 K-Fold Cross-Validation

K-fold cross-validation is a model evaluation technique on a limited data sample, taking 10 folds as an example, the flowchart is shown in Fig. 1. General procedures for k-fold cross-validation to evaluate a model are as follows:

(1) Randomly disrupt the original data D into k datasets D_1, ..., D_k. Suppose D has n data, the number of data in each dataset is n/k;
(2) Select one dataset as the testing data D_j and the other $k-1$ datasets as the training data $D_{-j} = D \backslash D_j$;
(3) Use the training data training a model and the accuracy of the model is evaluated on the testing data;
(4) Repeat step (2) and step (3), until all the datasets are used as training data. The true accuracy of the model is the average of all k accuracies.

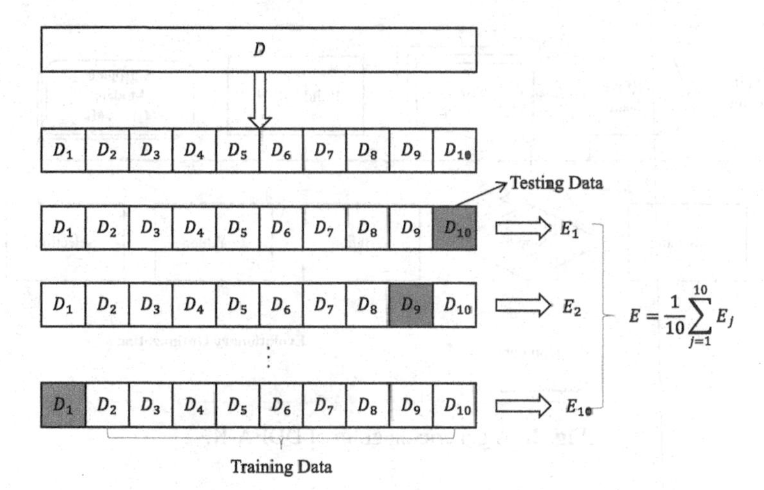

Fig. 1. Flowchart of 10-fold cross-validation.

2.2 Radial Basis Function Network

RBFN is a widely used interpolation method, which has three layers: input layer, hidden layer, and output layer. The transformations from the input layer to the hidden layer are nonlinear, while the transformations from the hidden layer to the output layer are linear. The hidden layer uses the radial basis function (RBF) to map the input to a higher dimension so that linearly indistinguishable in low-dimensional space can be linearly distinguishable. In RBFN, each hidden layer neuron is the center of the RBF, and the hidden layer neuron takes the distance between the input and the center point as the independent variable of the function. Four kernel functions are widely-used in RBFN [14]: gaussian function, reflected sigmoid function, inverse multiquadric function, and multiquadric function, as shown in Eq. (1)–(4).

$$r(d) = exp\left(-\frac{d^2}{2\sigma^2}\right). \tag{1}$$

$$r(d) = \frac{1}{1 + \exp\left(\frac{d^2}{\sigma^2}\right)}. \tag{2}$$

$$r(d) = \frac{1}{\sqrt{d^2 + \sigma^2}}. \tag{3}$$

$$r(d) = \frac{1}{\sqrt{d^2 + \sigma^2}}. \tag{4}$$

where d indicates the distance between the input x and the center point c, σ is the width parameter of the function, which controls the radial range of function.

3 Proposed Algorithm

The flowchart of DDEA-K is shown in Fig. 2, and the algorithm consists of two parts: surrogate modeling and evolutionary optimization. To reduce the size of the ensemble,

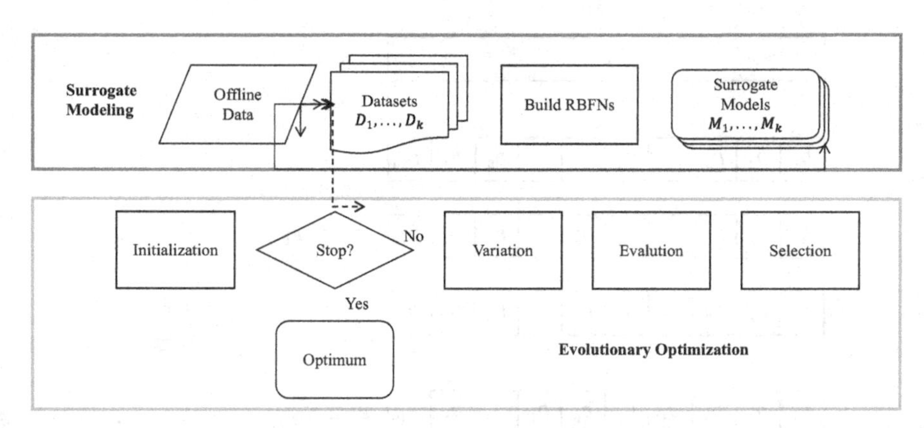

Fig. 2. A generic diagram of DDEA-K.

DDEA-K refers to k-fold cross-validation, uses the k-fold cross method to train K surrogate models, then uses the K surrogate models to form an ensemble surrogate. DDEA-K uses RBFNs as surrogate models. The part of evolutionary optimization in DDEA-K is similar to the traditional genetic algorithm (GA) and the difference between them is that the objective values in DDEA-K are evaluated by surrogate models.

Algorithm 1. Pseudo code of DDEA-K

Input: $Data$: the offline data; N: the size of the population; K: the number of k-fold cross; gen: the maximum number of generations.
Output: v_{best}: Optimum.
1: Divide the $Data$ equally into K datasets, $D_1, D_2, \ldots, D_K$.
2: Initialize the ensemble surrogate $M = \{\}$
3: **for** $k = 1 \rightarrow K$ **do**
4: Choose $K-1$ datasets as training data, $Tran = Data \backslash D_k$.
5: Use $Tran$ training surrogate model M_k.
6: $M = M \cup M_k$.
7: **end for**
8: Initialize the population P.
9: **for** $g = 1 \rightarrow gen$ **do**
10: Generate the offspring population NP from P using selection, crossover and mutation.
11: Combine parent and offspring populations, $P = [P, NP]$.
12: Use M to calculate fitness values of individuals in P.
13: Update the population, sorted by fitness values, $P = P(rank(1{:}N))$.
14: **end for**
15: Choose the decision variables of the first individual in P as x_{best}.
16: $v_{best} = f(x_{best})$.

Before running the optimizer, $11D$ (D is the problem dimension) offline data are created using the Latin hypercube sampling (LHS) [15] and their real objective function. When the optimizer starts running, there is no more real objective function evaluation and the objective function values are predicted using the surrogate models. The pseudo-code for DDEA-K is shown in Algorithm 1.

Steps 1–7 are the construction of the ensemble surrogate. The k-fold cross method is used to construct the ensemble surrogate. The ensemble surrogate consists of K surrogate models and the output of the ensemble is the plain average of the outputs of K surrogate models. Steps 8–14 are the evolutionary optimization process. This part is the same as GA, the only difference between them is that the objective function values of the individuals are not calculated using the real objective function, but are predicted using the ensemble surrogate. Step 15 chooses the best individual in the population and step 16 outputs the final optimal solution. The solution is calculated using the real objective function, since in real life, the goodness of the solution is evaluated in the real situation.

4 Experimental Results

In the experiments, five commonly used benchmark problems [16] are choose to test the algorithms, as presented in Table 1.

Table 1. Benchmark problems.

Problems	D	Global Optimum	Characteristics
Ellipsoid	10,30,50,100	0.0	Uni-modal
Rosenbrock	10,30,50,100	0.0	Multi-modal
Ackley	10,30,50,100	0.0	Multi-modal
Griewank	10,30,50,100	0.0	Multi-modal
Rastrigin	10,30,50,100	0.0	Multi-modal

In DDEA-K, the GA is real-coded and uses the simulated binary crossover (SBX) ($\eta = 15$), polynomial mutation ($\eta = 15$), and tournament selection. The population size of GA is set to 100, the crossover probability is set to 1.0 and the mutation probability is set to $1/D$ (D is the problem dimension). The optimization problems dimensions are 10, 30, 50, and 100, and the corresponding maximum numbers of generations are set to 98, 96, 94, and 88. The offline data is generated using LHS, and the number of data is $11D$. Further, the number of fold crosses is 10 and choosing RBFNs as surrogate models. The kernel function used in RBFN is the multiquadric function and there are $0.1D$ neurons in the hidden layer. Other parameters of RBFN are the same as [3].

All the algorithms are run on each test problem 20 independent times to increase data credibility. All the experiments are run on PlatEMO v3.4 [17], CPU is AMD Ryzen 5 3400G 4-Core 3.7 GHz, System OS is Windows 11, and MATLAB version is R2020b 64-bit.

4.1 Influence of Kernel Function

This part investigates the influence of kernel function in RBFN. DDEA-K variants with different kernel functions are compared on 20 test problems. DDEA-KI, DDEA-KS,

Table 2. Optimization results obtained by DDEA-K with different kernel functions

P	D	DDEA-KS	DDEA-KI	DDEA-KM	DDEA-KG
E	10	9.55e−1(7.61e−1)	7.40e−1(4.86e−1)	5.51e−1(4.08e−1)	**4.09e−1(3.44e−1)**
Ro	10	**2.39e + 1(7.38e + 0)**	2.92e + 1(9.84e + 0)	2.52e + 1(6.89e + 0)	3.58e + 1(1.01e + 1)
A	10	5.34e + 0(1.10e + 0)	5.68e + 0(1.30e + 0)	**4.67e + 0(7.73e−1)**	4.76e + 0(1.12e + 0)
G	10	1.26e + 0(1.37e−1)	1.16e + 0(1.39e−1)	1.06e + 0(8.33e−2)	**9.67e−1(9.38e−2)**
Ra	10	5.04e + 1(2.59e + 1)	5.09e + 1(1.92e + 1)	**3.92e + 1(1.95e + 1)**	4.19e + 1(2.51e + 1)
E	30	4.36e + 0(2.45e + 0)	3.22e + 0(9.49e−1)	3.96e + 0(1.65e + 0)	**3.15e + 0(1.27e + 0)**
Ro	30	4.78e + 1(8.70e + 0)	5.59e + 1(1.01e + 1)	**4.33e + 1(7.93e + 0)**	5.81e + 1(1.19e + 1)
A	30	4.30e + 0(4.16e−1)	4.41e + 0(7.40e−1)	**4.16e + 0(5.79e−1)**	4.28e + 0(4.66e−1)
G	30	1.79e + 0(4.38e−1)	1.30e + 0(1.11e−1)	1.56e + 0(2.28e−1)	**1.26e + 0(7.65e−2)**
Ra	30	8.11e + 1(2.64e + 1)	8.81e + 1(2.14e + 1)	**6.66e + 1(1.83e + 1)**	7.08e + 1(2.71e + 1)
E	50	2.03e + 1(4.20e + 0)	1.75e + 1(3.97e + 0)	**1.68e + 1(3.93e + 0)**	1.85e + 1(3.58e + 0)
Ro	50	**6.70e + 1(4.67e + 0)**	7.95e + 1(9.28e + 0)	6.98e + 1(6.42e + 0)	8.33e + 1(1.22e + 1)
A	50	**4.71e + 0(3.22e−1)**	4.91e + 0(3.18e−1)	4.77e + 0(2.61e−1)	4.98e + 0(4.20e−1)
G	50	3.64e + 0(9.50e−1)	**3.04e + 0(4.24e−1)**	3.38e + 0(4.61e−1)	3.11e + 0(5.51e−1)
Ra	50	1.61e + 2(2.91e + 1)	1.56e + 2(2.82e + 1)	1.51e + 2(3.38e + 1)	**1.41e + 2(3.32e + 1)**
E	100	7.26e + 2(1.43e + 2)	7.71e + 2(1.99e + 2)	7.38e + 2(1.65e + 2)	**6.93e + 2(1.75e + 2)**
Ro	100	3.56e + 2(6.56e + 1)	3.40e + 2(5.14e + 1)	**3.31e + 2(5.62e + 1)**	3.46e + 2(3.77e + 1)
A	100	9.34e + 0(7.47e−1)	9.33e + 0(5.46e−1)	9.39e + 0(7.30e−1)	**9.22e + 0(4.69e−1)**
G	100	4.96e + 1(1.00e + 1)	5.07e + 1(9.50e + 0)	**4.68e + 1(9.78e + 0)**	5.14e + 1(1.17e + 1)
Ra	100	9.37e + 2(8.16e + 1)	9.10e + 2(4.19e + 1)	**9.03e + 2(6.28e + 1)**	9.26e + 2(6.75e + 1)

(continued)

Table 2. (*continued*)

P	D	DDEA-KS	DDEA-KI	DDEA-KM	DDEA-KG
Rank		2.95	2.95	1.85	2.25

DDEA-KM, and DDEA-KG are DDEA-K with the kernel functions of inverse multiquadric function, sigmoid function, multiquadric function, and gaussian function, the numbers of hidden neurons in the RBFN are set to $\sqrt{1+D}+3$ [12]. The optimization results are shown in Table 2. In the table, the number outside the parentheses is the average of the optimization results obtained from 20 independent runs, the number inside the parentheses is the standard deviation, and the last row in the table shows the average ranking values of the optimization results obtained by the current algorithm. Furthermore, the name of the test problem in the table is replaced by the first letter or the first two letters.

As we can see from the table, DDEA-KM has the best performance with the 9 best results out of 20 test problems and an average rank of 1.85. DDEA-KG has the second-best performance with the 7 best results out of 20 test problems and an average rank of 2.25. DDEA-KS and DDEA-KI obtained 3 and 1 best results respectively with the same average rank. Therefore, the default kernel function of RBFN in DDEA-K chooses the multiquadric function.

4.2 Influence of the Number of Neurons in Hidden Layer

This part investigates the influence of the number of hidden neurons nc in RBFN. DDEA-K variants with different neuron numbers, such as $0.1D$, $0.2D$, $0.3D$, $0.4D$, are compared on 20 test problems. As we can see from Table 3, compared with other nc values, DDEA-K obtains 15 best results on 20 test problems with the lowest average ranking of optimization results while nc is set to $0.1D$. Therefore, the default value of nc of RBFN in DDEA-K is set to $0.1D$.

Table 3. Optimization results obtained by DDEA-K with different numbers of hidden neurons

P	D	$nc = 0.1D$	$nc = 0.2D$	$nc = 0.3D$	$nc = 0.4D$
E	10	**6.10e−3(3.21e−3)**	7.41e−2(8.73e−2)	1.70e−1(1.92e−1)	2.32e−1(1.52e−1)
Ro	10	**8.93e + 0(7.04e−2)**	1.13e + 1(3.14e + 0)	1.76e + 1(7.63e + 0)	2.52e + 1(1.08e + 1)
A	10	**5.78e−1(2.04e−1)**	2.27e + 0(9.27e−1)	3.50e + 0(7.89e−1)	4.11e + 0(7.15e−1)
G	10	**8.61e−1(1.24e−1)**	9.50e−1(7.54e−2)	9.47e−1(1.35e−1)	9.67e−1(8.47e−2)
Ra	10	**2.00e−1(1.26e−1)**	7.06e + 0(7.33e + 0)	9.15e + 0(8.61e + 0)	3.27e + 1(2.05e + 1)

(*continued*)

Table 3. *(continued)*

P	D	$nc = 0.1D$	$nc = 0.2D$	$nc = 0.3D$	$nc = 0.4D$
E	30	**1.35e + 0(3.61e−1)**	2.28e + 0(7.45e−1)	3.42e + 0(1.32e + 0)	4.56e + 0(1.83e + 0)
Ro	30	**3.50e + 1(3.87e + 0)**	4.34e + 1(6.56e + 0)	6.02e + 1(1.11e + 1)	6.95e + 1(9.17e + 0)
A	30	**3.19e + 0(3.70e−1)**	3.84e + 0(5.19e−1)	4.51e + 0(5.93e−1)	4.88e + 0(5.55e−1)
G	30	1.25e + 0(6.11e−2)	**1.23e + 0(7.61e−2)**	1.24e + 0(1.02e−1)	1.24e + 0(6.24e−2)
Ra	30	**2.09e + 1(7.52e + 0)**	4.96e + 1(1.23e + 1)	7.42e + 1(2.53e + 1)	1.05e + 2(3.55e + 1)
E	50	**1.63e + 1(3.71e + 0)**	1.86e + 1(3.62e + 0)	1.90e + 1(4.93e + 0)	2.33e + 1(4.23e + 0)
Ro	50	**7.07e + 1(1.05e + 1)**	8.70e + 1(1.23e + 1)	1.09e + 2(1.49e + 1)	1.22e + 2(1.63e + 1)
A	50	**4.37e + 0(3.50e−1)**	4.84e + 0(3.81e−1)	5.14e + 0(2.69e−1)	5.61e + 0(2.89e−1)
G	50	2.96e + 0(5.65e−1)	**2.82e + 0(6.16e−1)**	2.84e + 0(4.11e−1)	3.03e + 0(5.02e−1)
Ra	50	**1.15e + 2(2.65e + 1)**	1.49e + 2(3.14e + 1)	1.87e + 2(4.44e + 1)	2.26e + 2(4.49e + 1)
E	100	**6.84e + 2(1.33e + 2)**	7.15e + 2(1.75e + 2)	7.46e + 2(1.52e + 2)	7.23e + 2(1.48e + 2)
Ro	100	**3.41e + 2(5.10e + 1)**	3.74e + 2(4.38e + 1)	3.97e + 2(6.42e + 1)	4.13e + 2(4.73e + 1)
A	100	9.47e + 0(5.80e−1)	**9.28e + 0(7.71e−1)**	9.43e + 0(4.90e−1)	9.73e + 0(6.11e−1)
G	100	5.16e + 1(1.25e + 1)	**4.61e + 1(8.91e + 0)**	5.02e + 1(1.18e + 1)	5.13e + 1(9.17e + 0)
Ra	100	9.30e + 2(8.41e + 1)	9.19e + 2(7.04e + 1)	**9.12e + 2(6.81e + 1)**	9.34e + 2(7.36e + 1)
Rank		1.6	1.85	2.75	3.8

Analyzing the results from the perspective of the dimension, we can see that the margin of leading in the optimization results obtained when nc is $0.1D$ is gradually decreasing as D increases. Taking the Ellipsoid problem as an example, the ratios of the best results divided by the second-best results are 0.08, 0.59, 0.88 and 0.96. In addition, on some Griewank, Ackley, and Rastrigin problems, the best results are obtained when nc is set to $0.2D$ and $0.3D$. Therefore, when solving optimization problems, the value of nc in DDEA-K needs to be adjusted timely according to the dimensions of the problems.

4.3 Comparison with Offline Data-Driven EAs

Table 4. Optimization results obtained by DDEA-SE, DDEA-PSE and DDEA-K.

Problem	D	DDEA-SE	DDEA-PSE	DDEA-K
Ellipsoid	10	8.75e−1 (5.80e−1)	1.03e + 0 (5.51e−1)	**6.10e−3 (3.21e−3)**
Rosenbrock	10	2.40e + 1 (6.40e + 0)	2.93e + 1 (7.42e + 0)	**8.93e + 0 (7.04e−2)**
Ackley	10	5.46e + 0 (7.42e−1)	5.64e + 0 (6.50e−1)	**5.78e−1 (2.04e−1)**
Griewank	10	1.24e + 0 (1.76e−1)	1.23e + 0 (2.40e−1)	**8.61e−1 (1.24e−1)**
Rastrigin	10	5.34e + 1 (2.35e + 1)	5.47e + 1 (1.01e + 1)	**2.00e + 1 (1.26e−1)**
Ellipsoid	30	4.36e + 0 (1.73e + 0)	6.58e + 0 (1.51e + 0)	**1.35e + 0 (3.61e−1)**
Rosenbrock	30	5.50e + 1 (3.99e + 0)	6.46e + 1 (6.14e + 0)	**3.50e + 1 (3.87e + 0)**
Ackley	30	4.81e + 0 (4.91e−1)	5.13e + 0 (5.01e−1)	**3.19e + 0 (3.70e−1)**
Griewank	30	1.43e + 0 (9.53e−2)	1.51e + 0 (1.73e−1)	**1.25e + 0 (6.11e−2)**
Rastrigin	30	1.06e + 2 (2.00e + 1)	1.43e + 2 (3.03e + 1)	**2.09e + 1 (7.52e + 0)**
Ellipsoid	50	2.03e + 1 (5.69e + 0)	2.73e + 1 (7.53e + 0)	**1.63e + 1 (3.71e + 0)**
Rosenbrock	50	8.68e + 1 (6.41e + 0)	1.11e + 2 (7.91e + 0)	**7.07e + 1 (1.05e + 1)**
Ackley	50	5.09e + 0 (3.31e−1)	5.60e + 0 (3.56e−1)	**4.37e + 0 (3.50e−1)**
Griewank	50	**1.84e + 0 (2.21e−1)**	3.18e + 0 (4.32e−1)	2.96e + 0 (5.65e−1)
Rastrigin	50	1.90e + 2 (2.87e + 1)	2.46e + 2 (3.16e + 1)	**1.15e + 2 (2.65e + 1)**
Ellipsoid	100	7.68e + 2 (5.52e + 1)	8.78e + 2 (1.74e + 2)	**6.84e + 2 (1.33e + 2)**
Rosenbrock	100	3.83e + 2 (6.21e + 1)	5.78e + 2 (1.21e + 2)	**3.41e + 2 (5.10e + 1)**
Ackley	100	9.75e + 0 (1.22e + 0)	9.94e + 0 (4.69e−1)	**9.47e + 0 (5.80e−1)**
Griewank	100	**4.74e + 1 (1.45e + 1)**	5.16e + 1 (9.74e + 0)	5.16e + 1 (1.25e + 1)
Rastrigin	100	9.30e + 2 (8.57e + 1)	9.49e + 2 (6.00e + 1)	**9.30e + 2 (8.41e + 1)**

This part compares the proposed algorithm with two offline data-driven EAs: DDEA-SE [3] and DDEA-PSE [10]. Both of them use RBFNs as surrogate models, with 2000 and 200 RBFNs trained respectively during the optimization process. The GA parameters

Table 5. Running time on 100 dimensional problems.

Problem	DDEA-SE	DDEA-PSE	DDEA-K
Ellipsoid	5.03e + 2 (4.43e + 1)	2.69e + 2 (6.68e + 0)	**6.49e + 0 (4.63e−1)**
Rosenbrock	5.07e + 2 (4.36e + 1)	2.72e + 2 (3.85e + 0)	**6.41e + 0 (7.01e−1)**
Ackley	5.18e + 2 (4.57e + 1)	2.73e + 2 (4.34e + 0)	**6.34e + 0 (4.31e−1)**
Griewank	5.04e + 2 (6.31e + 0)	2.65e + 2 (3.78e + 0)	**6.60e + 0 (4.75e−1)**
Rastrigin	5.24e + 2 (8.01e + 0)	2.71e + 2 (4.91e + 0)	**6.45e + 0 (5.77e−1)**

are the same as DDEA-K, and the other parameters choose the best configuration from their original paper. The optimization results obtained by DDEA-SE, DDEA-PSE, and DDEA-K on the 20 test problems are shown in Table 4.

As we can see from Table 4, besides 50– and 100–dimensional Griewank problems, DDEA-K obtains the best results on all the remaining 18 test problems. In addition, during the optimization process, DDEA-K only needs to train 10 RBFNs, which makes its running time significantly less than DDEA-SE and DDEA-PSE. For example, the running times of the three algorithms on 100-dimensional test problems are shown in Table 5, the unit of time is second. From the table, we can see that DDEA-K takes a very short time to complete the optimization compared to DDEA-SE and DDEA-PSE. From the results in Table 4 and Table 5, it is obvious that DDEA-K not only has better performance but also has significantly lower time complexity than the comparison algorithms. Therefore, DDEA-K is an effective algorithm.

4.4 Analysis of the Universality of RBFN Parameters

Table 6. Optimization results obtained by DDEA-SE and DDEA-SE-alter

Problem	D	DDEA-SE	DDEA-SE-alter
Ellipsoid	10	8.75e−1 (5.80e−1)	**1.60e−2 (8.54e−3)**
Rosenbrock	10	2.40e + 1 (6.40e + 0)	**8.97e + 0 (6.76e−2)**
Ackley	10	5.46e + 0 (7.42e−1)	**1.08e + 0 (2.96e−1)**
Griewank	10	1.24e + 0 (1.76e−1)	**9.36e−1 (1.17e−1)**
Rastrigin	10	5.34e + 1 (2.35e + 1)	**5.54e−1 (2.69e−1)**
Ellipsoid	30	4.36e + 0 (1.73e + 0)	**1.21e + 0 (4.28e−1)**
Rosenbrock	30	5.50e + 1 (3.99e + 0)	**3.18e + 1 (7.71e−1)**
Ackley	30	4.81e + 0 (4.91e−1)	**3.03e + 0 (2.97e−1)**
Griewank	30	1.43e + 0 (9.53e−2)	**1.27e + 0 (6.94e−2)**
Rastrigin	30	1.06e + 2 (2.00e + 1)	**1.89e + 1 (5.75e + 0)**
Ellipsoid	50	2.03e + 1 (5.69e + 0)	**1.41e + 1 (2.55e + 0)**
Rosenbrock	50	8.68e + 1 (6.41e + 0)	**6.07e + 1 (1.93e + 0)**
Ackley	50	5.09e + 0 (3.31e−1)	**4.46e + 0 (2.40e−1)**
Griewank	50	**1.84e + 0 (2.21e−1)**	2.81e + 0 (4.07e−1)
Rastrigin	50	1.90e + 2 (2.87e + 1)	**1.01e + 2 (1.87e + 1)**
Ellipsoid	100	7.68e + 2 (5.52e + 1)	**7.06e + 2 (1.81e + 2**
Rosenbrock	100	3.83e + 2 (6.21e + 1)	**3.11e + 2 (4.23e + 1)**
Ackley	100	9.75e + 0 (1.22e + 0)	**9.26e + 0 (5.85e−1)**

(continued)

Table 6. (*continued*)

Problem	D	DDEA-SE	DDEA-SE-alter
Griewank	100	4.74e + 1 (1.45e + 1)	**4.73e + 1 (1.01e + 1)**
Rastrigin	100	9.30e + 2 (8.57e + 1)	**9.19e + 2 (6.59e + 1)**

Theoretically, the RBFN parameters in DDEA-K can be used in algorithms that also use RBFNs as surrogate models. Therefore, to verify the universality of the RBFN parameters in DDEA-K, the RBFN parameters in DDEA-SE are changed to those in DDEA-K, and the changed DDEA-SE is called DDEA-SE-alert. The kernel function in the original DDEA-SE is the gaussian function, the number of the hidden neurons is *D*, and the rest parameters of RBFN are the same as those in DDEA-K. Comparing DDEA-SE-alert with DDEA-SE on 20 test problems, the optimization results are shown in Table 6.

As we can see from the table, after using the new parameters, the performance of DDEA-SE on all the tested problems is improved besides the 50-dimensional griewank problem, and the margin of improvement decreases with the increase of dimension. Therefore, we can conclude that the RBFN parameters in DDEA-K have some universality and can be used in other DDEAs which use RBFNs as surrogate models to improve the performance of the algorithms.

5 Conclusion

This paper aims to address offline data-driven optimization problems and proposes an offline data-driven evolutionary optimization algorithm using k-fold cross (DDEA-K). DDEA-K uses the k-fold cross method to build the ensemble surrogate, which saves the running time of the algorithm; uses RBFNs as surrogate models, and the number of the hidden neurons and the kernel function in RBFN are determined by analyzing the influence of the parameters on the performance of the algorithm, which improves the performance of the algorithm. The experimental results on benchmark problems show that DDEA-K is an algorithm with good performance and low time cost. The RBFN parameters in DDEA-K have some universality, which can improve the performance of similar algorithms that also use RBFNs as surrogate models.

Although DDEA-K builds surrogate models with reference to k-fold cross-validation, it does not use the validation part. In future work, researchers can use the model accuracy information obtained from the validation part to help select better models or generate synthetic data to further improve the accuracy of surrogate models.

Acknowledgements. This work is supported by the National Natural Science Foundation of China (62072201).

References

1. Chugh, T., Chakraborti, N., Sindhya, K., et al.: A data-driven surrogate-assisted evolutionary algorithm applied to a many-objective blast furnace optimization problem. Mater. Manuf. Processes **32**(10), 1172–1178 (2017)
2. Wang, H., Jin, Y., Jansen, J.O.: Data-driven surrogate-assisted multiobjective evolutionary optimization of a trauma system. IEEE Trans. Evol. Comput. **20**(6), 939–952 (2016)
3. Wang, H., Jin, Y., Sun, C., et al.: Offline data-driven evolutionary optimization using selective surrogate ensembles. IEEE Trans. Evol. Comput. **23**(2), 203–216 (2018)
4. Jin, Y., Wang, H., Sun, C.: Data-driven evolutionary optimization. Springer Cham, Switzerland (2021). https://doi.org/10.1007/978-3-030-74640-7
5. Song, Z., Wang, H., He, C., et al.: A Kriging-assisted two-archive evolutionary algorithm for expensive many-objective optimization. IEEE Trans. Evol. Comput. **25**(6), 1013–1027 (2021)
6. Pan, L., He, C., Tian, Y., et al.: A classification-based surrogate-assisted evolutionary algorithm for expensive many-objective optimization. IEEE Trans. Evol. Comput. **23**(1), 74–88 (2018)
7. Zhou, Z., Ong, Y.S., Nguyen, M.H., et al.: A study on polynomial regression and Gaussian process global surrogate model in hierarchical surrogate-assisted evolutionary algorithm. In: 2005 IEEE Congress on Evolutionary Computation, vol. 3, pp. 2832–2839 IEEE (2005)
8. Li, J.Y., Zhan, Z.H., Wang, C., et al.: Boosting data-driven evolutionary algorithm with localized data generation. IEEE Trans. Evol. Comput. **24**(5), 923–937 (2020)
9. Shan, Y., Hou, Y., Wang, M., Xu, F.: Trimmed data-driven evolutionary optimization using selective surrogate ensembles. In: Pan, L., Pang, S., Song, T., Gong, F. (eds.) BIC-TA 2020. CCIS, vol. 1363, pp. 106–116. Springer, Singapore (2021). https://doi.org/10.1007/978-981-16-1354-8_10
10. Li, J.Y., Zhan, Z.H., Wang, H., et al.: Data-driven evolutionary algorithm with perturbation-based ensemble surrogates. IEEE Trans. Cybern. **51**(8), 3925–3937 (2020)
11. Huang, P., Wang, H., Jin, Y.: Offline data-driven evolutionary optimization based on tri-training. Swarm Evol. Comput. **60**, 100800 (2021)
12. Guo, D., Jin, Y., Ding, J., et al.: Heterogeneous ensemble-based infill criterion for evolutionary multiobjective optimization of expensive problems. IEEE Trans. Cybern. **49**(3), 1012–1025 (2018)
13. Zhou, Z.H.: Ensemble Methods: Foundations and Algorithms. CRC press, Boca Raton (2012)
14. Huang, P., Wang, H., Ma, W.: Stochastic ranking for offline data-driven evolutionary optimization using radial basis function networks with multiple kernels. In: 2019 IEEE Symposium Series on Computational Intelligence, pp. 2050–2057. IEEE (2019)
15. Stein, M.: Large sample properties of simulations using Latin hypercube sampling. Technometrics **29**(2), 143–151 (1987)
16. Liu, B., Zhang, Q., Gielen, G.G.: A Gaussian process surrogate model assisted evolutionary algorithm for medium scale expensive optimization problems. IEEE Trans. Evol. Comput. **18**(2), 180–192 (2013)
17. Tian, Y., Cheng, R., Zhang, X., et al.: PlatEMO: a MATLAB platform for evolutionary multi-objective optimization [educational forum]. IEEE Comput. Intell. M. **12**(4), 73–87 (2017)

Fireworks Algorithms

A Micro-population Evolution Strategy for Loser-Out Tournament-Based Firework Algorithm

Mingzhang Han, Mingjie Fan, Ning Han, and Xinchao Zhao(✉)

School of Science, Beijing University of Posts and Telecommunications, Beijing 100876, China
zhaoxc@bupt.edu.cn

Abstract. The loser-out tournament-based firework algorithm (LoTFWA) is a new baseline among firework algorithm (FWA) variants due to its outstanding performance in multimodal optimization problems. LoTFWA successfully achieves information-interaction among populations by introducing a competition mechanism, while information-interaction within each sub-population remains insufficient. To solve this issue, this paper proposes a micro-population evolution strategy and a hybrid algorithm LoTFWA-microDE. Under the proposed strategy, sparks generated by one firework make up a micro-population which is taken into the differential evolution procedure. The proposed algorithm is tested on the CEC'13 benchmark functions. Experimental results show that the proposed algorithm attains significantly better performance than LoTFWA and DE in multimodal functions, which indicates the superiority of the proposed micro-population evolution strategy.

Keywords: Firework algorithm · Evolutionary algorithm · Multimodal optimization · Micro-population algorithms

1 Introduction

Unimodal optimization problems can be solved easily by mathematical methods or simple heuristic algorithms, while for multimodal optimization problems researchers often apply swarm intelligence or evolutionary algorithms, such as evolution strategy (ES) [1], differential evolution (DE) [2], and particle swarm optimization (PSO) [3]. These algorithms use one or more populations for searching, which allows them to explore better local areas and exploit the local solutions more efficiently.

FWA [4] is a newly proposed optimization algorithm inspired by the phenomenon of fireworks explosion. With the core idea of generating sparks around fireworks, FWA has achieved much progress and been proven useful for real-world applications. Previous works introduced new mechanisms including dynamic explosion amplitude [5], elitism selection [6], and guiding spark generation [7], which had significantly enhanced the search efficiency of FWA. Many studies also show that the search performance of FWA can be improved by focusing on exploration [8–10] or exploitation [11–13].

Y. Tan et al. (Eds.): ICSI 2022, LNCS 13344, pp. 319–328, 2022.
https://doi.org/10.1007/978-3-031-09677-8_27

The loser-out tournament-based firework algorithm (LoTFWA) [14] is one of the best FWA variants and has outperformed several state-of-the-art evolutionary algorithms. LoTFWA innovatively adopts a competition mechanism that greatly improves the algorithm's ability of exploration, thus saving valuable computing resources for exploitation. LoTFWA also suggests that exploration and exploitation can help each other.

Several successful FWA variants are based on LoTFWA. [8] and [9] introduce cooperation among fireworks, enhancing both exploration and exploitation. EDFWA [16] improves the explosion operator with a series of guided explosion in an exponentially decaying manner. [17] uses covariance matrix adaptation method to improve explosion operator of LoTFWA. Further enhancement can be made to the exploitation of LoTFWA.

Similar to many FWA variants, LoTFWA adopts a relatively inefficient local search method, where sparks are generated independently with random sampling. Without information-interaction, the sparks search blindly and spend too much computing resources on finding a better solution. Aiming at solving the issue above, we propose a micro-population evolution strategy for LoTFWA to further strengthen its ability of local search.

The rest of this paper is organized as follows. Section 2 introduces the related work about the framework of LoTFWA and its competition mechanism in detail. Section 3 describes the proposed strategy and hybrid algorithm. Experimental results are shown in Sect. 4 and the last section concludes this paper.

2 Related Work

2.1 Explosion Operator

Algorithm 1 explosion operator for ith firework

Require: λ_i, A_i
1: **for** $j = 1$ **to** λ_i **to**
2: **for** each dimension $k = 1$ **to** d **do**
3: randomly sample η from $\mathcal{U}(-1,1)$
4: $S_{jk} = X_{ik} + A_i \cdot \eta$
5: **end for**
6: **end for**
7: **return** all sparks S_j

FWA can be considered as a $(\mu + \lambda)$ ES. For each generation, the algorithm retains μ fireworks, and λ sparks are generated around them with explosion operator.

In LoTFWA, the number of explosion sparks for each firework is determined by power law distribution. That is,

$$\lambda_r = \hat{\lambda}\frac{r^{-\alpha}}{\sum_{r=1}^{\mu} r^{-\alpha}} \tag{1}$$

where r is the fitness rank of the firework, α is the parameter of power law distribution, and $\hat{\lambda}$ is the total number of sparks generated in generation g.

Explosion sparks are generated with random sampling as described in Algorithm 1, where X_i is the ith firework, A_i is the explosion amplitude of X_i, and $\mathcal{U}$ is the uniform distribution. Besides, LoTFWA adopts a guiding spark mechanism [7], which generates a guiding spark based on the information of explosion sparks. This mechanism aims to help the firework escape local optimum and explore beyond the explosion amplitude.

2.2 Dynamic Explosion Amplitude and Selection Operator

In LoTFWA, an independent dynamic explosion amplitude is adopted for each firework [5]. That is,

$$A_i^g = \begin{cases} A_1, if \ g = 1 \\ C_a A_i^{g-1}, if \ f\left(X_i^g\right) < f\left(X_i^{g-1}\right) \\ C_r A_i^{g-1}, if \ f\left(X_i^g\right) = f\left(X_i^{g-1}\right) \end{cases} \quad (2)$$

where A_i^g is the explosion amplitude of the ith firework in the g generation, C_a and C_r are coefficients to dynamically control explosion amplitude, and A_1 is the initial explosion amplitude, which is usually radius of the search space.

A selection operator is performed on each firework, and the best individual among itself and all the sparks it generates is selected to be the firework of the next generation. Such operator aims to ensure that each firework remains optimum in the subpopulation.

2.3 Loser-Out Tournament

LoTFWA innovatively introduces a competition mechanism to assess the potential of each firework by considering the extent of achieved improvement and predicting their final fitness. The improvement of ith firework X_i in generation g is calculated as:

$$\delta_i^g = f\left(X_i^{g-1}\right) - f\left(X_i^g\right) \geq 0 \quad (3)$$

Then a prediction of the ith firework's fitness in the final generation g_{max} is made as:

$$\widehat{f\left(X_i^{g_{max}}\right)} = f\left(X_i^g\right) - \delta_i^g\left(g_{max} - g\right) \quad (4)$$

If the prediction is still worse than the current best firework, i.e., $\widehat{f\left(X_i^{g_{max}}\right)} > min_j\left\{f\left(X_j^g\right)\right\}$, the ith firework is considered a loser and will be reinitialized. Note that when $\delta_i^g = 0$ the loser-out tournament is not triggered, which is discussed in LoTFWA.

The loser-out tournament is the key mechanism in LoTFWA and main reason for its outstanding performance. It identifies bad areas of search space, thus saving computing resources for fireworks that are more promising.

3 Proposed Method

3.1 Motivation

Multi-population algorithms are inherently suitable for multi-modal optimization problems. FWA is a multi-population algorithm because each firework and its sparks naturally make up a population [14]. In swarm intelligence, main mechanisms of interaction are cooperation and competition. LoTFWA successfully achieves information-interaction among sub-populations by introducing a competition mechanism, while information-interaction within each sub-population remains insufficient. On one hand, sparks of one sub-population are generated all at once, which means that information-interaction only occurs between firework and each spark, but not among the sparks. On the other, sparks in LoTFWA are not generated iteratively. Yu etc. [11–13] proved that in iterative generations, new sparks search more effectively with history information. Therefore, the simple explosion operator adopted in LoTFWA lacks information-interaction among sparks.

Algorithm 2 micro-population evolution with DE

Require: φ, g^{max}, F, CR

1: initialize $g = 1$, randomly generate φ sparks $S_1^g, S_2^g, \dots, S_\lambda^g$ around firework X_i according to Algorithm 1
2: **for** $g = 2$ **to** g^{max} **do**
3: **for** $i = 1$ **to** φ **do**
4: generate random indexes r_1, r_2, r_3 different from i and each other
5: generate mutant vector $V_i^g = S_{r_1}^{g-1} + F \cdot (S_{r_2}^{g-1} - S_{r_3}^{g-1})$
6: generate trial vector $U_i^g = S_i^{g-1}$
7: generate random index d' from $1, 2, \dots, d$
8: **for** each dimension $j = 1$ **to** d **do**
9: randomly sample η from $\mathcal{U}(0,1)$
10: **if** $\eta < CR$ or $j = d'$ **then**
11: $U_{ij}^g = V_{ij}^g$
12: **end if**
13: **end for**
14: **if** $f(S_i^g) < f(U_i^g)$ **then**
15: $S_i^g = U_i^g$
16: **end if**
17: **end for**
18: **end for**

In order to solve the issue above, this paper proposes an iterative information-interaction strategy as described in the next subsection. To achieve the idea, this paper introduces DE operators as the information-interaction mechanism. Because DE operators have been proven to be successful as interaction mechanism and DE holds stable and balanced performance for both exploitation and exploration.

3.2 Micro-population Evolution Strategy

This paper proposes a micro-population evolution strategy, introducing information-interaction to the sparks in the form of cooperation. Under the proposed strategy, φ sparks generated by one firework make up an independent population, where φ is a rather small number because a small population can easily converge. The micro-population is then taken into the differential evolution procedure. Algorithm 2 describes the strategy in detail, where parameter g^{max} is the maximum number of generations of micro-populations, and F, CR are parameters of DE.

The proposed strategy is considered as a new explosion operator. At first, φ sparks are generated by random sampling (line 1). Then the sparks are taken into an evolution procedure (lines 2–18), where DE operators give them the ability to search iteratively in cooperation. During the update of every iteration, the search space of sparks is not limited. When the micro-population reaches its maximum generation g^{max}, it is considered converged and the procedure ends.

Algorithm 3 LoTFWA-microDE

1: Randomly initialize μ fireworks in the search space and evaluate their fitness
2: Initialize fireworks' explosion amplitude
3: **repeat**
4:   **for** $i = 1$ **to** μ **do**
5:     Perform micro-population evolution to the ith firework X_i according to Algorithm 2
6:   **End for**
7:   **for** $i = 1$ **to** μ **do**
8:     Select the best individual from the ith firework and its explosion sparks as the ith firework of the next generation
9:     Calculate A_i according to (2)
10:   **End for**
11:   Perform the loser-out tournament according to subsection 2.4
12: **until** termination criterion is met
13: **return** the best firework

3.3 LoTFWA-microDE

Applying the micro-population evolution strategy to LoTFWA, this paper proposes a hybrid algorithm LoTFWA-microDE. Algorithm 3 gives the complete process of the proposed algorithm.

In its initial phase (lines 1–2), μ fireworks are randomly generated in the search space. Since micro-populations work better in a small search space, the initial explosion amplitudes A_1 of the μ fireworks are set to $1/2$ radius of the search space.

In the first part of its iteration phase (lines 4–6), fireworks first generate their sparks under micro-population evolution strategy. During evolution, when a spark is out of boundaries, it will be bounced back with the following mapping rule:

$$x_i = \begin{cases} ub_i - mod(x_i - ub_i, ub_i - lb_i), if \ x_i > ub_i \\ lb_i + mod(lb_i - x_i, ub_i - lb_i), if \ x_i < lb_i \end{cases} \tag{5}$$

In the second part of its iteration phase (lines 7–10), the fireworks of the next generation are selected and their explosion amplitudes are updated.

In the final part of its iteration phase (line 11), the loser-out tournament is performed and the loser fireworks are reinitialized.

4 Experiments

In order to test the performance of the proposed algorithm, numerical experiments are conducted on the CEC 2013 benchmark suite [15] including 5 unimodal functions and 23 multimodal functions. According to the instructions of the benchmark suite, all the test functions are repeated for 51 times. The number of dimensions of all functions are set to $d = 30$, and the maximal number of function evaluations in each run is 10000 d.

4.1 Parameter Settings

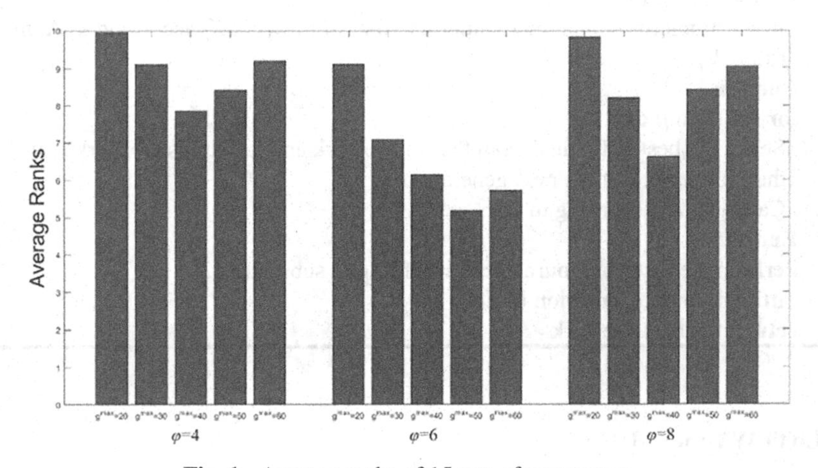

Fig. 1. Average ranks of 15 sets of parameters.

Since LoTFWA-microDE is a hybrid algorithm, it holds 2 sets of parameters from LoTFWA and DE, and a set of parameters of micro-population evolution strategy. There are two parameters of micro-population evolution strategy, micro-population size φ, and maximum iterations for one explosion g^{max}. φ and g^{max} have a strong correlation. A smaller φ means the micro-population converges faster and covers less areas, while a bigger φ means a slower convergence with more areas covered. g^{max} needs to cooperate with φ so that the micro-populations consume enough function evaluations to search and do not waste computing resources after convergence.

Fine-tuned φ and g^{max} are given based on 15 sets of experiments on CEC 2013 benchmark suite as illustrated in Fig. 1. Effect of parameters can be analyzed from the experiment. On one hand, when the number of iterations is not enough, that is, smaller than 30, the algorithm does not show its advantages. On the other hand, number of iterations should not exceed a certain number around 40 or 50, which is related to the value of φ. When suitable iterations are consumed, the performance of the algorithm is more sensitive to φ. Form the view of total number of function evaluations, which is $\varphi \times g^{max}$, the number changes more obviously when φ changes. Under the set of $\varphi = 6$ and $g^{max} = 30$, the proposed algorithm shows stable improvement on LoTFWA with 17 wins and 3 losses. However, this paper chooses the best set of the experiments, which is $\varphi = 6$ and $g^{max} = 50$.

For other parameters, this paper follows the suggestions in [14] and set dynamic amplitude explosion coefficients $C_a = 1.2$, $C_r = 0.9$, the number of fireworks $\mu = 5$. DE parameters $F = 0.5$, $CR = 0.5$ are set as recommended in [2].

4.2 Experimental Result

In this section we compare the proposed algorithm with LoTFWA [14] and DE [2], on which the proposed hybrid algorithm is based. We also compare the proposed algorithm with two other traditional algorithms PSO [3] and CMAES [18]. The mean errors of experimental results are presented in Table 1 and the Wilcoxon signed-rank tests with confidence level 95% are conducted between LoTFWA-microDE and each other algorithm. The P-value of Wilcoxon signed-rank tests between LoTFWA-microDE and LoTFWA is shown in the last column of Table 1 (the algorithms are significantly different when P-value is smaller than 5%). The "−" indicates LoTFWA-microDE performs significantly better, the "+" indicates its opponent performs significantly better, while the "≈" indicates that their performances are not significantly different.

According to the results, LoTFWA-microDE outperforms LoTFWA on 17 functions, and is worse than LoTFWA on 7 functions, indicating that the proposed strategy successfully enhances exploitation by introducing information-interaction to sparks. With efficient exploitation, there are more computing resources for exploration. Therefore, the overall performance of the algorithm is improved.

Table 1. Comparison among LoTFWA-microDE, LoTFWA and DE.

f	LoTFWA-microDE	LoTFWA	DE	PSO	CMAES	P-value
f_1	3.11E−03	**0.00E+00**$^{+}$	**0.00E+00**$^{+}$	2.95E−02^{-}	**0.00E+00**$^{+}$	0
f_2	1.26E+06	1.21E+06$^{\approx}$	1.02E+08^{-}	1.21E+07^{-}	**0.00E+00**$^{+}$	0.63
f_3	3.64E+06	2.39E+07^{-}	5.37E+05^{+}	5.71E+09^{-}	**2.33E+02**$^{+}$	0
f_4	9.57E+02	1.93E+03^{-}	5.49E+04^{-}	7.54E+04^{-}	**0.00E+00**$^{+}$	0
f_5	3.83E−02	3.58E−03^{+}	**0.00E+00**$^{+}$	6.32E−01^{-}	**0.00E+00**$^{+}$	0
f_6	1.46E+01	1.31E+01$^{\approx}$	1.62E+01^{-}	7.67E+01^{-}	**1.63E+00**$^{+}$	0.30

(continued)

Table 1. *(continued)*

f	LoTFWA-microDE	LoTFWA	DE	PSO	CMAES	P-value
f_7	**6.51E+00**	5.02E+01$^-$	2.29E+01$^-$	1.42E+02$^-$	1.35E+01$^-$	0
f_8	2.09E+01	**2.09E+01**$^+$	2.09E+01$^\approx$	2.10E+01$^-$	2.14E+01$^-$	0
f_9	**8.72E+00**	1.45E+01$^-$	3.93E+01$^-$	2.88E+01$^-$	4.26E+01$^-$	0
f_{10}	6.65E−01	4.04E−02$^+$	7.70E+00$^-$	3.94E+01$^-$	**1.57E−02**$^+$	0
f_{11}	**2.91E+01**	6.40E+01$^-$	1.10E+02$^-$	1.47E+02$^-$	9.67E+01$^-$	0
f_{12}	**4.90E+01**	6.96E + 01$^-$	1.96E+02$^-$	1.91E+02$^-$	1.19E+03$^-$	0
f_{13}	**9.03E+01**	1.31E+02$^-$	1.93E+02$^-$	3.02E+02$^-$	1.55E+03$^-$	0
f_{14}	**1.12E+03**	2.42E+03$^-$	4.29E+03$^-$	3.40E+03$^-$	5.18E+03$^-$	0
f_{15}	**2.27E+03**	2.56E+03$^-$	7.34E+03$^-$	4.36E+03$^-$	5.14E+03$^-$	0
f_{16}	**3.88E−02**	5.74E−02$^-$	2.41E+00$^-$	1.04E+00$^-$	7.91E−02$^-$	0
f_{17}	6.58E+01	**6.31E+01**$^\approx$	1.41E+02$^-$	2.65E+02$^-$	4.17E+03$^-$	0.10
f_{18}	8.70E+01	**6.33E+01**$^+$	2.23E+02$^-$	2.83E+02$^-$	4.15E+03$^-$	0
f_{19}	**2.58E+00**	3.17E+00$^-$	1.37E+01$^-$	1.52E+01$^-$	3.49E+00$^-$	0
f_{20}	1.33E+01	1.34E+01$^\approx$	1.28E+01$^\approx$	1.47E+01$^-$	**1.26E+01**$^\approx$	0.44
f_{21}	2.21E+02	**2.00E+02**$^+$	2.94E+02$^-$	3.08E+02$^-$	3.07E+02$^-$	0
f_{22}	**1.21E+03**	2.84E+03$^-$	5.26E+03$^-$	3.87E+03$^-$	7.04E+03$^-$	0
f_{23}	**2.40E+03**	3.11E+03$^-$	7.88E+03$^-$	5.26E+03$^-$	6.67E+03$^-$	0
f_{24}	**2.04E+02**	2.40E+02$^-$	2.16E+02$^\approx$	2.88E+02$^-$	7.10E+02$^-$	0
f_{25}	**2.47E+02**	2.76E+02$^-$	3.00E+02$^-$	3.01E+02$^-$	3.23E+02$^\approx$	0
f_{26}	**1.94E+02**	2.00E+02$^-$	2.08E+02$^-$	3.21E+02$^-$	4.62E+02$^-$	0
f_{27}	**4.87E+02**	6.96E+02$^-$	1.10E+03$^-$	1.11E+03$^-$	5.89E+02$^-$	0
f_{28}	2.92E+02	**2.69E+02**$^+$	3.00E+02$^-$	1.19E+03$^-$	1.47E+03$^-$	0
AR	1.79	2.14	3.39	4.21	3.32	
	+	7	3	0	7	
	−	17	22	28	19	
	≈	4	3	0	2	

5 Conclusion

This paper proposes a micro-population evolution strategy for the loser-out tournament-based fireworks algorithm to enhance its local search capability. Under the proposed strategy, the explosion sparks make up a micro-population and are taken into the differential evolution procedure. Experimental results on the CEC 2013 benchmark suite indicates that the proposed strategy can improve the overall performance of the algorithm.

Further investigation is worthwhile on two aspects. On the aspect of the proposed strategy itself. More improvements can be achieved with adaptive φ and g^{max} considering the search pattern of micro-populations. Besides, other evolutionary algorithms can be implemented considering their advantages when facing different type of problems, such as combinatorial optimizations and noisy optimizations. On the aspect of the algorithm. Performance of the algorithm can further improve when utilizing the proposed strategy for exploitation, and other mechanisms that enhance exploration.

Acknowledgements. This work is supported by Beijing Natural Science Foundation (1202020), National Natural Science Foundation of China (61973042) and BUPT innovation and entrepreneurship support program (2022-YC-A287). Awfully thanks will be given to Swarm Intelligence Research Team of BeiYou University.

References

1. Rechenberg, I.: Evolution strategy. Computational intelligence: Imitating life (1994)
2. Storn, R., Price, K.: Differential evolution–a simple and efficient heuristic for global optimization over continuous spaces. J. Global Optim. **11**(4), 341–359 (1997)
3. Shi, Y., Eberhart, R.: A modified particle swarm optimizer. In: 1998 IEEE International Conference on Evolutionary Computation Proceedings. IEEE World Congress on Computational Intelligence (Cat. No.98TH8360), pp. 69–73 (1998)
4. Tan, Y., Zhu, Y.: Fireworks algorithm for optimization. In: Tan, Y., Shi, Y., Tan, K.C. (eds.) ICSI 2010. LNCS, vol. 6145, pp. 355–364. Springer, Heidelberg (2010). https://doi.org/10.1007/978-3-642-13495-1_44
5. Zheng, S., Janecek, A., Li, J., et al.: Dynamic search in fireworks algorithm. In: 2014 IEEE Congress on Evolutionary Computation (CEC), pp. 3222–3229. IEEE (2014)
6. Zheng, S., Li, J., Janecek, A., et al.: A cooperative framework for fireworks algorithm. IEEE/ACM Trans. Comput. Biol. Bioinform. **14**(1), 27–41 (2015)
7. Li, J., Zheng, S., Tan, Y.: The effect of information utilization: introducing a novel guiding spark in the fireworks algorithm. IEEE Trans. Evol. Comput. **21**(1), 153–166 (2016)
8. Li, Y., Tan, Y.: Multi-scale collaborative fireworks algorithm. In: 2020 IEEE Congress on Evolutionary Computation (CEC), pp. 1–8. IEEE (2020)
9. Hong, P., Zhang, J.: Using population migration and mutation to improve loser-out tournament-based fireworks algorithm. In: Tan, Y., Shi, Y. (eds.) ICSI 2021. LNCS, vol. 12689, pp. 423–432. Springer, Cham (2021). https://doi.org/10.1007/978-3-030-78743-1_38
10. Zheng, Y.J., Xu, X.L., Ling, H.F., et al.: A hybrid fireworks optimization method with differential evolution operators. Neurocomputing **148**, 75–82 (2015)
11. Yu, J., Takagi, H.: Acceleration for fireworks algorithm based on amplitude reduction strategy and local optima-based selection strategy. In: Tan, Y., Takagi, H., Shi, Y. (eds.) ICSI 2017. LNCS, vol. 10385, pp. 477–484. Springer, Cham (2017). https://doi.org/10.1007/978-3-319-61824-1_52
12. Yu, J., Takagi, H., Tan, Y.: Multi-layer explosion-based fireworks algorithm. Int. J. Swarm Intell. Evol. Comput. **7**(3), 1–9 (2018)
13. Yu, J., Tan, Y., Takagi, H.: Scouting strategy for biasing fireworks algorithm search to promising directions. In: Proceedings of the GECCO 2018, pp. 99–100 (2018)
14. Li, J., Tan, Y.: Loser-out tournament-based fireworks algorithm for multimodal function optimization. IEEE Trans. Evol. Comput. **22**(5), 679–691 (2017)

15. Liang, J.J., Qu, B.Y., Suganthan, P.N., et al.: Problem definitions and evaluation criteria for the CEC 2013 special session on real-parameter optimization. Comput. Intell. Lab., Zhengzhou Univ., Zhengzhou, China Nanyang Technol. Univ., Singapore, Tech. Rep. **201212**(34), 281–295 (2013)
16. Chen, M., Tan, Y.: Exponentially decaying explosion in fireworks algorithm. In: 2021 IEEE Congress on Evolutionary Computation (CEC), pp. 1406–1413. IEEE (2021)
17. Li, Y., Tan, Y.: Enhancing fireworks algorithm in local adaptation and global collaboration. In: Tan, Y., Shi, Y. (eds.) ICSI 2021. LNCS, vol. 12689, pp. 451–465. Springer, Cham (2021). https://doi.org/10.1007/978-3-030-78743-1_41
18. Hansen, N., Müller, S.D., Koumoutsakos, P.: Reducing the time complexity of the derandomized evolution strategy with covariance matrix adaptation (CMA-ES). Evol. Comput. **11**, 1–18 (2003)

An Improved Fireworks Algorithm for Integrated Flight Timetable and Crew Schedule Recovery Problem

Xiaobing Gan[1], Tianwei Zhou[1], Yuhan Mai[1], Huifen Zhong[1,2](✉), Xiuyun Zhang[3], and Qinge Xiao[1]

[1] College of Management, Shenzhen University, Shenzhen 518060, China
zhonghf510520@163.com
[2] Faculty of Business and Administration, University of Macau, Macau 999078, China
[3] School of Automation and Electrical Engineering, Tianjin University of Technology and Education, Tianjin 300222, China

Abstract. Abnormal flights, which deviate from their scheduled plans, incurred huge costs for airlines and serious inconvenience for passengers. This phenomenon occurs frequently, especially under the influence of COVID-19 and requires high-quality solution within short time limits. To mitigate these negative effects, first, an integrated flight timetable and crew schedule recovery model with the aim of minimizing total cost is constructed in this paper. Second, an improved fireworks algorithm is proposed to effectively solve the model. Finally, an unscheduled temporary aircraft maintenance scenario is obtained to illustrate the superiority of the proposed algorithm in terms of computing time and solution quality.

Keywords: Integrated flight timetable and crew · Abnormal flight · Recovery · fireworks algorithm

1 Introduction

Abnormal flights are flights that deviate from their scheduled plans. They are very common in the airline industry and influenced by various unforeseen factors, such as poor weather, congestion, unscheduled maintenance, and so on. According to the Statistical Communique on the Development of the Civil Aviation Industry, in 2020, 404,164 flights are suffering from disruptions, accounting for hundreds of billions in economic losses for airlines [1].

Airlines typically recover from disruptions in stages, which are broadly categorized as timetable, aircraft, crew, and passenger recovery [2]. References [3–6] are some recent research for these four sub-stage recovery problems and readers are referred to Hassan for a more comprehensive review [7]. Since Teodorović pioneering focused on the abnormal flight recovery problem in 1984, research has announced that timetable recovery is the basic core step, and crew resource is the largest controllable cost for the airlines [8]. Thus, this paper focuses on integrated flight timetable and crew schedule recovery (I-FTCSR) problem, i.e., to determine new flight takeoff time and crew task.

Y. Tan et al. (Eds.): ICSI 2022, LNCS 13344, pp. 329–338, 2022.
https://doi.org/10.1007/978-3-031-09677-8_28

An integrated method can overcome the suboptimality of the recovery solution, however, solving it is intractable in computational efficiency because of the huge solution space and complex problem structure. Petersen et al. employed Bender's decomposition to decompose the integrated recovery model into a master problem and three sub-problems [9]. Le and Wu designed a heuristic approach to solve the integrated airline recovery problem [10]. Zhang et al. proposed an interactive two-stage heuristic, which runs iteratively until the optimal solution was achieved [11]. Maher presented a column-and-row generation technique to solve the integrated airline problem [12]. These methods indeed decrease computation times, while are still too long for airlines' operational implementation, especially given the NP-hard complexity.

To acquire an efficient and accurate solution within a short time limit, swarm evolutionary algorithm may be a good attempt for its easy implementation, excellent performance, and high efficiency in solving complex problems. In this paper, one of the popular swarm evolutionary algorithms, fireworks algorithm (FWA) proposed by Tan and Zhu, is improved with two mechanisms to apply to the I-FTCSR for its global exploration and local exploitation abilities [13].

Inspired by the above analysis, this paper aims to acquire an efficient and accurate solution for the I-FTCSR using a tailored swarm evolutionary algorithm. Main contributions are listed below:

- Construct the model of integrated flight timetable and crew schedule recovery (I-FTCSR) problem with the objective of minimizing total recovery cost.
- Design an improved fireworks algorithm (I-FWA) for the I-FTCSR and propose a corresponding coding scheme between the model and algorithm.
- Conduct experiments with unscheduled maintenance scenario and provide its competitive recovery solution.

The remainder of this paper is organized as follows. Section 2 shows the detailed model description for considered I-FTCSR. Section 3 introduces the coding schema and the improved I-FWA in detail. Section 4 designs experiments and provides the recovery scheme. Finally, Sect. 5 concludes the paper and discusses the future work.

2 Problem Formulation and Model Construction

2.1 Problem Description

This paper focuses on I-FTCSR, which suffers temporarily unscheduled maintenance. Therefore, how to re-schedule all planned flight tasks using some repair methods (e.g., flight cancellation, takeoff time rearrangement, and crew duty reallocation) under various time and space constraints with minimum recovery cost, is our main concern.

For simplicity, this model considers only one aircraft type and assumes the recovery period is within one day.

2.2 Model Construction

Model Parameters. Referencing the latest aviation industry-government document CCAR121-R5 in China, Table 1 presents the sets, parameters, and variables used in I-FTCSR model [14].

Table 1. Definition of sets, parameters, and variables.

Sets			
Set	Definition	Set	Definition
S	airport set, $s \in S$	F	flight set, $f \in F$
E	crew set, $e \in E$		
Parameters			
Name	Definition	Name	Definition
c_{fd}	flight unit delay cost	c_{fc}	flight cancellation cost
c_{es}	unit cost of using standby crew	t_f	landing time of flight f
f_e	crew arrangement for flight f	a_f	landing airport of flight f
w_e	total working time of crew e	d_e	total duty time of crew e
u_1	crew maximum flight time	u_2	crew maximum duty time
t	transit time between flights	t_e	most recent landing time of crew e
q_f	departure time of flight f's subsequent flight		
q_e	departure time of crew e's subsequent task		
v_f	departure airport of flight f's subsequent flight		
v_e	departure airport of crew e's subsequent task		
Variables			
Name	Definition		
x_f	delay time of flight f compared to its planned departure time		
y_f	whether flight f is canceled or not. If x_f larger than 8 hours, it is cancelled		
z_f	whether the crew allocation on flight f is a standby crew or not		

Model Formulation. With the parameter definitions presented in Table 1, the specific objective and constraint formulations of the I-FTCSR are defined below.

$$\min \sum_{f \in F} c_{fd} x_f (1 - y_f) + \sum_{f \in F} c_{fc} y_f + \sum_{f \in F} c_{es} z_f \tag{1}$$

$$\text{s.t.} q_f > t_f + t. \quad \forall f \in F \tag{2}$$

$$q_e > t_e + t. \quad \forall f \in F \tag{3}$$

$$a_f = v_f = v_e. \quad \forall f \in F \tag{4}$$

$$w_e < u_1. \quad \forall e \in E \tag{5}$$

$$d_e < u_2. \quad \forall e \in E \tag{6}$$

$$\sum_e f_e \leq 1. \quad \forall f \in F \tag{7}$$

$$x_f \geq 0. \quad \forall f \in F \tag{8}$$

$$y_f \in \{0, 1\}. \quad \forall f \in F \tag{9}$$

$$z_f \in \{0, 1\}. \quad \forall f \in F \tag{10}$$

Formula (1) is the objective function, aiming to minimize the sum of flight delay cost, flight cancellation cost, and the cost of using standby crew under a temporarily unscheduled maintenance case. Formulas (2) and (3) ensure minimum transit time between adjacent flights and adjacent crew tasks, respectively. Formula (4) guarantees space feasibility for both adjacent flights and crew tasks. Formulas (5) and (6) make each crew's daily flying and duty times not exceed their prescriptive maximum. Formula (7) ensures that no crew can execute two tasks at the same time. The remaining three formulas are the ranges of the variables. The remaining formulas define the range of variables.

3 Solution Procedure Based on Improved Fireworks Algorithm

3.1 Coding Design for I-FTCSR Model

To illustrate the coding scheme, in Table 2, an example comprising 10 abnormal flights is presented. Because there are 2 decision variables (y_f is an auxiliary variable that can be calculated by x_f) in our I-FTCSR, the coding dimensions are 20 (flights number × 2 dimensions). Every 10 dimensions represent decision variables x_f and z_f for each flight, respectively.

Table 2. I-FTCSR coding design.

First 10 dimensions	Last 10 dimensions
Delay time of flight f compared to its planned departure time	Crew arrangement for each flight

3.2 Improved Fireworks Algorithm (I-FWA) Steps

The basic fireworks algorithm originated from observing the fireworks explosion process and is a kind of new evolutionary optimization technique. There are two main operators, explosion and Gauss mutation operators, which serve for optimization evolutionary and diversity keeping, respectively.

For reading convenience, we firstly introduce formulations for the main parameters, i.e., maximum explosion radius A_i and spark number $\hat{s}_i$ for fireworks i as below.

$$A_i = \hat{A} \cdot \frac{f(x_i) - y_{min} + \xi}{\sum_{i=1}^{n} (f(x_i) - y_{min}) + \xi}. \tag{11}$$

where $\hat{A}$ represents maximum explosion radius (this is generated by our first improved mechanism as formula (14) shown), $f(x_i)$ is the fitness of firework i, n is firework quantities, y_{min} is the best solution among n fireworks, and ξ is the smallest constant in the computer to avoid zero-division-error.

$$s_i = m \cdot \frac{y_{max} - f(x_i) + \xi}{\sum_{i=1}^{n} (y_{max} - f(x_i)) + \xi}. \tag{12}$$

$$\hat{s}_i = \begin{cases} round(am) \text{ } if \text{ } s_i < am \\ round(bm) \text{ } if \text{ } s_i > bm, a < b < 1 \\ round(s_i). \text{ } otherwise \end{cases} \tag{13}$$

where $\hat{s}_i$ is the bound for s_i, a and b are constant parameters, m is a customized spark generated parameter, y_{max} is the worst solution, and *round(..)* represents the rounding symbol.

In order to obtain a better solution for I-FTCSR model, this paper proposes a novel I-FWA algorithm. Its detailed steps are presented in Algorithm 1, which comprises the following two improvements:

(1) Nonlinear Decreasing Maximum Explosion Radius. The maximum explosion radius in the basic FWA is determined, which is not conducive to the convergence in the later iteration stage. Therefore, a decreasing non-linearly maximum explosion radius as formula (14) shown is designed to keep strong global exploration firstly and improve local exploration capability later.

$$\hat{A} = \hat{A} \times \left(1 - \frac{T}{T_{Max}}\right). \tag{14}$$

where T is the iteration number and T_{Max} is the maximum iteration number.

(2) Cauchy Mutation Operator. The basic Gaussian distribution mutation allows only a small number of sparks to be far away from the current fireworks, which weakens the algorithm ability to find the global optimal solution. Therefore, a more flat curve, Cauchy distribution, is utilized instead, to make the algorithm find global optimization ability effectively, jump out of the local optimal quickly, and enhance the diversity of the population. Figure 1 shows the two standard probability density functions intuitively.

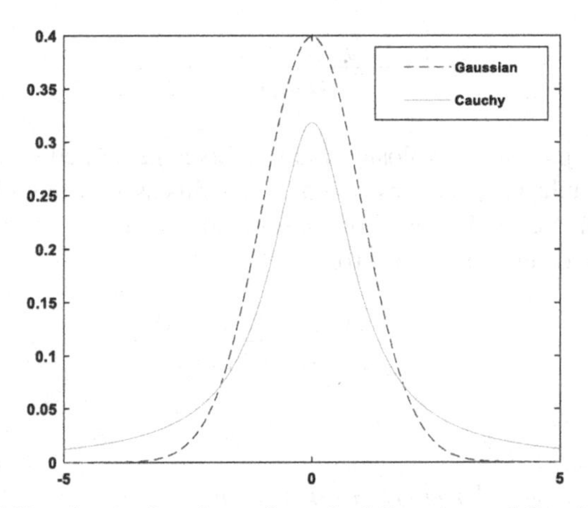

Fig. 1. Probability density functions of standard Gaussian and Cauchy distribution.

Algorithm 1. Detailed steps of I-FWA

Input: Parameters including population size *n*, and termination criteria.
Output: Optimal solution.
1: Initial population generated randomly for *n* fireworks.`
2: **While** termination criteria=false **do**
3: **for** each firework *i* **do**
4: Calculate explosion radius A_i and spark number $\hat{s}_i$ by formulas (11)-(14).
5: Obtain the location of all sparks by referencing Algorithm 1 in Tan and Zhu [13]. The solution should be within the feasible region.
6: Evaluate the fitness of fireworks and sparks, and flag the best one.
7: Execute Cauchy mutation operator for the fireworks.
end for
8: Select the best location and keep it for the next explosion generation.
9: Randomly select $n - 1$ locations from all current sparks and fireworks.
end while

---

4 Experimental Analysis

In this part, an actual unscheduled maintenance case, in which 2 aircraft are breakdown before takeoff and will be recovered after 120 min, is adopted. Since particle swarm optimization (PSO) has already demonstrated significant success in many scheduling problems, it is chosen as a peer algorithm to investigate the performance of our proposed I-FWA in solving the complex I-FTCSR. Some available information, such as instance information, I-FTCSR model parameters, and algorithm parameters, are listed in Tables 3 and 4, respectively.

Table 3. Disrupted flights information.

No	flights	Crew duty	Departure airport	Land airport	Departure time	Land time
1	1	1	Shanghai	Chengdu	08:20	11:40
	2		Chengdu	Shanghai	12:30	15:10
	3		Shanghai	Sanya	17:05	20:00
	4		Sanya	Shanghai	21:05	23:50
2	1	2	Shanghai	Taipei	09:20	11:00
	2		Taipei	Shanghai	12:00	13:30
	3		Shanghai	Harbin	14:25	17:05
	4		Harbin	Shanghai	18:00	20:40
	5	3	Shanghai	Jinan	21:30	22:25
	6		Jinan	Shanghai	23:05	24:00

Since our two proposed strategies in Sect. 3.2 only enhance evolution mechanisms on the local search and mutation processes, the basic input parameters of algorithms I-FWA and FWA are consistent. To make a fair comparison, all experiments are run 30 times independently under the same maximum iterations T = 5000, and the best results are highlighted in boldface.

Table 5 firstly gives the comparison results obtained by different algorithms for solving the I-FTCSR. It is clear that the I-FWA is the best solver with the shortest computing time to obtain the optimal solution. What's more, I-FWA can always find the minimum cost. It is because the nonlinear decreasing maximum explosion radius mechanism helps in global detection capability firstly and ensures relatively speed in the middle stage. The introduction of Cauchy mutation operator enhances the mutation probability of the population, which reduces the possibility of falling into local optimization.

The average convergence processes of the three compared algorithms are given intuitively in Fig. 2. Although PSO can find a smaller cost recovery schema at the very beginning, it falls into local sub-optimal at the same time. FWA and I-FWA have similar convergence effectiveness, while the I-FWA algorithm always finds a lower-cost scheme, which indices the effectiveness of our proposed improved strategies.

Table 4. Model and algorithm parameters.

Definition	Value	
Flight unit delay cost (yuan / min)	100	I-FTCSR model
Flight cancellation cost (yuan)	10000	
Unit cost of using standby crew (yuan)	5000	
Initial individual number of fireworks	5	I-FWA, FWA
Sparks generated parameter *m*	50	
Maximum number of sparks	40	
Minimum number of sparks	2	
Initial population size	50	PSO
Inertia weight	0.5	
Self-learning factor	0.5	
Group-learning factor	0.5	

Table 5. Comparison results obtained by different algorithms on solving the I-FTCSR.

Algorithms	Average computing time/s	Solution cost
FWA	0.3089700	71276
I-FWA	*0.2094590*	*69450*
PSO	5.6175748	138746

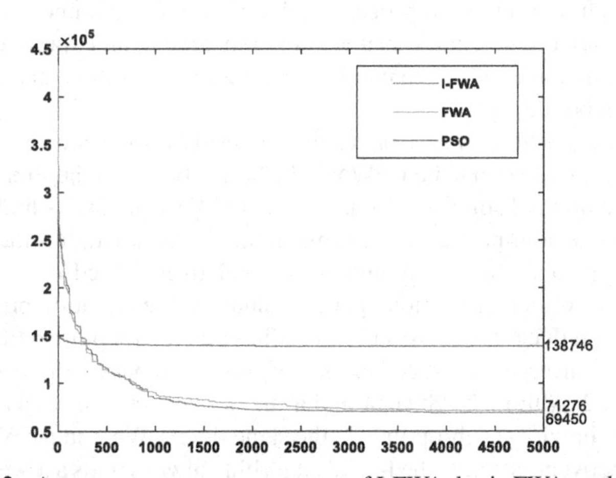

Fig. 2. Average convergence processes of I-FWA, basic FWA, and PSO.

Finally, the best recovery scheme provided by the I-FWA algorithm is presented in Table 6. All flight takeoff timetables are subject to various degrees of adjustment, for example, the 1st flight of aircraft 1 will delay 120 min, and the 2nd flight delays 100 min. The crew arrangements keep the same in this instance, it is because all crews' industry requirements can be ensured by these effective timetable adjustments. Table 6 can be used by the decision-maker directly.

Table 6. Abnormal I-FTCSR recovery solution.

No	flights	Crew duty	Departure airport	Land airport	Departure time	Land time
1	1	1	Shanghai	Chengdu	10:20	13:40
	2		Chengdu	Shanghai	14:10	16:50
	3		Shanghai	Sanya	17:20	20:15
	4		Sanya	Shanghai	21:05	23:50
2	1	2	Shanghai	Taipei	11:20	13:00
	2		Taipei	Shanghai	13:30	15:00
	3		Shanghai	Harbin	15:30	18:10
	4		Harbin	Shanghai	18:40	21:20
	5	3	Shanghai	Jinan	21:50	22:45
	6		Jinan	Shanghai	23:15	00:35

5 Conclusion and Future Research

This research establishes an integrated flight timetable and crew schedule recovery problem under a temporarily unscheduled maintenance scenario. A tailored improved fireworks algorithm is proposed to solve this problem efficiently. Experiment results demonstrate the effectiveness of this proposed algorithm in terms of computing time and solution quality.

In the future study, we will consider multi-aircraft types to overcome its single assumption. The passenger recovery problem also deserves consideration in conjunction with the problem considered in this research.

Acknowledgment. The study was supported in part by the Natural Science Foundation of China Grant No.62103286, No.71971143, No.62001302, in part by Social Science Youth Foundation of Ministry of Education of China under Grant 21YJC630181, in part by Guangdong Basic and Applied Basic Research Foundation under Grant 2021A1515011348, 2019A1515111205, 2019A1515110401, 2020A1515010752, in part by Natural Science Foundation of Guangdong Province under Grant 2020A1515010749, 2020A1515010752, in part by Key Research Foundation of Higher Education of Guangdong Provincial Education Bureau under Grant 2019KZDXM030, in part by Natural Science Foundation of Shenzhen under Grant JCYJ20190808145011259, in part by Shenzhen Science and Technology Program under Grant

RCBS20200714114920379, in part by Guangdong Province Innovation Team Intelligent Management and Interdisciplinary Innovation under Grant 2021WCXTD002, Tianjin Nature Science Foundation under Grant 20JCQNJC00370.

References

1. China Civil Aviation Network Statistical Announcement on Development of Civil Aviation Industry in 2020. http://www.caac.gov.cn/XXGK/XXGK/TJSJ/202106/t20210610_207915.html
2. Clausen, J., Larsen, A., Larsen, J., Rezanova, N.J.: Disruption management in the airline industry—concepts, models and methods. Comput. Oper. Res. **37**(5), 809–821 (2010)
3. Zhou, T., Lu, J., Zhang, W., He, P., Niu, B.: Irregular flight timetable recovery under covid-19: an approach based on genetic algorithm. In: Tan, Y., Shi, Y., Zomaya, A., Yan, H., Cai, J. (eds.) DMBD 2021. CCIS, vol. 1453, pp. 240–249. Springer, Singapore (2021). https://doi.org/10.1007/978-981-16-7476-1_22
4. Evler, J., Lindner, M., Fricke, H., Schultz, M.: Integration of turnaround and aircraft recovery to mitigate delay propagation in airline networks. J. Comput. Oper. Res. **138**, 105602 (2022)
5. Wen, X., Sun, X., Sun, Y., Yue, X.: Airline crew scheduling: models and algorithms. J. Transp. Res. Part E: Logist. Transp. Rev. **149**, 102304 (2021)
6. McCarty, L.A., Cohn, A.E.: Preemptive rerouting of airline passengers under uncertain delays. J. Comput. Oper. Res. **90**, 1–11 (2018)
7. Hassan, L.K., Santos, B.F., Vink, J.: Airline disruption management: a literature review and practical challenges. Comput. Oper. Res. **127**, 105137 (2021)
8. Teodorović, D., Guberinić, S.: Optimal dispatching strategy on an airline network after a schedule perturbation. Eur. J. Oper. Res. **15**(2), 178–182 (1984)
9. Petersen, J.D., Sölveling, G., Clarke, J.P., Johnson, E.L., Shebalov, S.: An optimization approach to airline integrated recovery. J. Transp. Sci. **46**(4), 482–500 (2012)
10. Le, M.L., Wu, C.C.: Solving airlines disruption by considering aircraft and crew recovery simultaneously. J. Shanghai Jiaotong Univ. (Sci.) **18**(2), 243–252 (2013)
11. Zhang, D., Lau, H.H., Yu, C.: A two stage heuristic algorithm for the integrated aircraft and crew schedule recovery problems. J. Comput. Ind. Eng. **87**, 436–453 (2015)
12. Maher, S.J.: Solving the integrated airline recovery problem using column-and-row generation. J. Transp. Sci. **50**(1), 216–239 (2016)
13. Tan, Y., Zhu, Y.: Fireworks algorithm for optimization. In: Tan, Y., Shi, Y., Tan, K.C. (eds.) ICSI 2010. LNCS, vol. 6145, pp. 355–364. Springer, Heidelberg (2010). https://doi.org/10.1007/978-3-642-13495-1_44
14. Ministry of Transport of the People's Republic of China. 14th Ministerial Meeting (Aug. 29, 2017). Large Aircraft Public Air Transport Carrier Operation Certification Rules. http://www.caac.gov.cn/XXGK/XXGK/MHGZ/201710/P020171009385743667633.pdf

Economic Dispatch Optimization for Microgrid Based on Fireworks Algorithm with Momentum

Mingze Li and Ying Tan(✉)

Key laboratory of Machine Perception (MOE), School of Artificial Intelligence, Institute for Artificial Intelligence, Peking University, Beijing, China
{mingzeli,ytan}@pku.edu.cn

Abstract. As an efficient organization form of distributed energy resources with high permeability, microgrid (MG) is recognized as a promising technology with the promotion of various clean renewable sources. Due to uncertainties of renewable sources and load demands, optimizing the dispatch of controllable units in microgrid to reduce economic cost has become a critical issue. In this paper, an economic dispatch optimization model for microgrid including distributed generation and storage is established with the considering of inherent links between intervals, which aims to minimize the economic and environmental costs. In order to solve the optimization problem, a novel swarm intelligence algorithm called fireworks algorithm with momentum (FWAM) is also proposed. In the algorithm, the momentum mechanism is introduced into the mutation strategy, and the generation of the guiding spark is modified with the historical information to improve the searching capability. Finally, in order to verify the rationality and effectiveness of the proposed model and algorithm, a microgrid system is simulated with open data. The simulation results demonstrate FWAM lowers the economic cost of the microgrid system more effectively compared with other swarm intelligence algorithms such as GFWA and CMA-ES.

Keywords: Fireworks algorithm · Swarm intelligence · Microgrid · Smart grid · Economic dispatch

1 Introduction

Facing the increasing environment protection needs, clean energy with remarkable renewable and environment-friendly characteristics, such as photovoltaic (PV) power and wind power (WP), is gradually replacing the traditional thermal power which has harmful environmental effects. In the relevant case study, the global renewable energy consumption has already accounted for 15% of global energy consumption in 2020 and will further increase to 27% in 2050 [15]. However, due to the randomness of natural conditions, renewable clean energy usually shows significant intermittent and irregularity. Directly injecting the renewable

Y. Tan et al. (Eds.): ICSI 2022, LNCS 13344, pp. 339–353, 2022.
https://doi.org/10.1007/978-3-031-09677-8_29

power into the utility grid will lead to the power mismatching and seriously affect the power quality [1]. The technique of energy decentralization like microgrid can effectively alleviate the problem by maintaining a stable power demand and supply ratio.

Microgrid is a system concept including multiple coordinated loads and distributed energy resources (DER), operating as a controllable structure to the utility grid with well defined electrical boundaries [14]. In addition, the microgrid is also equipped with necessary control device which can manage the power output of the controllable unit to maintain the power balance and control interaction with the utility grid under the grid-tied mode, so as to downscale the fluctuation and boost the overall economic benefits of grid and users [16]. On the basis of meeting the above requirements, making a reasonable day ahead dispatch schedule to minimize the economic cost is of great significance in the microgrid and smart grid.

Due to the complex form of objective functions of the microgrid economic dispatch optimization, various swarm intelligence algorithms are introduced to solve the optimization problem, which have already achieved notable success on some real-world problems like spam detection [12], multiple targets search [17] and multi-objective optimization [3,8,18]. Fireworks algorithm (FWA) is a novel swarm intelligence algorithm proposed by Tan et al. in 2010 [13]. FWA has a double-layer structure, one layer is the global coordination between the populations represented by fireworks, and the other one is the independent search of each firework. This hierarchical structure ensures that FWA can adapt to a variety of optimization problems with different characteristics. In recent years, some variants of fireworks algorithms such as guided FWA (GFWA [7]) and loser-out tournament FWA (LoTFWA [5]) further enhance the search ability of FWA. The superiority of the those variants on the optimization of multi-modal test functions prove that FWA has great potential in real-world optimization problems like multi-objective.

Based on the comprehensive consideration of the power characteristics and constraints of renewable energy and energy storage, a dynamic economic dispatch optimization model for the microgrid is built with the goal of minimizing the overall costs and simulated with the open datasets in this paper. This work also improves the mutation operator of FWA by introducing momentum mechanism and the resulting algorithm is called FWA with momentum (FWAM). In GFWA, the guidance vector (GV) is determined by the difference between the centroids of two certain groups of sparks in the current population, and a guidance spark (GS) are generated accordingly as the elite individual. Meanwhile, FWAM additionally introduces its own historical information when calculating the guiding vector to reduce the randomness of guidance spark generation. Simulation result shows that FWAM exhibits more powerful exploration and exploitation ability than previous FWA variants and other swarm intelligence algorithms like CMA-ES.

The remaining chapters of this paper is organized as follows. Section 2 introduces the essential background information and related works. Section 3 describes our proposed dispatch model in detail. Section 4 explains and analyzes

FWAM and its improved mutation operator. Then Sect. 5 presents the simulation results to show the performance of FWAM on the dispatch problem. And Sect. 6 gives the conclusion.

2 Related Works

2.1 Economic Dispatch Optimization for Microgrid

Microgrid economic dispatch can be roughly divided into static dispatch and dynamic dispatch. The static dispatch strategy obtains the optimal value of the objective function for each time interval, and adds the results together to obtain the global optimal. The static strategy ignores the inherent links between intervals, and thus cannot meet the actual requirements. In addition, some studies also simplify the architecture, constraint and objective function of the microgrid system. Peng et al. built an economic dispatch model for microgrid under the island mode without the state of charge constraints of the energy storage [9]. Ding et al. proposed a similar dispatch model with the goal of minimizing operating cost of distributed generation system, but ignore the environmental cost [2]. Recently, some studies try to simulate a more realistic microgrid system model, and put forward more meaningful and useful dispatch strategies on this basis, which leads to a sharp increase in the complexity of the microgrid dispatch optimization, and a variety of swarm intelligence algorithms are introduced to solve the problems. Tan et al. proposed a hybrid non-dominated sorting genetic algorithm (NSGA) and adopted it on the multi-objective dispatch optimization for microgrid [11]. Lezama et al. optimized bidding in local energy market with particle swarm algorithm (PSO) [4].

This paper describes the constraints of each DER and the objective function of the entire system in detail, and establishes the links between intervals. In order to solving the optimization problem, a novel FWA is proposed and introduced.

2.2 Guided Fireworks Algorithm

FWA conducts explosion and selection iteratively to search the global optimum. In the explosion operation, each firework would generate several sparks in a hypersphere centered on itself, where the radius of hypersphere is called explosion amplitude. Then, firework of next iteration would be selected from the candidate pool formed by firework and its sparks. Variants like adaptive FWA (AFWA [6]) and dynamic search FWA (dynFWA [16]) improve the explosion operator by adjusting the explosion amplitude adaptively in each iteration. LoTFWA and Fireworks Algorithm based on search space partition (FWASSP [8]) attempted to design a more efficient collaboration mechanism.

GFWA introduced a landscape information utilization-based elite steategy [6]. In GFWA, the firework and its sparks are sorted according to their fitness after the explosion operation in each iteration. Then, the guiding vector (GV) is calculated as the difference between the centroids of the top $\sigma\lambda_i$ sparks and the bottom $\sigma\lambda_i$ sparks, where σ is a super parameter to control the size of the two subsets. By adding the GV to the firework, a elite individual named guiding spark (GS) is generated. And GS will be selected together with other individuals in the candidate pool to select new firework. The experimental result shows this novel mutation operator can enhance the convergence speed and the local search ability significantly.

3 Dynamic Economic Dispatch Optimization Model for Microgrid

A dynamic economic dispatch optimization model for microgrid under the grid-tied mode is established in this paper. The microgrid system is mainly consist of the distribution energy resources and the load, where the distribution energy resources include photovoltaic (PV) system, wind turbine (WT), micro turbine(MT) and energy storage (ES) devices. And there is also a control device to control the power output of the DER and the interaction with the utility grid. The power generated by DER gives the priority to meeting the load demand, and the excess energy will be transmitted to ES and the utility grid according to the electricity price. Figure 1 illustrates the structure of the entire microgrid system.

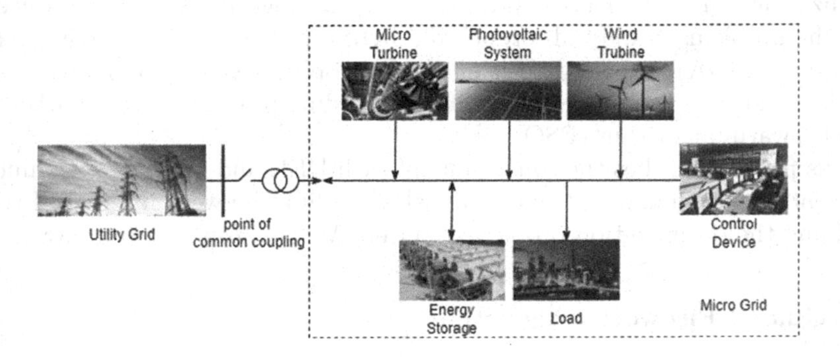

Fig. 1. The structure of the microgrid system.

The dynamic economic dispatch optimization model is usually regarded as a dynamic system. Taking the state of charge (SoC) of the ES as the system state variable, the model can be described as the following dynamic equation:

$$SoC_{t+1} = SoC_t + \sum_{i=1}^{N} P_t^{(i)} + P_{g,t}, \quad t = 0, 1, ..., T-1, \tag{1}$$

where N is the number of DER, and P_g is the power interaction with the utility grid. The dispatch in this paper is a day-ahead hourly scheduling, thus T is set as 24.

3.1 Objective Function of the Dispatch Model

The main objective of the dispatch is maximizing the utilization of renewable resources to reduce the pollution emissions, while minimizing the economic cost of the entire system. The objective function is expressed as the following:

$$\min f = f_{eco} + f_{env}, \tag{2}$$

$$f_{eco} = \sum_{t=0}^{T-1}(\sum_{i=1}^{N_W} C_W(P_{W,t}^{(i)}) + \sum_{i=1}^{N_P} C_P(P_{P,t}^{(i)}) + \sum_{i=1}^{N_M} C_M(P_{M,t}^{(i)}) + \sum_{i=1}^{N_S} C_S(P_{S,t}^{(i)}) + C_G(P_{G,t})), \tag{3}$$

$$f_{env} = \sum_{t=0}^{T-1}(\sum_{i=1}^{N_M} E_M(P_{M,t}^{(i)}) + E_G(P_{G,t})), \tag{4}$$

where f_{eco} and f_{env} are functions of the economic cost and environmental cost. C_W, C_P, C_M and C_S are the operation cost of WT, PV, MT and ES. N and P represent the number and power of the corresponding units. C_G is the transaction cost with utility grid. E_M and E_G represent the environmental compensation expense of MT and grid. Detailed definitions of the cost function above are introduced as follows.

Cost Function of Wind Turbine. Wind power is one of the main clear energy resources with the well established technology at present, which could lower the pollution emissions effectively. The maintenance cost of the wind turbines can be abstracted as a linear relation with the active output:

$$C_W = \alpha_W P_W, \tag{5}$$

where α_W is the coefficient of the maintenance cost of WT.

Cost Function of Photovoltaic System. Photovoltaic power is also a important clean alternative energy, and it has a more extensive application scenarios compared with the wind power. The maintenance cost of photovoltaic system can also be expressed as a linear relation:

$$C_P = \alpha_P P_P, \tag{6}$$

where α_P is the coefficient of the maintenance cost of PV.

Cost Function of Micro Turbine. Due to the stable and controllable power output, the micro turbine can relieve the short-term power shortage and stabilize the fluctuation of voltage and frequency caused by the randomness of clean energy. The operation cost of micro turbine mainly includes the maintenance cost and the fuel cost, which are defined as the followings:

$$C_M = C_{mt} + C_f, \tag{7}$$

$$C_{mt} = \alpha_M P_M, \tag{8}$$

$$C_f = \alpha_f P_M^2 + \beta_f P_M + \gamma_f, \tag{9}$$

where α_M, α_f, β_f and γ_f are parameters determined by the type of the micro turbine.

Cost Function of Energy Storage. ES can be charged and discharged according to the electricity price and power surplus, so as to effectively mitigate the negative impact of fluctuation of the load and reduce the operation cost of microgrid system. The maintenance cost of ES are usually expressed as a linear relation with the power of charging and discharging:

$$C_S = \alpha_S |P_S|, \tag{10}$$

where α_S is the coefficient of the maintenance cost of ES, and P_S represents the charging power of discharging power of ES. For convenience, the charging power is set to negative, and the discharging power is set to positive.

Cost Function of Energy Transaction. Under the grid-tied mode, the microgrid system can establish the energy transaction between the utility grid. If the power of DER cannot meet the load demand, the microgrid can purchase energy from the utility grid. If there is a power surplus, the excess energy can be transmitted to ES or sell to the utility grid according to the real-time electricity price. Thus, the cost of energy transaction can be expressed as the following:

$$C_G = \begin{cases} p_b P_G, & P_G \geq 0 \\ p_s P_G, & P_G < 0, \end{cases} \tag{11}$$

where p_b and p_s are the purchase price and the selling price respectively, and P_G is the interactive power. The interactive power P_G is set to negative when the microgrid purchase the electricity form the utility grid, otherwise it is set to positive.

Environmental Cost. Thermal power generation like micro turbine usually emits certain polluting gases, among which sulfide and nitride have a relatively strong negative impact on the environment. This would require certain environmental compensation for the pollution prevention and control. It's worth noting that thermal power also account for a significant portion in the utility grid today. Thus, when purchasing the electricity from the utility grid, the microgrid system still need paying the environmental compensation. The environmental compensation expense of MT can be abstracted as the following:

$$E_M = \alpha_n \beta_n^M P_M + \alpha_s \beta_s^M P_M, \tag{12}$$

where α_n and α_s are the compensation expense of nitride and sulfide. β_n^M and β_s^M are emission parameters of nitride and sulfide. And the compensation expense of the utility grid is similar as the MT:

$$E_G = \begin{cases} \alpha_n \beta_n^G P_G + \alpha_S \beta_s^G P_G, & P_G > 0 \\ 0, & P_G \leq 0, \end{cases} \tag{13}$$

where β_n^G and β_s^G are emission parameters of nitride and sulfide of the utility grid.

3.2 Constraints of the Dispatch Model

For the stability and safety of microgrid operation, it is necessary to enforce certain constraints on each unit in the microgrid. The constraints can be divided into equality constraint and the inequality constraints in this paper, where the equality constraint describe the power balance. And the inequality constraints are mainly the power constraints of the distributed generation and storage. Detailed constraints are listed as follows.

The Power Balance of the Microgrid System. There must be a balance between the power supply and demand in each time interval:

$$P_L(t) = P_W(t) + P_P(t) + P_M(t) + P_E(t) + P_G(t), \tag{14}$$

where P_L represents the power of all loads int the microgrid system.

The Constraints of Distributed Generations.

$$P_W^{min} \leq P_W(t) \leq P_W^{max}, \tag{15}$$

$$P_P^{min} \leq P_P(t) \leq P_P^{max}, \tag{16}$$

$$P_M^{min} \leq P_M(t) \leq P_M^{max}, \tag{17}$$

$$P_M(t) - P_M(t-1) \leq R_{up}, \tag{18}$$

$$P_M(t) - P_M(t-1) \geq R_{down}, \tag{19}$$

where P_W^{min}, P_P^{min}, P_M^{min}, P_W^{max}, P_P^{max} and P_M^{max} are the minimum and the maximum of active output power of WT, PV and MT respectively. R_{up} and R_{down} are limitations of the ramp rate of MT.

The Constraints of Energy Storage. Both the capacity and power of ES need to be limited, where the capacity is usually described by the state of charge SoC, that is the ratio of the residual capacity to the rated capacity:

$$SoC = \frac{Q_0 - \int_0^t I(t)dt}{Q_m}, \tag{20}$$

where Q_0 is the initial capacity of ES, and Q_m is the rated capacity of ES. Then, the main constraints of ES can be given as follows:

$$\begin{cases} SoC(t+1) = SoC(t) + \eta_{in} P_S(t)/Q_m \\ P_{min}^{in} \leq P_S(t) \leq P_{max}^{in} \\ SoC_{min} \leq SoC(t) \leq SoC_{max} \end{cases}, P_S(t)<0, \tag{21}$$

$$\begin{cases} SoC(t+1) = SoC(t) + \eta_{out} P_S(t)/Q_m \\ P_{min}^{out} \leq P_S(t) \leq P_{max}^{out} \\ SoC_{min} \leq SoC(t) \leq SoC_{max} \end{cases}, P_S(t) \geq 0. \tag{22}$$

where η_{in} and η_{out} are the charging and discharging efficiency of ES. P_{min}^{in}, P_{max}^{in}, P_{min}^{out} and P_{max}^{out} are limitations of charging and discharging power. SoC_{min} and SoC_{max} are limitations of SoC.

4 Fireworks Algorithm with Momentum

4.1 Principle

GFWA improves the local search ability of fireworks algorithm by further utilizing the information of population and landscape. In GFWA, the guiding vector (GV) can be seem as an estimator of the gradient of the objective function, especially when the explosion amplitude is short. Thus, a GV with the accurate direction and length could generate a guiding spark (GS) on a promising position, which is more likely to be selected as the firework of the population in the next iteration.

For reducing the randomness, GFWA calculate the GV by the centroids of the top and bottom sparks instead of the best and worst spark. The technique could extract the common qualities of the top sparks (the bottom sparks), and cancels out the random noise on the irrelevant directions. However, there are still several weaknesses in the technique: (1) When the firework locates in a local/global optimum area, the explosion amplitude is usually shortened dramatically, which means that the length of GV would also be too short to generate a GS on the promising position. And the effect of the elite strategy would be weakened. (2) The stability of the guiding spark mechanism is sensitive to the change of super parameter σ. If the guiding mutation ratio σ is too large, some moderate sparks would be selected to calculate the centroid, and this would lead to the vague of common qualities of the top sparks/bottom sparks. While if σ is too small, the random noise could not be cancelled out.

To solve the problems above, FWAM introduces a momentum mechanism to generate GS with the historical information. Specifically, in each iteration, the calculation of GV is not only determined by the difference in the current iteration, but also the GV in the previous iteration:

$$\Delta_{i,t} = \frac{1}{\sigma\lambda_i}\left(\sum_{j=1}^{\sigma\lambda_i} \mathbf{s}_{\mathbf{ij,t}} - \sum_{\mathbf{j}=\lambda_\mathbf{i}-\sigma\lambda_\mathbf{i}+1}^{\lambda_\mathbf{i}} \mathbf{s}_{\mathbf{ij,t}}\right), \tag{23}$$

$$\mathbf{v}_{i,t} = \gamma \mathbf{v}_{i,t-1} + \eta \Delta_{i,t}, \tag{24}$$

where $\mathbf{v_t}$ is defined as the GV in FWAM, and $\mathbf{v_{t-1}}$ can be seem as a momentum term. γ is a momentum parameter to control the ratio of current difference Δ_t and the historical information, and a larger γ means that GV would sotre more historical information in the current GV. Compared with GFWA, a obvious advantage is that even if the random noise in the difference between the top and bottom sparks affects GV's estimation of gradient, GV could still be corrected and compensated by the historical information. Thus, when the firework locates near the optimum, this improved GV could also has a promising direction. Another important advantage is that the momentum mechanism can lengthen the GV on the relevant direction, which would accelerate the convergence of the firework. Algorithm 1 gives the description of FWAM. The next subsection will give analysis of this mechanism in detail.

Algorithm 1. Framework of Fireworks Algorithm with Momentum

Input: Firework num μ, spark num λ, mutation ratio σ, momentum params γ, η
Output: Optimal solution
Initialize μ fireworks randomly within the feasible region Ω
while termination condition not satisfied **do**
  for $Fireworks_i$ in $Fireworks$ **do**
    Explosion:
    Generate λ_i spark randomly around $Fireworks_i$ within amplitude A_i
    Mutation:
    Sort sparks according to their fitness in ascending order
    Calculate guiding vector $\mathbf{v}_{i,t} = \gamma \mathbf{v}_{i,t-1} + \eta \frac{1}{\sigma\lambda_i}(\sum_{j=1}^{\sigma\lambda_i} \mathbf{s_{ij,t}} - \sum_{\mathbf{j}=\lambda_\mathbf{i}-\sigma\lambda_\mathbf{i}+1}^{\lambda_\mathbf{i}} \mathbf{s_{ij,t}})$
    Generate guiding spark $G_{i,t} = \mathbf{v}_{i,t} + Firework_{i,t}$
    Selection:
    Evaluate $Firework_i$ and all sparks' fitness
    Select the best candidate as the $Firework_i$ of the next iteration
    Adjust A_i adaptively
  end for
end while

4.2 Analysis

Considering the relation between the previous GV (momentum term) $\mathbf{v_{t-1}}$ and current difference Δ_t, the effect of momentum mechanism can be analyzed in the following two possible situations.

If current difference Δ_t has a direction consistent with $\mathbf{v_{t-1}}$ (see Fig. 2(a)), the projection of $\mathbf{v_{t-1}}$ on Δ_t would be relatively large and current GV $\mathbf{v_t}$ would be lengthened on the relevant direction. Actually, the momentum mechanism can be regarded as a weighted average method, and thus the lengthening effect would be more significant if the direction of Δ_t always keeps consistent with the historical GV in recent iterations. Due to the characteristic of adaptive strategy in FWA, there is always the same change trend of amplitudes on different dimensions, and this would lead to premature convergence on some key dimensions. As a result, FWA might underperform when the global optimum locates on the boundaries of feasible region. While the momentum mechanism can help the GS keep sufficient distance separation with the firework even when the population locates near the optimum or boundaries, which ensure the guiding spark can still make sense in this situation.

If the direction of current difference Δ_t has a obvious divergence with the promising direction (see Fig. 2(b)), GV in GFWA tends to have a large oscillation on the irrelevant direction and the GS is likely not going to seek a better position, which means that GS would not be selected as the firework. While $\mathbf{v_{t-1}}$ represents the accumulation of historical GV information in FWAM, and thus there is a higher probability for GV to approach the relevant direction. As the sum of $\mathbf{v_{t-1}}$ and Δ_t, components of GV on the irrelevant direction in FWAM would be shortened and the GS would be closer to the optimum. From another point of view, Δ_t also has a significant effect on $\mathbf{v_t}$, and such a "compromise" strategy ensure that the GS in FWAM still can lead the population to get rid of the local optimum.

In summary, by introducing the momentum mechanism, the variance of GV in FWAM can be reduced effectively and the GV would be more aggressive when it approach the promising direction, which could accelerate the convergence of the algorithm and enhance the tolerance of the selection of parameter σ.

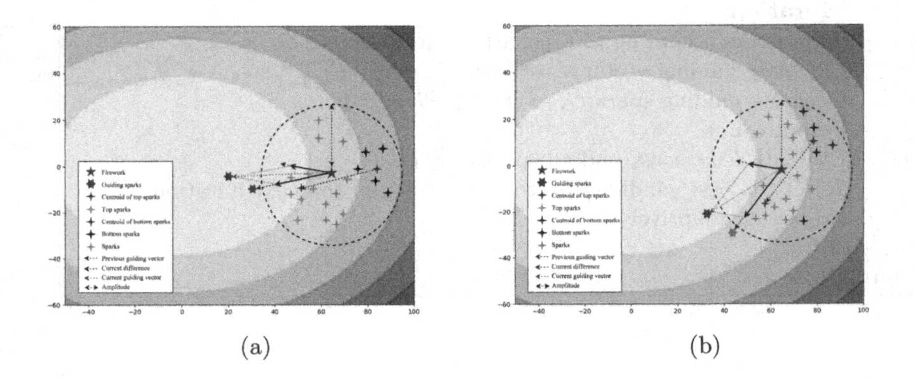

Fig. 2. Illustration of two possible situations of FWAM in the search process.

5 Case Study

For verifying the rationality and effectiveness of the proposed dispatch model and the improved algorithm, this paper simulates a dispatch optimization model under the grid-tied mode as the definition in Sect. 3 , and conducts the FWAM on the model to get a day-ahead dispatch schedule for each controllable components. Detailed description are given as the followings.

5.1 Simulation Settings

The distributed generations in the system consist of 4 micro turbines, 100 wind turbines, 20 PV arrays (each PV array has 40 PV panels) and 2 energy storage systems. The detailed parameters of each DER units above are listed in Table 1 and Table 2. Loads is mainly composed of 3 office buildings and 10 personal residential houses. And the real-world power profiles of WT, PV and loads are selected form the open datasets available in PES ISS website [10]. Time-of-use (TOU) price of the electricity transaction with utility grid is shown as Table 3.

In order to fully verify the reliability of the proposed algorithm, the power profiles from 1 June to 30 June are chosen to conduct the simulation 30 times repeatedly with a maximum evaluation number of 100000. Some FWA variants, such as GFWA and FWA based on search space partition (FWASSP), and other swarm intelligence algorithm like CMA-ES and PSO are selected as the baseline.

Table 1. Parameters of distributed generations in the microgrid system.

DG	Minimum Output (kW)	Maximum Output (kW)	Minimum Ramp Rate (kW/h)	Maximum Ramp Rate (kW/h)	Maintenance Parameters (CNY/(kW·h))
WT	0	50.0	–	–	0.12
PV	0	80.0	–	–	0.02
MT1	0	35.0	−15.0	15.0	0.03
MT2	0	35.0	−15.0	15.0	0.02
MT3	0	35.0	−15.0	15.0	0.04
MT4	0	35.0	−15.0	15.0	0.01

Table 2. Parameters of distributed storage systems in the microgrid system.

DS	rated Capacity (kW·h)	Minimum SoC (%)	Maximum SoC (%)	Maximum Discharing Power (kW)	Maximum Charging Power (kW)	Maintenance Parameters (CNY/(kW·h))
ES1	30.0	0.1	0.9	10.0	−5.0	0
ES2	30.0	0.1	0.9	10.0	−5.0	0

Table 3. Time-of-use price of electricity transaction.

Period	Time	Purchase price (CNY)	Sell price (CNY)
Peak	11:00–15:00, 19:00–21:00	0.83	0.65
Peace	8:00–10:00, 16:00–18:00, 22:00–23:00	0.49	0.38
Valley	0:00–7:00	0.17	0.13

5.2 Encoding of Solutions

The object of the dispatch is reducing the overall cost of the microgrid system by adjust the active output of the controllable DER and the transaction with the utility grid. Thus, the solutions in this paper is 144-dimensional, which consists of the scheduled hourly active output of 4 MT and 2 ES in the next day. And the transaction with the utility grid can be determined according to the constraint of power balance.

5.3 Cost Analysis

The simulation results are shown as Table 4. Momentum parameters γ and η are set as 0.9 and 0.6 by grid search. The results indicate that the average rank of mean cost of FWAM is 1.57, which is the best compared with other baseline algorithms. And the standard deviation of FWAM also indicates that the momentum mechanism improve the stability of the algorithm.

Table 4. Comparing FWAM with baseline algorithms on the simulation datasets.

Day	FWAM		GFWA		FWASSP		PSO		CMA-ES	
	mean	std	mean	std	mean	std	mean	std	mean	std
1	**3.41e+02**	3.46e+00	3.84e+02	1.56e+01	4.29e+02	9.67e+00	4.21e+02	1.48e+01	3.87e+02	1.26e+00
2	**1.44e+02**	9.01e+00	1.66e+02	1.05e+01	2.14e+02	1.90e+01	1.74e+02	3.76e+00	1.50e+02	7.07e+00
3	**−2.16e+02**	3.26e+00	−2.08e+02	7.35e+00	−1.75e+02	7.60e+00	−1.68e+02	1.64e+01	−2.12e+02	2.22e+01
4	5.32e+02	1.04e+01	5.46e+02	8.43e+00	5.64e+02	2.81e+01	5.64e+02	1.34e+01	**5.28e+02**	3.81e+00
5	**2.27e+02**	6.36e+00	2.53e+02	1.23e+01	3.05e+02	2.49e+01	3.18e+02	9.54e+00	2.52e+02	6.19e+00
6	**5.05e+02**	8.79e+00	5.14e+02	8.76e+00	5.62e+02	1.69e+01	5.85e+02	5.64e+00	5.07e+02	1.09e+01
7	2.60e+02	6.17e+00	2.61e+02	5.02e+00	3.40e+02	3.25e+01	3.30e+02	1.28e+01	**2.56e+02**	5.20e+00
8	**4.80e+02**	3.44e+00	5.01e+02	1.32e+01	5.16e+02	2.02e+01	5.28e+02	2.50e+01	4.88e+02	1.30e+01
9	6.17e+02	8.51e+00	6.23e+02	1.03e+01	6.43e+02	1.61e+01	6.64e+02	1.65e+01	**6.16e+02**	6.03e+00
10	3.95e+02	7.38e+00	**3.93e+02**	4.15e+00	4.63e+02	3.56e+01	4.37e+02	1.37e+01	4.03e+02	1.42e+01
11	8.21e+02	1.90e+00	8.24e+02	7.87e+00	8.29e+02	7.31e+00	8.34e+02	1.63e+01	**8.08e+02**	1.14e+01
12	**−4.88e+01**	5.32e+00	−4.75e+01	6.38e+00	3.15e+01	3.29e+01	1.01e+00	1.47e+01	−4.45e+01	6.59e+00
13	6.49e+02	6.63e+00	6.58e+02	5.34e+00	6.89e+02	1.46e+01	6.86e+02	7.86e+00	**6.37e+02**	1.60e+00
14	**6.04e+02**	9.69e+00	6.15e+02	7.59e+00	6.60e+02	2.64e+01	6.45e+02	1.16e+01	6.15e+02	1.85e+00
15	4.30e+02	7.84e+00	4.35e+02	1.43e+01	4.59e+02	2.39e+01	4.87e+02	1.61e+01	**4.26e+02**	8.33e+00
16	7.32e+02	4.94e+00	7.49e+02	5.81e+00	7.82e+02	5.53e+00	7.58e+02	1.21e+01	**7.26e+02**	1.00e+01
17	**6.66e+02**	8.11e+00	6.81e+02	1.15e+01	7.03e+02	1.36e+01	7.14e+02	1.68e+01	6.68e+02	5.36e+00
18	**3.43e+02**	3.90e+00	3.67e+02	9.25e+00	4.00e+02	3.20e+01	4.10e+02	2.06e+01	3.61e+02	1.06e+01
19	**2.89e+02**	3.90e+00	3.15e+02	9.42e+00	3.71e+02	1.59e+01	3.78e+02	2.10e+01	2.91e+02	5.56e+00
20	3.97e+02	5.04e+00	3.98e+02	7.06e+00	4.46e+02	8.13e+00	4.59e+02	4.91e+00	**3.85e+02**	4.31e+00
21	−6.97e+02	2.06e+00	−6.95e+02	9.17e+00	−6.39e+02	2.95e+01	−6.30e+02	1.35e+01	**−7.02e+02**	5.31e+00
22	**−1.09e+02**	9.29e+00	−1.03e+02	1.12e+01	−6.35e+01	1.70e+01	−4.20e+01	1.40e+00	−1.04e+02	7.32e+00
23	−3.72e+02	3.18e+00	**−3.76e+02**	1.61e+01	−2.60e+02	1.80e+01	−3.13e+02	1.95e+01	−3.69e+02	8.32e+00
24	−6.45e+01	1.59e+00	**−6.70e+01**	1.34e+01	−7.80e+00	1.47e+01	−9.13e+00	1.42e+01	−5.97e+01	1.48e+01
25	3.99e+02	4.44e+00	3.94e+02	1.08e+01	4.75e+02	9.55e+00	4.68e+02	9.14e+00	**3.93e+02**	1.43e+01
26	4.33e+02	5.54e+00	4.54e+02	1.59e+01	4.77e+02	2.03e+01	4.92e+02	1.09e+01	**4.30e+02**	9.49e+00
27	**4.37e+02**	6.53e+00	4.49e+02	5.98e+00	5.08e+02	1.07e+01	4.82e+02	2.02e+01	4.40e+02	3.83e+00
28	8.51e+02	2.24e+00	8.54e+02	6.15e+00	8.67e+02	3.05e+01	8.71e+02	9.58e+00	**8.35e+02**	3.06e+00
29	8.18e+02	5.48e+00	8.21e+02	9.71e+00	8.39e+02	1.68e+01	8.44e+02	1.32e+01	**8.11e+02**	5.27e+00
30	**3.04e+02**	9.95e+00	3.16e+02	1.14e+01	3.63e+02	1.71e+01	3.50e+02	1.80e+01	3.20e+02	6.14e+00
AR	1.57	1.83	2.67	2.83	4.47	4.33	4.53	3.73	1.77	2.27

Here, we take the simulation of 1 June as the case to analyze the internal logic of the dispatch schedule. The scheduled output of each component is shown as the Fig. 3. During 1:00–8:00, the output of PV and WT is in a relatively low range and cannot meet the load demand. Owing to that the cost of electricity transaction is lower than the MT's in this time, purchasing electricity from the utility grid accounts for a large proportion in the power supply. As the growth of load demand and electricity price, the output of MT increases gradually during 9:00–13:00. And If there is power supply surplus, the microgrid system would sell the extra power to the utility grid. When the clean energy covers the most power demands in the daytime, the power of MT would decrease accordingly to reduce the pollution emission. The output schedule during 16:00–24:00 follows the same logic as the daytime. ES generally tends to store the energy while the load demand in a low range, and discharges during the peak time to reduce the power fluctuation of the system. Compared with the dispatch schedule obtained by GFWA during 13:00–19:00 and 23:00–24:00, FWAM can response to the changes and adjust the output of the controllable units more timely. Besides, although there is a differences of parameters between MTs, the cost of each MT is still maintained at a relatively same level under the dispatch of FWAM, which means that FWAM could find a better solution to balance the output of different MTs.

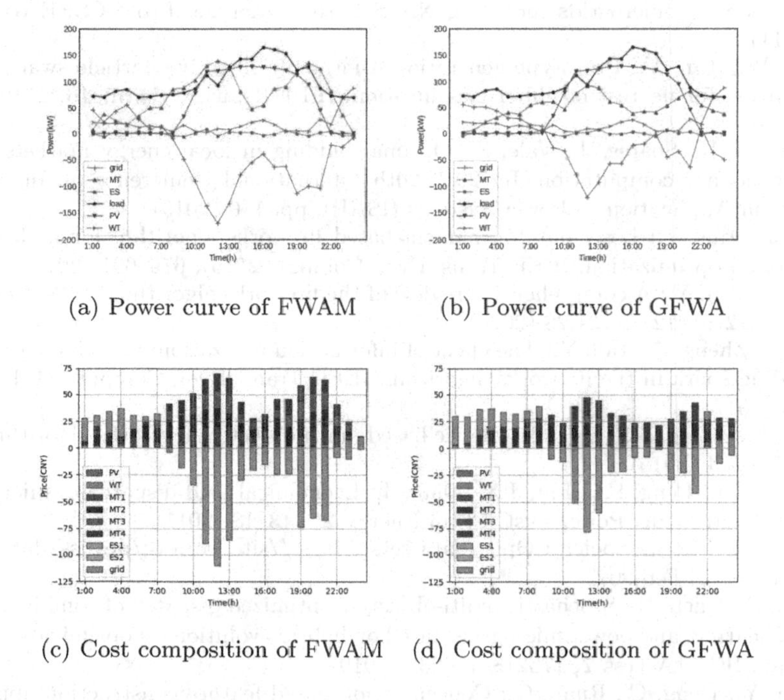

(a) Power curve of FWAM (b) Power curve of GFWA

(c) Cost composition of FWAM (d) Cost composition of GFWA

Fig. 3. Power curve and overall cost of 1 June obtained by FWAM and GFWA.

6 Conclusion

This paper establishes an economic dispatch optimization model for microgrid system with the objective of minimizing the economic cost and environment cost, and proposes an improved GFWA with the momentum mechanism to improve the search ability and mitigate the instability caused by the randomness of guiding spark generation. The simulation results indicate that FWAM is competitive against other swarm intelligence algorithms on the grid application. There are plenty of application scenarios for FWA, and we expect this work could be a inspiration for more application researches of FWA on the real-world problems.

Acknowledgments. This work is supported by the National Natural Science Foundation of China (Grant No. 62076010), and partially supported by Science and Technology Innovation 2030 - "New Generation Artificial Intelligence" Major Project (Grant Nos.: 2018AAA0102301 and 2018AAA0100302). (Ying Tan is the corresponding author.)

References

1. Blaabjerg, F., Teodorescu, R., Liserre, M., Timbus, A.V.: Overview of control and grid synchronization for distributed power generation systems. IEEE Trans. Ind. Electron. **53**(5), 1398–1409 (2006)
2. Ding, M., Zhang, Y.Y., Mao, M.Q., Liu, X.P., Xu, N.Z.: Economic operation optimization for microgrids including NA/S battery storage. Proc. CSEE **31**(4), 8 (2011)
3. Hu, W., Tan, Y.: Prototype generation using multiobjective particle swarm optimization for nearest neighbor classification. IEEE Trans. Cybern. **46**, 2719–2731 (2015)
4. Lezama, F., Soares, J., Vale, Z.: Optimal bidding in local energy markets using evolutionary computation. In: 2019 20th International Conference on Intelligent System Application to Power Systems (ISAP), pp. 1–6 (2019)
5. Li, J., Tan, Y.: Loser-out tournament-based fireworks algorithm for multimodal function optimization. IEEE Trans. Evol. Comput. **22**(5), 679–691 (2018)
6. Li, J., Tan, Y.: A comprehensive review of the fireworks algorithm. ACM Comput. Surv. **52**(6), 121:1-121:28 (2020)
7. Li, J., Zheng, S., Tan, Y.: The effect of information utilization: introducing a novel guiding spark in the fireworks algorithm. IEEE Trans. Evol. Comput. **21**(1), 153–166 (2017)
8. Liu, L., Zheng, S., Tan, Y.: S-metric based multi-objective fireworks algorithm, pp. 1257–1264 (2015)
9. Peng, X.I., Peng, L.I., Liu, J.X., Qian, J.: Economical load dispatch of microgrids in isolated mode. Power Syst. Clean Energy **27**, 13–18 (2011)
10. Power & Energy Society: Open data sets. https://site.ieee.org/pes-iss/data-sets/ Accessed 23 Feb 2022
11. Tan, B., Chen, H.: Stochastic multi-objective optimized dispatch of combined cooling, heating, and power microgrids based on hybrid evolutionary optimization algorithm. IEEE Access **7**, 176218–176232 (2019)
12. Tan, Y., Deng, C., Ruan, G.: Concentration based feature construction approach for spam detection. In: International Joint Conference on Neural Networks, IJCNN 2009, Atlanta, Georgia, USA, 14–19 June 2009, pp. 3088–3093. IEEE Computer Society (2009)

13. Tan, Y., Zhu, Y.: Fireworks algorithm for optimization. In: Tan, Y., Shi, Y., Tan, K.C. (eds.) ICSI 2010. LNCS, vol. 6145, pp. 355–364. Springer, Heidelberg (2010). https://doi.org/10.1007/978-3-642-13495-1_44
14. Ton, D.T., Smith, M.A.: The U.S. department of energy's microgrid initiative. Electricity J. **25**(8), 84–94 (2012)
15. U.S. Energy Information Administration: International energy outlook 2021. https://www.eia.gov/outlooks/ieo/ Accessed 23 Feb 2022
16. Wang, B., Sechilariu, M., Locment, F.: Intelligent dc microgrid with smart grid communications: control strategy consideration and design. IEEE Trans. Smart Grid **3**(4), 2148–2156 (2012). https://doi.org/10.1109/TSG.2012.2217764
17. Zheng, Z., Tan, Y.: Group explosion strategy for searching multiple targets using swarm robotic. In: 2013 IEEE Congress on Evolutionary Computation, pp. 821–828 (2013)
18. Zhou, Y., Tan, Y.: GPU-based parallel multi-objective particle swarm optimization. Int. J. Artif. Intell. **7**(11 A), 125–141 (2011)

UAV Path Planning Based on Hybrid Differential Evolution with Fireworks Algorithm

Xiangsen Zhang[1,2] and Xiangyin Zhang[1,2](✉)

[1] Faculty of Information Technology, Beijing University of Technology, Beijing 100124, China
xy_zhang@bjut.edu.cn

[2] Engineering Research Center of Digital Community, Ministry of Education, Beijing 100124, China

Abstract. This paper considers the unmanned aerial vehicle (UAV) global path planning as an optimization problem with constraints and proposes a hybrid differential evolution with firework algorithm (HDEFWA) to generate the optimal feasible path. The multiple constraints based on the realistic scenarios are taken into account, including terrain and threat area constraints. The hybrid algorithm integrates the differential evolution operator into the mechanism of optimizing the fireworks algorithm (FWA) and uses the ideas of mutation, crossover, and selection to transform the spark particles generated by the explosion. The source of the differential mutation operator is the excellent particles in the iterative population. This mechanism makes up for the basic firework algorithm's neglect of the excellent solution resources in the population, which greatly improves the information sharing among the solutions. Experiments show that the proposed hybrid algorithm is superior to other intelligent algorithms in UAV path planning.

Keywords: Path planning · Unmanned air vehicle (UAV) · Fireworks algorithm (FWA) · Differential evolution (DE) · Global route planning

1 Introduction

Benefitting from the development of modern aviation technology and radio technology, unmanned aerial vehicle (UAV) are widely used in military and civil fields with their low price and flexible operation [1–3]. The UAV path planning is an important part and the basis for performing complex tasks. Path planning refers to planning a route from the start point to the endpoint within a specified time. The optimal or sub-optimal flight path should consider the constraints of UAV performance and environment in traveling to avoid the threat area that affects the flight safety of the UAV.

The path planning problem of UAV is usually regarded as an optimization problem with high-dimensional equality and inequality constraints [4]. Some traditional methods such as the A* algorithm [5], artificial potential field [6], and rapidly-exploring random tree (RRT) [7] are used to solve this problem. To reduce the complexity, some intelligent population-based algorithms are proposed to solve the path planning problem by

Y. Tan et al. (Eds.): ICSI 2022, LNCS 13344, pp. 354–364, 2022.
https://doi.org/10.1007/978-3-031-09677-8_30

the researcher, for example: Fu proposed a hybrid differential evolution and quantum-behaved particle swarm optimization (DEQPSO) to solve the constraint problem of UAV path planning on the sea [8]; Zhang proposed an improved differential evolution algorithm (DE) to deal with the UAV path planning problem in 3D environment, and used the method of level comparison to improve the efficiency of processing constraints [9]; Qi uses the firework algorithm (FWA) to solve the path planning problem of amphibious robots and considers switching energy consumption in amphibious environments [10]. FWA is a swarm intelligence algorithm proposed by Tan in 2010 [11]. To improve the performance of the firework algorithm, Zheng et al. proposed an enhanced firework algorithm (EFWA) [12] in 2013. Li and Tan proposed the bare bones fireworks algorithm (bbFWA) [13] in 2018, which only uses a minimalist basic algorithm framework, dynamically adjusts the search range, also shows good performance. Mixing other intelligent algorithms has also gained widespread attention, such as the cultural fireworks algorithm (CFWA) [14] and the hybrid particle swarm with fireworks algorithm (FWPS) [15].

In FWA only the optimal spark enters the next iteration in each iteration, and other fireworks are obtained through a distance-based random mechanism. In the sense of swarm intelligence, the other excellent solutions are not well-informed by the whole swarm. Inspired by this observation, we propose a hybrid differential evolution with fireworks algorithm (HDEFWA). The hybrid algorithm uses the mutation, crossover, and selection mechanism of differential evolution operators to transform explosion sparks. The source of the differential mutation material is the excellent fireworks or sparks of the previous generation. These operators are applied to guide the generation of new solutions, which improve the diversity of the population and avoid falling into local optimum in the early stage. Experiments show that our proposed algorithm performs better than other existing algorithms in obtaining high-quality solutions.

The rest of this paper is organized as follows. Section 2 conducts mathematical modeling for UAV path planning. Section 3 describes the hybrid differential evolution algorithm in detail and gives the detailed implementation process of UAV path planning. Section 4 designs experiments to verify the feasibility of the algorithm, and verifies the superiority of the proposed HDEFWA by comparing it with other existing constraint algorithms. Section 5 briefly summarizes this paper.

2 Mathematical Model of UAV Path Planning

2.1 Path Representation

For the UAV path planning problem described in this paper, we propose two assumptions as follows. The locations of threat areas in the whole flight environment are known. UAVs are represented by a point mass that maintains constant flight altitude and speed during the flight [16]. According to the above assumptions, the problem can be simplified to the path planning problem of UAVs in two-dimensional space. The UAV path is described in order of the flight starting point S (x_s, y_s), N waypoints, and target point T (x_t, y_t) as path = $\{S, P_1, P_2, \ldots, P_D, T\}$, as is shown in Fig. 1.

The rotated coordinate frame $X'O'Y'$ shown in Fig. 1 is established using $\overline{ST}$ as the new *X-axis*. The coordinate transformation between two reference systems by Eq. (1)

$$\begin{bmatrix} x' \\ y' \end{bmatrix} = \begin{bmatrix} \cos\theta & \sin\theta \\ -\sin\theta & \cos\theta \end{bmatrix} \begin{bmatrix} x - x_s \\ y - y_s \end{bmatrix}, \tag{1}$$

where θ is the angle between the line $\overline{ST}$ and the X-axis. Divide $\overline{ST}$ into $(D+1)$ equal parts with the length of Δl by D vertical lines, denoted as $l_1, l_2, \ldots l_D$. The vertical coordinate of the discrete point at the vertical line L_k is limited in $[S_{min}, S_{max}]$. S_{max} and S_{min} are the maximum value $+\Delta d$ and the minimum value $-\Delta d$ of all threats in the transformed coordinate system.

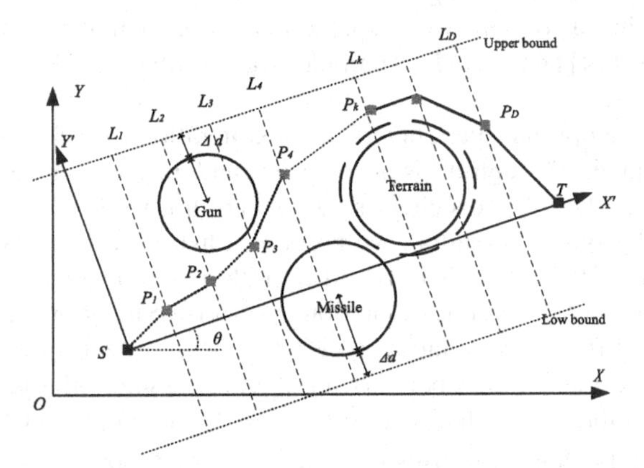

Fig. 1. Schematic diagram of UAV path

2.2 Threat Constraint

The safety of flight is a primary requirement for UAV path planning. It is usually necessary to use various methods to analyze and process the obtained flight geographic information and threat information and then establish a suitable mathematical model to enable the UAV to avoid various threats while flying. During the flight, the UAV is mainly threatened by terrain, radar, missiles, and anti-aircraft guns. The threat probability to any point on the flight path is calculated by Eq. (2) [17].

$$\begin{aligned}
P_{R,k} &= \frac{R_{R\max}^4}{d^4 + R_{R\max}^4}, d \le R_{R\max} \\
P_{p,k} &= \begin{cases} \frac{1}{d}, \; 2 + R_{p\max} \le d \le 10 + R_{p\max} \\ 1, \; d < 2 + R_p \end{cases} \\
P_{M,k} &= \frac{R_{M\max}^4}{d^4 + R_{M\max}^4}, d \le R_{M\max} \\
P_{G,k} &= \frac{R_{G\max}^4}{d^4 + R_{G\max}^4}, d \le R_{G\max},
\end{aligned} \tag{2}$$

where $P_{R,k}$, $P_{P,k}$, $P_{M,k}$, $P_{G,k}$ represent the threat values of radar, terrain, missiles, and anti-aircraft guns, respectively. R_{max} represents the scope of threats. d represents the distance of the UAV relative to its center.

The threat cost J_T is used to punish the path that allows the UAV to enter the danger areas. To simplify the calculation, the threat probabilities at five points on each path segment between two waypoints are usually selected to calculate the threat cost of the entire path segment, and the threat cost is the weighted sum of all threat probabilities by Eq. (3):

$$J_T = \sum_{j=1}^{n_T} \sum_{k=0}^{D} (P_{0.1,j,k} + P_{0.3,j,k} + P_{0.5,j,k} + P_{0.7,j,k} + P_{0.9,j,k}), \tag{3}$$

where n_T is the number of threats; $P_{0.1,j,k}$ is the threat probability from the *j-th* threat at the 1/10 point on the path segment $p_k\ p_{k+1}$.

2.3 Constrained Objective Function

The performance indicators of the UAV path planning mainly include the length cost and the threat cost. The length cost J_L of the path can be calculated as follows. The path planning of UAV is regarded as a constrained optimization problem, and a penalty function method is adopted to deal with the constraints. Therefore, the optimization model is established as follows:

$$\begin{aligned} \min\ J_L &= \sum_{n=1}^{r} \sum_{k=0}^{D} \sqrt{(x_{k+1} - x_k)^2 + (y_{k+1} - y_k)^2} \\ s.t.\ J_T &= 0. \end{aligned} \tag{4}$$

3 Global Path Planning Constraints Based on HDEFWA

3.1 Standard Firework Algorithm

As one swarm intelligence algorithm, the fireworks algorithm is inspired by the phenomenon of fireworks explosion in the night sky. The general framework of Standard FWA is described in Algorithm 1. For a detailed introduction to the standard fireworks algorithm, please refer to [11].

Algorithm 1 Standard firework algorithm.

Set the algorithm parameters and the generation counter $N_C = 1$
While termination criteria are not satisfied **do**
  For $i = 1: N_F$ **do**
    Calculate the number of explosion sparks S_i according to
    $S_i = M \times \frac{f_{\max} - f(X_i) + \varepsilon}{\sum_{i=1}^{N_F}(f_{\max} - f(X_i)) + \varepsilon}$.
    Calculate the explosion amplitude A_i according to
    $A_i = \hat{A} \times \frac{f(X_i) - f_{\min} + \varepsilon}{\sum_{i=1}^{N_F}(f(X_i) - f_{\min}) + \varepsilon}$.
    Generate the explosion sparks X_p according to
    $X_p^k = X_i^k + rand(-1,1) \times A_i$.
  End for
  For $i = 1: N_F$ **do**
    Generate the *Gaussian* spark X_q according to $X_q^k = X_i^k \cdot Gaussion\ (0,1)$.
  End for
  Select the best individual and keep it for the next explosion generation.
  Randomly select $N_F - 1$ individuals based on distance probabilities. $N_C = N_C + 1$.
End while

where N_F is the number of fireworks per generation; M is a parameter that controls the total number of sparks generated by the fireworks; $\hat{A}$ is a parameter that controls the maximum explosion amplitude; $f_{\min} = \min(f(X_i))$ stands for the best fitness value of all the fireworks. ε is defined as the smallest positive constant to avoid the zero-division error. *Gaussian* (0,1) is a function that generates Gaussian random numbers.

3.2 Hybrid Differential Evolution with Fireworks Algorithm

Differential evolution is a simple and powerful stochastic evolution algorithm. Here we introduce the DE operators to the FWA to improve the diversification strategy. The detailed steps of the HDEFWA are as follows.

Generate Explosion Sparks. In this paper, the minimum explosion radius detection mechanism is set regarding the EFWA, and the minimum radius changes linearly with the number of iterations to utilize the search resources of the optimal fireworks.

$$A_i = \hat{A} \times \frac{f(X_i) - f_{\min} + \varepsilon}{\sum_{i=1}^{N_F}(f(X_i) - f_{\min}) + \varepsilon}.$$
$$with \tag{5}$$
$$A_i = \begin{cases} A_{\min}, if A_i < A_{\min} \\ A_i, if A_i \geq A_{\min} \end{cases}.$$

$$A_{\min} = A_{\min start} - \frac{(A_{\min start} - A_{\min final}) * NC}{NC_{\max}}, \tag{6}$$

where $A_{\min start}$ and $A_{\min final}$ are the maximum and minimum values of the A_{min}.

Generation of Gaussian Sparks. The new Gaussian spark is obtained by Eq. (7)

$$X_q = X_i + Gaussion(0, 1) \cdot (X_{best} - X_i). \tag{7}$$

Generation of the Optimal Table (OT). Combined with the structural characteristics of the differential evolution algorithm, an *OT* with a capacity of 2*N_F ($X_g \in OT$) is built to store the excellent sparks of the current iteration population.

Generation of DE-explosion Sparks. Differential evolution operation is introduced to transform explosive sparks to generate *DE-explosion* sparks, in the first half of the iterations. The variation term is derived from the *OT* of the parent. The mutation operation is as shown in Eq. (8)

$$\begin{aligned} V_i^{N_C} &= X_{p,i}^{N_C} + F \cdot (x_{r2}^{N_C - 1} - x_{r3}^{N_C - 1}) \\ F &= F_{\max} - \frac{F_{\max} \cdot N_C}{(F_{\max} - F_{\min}) \cdot NC_{\max}}, \end{aligned} \tag{8}$$

where r_2, r_3 are the different random numbers, $F > 0$ is called the scaling factor.

The experimental vector $U_i = [U_{i,1}, U_{i,2}, \ldots U_{i,D}]$ is obtained by mixing the basis vector and the differential mutation vector by Eq. (9)

$$U_{i,j} = \begin{cases} V_{i,j}, \text{ if } rand(0, 1) \le C_r \vee j = r_n \\ X_{i,j}, \text{ otherwise} \end{cases}, \tag{9}$$

where C_r is the crossover rate between 0 and 1, r_n is a randomly selected index from {1, 2, ..., D}, which ensures that at least one component is taken from the donor vector.

After the differential mutation operation, if the mutation spark is better than the original spark, it will be retained and replaced by the original spark to become a member of the population, otherwise, the original explosion spark particles will be retained. The greedy selection operator can be described:

$$X_{p,i} = \begin{cases} U_i, \text{ if } f(Ui) < f(X_{p,i}) \\ X_{p,i}, \text{ otherwise} \end{cases}. \tag{10}$$

Population Update Operation. The update operation consists of two parts: one is to sort all fireworks and spark particles according to the fitness values in the current population, and update the *OT* to prepare for the next differential operation; the other is to select the next generation of fireworks, here we use elite random operator to avoid more computational cost, except for optimal particles the remaining N_F-1 fireworks will be randomly selected.

3.3 Path Planning for UAV

The pseudo-code of UAV path planning based on HDEFWA is shown as follows.

Algorithm 2 The pseudo-code for UAV path planning based on HDEFWA.

/**Modeling**/
Set the start point $(x_S, y_S)^T$ and target point $(x_T, y_T)^T$ of the UAV and the threat centers, radius, levels. Determine the transformation coordinate system $X'O'Y'$ based on the start and target points.
Determine the upper and low bounds S_{max}, S_{min}
/**Input for HDEFWA**/
Set the NC_{max} and the algorithm parameters N_F, D, M, $A_{min\ start}$, $A_{min\ final}$, F_{max}, F_{min}, C_r.
/**Initialization**/
Set the generation number $N_C = 1$; Initialize the location X_i of N_F fireworks randomly
/**Evaluation**/
Generate the $Path = \{p_0, p_1, p_2, \ldots, p_{D+1},\}$ Calculate the length cost by Eq. (4), threat cost by Eq. (3)
/**Iteration computation**/
While $N_C < NC_{max}$ **do**
    For $i = 1: N_F$ **do**
        Calculate the number of explosion sparks S_i according to **Algorithm 1**
        Calculate the explosion amplitude A_i according to Eq. (5).
        Generate the Gaussian spark X_q according to Eq. (7)
    End for
        Evaluate the fitness value of all spark particles $f(X_P)$.
    If $1 < N_C < NC_{max} / 2$ **do**
        For $j = 1: M$ **do**
            Generate DE-firework sparks U_{Pj}.
        If $f(U_{Pj}) < f(X_{Pj})$
            Preserves U_{Pj} and replace exploding spark X_{Pj}.
        Else
            Preserves the original explosion spark particles.
        End if
        End for
    End if
    Select the best individual and keep it for the next explosion generation. Randomly select $N_F - 1$ individuals from the remaining population. $N_C = N_C + 1$. Update the globally optimal path.
End while
/**Output**/
Output the best path for UAV.

---

4 Experiment Evaluation and Comparison

To evaluate the performance of HDEFWA in solving the UAV path planning problem, we design two experiments to compare the performance of HDEFWA with other algorithms, including FWA, PSO, and DE. The algorithm is implemented by Matlab-R2021a, and no commercial algorithm tools are used. Each algorithm runs 30 times independently. The information for test cases is listed in Table 1.

The main parameters of HDEFWA are set as follows: $NCmax = 500$, $N_F = 10$, $A_{min\ start} = 20$, $A_{min\ final} = 5$, $M = 50$, $F_{max} = 0.2$, $F_{\min} = 0.9$. The other algorithm parameters are set as follows: for FWA, $N = 10$, $M = 50$, for PSO, $N = 50$, $w_{max} = 0.9$, $w_{min} = 0.1$, $c_1 = c_2 = 0.9$; for DE, F is set as a random value in [0.2, 0.9], $N = 50$, $C_r = 0.9$.

In test Case I, the start and end points are set to [50, 50] and [950, 450]. Figure 2 shows the optimal path generated in Case I. In Case II the start and end points are [50, 200] and [950, 450]. Figure 4 shows the optimal path in Case II.

The statistical results of HDEFWA, FWA, DE, and PSO during 30 runs on Case I are listed in Table 2, and Case II are listed in Table 3, where the best, median, mean, worst and standard deviation of the optimal path length are recorded. These statistics listed in columns 2–5 only count successful runs with $J_T = 0$.

Table 1. Parameters of the threats.

	Case I			Case II		
Threat	Center	Radius	Level	Center	Radius	Level
Terrain	[400, 450]	95	8	[700, 200]	90	6
Radar	[700, 180] [650, 450]	142 100	5 5	[480, 230] [650, 450]	105 100	5 5
Missile	[200, 240]	85	9	[200, 210]	85	9
Anti-aircraft gun	[420, 220] [850, 430]	75 70	6 6	[300, 450] [850, 370]	120 70	8 10

It is observed from Table 2 and Fig. 3 that the HDEFWA is better than other algorithms in the globally optimal path planning in case I. It is better than other algorithms in the average and median path length of the optimal path. Although the algorithm ranks second in the standard deviation and worst length of the planned path length, it still maintains the advanced level with a small gap with the first. From the evolution curve of the average optimal path length, it can be seen that HDEFWA can always remain optimal at the end, although it may be overtaken by PSO in the early iterations. In general, HDEFWA is superior to the other three algorithms in global UAV path planning.

In the Case II we modified the environmental parameters and the start point of the UAV. As can be seen in Fig. 4, each algorithm can be found or close to the global optimal solution in 30 independent operations. In the second column in Table 3, it can be seen that the length of the global best path is not much different, but the HDEFWA maintains optimal parameters of the average and median parameters of the length of the planning path in 30 independent operations (Fig. 5).

Table 2. Performance comparison of various algorithms on Case I

Algorithm	Best	Mean	Median	Std	Worst
HDEFWA	**1021.214**	**1047.136**	**1028.554**	27.44116	1120.183
FWA	1060.957	1105.376	1100.792	31.2495	1158.744
PSO	1029.041	1068.274	1071.196	**24.81158**	**1115.781**
DE	1022.867	1149.094	1149.091	123.4526	1390.799

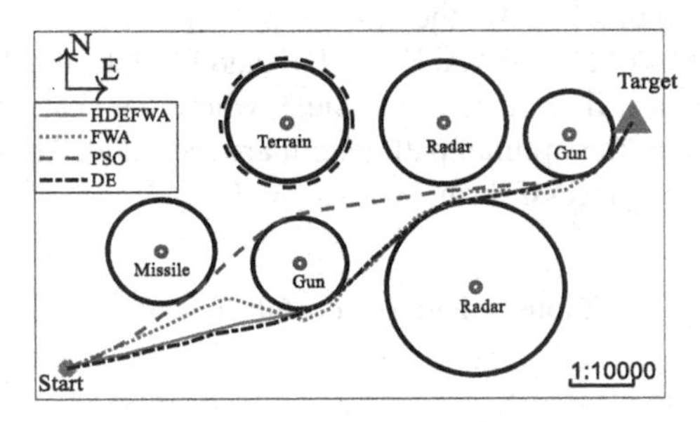

Fig. 2. The best UAV path of Case I

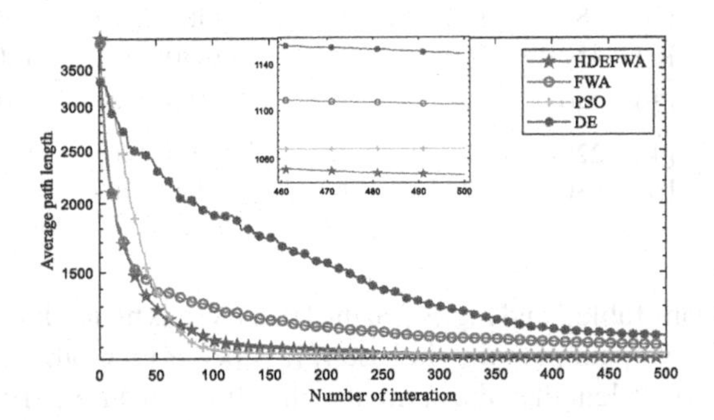

Fig. 3. Average path length evolution curve in Case I

Table 3. Performance comparison of various algorithms on Case II

Algorithm	Best	Mean	Median	Std	Worst
HDEFWA	959.4938	**974.6509**	**962.0771**	**26.48598**	**1032.565**
FWA	961.9972	980.9554	965.0887	31.12052	1044.86
PSO	960.1708	1001.065	963.9262	72.26618	1204.19
DE	**958.9633**	1002.261	1027.144	33.98229	1035.973

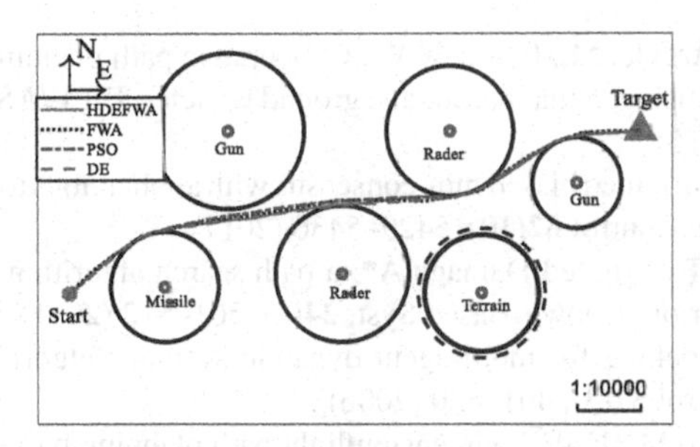

Fig. 4. The best UAV path of Case II

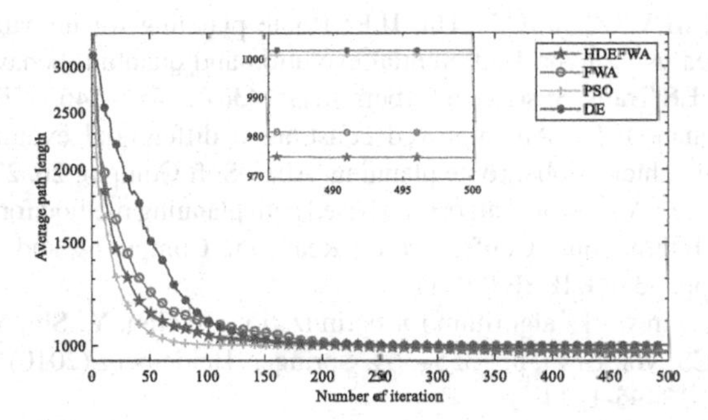

Fig. 5. Average path length evolution curve in Case II

5 Conclusion

Aiming at the inefficiency of the FWA in solving the UAV path planning, the optimal fitness table in each iteration is screened, and the explosive particles generated in the population are mutated by the differential evolution operator. Combining the search advantages of the differential evolution algorithm and fireworks algorithm, it can better fulfill the task requirements for UAV path planning.

The HDEFWA is tested in UAV path planning scenarios. At the same time, compare with other existing algorithms. The path planned by the algorithm allows the UAVs to reach their target faster. Safer without violating any performance constraints, which proves its effectiveness and efficiency. Experiments show that the algorithm has better performance in solving the UAVs' path planning problem. In the future, we will further improve the proposed algorithm in some aspects, including the initialization of the population, the adjustment of parameters in the algorithm, and so on. In addition, path planning for large-scale UAV swarms is also our next research direction.

References

1. Zhao, Y., Zheng, Z., Liu, Y.: Survey on computational-intelligence-based UAV path planning. Knowl.-Based Syst. **158**, 54–64 (2018)
2. Mac, T.T., Copot, C., Tran, D.T., Keyser, R.D.: Heuristic approaches in robot path planning: a survey. Robot. Auton. Syst. **86**, 13–28 (2016)

3. Yu, H., Meier, K., Argyle, M., Beard, R.W.: Cooperative path planning for target tracking in urban environments using unmanned air and ground vehicles. IEEE/ASME Trans. Mechatron. **20**(2), 541–552 (2015)
4. Zhang, Y., Li, S.: Distributed biased min-consensus with applications to shortest path planning. IEEE Trans. Autom. Control **62**(10), 5429–5436 (2017)
5. Bayilia, S., Polatb, F.: Limited-Damage A*: a path search algorithm that considers damage as a feasibility criterion. Knowl.-Based Syst. **24**(4), 501–512 (2011)
6. Olfati-Saber, R.: Flocking for multi-agent dynamic systems: algorithms and theory. IEEE Trans. Autom. Control **51**(3), 401–420 (2006)
7. Li, M., Sun, Q., Zhu, M.: UAV 3-dimensionflight path planning based on improved rapidly-exploring random tree. In: 31st Chinese Control and Decision Conference, Nanchang, China, pp. 921–925. IEEE (2019)
8. Fu, Y.G., Ding, M.Y., Zhou, C.P., Hu, H.P.: Route planning for unmanned aerial vehicle (UAV) on the sea using hybrid differential evolution and quantum-behaved particle swarm optimization. IEEE Trans. Syst. Man Cybern.-Syst. **43**(6), 1451–1465 (2013)
9. Zhang, X.Y., Duan, H.B.: An improved constrained differential evolution algorithm for unmanned aerial vehicle global route planning. Appl. Soft Comput. **26**, 270–284 (2015)
10. Qi, Y., Liu, J., Yu, J.: A fireworks algorithm based path planning method for amphibious robot. In: 2021 IEEE International Conference on Real-time Computing and Robotics (RCAR), Xining, China, pp. 33–38. IEEE (2021)
11. Tan, Y., Zhu, Y.: Fireworks algorithm for optimization. In: Tan, Y., Shi, Y., Tan, K.C. (eds.) ICSI 2010. LNCS, vol. 6145, pp. 355–364. Springer, Heidelberg (2010). https://doi.org/10.1007/978-3-642-13495-1_44
12. Zheng, S., Janecek, A., Tan, Y.: Enhanced fireworks algorithm. In: 2013 IEEE Congress on Evolutionary Computation, Cancun, Mexico, pp. 2069–2077. IEEE (2013)
13. Li, J., Tan, Y.: The bare bones fireworks algorithm: a minimalist global optimizer. Appl. Soft Comput. **62**, 454–462 (2018)
14. Gao, H., Diao, M.: Cultural firework algorithm and its application for digital filters design. Int. J. Modell. Ident. Control **14**(4), 324–331 (2011)
15. Zhang, X.Y., Xia, S.: Hybrid FWPS cooperation algorithm based unmanned aerial vehicle constrained path planning. Aerosp. Sci. Technol. **118**(1), 107004 (2021)
16. Xu, C.F., Duan, H.B., Liu, F.: Chaotic artificial bee colony approach to uninhabited combat air vehicle (UCAV) path planning. Aerosp. Sci. Technol. **14**(8), 535–541 (2010)
17. Besada-Portas, E., et al.: On the performance comparison of multi-objective evolutionary UAV path planners. Inf. Sci. **238**, 111–125 (2013)

Brain Storm Optimization Algorithm

Comparing the Brain Storm Optimization Algorithm on the Ambiguous Benchmark Set

Jakub Kudela(✉), Tomas Nevoral, and Tomas Holoubek

Institute of Automation and Computer Science, Brno University of Technology, Brno, Czech Republic
Jakub.Kudela@vutbr.cz

Abstract. In the field of evolutionary computation, benchmarking has a pivotal place in both the development of novel algorithms, and in performing comparisons between existing techniques. In this paper, the computational comparison of the Brain Storm Optimization (BSO) algorithm (a swarm intelligence paradigm inspired by the behaviors of the human process of brainstorming) was performed. A selected representative of the BSO algorithms (namely, BSO20) was compared with other selected methods, which were a mix of canonical methods (both swarm intelligence and evolutionary algorithms) and state-of-the-art techniques. As a test bed, the ambiguous benchmark set was employed. The results showed that even though BSO is not among the best algorithms on this test bed, it is still a well performing method comparable to some state-of-the-art algorithms.

Keywords: Brain Storm Optimization · Ambiguous benchmark set · Benchmarking · Numerical optimization · Single objective problems

1 Introduction

Swarm intelligence (SI) and evolutionary algorithms (EAs) are effective metaheuristics used for global optimization. EAs are inspired by the processes of biological evolution, such as reproduction, recombination, mutation, and natural selection. SI methods are also biology-based, as they aim to mimic the intelligent behavior of various social animals in the ecosystem. The most widely known examples of these techniques are genetic algorithms, differential evolution, and evolutionary strategy (for the EAs), and ant colony optimization, particle swarm optimization, and artificial bee colony optimization (for the SI methods). These methods were successfully used in the optimization of various complex problems such as the hyperparameter optimization in deep learning [25], stabilization of chaos [14], difficult assignment problems [15], or design of quantum operators [27].

Y. Tan et al. (Eds.): ICSI 2022, LNCS 13344, pp. 367–379, 2022.
https://doi.org/10.1007/978-3-031-09677-8_31

In the SI methods, different learning strategies were proposed for the exchange of information between individuals in the population in the attempt to simulate the behaviour of cooperative intelligence. One of the promising SI techniques is the Brain Storm Optimization (BSO) algorithm [20] that is inspired by human thinking and behavior. Various studies has shown that BSO is a robust and fast converging, with a solid global search capabilities.

As most metaheuristic methods are quite difficult (or impossible) to compare on a theoretical level, benchmarking became a central method in the development of new methods [4] as well as in the comparison and assessment of the already used ones [8]. In this paper, we compare the abilities of a representative BSO variant with other biology-based metaheuristics on a newly proposed set of benchmark functions, called the ambiguous benchmark set [12]. These benchmark functions were designed with the help of various state-of-the-art (SOTA) techniques in order to select functions, on which there was a clear ranking between the algorithms, but for the entire set the ranking remained ambiguous.

The rest of the paper is structured as follows: Sect. 2 introduces the BSO algorithm and the chosen representative, Sect. 3 describes the benchmark functions used in the ambiguous benchmark set, Sect. 4 shows the comparison of BSO with other selected methods, in Sect. 5 conclusions are drawn and future research is outlined.

2 Brain Storm Optimization

BSO was first introduced in [20], as a method inspired by human being's brainstorming process. Since then, various modifications of the BSO have been proposed [5,6,24] improving its performance. The fundamental operations of one iteration of BSO are made up of the following parts:

- Clustering of solutions: A certain clustering strategy is utilized for partitioning the population – NP individuals in the population are grouped to k clusters. Modern metaheuristics incorporate various adaptive approaches for address the issue of premature convergence [18]. In BSO, the possibility of the method converging prematurely is avoided by using a replacing operator, which controls the initialization of a center of a given cluster with probability p_{init}.
- Generation of new solutions: In the original BSO, a new solution x_{new} is formed by using equation (1), where y is a base individual, D is the dimension, $\mathcal{N}^D(0,1)$ is a random number from the standard normal distribution, and $\xi(t)$ is the step size which depends on the iteration t (2).

$$x_{new} = y + \xi(t) \cdot \mathcal{N}^D(0,1). \tag{1}$$

$$\xi(t) = \text{logsig}\left(\frac{0.5T - t}{20}\right) \cdot \text{rand}. \tag{2}$$

where T is the maximum number of iterations, rand is a uniformly distributed random number between 0 and 1, and the function logsig($\cdot$) has the following form:

$$\text{logsig}(a) = \frac{1}{1 + \exp(-a)}. \tag{3}$$

In BSO, two mutation strategies are incorporated: the inter-cluster mutation and the intra-cluster mutation. The probability of the intra-cluster mutation, which is denoted by $p_{one\_cluster}$, leaving the probability of the inter-cluster mutation on the value $(1 - p_{one\_cluster})$ The intra-cluster mutation selects with the probability p_{one_best} an individual from a random cluster as the base individual y, where y is the center of the selected cluster. The inter-cluster mutation generates the base individual y by the following equation:

$$y = r \cdot x_{i1} + (1 - r) \cdot x_{i2}. \tag{4}$$

where r is a random number from the interval (0, 1), x_{i1} and x_{i2} are two different individuals which are chosen from two randomly selected clusters. Moreover, x_{i1} and x_{i2} are selected as the centers of the respective clusters with the probability p_{two_best}.

- Selection of the new population: The newly obtained solutions from the mutation operations are compared with the old solutions with the same index, and the better ones are kept in the new population.

As there are many possibilities for the clustering and mutation strategies, a high-performing representative of the family of the BSO algorithms had to be chosen for the computational comparison. For this purpose, the BSO20 algorithm [24] was selected. BSO20 utilizes a combination of two clustering strategies, namely the random grouping and nearest-better clustering. It also uses a modified mutation strategy for enhancing its exploration capabilities by sharing information within a cluster or among multiple clusters. A analysis comparing BSO20 with different other BSO variants on the 2017 CEC competition benchmark set [2] showed it is competitive and appropriate to serve as the representative of the BSO methods.

3 Ambiguous Benchmark Set

The ambiguous benchmark set is a newly proposed set of functions for comparing algorithms for single-objective bound-constraint optimization [12]. It was designed by selecting a subset of certain functions such the resulting problems introduce a statistically significant ranking among selected algorithms (a mix of SOTA algorithms and canonical techniques), but the ranking for the entire set is still ambiguous with no clear dominating relationship between the algorithms. The benchmark functions in this set are based upon a zigzag function [11], which has the following form:

$$z(x, k, m, \lambda) = \begin{cases} 1 - m + \frac{m}{\lambda}(|x|/k - \lfloor |x|/k \rfloor), & \text{if } |x|/k - \lfloor |x|/k \rfloor \leq \lambda \\ 1 - m + \frac{m}{1-\lambda}(1 - |x|/k + \lfloor |x|/k \rfloor), & \text{otherwise} \end{cases}. \tag{5}$$

where $x \in \mathcal{R}$ is the point at which it should be evaluated, $k > 0$ controls the period of the zigzag, $m \in [0, 1]$ controls the amplitude, and $\lambda \in (0, 1)$ controls the location of local minima. The construction of the benchmark functions starts with four 1-D functions $\phi_1, \ldots, \phi_4$, which are formulated in (6) and utilize the zigzag function:

$$\begin{aligned}
\phi_1(x, k, m, \lambda) &= 3{\cdot}10^{-9}|(x-40)(x-185)x(x+50)(x+180)|z(x, k, m, \lambda) \\
&\quad + 10|\sin(0.1x)|. \\
\phi_2(x, k, m, \lambda) &= \phi_1(\phi_1(x, k, m, \lambda), k, m, \lambda). \\
\phi_3(x, k, m, \lambda) &= 3|\ln(1000|x|+1)|z(x, k, m, \lambda) + 30 - 30|\cos(\frac{x}{10\pi})|. \\
\phi_4(x, k, m, \lambda) &= \phi_3(\phi_3(x, k, m, \lambda), k, m, \lambda).
\end{aligned} \tag{6}$$

To obtain the benchmark functions for a dimension D, a simple sum of the functions ϕ for the individual components is used, while the inputs are modified by a shift vector $\mathbf{s} \in [-100, 100]^D$ and a rotation/scaling matrix $\mathbf{M}$:

$$f_j(\mathbf{x}, k, m, \lambda) = \sum_{i=1}^{D} \phi_j(x_i, k, m, \lambda) \qquad j = 1, \ldots, 4. \tag{7}$$

$$\mathrm{F}j(\mathbf{x}, k, m, \lambda) = f_j(\mathbf{M}_j(\mathbf{x} - \mathbf{s}_j), k, m, \lambda) \qquad j = 1, \ldots, 4. \tag{8}$$

The optimization of these function should be carried out over the search space $[-100, 100]^D$, where also lie the global optima of these functions (each identically equal to 0). The ambiguous benchmark set is comprised of 32 functions whose parametrization is shown in Table 1. The particular values for the shift vectors and rotation/scaling matrices can be found in [12].

Table 1. Parametrizations used in the ambiguous benchmark set.

ID	Function	D	k	m	λ	ID	Function	D	k	m	λ
1	F1	5	16	1	0.01	17	F3	5	16	0.9	0.01
2	F1	10	16	1	0.01	18	F3	10	16	0.9	0.01
3	F1	15	16	1	0.01	19	F3	15	16	0.9	0.01
4	F1	20	16	1	0.01	20	F3	20	16	0.9	0.01
5	F1	5	8	0.5	0.01	21	F3	5	8	0.9	0.9
6	F1	10	8	0.5	0.01	22	F3	10	8	0.9	0.9
7	F1	15	8	0.5	0.01	23	F3	15	8	0.9	0.9
8	F1	20	8	0.5	0.01	24	F3	20	8	0.9	0.9
9	F2	5	2	0.5	0.99	25	F4	5	16	0.1	0.1
10	F2	10	2	0.5	0.99	26	F4	10	16	0.1	0.1
11	F2	15	2	0.5	0.99	27	F4	15	16	0.1	0.1
12	F2	20	2	0.5	0.99	28	F4	20	16	0.1	0.1
13	F2	5	1	1	0.1	29	F4	5	4	0.9	0.01
14	F2	10	1	1	0.1	30	F4	10	4	0.9	0.01
15	F2	15	1	1	0.1	31	F4	15	4	0.9	0.01
16	F2	20	1	1	0.1	32	F4	20	4	0.9	0.01

4 Comparison with Selected Methods

In this section, the behaviour of the BSO on the ambiguous benchmark set is compared with other selected methods. Two main methods were chosen for this comparison: generally known and used EAs, and the best performing methods from the CEC Competitions [16]. The selected algorithms were the following:

- Adaptive Gaining-Sharing Knowledge (AGSK) [17] – one of the best performing algorithms in the CEC'20 competition.
- Covariance Matrix Adaptation Evolution Strategy (CMAES) – a canonical algorithm that adapts the covariance matrix of a mutation distribution [1].
- Differential Evolution (DE) – a canonical algorithm, one of the most utilized ones for continuous optimization [21].
- Hybrid Sampling Evolution Strategy (HSES) [26] – winner of the CEC'18 Competition.
- Improved Multi-operator Differential Evolution (IMODE) [19] – winner of the CEC'20 Competition.
- Linear Population Size Reduction SHADE (LSHADE) [22]– one of the most popular versions of adaptive DE, successfully utilized as a basis for many of the best-performing methods in the CEC Competitions in past several years.
- Multiple Adaptation DE Strategy (MadDE) [3] – one of the best performing methods from the CEC'21 competition.
- Particle Swarm Optimization (PSO) [10] – a canonical method that simulates swarm behavior of social animals such as the fish schooling or bird flocking.

The selected algorithms were compared on the ambiguous benchmark set with $D = \{5, 10, 15, 20\}$ dimensions, and a search space of $[-100, 100]^D$. The maximum number of function evaluations were set to 50,000, 200,000, 500,000, and 1,000,000 fitness function evaluations for problems with $D = \{5, 10, 15, 20\}$, respectively. All algorithms were run 30 times to obtain representative results. In each run, if the objective function value of the obtained solution was $\leq$ 1E-8, it was considered as a zero. For all algorithms the same parameter setting used was the one reported in the corresponding publication [9] and all started from the same random seed [13]. The computations were done in a MATLAB R2020b, on a PC with 3.2 GHz Core I5 processor, 16 GB RAM, and Windows 10.

The detailed results of the computations are summarized in Table 2 and Table 3. The first thing to notice is that BSO is, indeed, a well performing metaheuristic, as it achieves results similar to the SOTA and canonical methods. First, when comparing the "min" values, one can see that on some of the problems from the ambiguous benchmark set BSO was able to find (out of the 30 independent runs) the global optimum (ID = [5,9,21,25]), the best solution out of all considered methods (ID = 19), or one of the best solutions (ID = [20, 29, 30]). On the other hand, there were numerous instances where it performed among the 3 worst in this metric (ID = [1, 2, 3, 4, 6, 7, 8, 10, 11, 12, 13, 14, 15, 22, 23]).

Table 2. Detailed statistics of the 30 runs of the selected algorithms on the ambiguous benchmark set. Function IDs 1–16.

ID		AGSK	CMAES	DE	HSES	IMODE	LSHADE	MadDE	PSO	BSO
1	Min	0.079	0	4.7E−05	0.148	0	0	0	0.148	0.148
	Median	0.382	0.079	1.810	0.148	0.192	0.016	0.098	0.690	0.380
	Mean	0.389	1.120	1.700	0.182	0.165	0.045	0.139	0.783	0.472
	Max	0.629	6.674	2.978	0.533	0.375	0.227	0.377	1.822	1.056
	Std	0.123	1.748	0.877	0.075	0.109	0.056	0.100	0.505	0.254
2	Min	1.257	0	0	0.306	0.237	0.002	0.080	0.306	0.690
	Median	2.407	6.692	0.341	0.464	0.805	0.139	0.475	0.751	1.228
	Mean	2.389	9.639	1.250	0.507	0.889	0.132	0.485	0.993	1.377
	Max	3.332	26.041	19.712	0.918	1.753	0.268	0.997	2.598	3.448
	Std	0.486	9.996	3.616	0.119	0.407	0.087	0.163	0.606	0.594
3	Min	2.534	0	0.158	0.079	0.621	0.254	0.158	1.066	0.384
	Median	3.643	39.538	0.582	0.237	1.228	0.482	0.380	2.587	0.612
	Mean	3.584	30.451	0.777	0.244	1.270	0.450	0.353	2.770	0.677
	Max	4.641	49.380	2.816	0.464	2.680	0.649	0.533	6.899	1.638
	Std	0.548	19.186	0.620	0.092	0.465	0.088	0.094	1.282	0.304
4	Min	6.394	0	0.227	0.909	1.835	0.200	1.136	0.988	1.214
	Median	7.750	0.237	1.105	1.210	3.430	0.526	1.882	2.292	2.611
	Mean	8.132	0.284	1.110	1.221	3.512	0.527	1.871	2.679	2.946
	Max	10.757	0.858	2.062	1.660	5.721	0.745	2.437	6.379	12.419
	Std	1.144	0.187	0.482	0.156	1.049	0.122	0.277	1.340	1.902
5	Min	0	0	0	0	0	0	0	0	0
	Median	0	0	0	2.159	0	0	0	1.148	3.307
	Mean	0.006	0.077	0.077	1.228	0.077	0	0.603	1.838	4.230
	Max	0.189	1.148	1.148	3.308	1.148	0	2.296	7.452	9.785
	Std	0.035	0.291	0.291	1.202	0.291	0	0.763	1.850	2.351
6	min	0	0	0	4.456	2.159	0	1.148	2.319	7.588
	Median	4.202	0	0	5.604	3.950	0	2.296	5.380	11.208
	Mean	3.818	1.779	0.790	5.796	3.870	2.4E−07	2.675	5.777	12.000
	Max	6.568	13.668	4.456	7.901	5.798	6.1E−06	4.593	10.896	22.776
	Std	1.645	3.678	1.252	0.803	1.212	1.1E−06	0.995	2.437	3.221
7	Min	3.445	0	0	1.148	5.0E−06	0.003	0	3.444	3.444
	Median	11.628	1.148	1.148	2.296	2.296	1.164	1.148	13.075	7.763
	Mean	11.302	3.697	1.412	2.717	2.586	1.487	1.526	12.509	7.567
	Max	14.614	35.888	5.741	5.741	4.456	3.455	3.445	20.485	13.192
	Std	2.284	8.487	1.515	0.976	1.054	0.942	0.747	3.695	2.378
8	Min	15.209	0	0	10.934	8.912	0.002	2.296	8.912	20.682
	Median	22.856	1.148	2.296	14.379	13.231	2.306	10.497	15.633	27.800
	Mean	23.114	1.297	2.797	14.180	13.583	1.877	9.913	15.192	28.690
	Max	29.106	4.456	8.912	17.550	19.534	4.466	12.744	23.990	44.931
	Std	3.309	1.258	2.218	1.731	2.636	1.186	2.273	4.007	6.241

(*continued*)

Table 2. (*continued*)

ID		AGSK	CMAES	DE	HSES	IMODE	LSHADE	MadDE	PSO	BSO
9	Min	0.351	0	0	0	0	0	0	0	0
	Median	2.876	1.886	6.1E−05	2.173	0	0.003	0.807	3.365	4.767
	Mean	2.908	2.396	0.865	1.695	0.254	0.497	0.925	3.274	4.608
	Max	5.554	9.906	8.781	2.881	1.961	1.995	1.886	9.186	8.538
	Std	1.355	2.719	2.037	0.882	0.659	0.832	0.942	2.023	2.186
10	Min	7.886	0	0	5.664	1.886	0.002	1.605	1.605	9.619
	Median	12.487	21.548	3.966	9.116	6.085	3.715	5.757	9.497	15.935
	Mean	12.865	16.344	10.284	9.108	5.989	3.283	5.745	9.717	16.157
	Max	17.286	37.279	28.568	14.860	9.641	6.727	9.033	17.369	23.290
	Std	2.040	13.365	10.621	2.360	2.127	1.710	1.909	3.814	3.616
11	Min	19.195	0	1.886	6.519	8.714	4.348	5.185	12.838	9.159
	Median	25.994	2.173	5.953	10.248	12.758	9.845	8.918	20.023	18.873
	Mean	25.629	2.807	8.006	10.219	12.961	9.169	8.968	22.022	18.573
	Max	32.357	8.416	34.796	14.028	17.685	13.648	12.851	43.048	28.811
	Std	3.492	2.678	6.217	1.690	2.222	2.560	1.848	7.163	4.566
12	Min	30.890	0	3.210	15.752	19.499	6.852	10.187	15.232	27.978
	Median	41.565	3.327	8.426	23.186	26.333	12.499	18.971	25.418	37.370
	Mean	41.338	4.129	9.680	23.674	26.167	12.471	18.161	26.322	38.153
	Max	50.257	10.308	21.103	30.395	36.428	18.686	24.438	43.032	51.271
	Std	4.713	3.007	4.534	3.463	4.169	2.740	3.826	7.099	5.040
13	Min	0	0	5.6E−08	0	0	0	0	0.199	0.199
	Median	1.842	0.637	1.917	0.637	0	0.014	0.469	1.620	1.106
	Mean	1.867	1.950	2.042	0.573	0.141	0.195	0.322	1.969	1.284
	Max	3.528	8.261	6.146	1.274	0.470	0.695	0.640	6.072	3.185
	Std	1.029	2.449	2.140	0.257	0.219	0.268	0.255	1.371	0.569
14	Min	5.961	0	4.0E−04	1.474	1.274	0.485	0	0.199	2.551
	Median	10.284	26.534	22.105	2.549	4.134	1.661	2.551	9.113	5.294
	Mean	10.248	20.786	19.752	2.534	4.004	1.722	2.797	10.249	5.828
	Max	14.187	30.922	30.011	3.385	7.986	3.130	9.179	26.090	13.299
	Std	2.151	11.369	8.803	0.519	1.881	0.723	1.762	6.107	2.745
15	Min	17.448	0	0.637	2.274	3.728	2.806	1.408	3.382	3.680
	Median	25.269	1.274	12.385	3.186	9.151	5.617	3.156	17.325	6.287
	Mean	25.000	1.396	23.789	3.311	10.639	6.248	3.687	18.817	7.056
	Max	32.461	3.018	53.380	4.460	20.560	13.618	8.005	57.909	18.895
	Std	3.840	0.807	22.321	0.605	4.464	3.025	1.503	12.130	3.127
16	Min	37.808	0.199	5.2E−07	4.993	14.945	4.194	1.878	16.267	8.031
	Median	47.335	1.374	20.635	6.416	35.142	16.225	5.674	24.982	14.182
	Mean	47.233	1.626	35.135	6.549	33.864	16.808	7.541	28.150	15.937
	Max	55.853	3.186	81.701	9.323	51.526	33.009	23.803	78.122	37.514
	Std	4.450	0.889	33.104	1.095	9.050	7.673	4.969	12.278	5.914

Table 3. Detailed statistics of the 30 runs of the selected algorithms on the ambiguous benchmark set. Function IDs 17–32.

ID		AGSK	CMAES	DE	HSES	IMODE	LSHADE	MadDE	PSO	BSO
17	Min	24.497	20.920	29.237	27.239	13.817	21.721	18.211	20.209	20.120
	Median	28.568	30.398	35.520	30.398	20.559	25.862	22.916	28.718	30.397
	Mean	28.468	32.684	35.650	30.532	21.078	25.313	23.226	28.147	31.353
	Max	31.597	51.998	41.569	33.557	27.393	27.175	27.811	34.698	37.223
	Std	1.978	8.097	3.210	1.310	3.177	1.664	2.499	3.940	3.283
18	Min	55.092	54.477	60.309	57.636	50.114	51.925	46.091	52.587	54.591
	Median	64.854	126.734	94.517	62.089	56.761	57.102	54.738	57.903	60.906
	Mean	64.553	107.265	90.495	62.136	56.653	57.323	54.832	59.166	61.351
	Max	72.476	156.234	116.656	63.955	63.579	61.972	61.251	67.202	68.391
	Std	3.653	38.127	15.928	1.761	3.729	2.392	2.863	4.328	3.696
19	Min	104.287	81.724	77.044	81.716	84.433	87.715	76.012	85.912	68.454
	Median	120.277	95.646	92.118	84.875	95.491	94.035	88.989	105.300	86.544
	Mean	120.845	153.290	91.115	85.651	94.512	94.278	88.154	104.025	87.486
	Max	137.601	264.754	111.408	90.047	102.123	102.945	93.989	127.015	109.494
	Std	8.426	75.605	7.885	2.583	4.458	3.957	4.213	10.525	7.675
20	Min	148.185	102.866	111.148	105.221	116.583	121.006	112.764	115.995	105.646
	Median	168.455	320.296	119.584	113.839	140.981	130.682	123.086	148.106	115.396
	Mean	166.773	261.881	119.236	113.669	139.884	131.311	123.438	148.521	116.597
	Max	185.485	377.771	132.276	122.457	154.659	139.613	135.300	183.048	135.890
	Std	10.307	105.028	4.649	4.272	10.055	5.071	5.052	15.519	7.651
21	Min	0	0	0	0	0	0	0	0	0
	Median	6.490	4.090	0	3.552	0	0	0	5.213	8.880
	Mean	6.351	11.857	7.5E−07	3.078	0.893	0.001	1.130	6.553	8.865
	Max	12.826	55.099	2.2E−05	3.552	5.138	0.037	3.762	15.020	15.993
	Std	3.770	14.631	4.1E−06	1.228	1.668	0.007	1.669	4.667	3.994
22	Min	33.289	0	0	3.552	8.699	3.387	3.552	7.649	11.050
	Median	42.885	14.208	0	3.552	21.159	11.961	13.212	25.586	20.848
	Mean	42.386	48.884	0.474	4.930	20.624	11.510	13.691	25.355	21.321
	Max	50.719	128.454	3.552	8.690	33.462	17.541	23.428	45.074	34.277
	Std	4.791	47.193	1.228	1.818	5.749	3.612	4.751	11.528	4.946
23	Min	65.592	0	3.552	17.760	27.757	20.753	12.042	26.357	27.951
	Median	94.560	14.378	8.067	18.100	52.652	29.963	33.199	59.561	42.100
	Mean	93.180	51.745	9.269	19.290	51.787	30.153	32.884	58.324	44.450
	Max	105.063	231.198	14.990	23.238	72.605	40.446	42.918	88.369	70.521
	Std	8.714	75.848	4.243	1.974	10.185	4.553	7.000	16.076	10.150
24	Min	90.555	0	0	22.898	70.394	43.377	31.612	57.931	45.442
	Median	135.687	20.019	14.548	26.450	91.950	59.616	50.779	90.813	63.857
	Mean	133.143	116.119	13.820	27.504	94.698	58.368	51.204	92.303	62.812
	Max	152.926	361.609	22.216	37.065	122.852	68.916	65.893	137.916	84.502
	Std	13.447	142.139	4.671	3.412	12.196	7.376	8.207	24.918	9.683
25	Min	60.686	0	0.183	0.120	53.576	0	26.523	29.946	0
	Median	104.889	0	37.116	0.279	86.615	62.168	76.352	84.601	64.246
	Mean	104.439	13.471	36.067	0.335	86.386	57.876	73.217	85.128	61.167
	Max	125.811	161.451	78.464	0.849	99.178	72.786	91.105	126.704	135.040
	Std	14.153	41.309	20.464	0.208	10.621	17.322	13.625	18.336	33.687

(*continued*)

Table 3. (*continued*)

ID		AGSK	CMAES	DE	HSES	IMODE	LSHADE	MadDE	PSO	BSO
26	Min	163.671	0	135.256	154.755	233.503	130.362	208.422	207.208	150.245
	Median	314.126	41.748	194.938	195.832	267.512	164.385	242.386	270.424	259.735
	Mean	295.113	74.707	187.533	195.388	265.412	161.362	239.153	268.315	257.513
	Max	331.847	329.923	222.756	237.361	300.395	187.579	261.766	330.771	326.756
	Std	40.180	106.332	27.320	18.559	19.013	12.448	13.686	31.063	51.762
27	Min	348.402	0	193.006	317.022	374.203	221.197	304.657	341.376	228.304
	Median	509.195	41.849	311.162	386.469	446.116	249.563	368.799	436.817	313.239
	Mean	488.725	54.362	301.460	382.444	441.602	247.384	362.159	436.950	338.338
	Max	528.011	494.286	354.265	413.896	467.535	271.327	391.294	529.460	507.306
	Std	50.179	89.360	43.729	18.832	24.525	13.193	25.341	38.330	81.328
28	Min	548.602	0	0	506.788	576.439	273.658	473.872	530.989	346.787
	Median	711.098	41.795	279.976	525.713	635.228	334.567	521.129	606.926	564.714
	Mean	692.511	65.597	276.620	528.635	632.042	329.717	518.021	614.407	544.894
	Max	727.464	580.101	527.866	554.467	672.459	364.781	563.720	715.581	636.883
	Std	48.104	108.293	201.979	13.152	21.947	24.451	24.890	43.689	71.110
29	Min	23.854	17.386	28.283	18.695	16.092	16.002	16.246	23.265	16.453
	Median	34.351	21.897	37.156	39.615	20.530	20.433	19.454	32.242	22.341
	Mean	33.905	25.213	36.893	36.677	20.750	20.953	19.828	32.490	22.968
	Max	42.422	49.843	43.193	51.209	26.747	26.529	22.987	44.599	31.926
	Std	4.011	9.400	3.599	9.305	2.768	2.137	1.618	5.791	3.975
30	Min	78.816	39.269	104.742	37.788	54.182	52.321	55.594	78.915	50.107
	Median	91.075	98.698	118.756	49.973	68.264	58.426	63.994	98.324	75.262
	Mean	90.372	87.538	115.562	51.413	68.231	58.990	63.439	97.430	75.640
	Max	102.035	123.785	122.909	77.245	83.318	65.581	73.942	114.089	100.108
	Std	6.501	29.765	6.035	8.978	7.479	3.394	3.731	9.774	13.272
31	Min	144.170	62.598	174.725	68.015	98.105	82.713	90.469	125.414	97.696
	Median	155.725	174.524	200.374	85.331	127.193	99.219	103.677	173.904	142.922
	Mean	156.636	145.368	200.810	84.046	128.558	99.454	104.418	174.087	139.695
	Max	181.204	199.811	219.169	104.426	149.310	111.173	117.778	202.850	168.708
	Std	9.036	51.971	10.301	10.350	12.461	5.827	7.043	17.711	20.268
32	Min	204.304	80.782	256.491	94.565	171.158	135.534	131.067	214.665	155.390
	Median	242.529	112.234	286.675	114.071	198.139	153.238	161.119	274.342	187.750
	Mean	242.741	170.763	285.455	117.005	197.970	152.829	158.040	268.152	192.019
	Max	275.858	271.524	306.145	138.448	223.666	164.808	177.022	309.456	239.120
	Std	15.207	82.614	11.193	12.941	14.681	6.469	10.930	24.682	19.291

In the "mean" metric, it ended up twice among the best three methods, 18 times in the middle (between fourth and sixth), and 12 times in the worst three (of which it was four times the worst). In the "max" metric, the results were similar to the "mean" one, with the main difference being that the BSO did not make it in the top 3 methods (21 times in the middle, 11 bottom three, 5 times the worst). The performance of BSO (when compared with the other methods) was a bit worse especially for the first 12 problems in the benchmark set.

On the other hand, its performance was very solid for the remaining 20 problems. The relatively worse performance of the studied BSO variant in the "max" metric might be caused by the other methods better ability for diversification.

To give a more balanced analysis of the result, and to give an understanding of the approximate ranking of the BSO method, a comparison of all the selected algorithms on the different benchmark functions together was performed. For this comparison, the IOHprofiler [7], which is a web-based benchmarking and profiling tool for (meta)heuristics used in optimization, was used. Within the IOHprofiler, the comparison based on a fixed-budget was selected (this comparison is defined by the maximum number of objective function evaluations) and compared the algorithms on each function from the ambiguous benchmark set. For this purpose, a so-called Glicko-2 rating system was utilized, as it was found to be appropriate for ranking evolutionary algorithms [23]. Glicko-2 is an Elo-base system that uses games between the methods (based on randomly selected runs). For the comparison, each pair of methods "played" 25 games for each of the 32 benchmark functions, resulting in 6400 games played (as each algorithm competed with the other eight). The results of this comparison are summarized in Fig. 1 and Table 4. Although it can be seen that BSO is not among the best methods in this ranking, it is still a method that has a solid performance (as more than a third of its games were wins). In this ranking, it is comparable to PSO (the other widely used SI method), IMODE (winner of the CEC'20 competition), and quite clearly outperforms AGSK (runner up in the CEC'20 competition).

Table 4. Glicko-2 ranking results.

Algorithm	Rating	Deviation	Volatility	Win	Draw	Loss
LSHADE	1675	12.9	0.043	4532	242	1626
MadDE	1607	12.6	0.044	4069	215	2116
HSES	1599	12.0	0.039	3913	176	2311
CMAES	1575	11.8	0.038	3774	236	2390
DE	1548	11.8	0.038	3474	232	2694
IMODE	1490	11.6	0.037	2919	223	3258
BSO	1413	12.2	0.040	2338	47	4015
PSO	1351	13.9	0.051	1863	76	4461
AGSK	1265	14.3	0.046	1124	141	5135

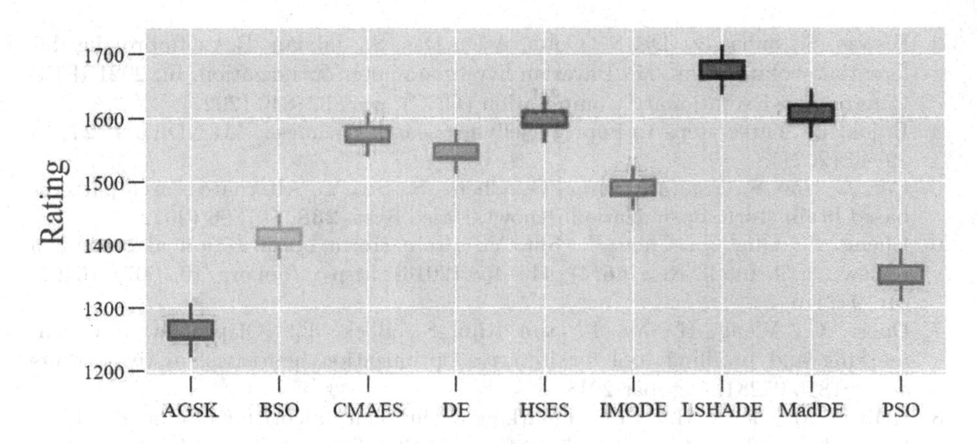

Fig. 1. Graphical representation of the resulting Glicko-2 rating.

5 Conclusions

This paper conducted a comparison of the selected representative of the BSO algorithms with other EA and SI techniques on the ambiguous benchmark set. The selected techniques were a mix of canonical methods as well as some of the SOTA algorithms. The results of the extensive computational comparison showed that BSO is a solid technique, as it ranked similarly to other well performing methods, such as PSO or IMODE. On the other hand, the computations also showed that there is still enough room for its improvements, as it lacked behind some of the recent DE and ES techniques.

As the behaviour of only one representative variant of the BSO was investigated, futures research will focus on comparing more variants of the BSO on the ambiguous benchmark set. Hyperparameter optimization of the BSO variants on the studied benchmark set could provide insight about the possible improvements of its performance. Additionally, more comparisons with similarly structured SI methods could provide valuable analysis of their strengths and weaknesses.

Acknowledgments. This work was supported by internal grant agency of BUT: FME-S-20-6538 "Industry 4.0 and AI methods", FIT/FSI-J-22-7980, and FEKT/FSI-J-22-7968.

References

1. Auger, A., Hansen, N.: A restart CMA evolution strategy with increasing population size. In: 2005 IEEE Congress on Evolutionary Computation, vol. 2, pp. 1769–1776 (2005)
2. Awad, N., Ali, M., Liang, J., Qu, B., Suganthan, P.: Problem definitions and evaluation criteria for the CEC 2017 special session and competition on single objective bound constrained real-parameter numerical optimization. Technical report, Nanyang Technological University, Singapore (2017)

3. Biswas, S., Saha, D., De, S., Cobb, A.D., Das, S., Jalaian, B.A.: Improving differential evolution through Bayesian hyperparameter optimization. In: 2021 IEEE Congress on Evolutionary Computation (CEC), pp. 832–840 (2021)
4. Bujok, P.: Three steps to improve jellyfish search optimiser. MENDEL J. **27**(1), 29–40 (2021)
5. Cai, Z., Gao, S., Yang, X., Yang, G., Cheng, S., Shi, Y.: Alternate search pattern-based brain storm optimization. Knowl.-Based Syst. **238**, 107896 (2022)
6. Cheng, S., Qin, Q., Chen, J., Shi, Y.: Brain storm optimization algorithm: a review. Artif. Intell. Rev. **46**(4), 445–458 (2016). https://doi.org/10.1007/s10462-016-9471-0
7. Doerr, C., Wang, H., Ye, F., van Rijn, S., Bäck, T.: IOHprofiler: a benchmarking and profiling tool for iterative optimization heuristics. arXiv e-prints arXiv:1810.05281, October 2018
8. Hellwig, M., Beyer, H.G.: Benchmarking evolutionary algorithms for single objective real-valued constrained optimization - a critical review. Swarm Evol. Comput. **44**, 927–944 (2019)
9. Kazikova, A., Pluhacek, M., Senkerik, R.: Why tuning the control parameters of metaheuristic algorithms is so important for fair comparison? MENDEL J. **26**(2), 9–16 (2020)
10. Kennedy, J., Eberhart, R.: Particle swarm optimization. In: Proceedings of ICNN 1995 - International Conference on Neural Networks, vol. 4, pp. 1942–1948 (1995)
11. Kudela, J.: Novel zigzag-based benchmark functions for bound constrained single objective optimization. In: 2021 IEEE Congress on Evolutionary Computation (CEC), pp. 857–862. IEEE (2021)
12. Kudela, J., Matousek, R.: New benchmark functions for single-objective optimization based on a zigzag pattern. IEEE Access **10**, 8262–8278 (2022)
13. Matousek, R., Dobrovsky, L., Kudela, J.: How to start a heuristic? Utilizing lower bounds for solving the quadratic assignment problem. Int. J. Ind. Eng. Comput. **13**(2), 151–164 (2022)
14. Matousek, R., Hulka, T.: Stabilization of higher periodic orbits of the chaotic logistic and Hénon maps using meta-evolutionary approaches. In: 2019 IEEE Congress on Evolutionary Computation (CEC), pp. 1758–1765. IEEE (2019)
15. Matousek, R., Popela, P., Kudela. J.: Heuristic approaches to stochastic quadratic assignment problem: VaR and CVar cases. Mendel. **23**, 73–78 (2017)
16. Mohamed, A.W., Hadi, A.A., Mohamed, A.K., Agrawal, P., Kumar, A., Suganthan, P.N.: Problem definitions and evaluation criteria for the CEC 2021 special session and competition on single objective bound constrained numerical optimization. Technical report, Cairo University, Egypt (2020)
17. Mohamed, A.W., Hadi, A.A., Mohamed, A.K., Awad, N.H.: Evaluating the performance of adaptive gaining sharing knowledge based algorithm on CEC 2020 benchmark problems. In: 2020 IEEE Congress on Evolutionary Computation (CEC), pp. 1–8 (2020)
18. Pluháček, M., Kazikova, A., Kadavy, T., Viktorin, A., Senkerik, R.: Relation of neighborhood size and diversity loss rate in particle swarm optimization with ring topology. MENDEL J. **27**(2), 74–79 (2021)
19. Sallam, K.M., Elsayed, S.M., Chakrabortty, R.K., Ryan, M.J.: Improved multi-operator differential evolution algorithm for solving unconstrained problems. In: 2020 IEEE Congress on Evolutionary Computation (CEC), pp. 1–8 (2020)
20. Shi, Y.: Brain storm optimization algorithm. In: Tan, Y., Shi, Y., Chai, Y., Wang, G. (eds.) ICSI 2011. LNCS, vol. 6728, pp. 303–309. Springer, Heidelberg (2011). https://doi.org/10.1007/978-3-642-21515-5_36

21. Storn, R., Price, K.: Differential evolution - a simple and efficient heuristic for global optimization over continuous spaces. J. Global Optim. **11**, 341–359 (1997)
22. Tanabe, R., Fukunaga, A.S.: Improving the search performance of shade using linear population size reduction. In: 2014 IEEE Congress on Evolutionary Computation (CEC), pp. 1658–1665 (2014)
23. Veček, N., Črepinšek, M., Mernik, M., Hrnčič, D.: A comparison between different chess rating systems for ranking evolutionary algorithms. In: 2014 Federated Conference on Computer Science and Information Systems, pp. 511–518. IEEE (2014)
24. Xu, P., Luo, W., Lin, X., Cheng, S., Shi, Y.: BSO20: efficient brain storm optimization for real-parameter numerical optimization. Complex Intell. Syst. **7**, 2415–2436 (2021)
25. Young, S.R., Rose, D.C., Karnowski, T.P., Lim, S.H., Patton, R.M.: Optimizing deep learning hyper-parameters through an evolutionary algorithm. In: Proceedings of the Workshop on Machine Learning in High-Performance Computing Environments, pp. 1–5 (2015)
26. Zhang, G., Shi, Y.: Hybrid sampling evolution strategy for solving single objective bound constrained problems. In: 2018 IEEE Congress on Evolutionary Computation (CEC), pp. 1–7 (2018)
27. Žufan, P., Bidlo, M.: Advances in evolutionary optimization of quantum operators. MENDEL J. **27**(2), 12–22 (2021)

Mine Ventilation Prediction Based on BSO-DG Optimized BP Neural Network

Junfeng Chen[1,2(✉)], Mao Mao[1], and Xueping Zhang[1]

[1] College of IOT Engineering, Hohai University, Changzhou 213022, Jiangsu, China
jfchen@hhu.edu.cn

[2] Jiangsu Key Laboratory of Power Transmission and Distribution Equipment, Hohai University, Changzhou 213022, Jiangsu, China

Abstract. This paper introduces the double grouping into the brain storm optimization algorithm, namely BSO-DG. The BSO-DG algorithm is used to optimize the weights and thresholds in the BP neural network. Then, the BP neural network is used to realize the mine ventilation prediction. Simulation and comparison experiments are carried out on the prediction model. The experimental results show that the BSO-DG algorithm can effectively improve the prediction accuracy of BP neural network.

Keywords: Brain Storm Optimization · Double grouping · BP neural network · Mine ventilation prediction

1 Introduction

The swarm intelligence optimization algorithm solves complex problems by simulating biological phenomena and has outstanding performance in solving complex optimization problems. Among the new swarm intelligence optimization algorithms, the Brain Storm Optimization (BSO) algorithm is a classic algorithm whose ideas come from human brainstorming meetings and are algorithms inspired by human social behavior. The algorithm searches for the individual optimal solution by gathering and dispersing the local optimum through the clustering operation and increasing the diversity of the population through divergent mutation based on the local optimum. The algorithm contains many parameters in the implementation process, and the convergence speed is not very fast.

In order to improve the performance of the BSO algorithm, various improvements have been made to the BSO algorithm. Cao et al. [1] designed a random grouping strategy BSO assisted by fitness value in the clustering process, reducing the load of parameter setting and balancing the detection and development at different search times. After that, they introduced the dynamic clustering method and proposed an improved BSO algorithm based on the dynamic clustering strategy [2], which reduced the time complexity. Zhu and Shi [3] proposed

Y. Tan et al. (Eds.): ICSI 2022, LNCS 13344, pp. 380–390, 2022.
https://doi.org/10.1007/978-3-031-09677-8_32

a variant of k-means clustering, k-medians clustering, which uses the median of individuals in the current class as the cluster center instead of the mean. Zhan et al. [4] introduced the Simple Grouping Method (SGM) based on the location information of individuals in the BSO algorithm, which makes the computational burden of the algorithm smaller. Guo et al. [5] used the maximum fitness clustering method to divide individuals into different subgroups instead of the primary k-means clustering to find different optimal solutions. Then, they established a simple clustering method [6] to improve the search speed and proposed an adaptive multi-objective BSO algorithm. Chen et al. [7] introduced Affinity Propagation (AP) clustering into the BSO algorithm, thus avoiding the fixed initial number of clusters in the k-means algorithm on the clustering results. In addition, they adopted the method of hierarchical clustering and formed clustering strategies with different precisions [8]. Xie and Wu [9] used the Density-Based Spatial Clustering of Applications with Noise (DBSCAN) in the BSO algorithm. Qiu and Duan proposed a new clustering method [10], classified by fitness value and applied to solve the optimization problem of unmanned aerial vehicle formation flight based on nonlinear rear horizon control mode. El-Abd designed a grouping method based on fitness value [11], which sorts and groups individuals according to fitness value. The probability of good and flawed individuals in different groups is the same. Zhou et al. [12] first modified the step size according to the dynamic range of individuals in each dimension. New individuals were generated in batch mode, and good individuals were selected for the next generation. Starting from the mechanism of quantum theory, Duan et al. [13] proposed a new quantum behavioral BSO algorithm to solve the optimization model of the Looney solenoid problem. Yang et al. [14] proposed an advanced discussion mechanism-based BSO algorithm, introducing intra-group and inter-group discussion mechanisms to control global and local search capabilities. Yang Z et al. [15] introduced chaotic technology in the basic BSO algorithm. They proposed an improved BSO algorithm, which uses chaotic technology to effectively solve the premature problem of the BSO algorithm and avoid falling into local optimum. Wang et al. [16] introduced the idea of graph theory. New individuals are generated when the BSO algorithm is in a poor state to replace some old ones.

This paper uses the modified BSO algorithm to optimize the weights and thresholds in the BP neural network in the prediction model of mine ventilation.

The remainder of this paper is organized as follows. In Sect. 2, the mine ventilation system is modeled. In Sect. 3, the BSO is modified based on Double Grouping (BSO-DG), and the BSO-DG algorithm is used to optimize the weights and thresholds in the BP neural network. Simulation experiments and comparisons are provided in Sect. 4. Finally, several conclusive remarks are given in Sect. 5.

2 Model Design of Mine Ventilation System

2.1 The Mine Ventilation System

A coal mine ventilation system usually needs to prevent safety emergencies such as carbon monoxide, coal dust, gas, and fire, so it is crucial to control the ventilation volume effectively. On the working face of the mine, its various parameters will change in real-time. The system needs to adjust the wind speed (or mine ventilation volume) according to the actual situation. Too high or too low wind speed will lead to unsafe conditions. Generally speaking, the draught fan of the current system uses constant power, and air volume is generally fixed and will not change. The operator can regulate the wind speed of the working face by controlling the fan and damper of the roadway or working face. Therefore, a problem worthy of study is reaching the optimal wind speed to control gas and coal dust on the working face in a reasonable range [17].

This paper proposes a BP neural network prediction model based on BSO-DG optimization. A forward neural network is used to establish a mine ventilation forecasting model. The global optimization ability of BSO-DG proposed above is employed to overcome the randomness of the initial weights and thresholds of the network. In this way, weights and thresholds of the network can be optimized, the approximation effect is the best, and the corresponding absolute error is the smallest.

Gas quantity, coal dust quantity, temperature, and humidity are selected as the indicators to measure the wind speed index. These four attributes are used as the input data of the network, and the wind speed is selected as the output data of it. Some data sheets are shown in Table 1 (only part data are shown here), and the data are normalized [18].

Table 1. Data points on a working station

Point of time	Wind speed (m/s)	Gas (%)	Coal dust (g/m^3)	Temperature (g/m^3)	Humidity (g/m^3)
1	2.4	0.71	8	17	15
2	2.56	0.62	8.4	15	13
3	2.24	0.66	7.61	16	14
4	2.27	0.62	7.69	15.5	13.2
5	2.41	0.63	7.99	15	14.1
6	2.4	0.66	7.87	14.6	13.5
7	2.27	0.78	7.75	14.3	15.3
8	2.22	0.71	7.71	13.9	14.3

2.2 The BP Neural Network for the Mine Ventilation

There are four input parameters and one output parameter in coal mine ventilation, so four input neurons and one output neuron are in the BP neural network. Since the determination of the hidden layer needs to be set artificially, an estimated value is regarded as the number of hidden layers according to the empirical formula. Then, the error value is compared in the experiment, and the parameter that makes the error minimum is selected by slowly increasing the number of hidden layers. Finally, the number of hidden layers is determined to be 7, so the BP neural network structure can be determined as 4-7-1. Therefore, there are $4*7+7*1 = 35$ weights and $7+1 = 8$ thresholds. The specific structure of the BP neural network is shown in Fig. 1.

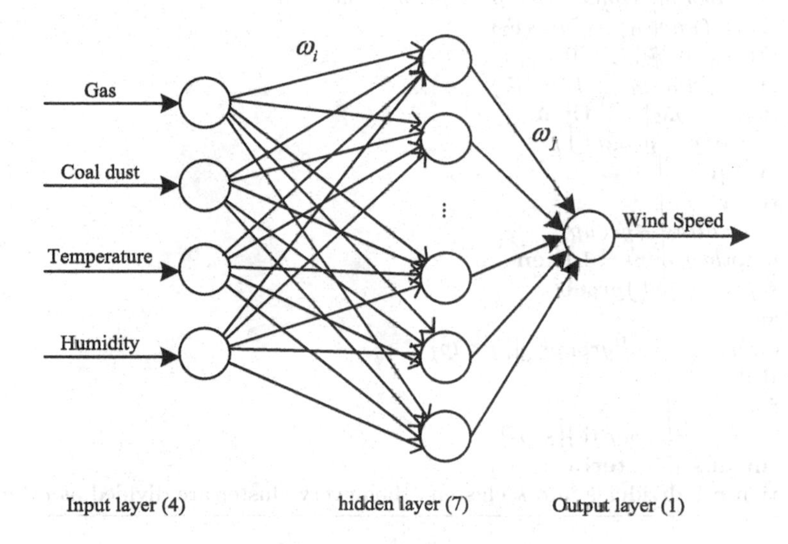

Fig. 1. Structure of BP neural network

3 BP Neural Network Based on Improved BSO Algorithm

3.1 BSO with Double Grouping (BSO-DG)

The BSO algorithm employs clustering, creating and selecting operators, which are all connected and have great impacts on the optimization performance. The original BSO and most of its variants employ k-means clustering. With the increase of dimension, there is some correlation information between each dimension of each individual. In this paper, we propose a double grouping, i.e., the differential grouping strategy [19] will be added to the k-means clustering strategy. The whole double grouping is as Algorithm 1.

Algorithm 1. Pseudocode of double grouping

Initialize parameters: a set of solutions $i \neq j$, where each solution is a m-dimensional real vector; clustering number $k = 2$; separable group $seps = \{\}$; all the subconponents $allgroups = \{\}$; dimension vector $dims = \{1, 2, 3, \cdots, m\}$
//differential grouping
for $i = 1$ to m **do**
  set vector $group = \{i\}$;
  for $j = 1$ to m and $i \neq j$ **do**
    $\vec{p_1} = lbound * ones(1, n)$, $\vec{p_2} = \vec{p_1}$, $\vec{p_2} = ubound$
    $\Delta_1 = func(\vec{p_1}) - func(\vec{p_2})$
    $\vec{p_1}(j) = 0$, $\vec{p_2}(j) = 0$
    $\Delta_2 = func(\vec{p_1}) - func(\vec{p_2})$
    if $|\Delta_1 - \Delta_2| > \varepsilon$ **then**
      $group = group \bigcup j$
    end if
  end for
  $dims = dims - group$
  if $length(group) = 1$ **then**
    $seps = seps \bigcup group$
  else
    $allgroups = allgroups \bigcup \{group\}$
  end if
end for
$allgroups = allgroups \bigcup \{seps\}$
// k-means Clustering
Partition n individuals into k clusters, then every cluster are divided as $allgroups$

3.2 Optimizing BP Neural Network Parameters Based on BSO-DG

In applying BP neural network to optimize mine ventilation, the threshold and weight are initialized to any random value between [−0.5, 0.5]. However, different weights and thresholds have a significant influence on the predictions. BSO can search the solution for the optimal global one as a global optimization algorithm. Therefore, BSO is employed to optimize the weight of the input and hidden layers, threshold of the hidden layer, weight of the hidden layer and output layer, and threshold of the output layer in this paper. Then the optimal weights and thresholds are assigned to the neural network, and the BP algorithm is used to achieve local optimization. Original data samples include training data and test ones. We consider the sum of the absolute value of output error as an individual fitness function. The smaller the fitness value of an individual is, the better the individual is. The specific steps of the algorithm are shown in Algorithm 2. The flow chart of the BP neural network based on BSO-DG algorithm is shown in Fig. 2.

Algorithm 2. BP neural network based on BSO-DG

Step 1. Initialize data, import data points, and normalize them.
Step 2. Initialize the algorithm's parameters, set the population size, variable range, Etc.
Step 3. Calculate individual fitness value, use BSO-DG to optimize weight and threshold, take the sum of absolute error as the fitness function, and get the optimal weight and threshold through continuous iteration.
Step 4. Use the optimized weight and threshold to train the training data and get the corresponding results.
Step 5. Employ the trained network to predict the data to be detected.
Step 6. Output the experimental data and draw the diagrams.

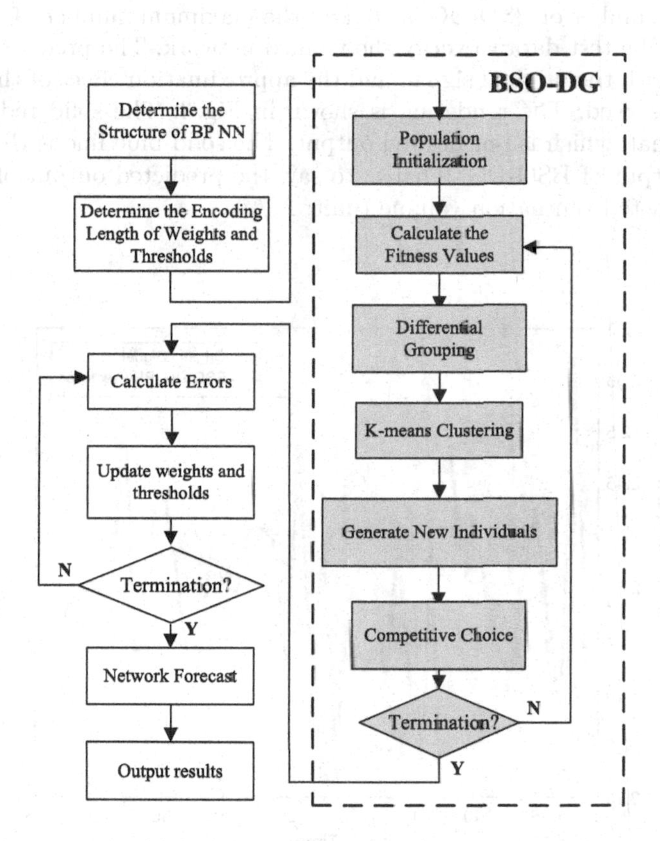

Fig. 2. BP neural network based on BSO-DG

It can be seen from Fig. 2 that the algorithm needs first to determine the BP neural network's structure. This paper's structure is 4-7-1, as shown in Fig. 1. Then, BSO-DG is applied to optimize the weights and thresholds. The system error is calculated after reaching the number of iterations optimized by BSO-DG.

Then the weights and thresholds are updated by the optimal solution obtained. Finally, the network is predicted, and the ultimate predictions are output.

4 Experimental Simulation and Result Analysis

In this paper, 43 groups of data of one mine working face are selected for experimental simulation, and the indexes include gas, coal dust, temperature, humidity, and wind speed. All the tests are carried out on a PC with Intel Core *i*7 4510u CPU, 2.6 GHz, and 8 GB memory. The wind speed prediction model based on BSO-DG and BP algorithm is established in MATLAB. In this experiment, the parameters are set as follows: the maximum number of iterations of BP neural network training is 1000, the learning rate is 0.1, the training goal is 0.0001, the population number of BSO-DG is 50, and the maximum number of iterations is 20. We use the test data to verify the trained network. The predicted output is compared with the original signal, and the approximation effect of the predicted result is observed. The rendering is shown in Fig. 3. The solid red line is the original signal, which is our desired output. The solid blue line is the optimized network output of BSO-DG. That is to say, the predicted output of the model is the forecasted ventilation volume (unit: M/s).

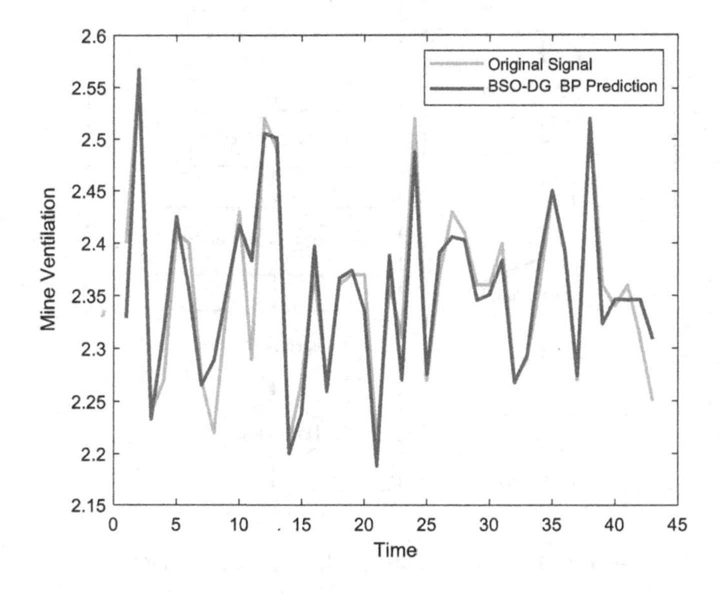

Fig. 3. Comparison of expected and predicted wind speed (BSO-DG)

In order to verify that BP neural network optimized by BSO-DG has a better prediction effect, comparative experiments are designed. We compare it with BP neural network prediction and BP neural network prediction based on Genetic Algorithm (GA). The prediction effect of BP is shown in Fig. 4. In addition, the comparison results between Mean Square Error (MSE) and Mean Absolute Error (MAE) of the three methods are shown in Table 2.

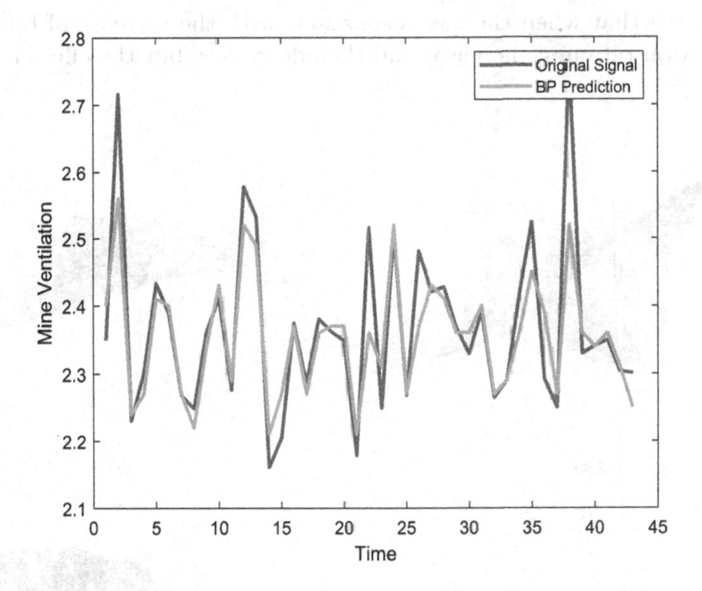

Fig. 4. Comparison of expected output wind speed and predicted output wind speed (BP)

Table 2. Comparison of prediction errors of three algorithms

Algorithm name	MSE	MAE
BP	0.0072	0.0613
GA-BP	0.0022	0.0279
BSO-DG-BP	**9.5404e−04**	**0.0236**

Compared with the predictions between BP and GA-BP, it is better to employ BSO-DG. The predictions estimated by BSO-DG are smooth, and MSE and mean error are much smaller than those predicted by BP and GA-BP. All predicted ventilation values are obtained through the global search of the system and taking the data of gas, coal dust, temperature, and humidity in their respective variable range. Finally, the minimum group of ventilation and the sum of gas and coal dust is selected as the optimal value. The results show that when gas is 0.5000, coal dust is 6.5000, the temperature is 13, humidity is 19, and the optimal wind speed is 2.1295, the condition is optimal.

Finally, we draw the surface diagram of the optimal ventilation volume, gas, coal dust, temperature, and humidity predicted by the BP neural network optimized by BSO-DG. The diagram is shown in Fig. 5.

When the amount of coal dust remains unchanged, the ventilation rate changes in an S-shaped curve with the gas increase in Fig. 5(a). It can be seen from Fig. 5(b) that when the gas is constant, with the increase of temperature, the ventilation rate first increases and then decreases, but the effect is relatively

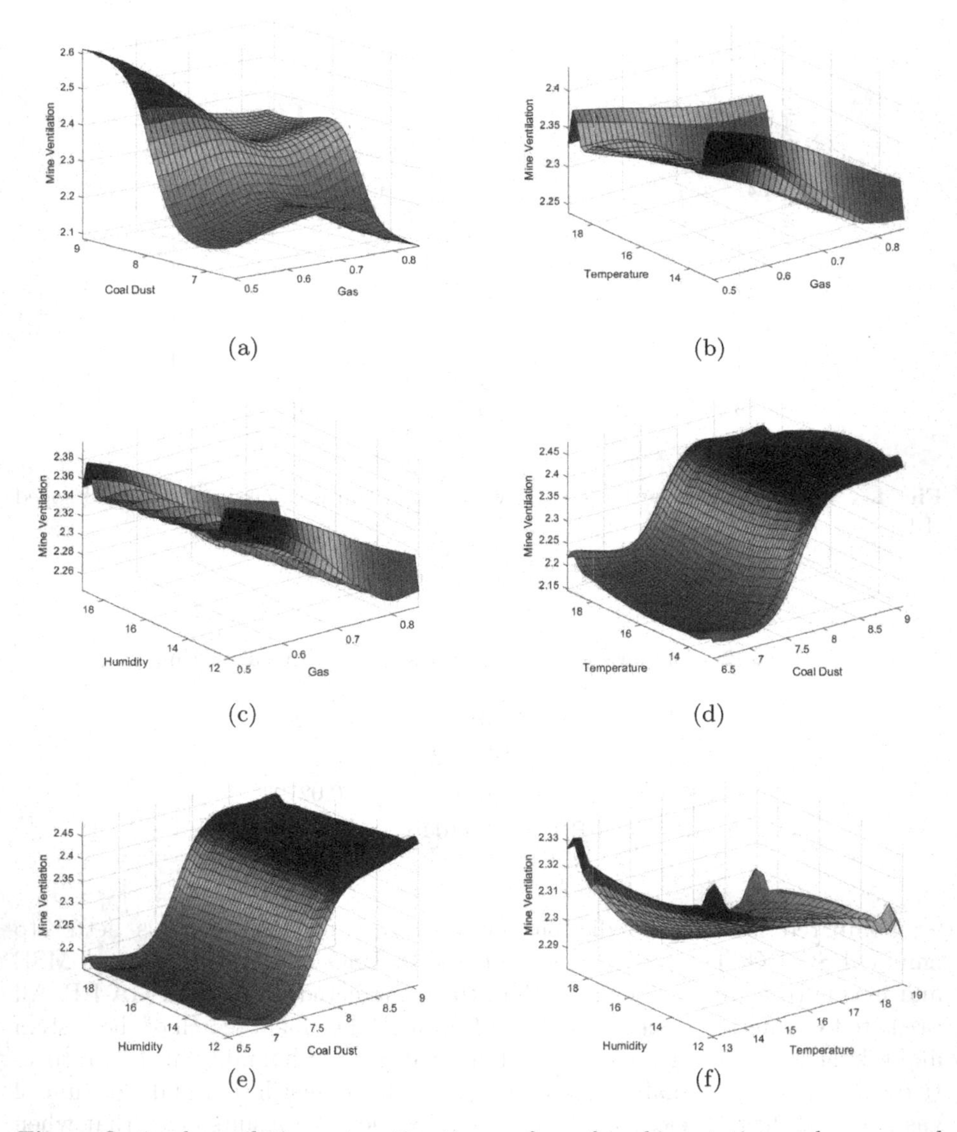

Fig. 5. Optimal ventilation rate estimation surface of working station under gas, coal dust, temperature and humidity

slow. In Fig. 5(c), when the gas is constant, the ventilation rate decreases slightly with the increase of humidity. From Fig. 5(d) and (e), we can see that when the temperature and humidity are constant, the ventilation rate also increases with the rise of coal dust. Figure 5(f) shows that the temperature and humidity change have a relatively slow ventilation rate. Therefore, we can conclude that coal dust and gas content is the main factor affecting ventilation.

5 Conclusions

Based on the advantages of BSO-DG in global optimization, the prediction model of mine ventilation based on the BP neural network optimized by BSO-DG is established. The model overcomes the defects of random initialization of weights and thresholds in the prediction process of the general BP neural network. Furthermore, it can improve the accuracy of network prediction. The optimal wind speed prediction model is established through gas, coal dust, temperature, and humidity, and the simulation research is carried out. The experimental results show that the BP neural network optimized by BSO-DG (BSO-DG-BP model) has higher accuracy than the BP model, and the GA-BP model, and is effective prediction model.

References

1. Cao, Z., Shi, Y., Rong, X., Liu, B., Du, Z., Yang, B.: Random grouping brain storm optimization algorithm with a new dynamically changing step size. In: Tan, Y., Shi, Y., Buarque, F., Gelbukh, A., Das, S., Engelbrecht, A. (eds.) ICSI 2015. LNCS, vol. 9140, pp. 357–364. Springer, Cham (2015). https://doi.org/10.1007/978-3-319-20466-6_38
2. Wang, H., Liu, J., Yi, W., Niu, B., Baek, J.: An improved brain storm optimization with learning strategy. In: Tan, Y., Takagi, H., Shi, Y. (eds.) ICSI 2017. LNCS, vol. 10385, pp. 511–518. Springer, Cham (2017). https://doi.org/10.1007/978-3-319-61824-1_56
3. Zhu, H., Shi, Y.: Brain storm optimization algorithms with k-medians clustering algorithms. In: 2015 Seventh International Conference on Advanced Computational Intelligence (ICACI), pp. 107–110. IEEE (2015)
4. Zhan, Z.-h., Zhang, J., Shi, Y.-h., Liu, H.-l.: A modified brain storm optimization. In: 2012 IEEE Congress on Evolutionary Computation, pp. 1–8. IEEE (2012)
5. Guo, X., Wu, Y., Xie, L.: Modified brain storm optimization algorithm for multimodal optimization. In: Tan, Y., Shi, Y., Coello, C.A.C. (eds.) ICSI 2014. LNCS, vol. 8795, pp. 340–351. Springer, Cham (2014). https://doi.org/10.1007/978-3-319-11897-0_40
6. Guo, X., Wu, Y., Xie, L., Cheng, S., Xin, J.: An adaptive brain storm optimization algorithm for multiobjective optimization problems. In: Tan, Y., Shi, Y., Buarque, F., Gelbukh, A., Das, S., Engelbrecht, A. (eds.) ICSI 2015. LNCS, vol. 9140, pp. 365–372. Springer, Cham (2015). https://doi.org/10.1007/978-3-319-20466-6_39
7. Chen, J., Cheng, S., Chen, Y., Xie, Y., Shi, Y.: Enhanced brain storm optimization algorithm for wireless sensor networks deployment. In: Tan, Y., Shi, Y., Buarque, F., Gelbukh, A., Das, S., Engelbrecht, A. (eds.) ICSI 2015. LNCS, vol. 9140, pp. 373–381. Springer, Cham (2015). https://doi.org/10.1007/978-3-319-20466-6_40

8. Chen, J., Wang, J., Cheng, S., Shi, Y.: Brain storm optimization with agglomerative hierarchical clustering analysis. In: Tan, Y., Shi, Y., Li, L. (eds.) ICSI 2016. LNCS, vol. 9713, pp. 115–122. Springer, Cham (2016). https://doi.org/10.1007/978-3-319-41009-8_12
9. Xie, L., Wu, Y.: A modified multi-objective optimization based on brain storm optimization algorithm. In: Tan, Y., Shi, Y., Coello, C.A.C. (eds.) ICSI 2014. LNCS, vol. 8795, pp. 328–339. Springer, Cham (2014). https://doi.org/10.1007/978-3-319-11897-0_39
10. Qiu, H., Duan, H.: Receding horizon control for multiple UAV formation flight based on modified brain storm optimization. Nonlinear Dyn. **78**(3), 1973–1988 (2014)
11. El-Abd, M.: Global-best brain storm optimization algorithm. Swarm Evol. Comput. **37**, 27–44 (2017)
12. Zhou, D., Shi, Y., Cheng, S.: Brain storm optimization algorithm with modified step-size and individual generation. In: Tan, Y., Shi, Y., Ji, Z. (eds.) ICSI 2012. LNCS, vol. 7331, pp. 243–252. Springer, Heidelberg (2012). https://doi.org/10.1007/978-3-642-30976-2_29
13. Duan, H., Li, C.: Quantum-behaved brain storm optimization approach to solving Loney's solenoid problem. IEEE Trans. Magn. **51**(1), 1–7 (2014)
14. Yang, Y., Shi, Y., Xia, S.: Advanced discussion mechanism-based brain storm optimization algorithm. Soft. Comput. **19**(10), 2997–3007 (2014). https://doi.org/10.1007/s00500-014-1463-x
15. Yang, Z., Shi, Y.: Brain storm optimization with chaotic operation. In: 2015 Seventh International Conference on Advanced Computational Intelligence (ICACI), pp. 111–115. IEEE (2015)
16. Wang, G.-G., Hao, G.-S., Cheng, S., Shi, Y., Cui, Z.: An improved brain storm optimization algorithm based on graph theory. In: 2017 IEEE Congress on Evolutionary Computation (CEC), pp. 509–515. IEEE (2017)
17. Su, Y., Go, L., Chen, S.: Mine ventilation wind speed prediction based on improved genetic algorithm and BP neural network. J. Henan Univ. Sci. Technol. (Nat. Sci. Ed.) **36**(4), 20–25 (2017)
18. Liu, X.: Research on normalization of data in the input layer of BP neural network. Mech. Eng. Autom. **3**, 122–123 (2010)
19. Chen, J., Xue, X., Ninjerdene, B.: An improved brain storm optimization algorithm based on maximum likelihood estimation. In: Tan, Y., Shi, Y. (eds.) ICSI 2021. LNCS, vol. 12689, pp. 492–501. Springer, Cham (2021). https://doi.org/10.1007/978-3-030-78743-1_44

Construction of Fuzzy Classifiers by a Brain Storm Optimization Algorithm

Marina Bardamova(✉), Ilya Hodashinsky(✉), and Mikhail Svetlakov

Tomsk State University of Control Systems and Radioelectronics, 40, prospect Lenina, 634050 Tomsk, Russia
722bmb@gmail.com, hodashn@rambler.ru

Abstract. The fuzzy classifier has such important advantages as an intuitive operation logic and high interpretability of the fuzzy rules base. The development of a fuzzy classifier includes three consecutive building steps: generating a fuzzy rules base, feature selection, and optimizing the parameters of membership functions. To create a rule base, clustering methods are most often used. Wrapper methods are used for feature selection, and parameter optimization is performed either by traditional optimization methods or by metaheuristic methods. In this paper, we use a metaheuristic called Brain Storm Optimization to construct a fuzzy classifier. The use of algorithms based on this metaheuristic made it possible to obtain comparable accuracy values comparable to counterparts such as D-MOFARC and FARC-HD, with a much smaller number of rules and features.

Keywords: Brain storm optimization · Fuzzy rule-based classifier · Membership function · Machine learning

1 Introduction

Classification is a major area of machine learning related to supervised learning. Among other classification methods, fuzzy rule-based classifiers occupy a special place. The fuzzy classifier has such important advantages as an intuitive operation logic and high interpretability of the fuzzy rule base [1]. Constructing a fuzzy classifier includes three stages: designing a fuzzy rules base, feature selection, and optimizing the parameters of membership functions. To create a rule base, clustering methods are most often used [2], wrapper methods are used for feature selection [3], and parameter tuning is performed either by traditional optimization methods or by metaheuristic methods [4].

Classification is an optimization problem, in which superior algorithms play a crucial role [5]. An increasing number of metaheuristic algorithms are being developed to find optimal solutions to optimization problems. Population-based metaheuristic algorithms can be divided into three categories based on what underlies their search engine; the first category consists of algorithms inspired by biological evolution, the second category consists of algorithms inspired by social behavior, and the third category consists of algorithms inspired by physics [6]. The concept of no universal optimization algorithm, formalized as "No Free Lunch Theorems" [7], states that there is no need to evaluate

Y. Tan et al. (Eds.): ICSI 2022, LNCS 13344, pp. 391–403, 2022.
https://doi.org/10.1007/978-3-031-09677-8_33

the effectiveness of optimization algorithms on average for all problems. The theorem encourages specialists to conduct new research in metaheuristic optimization algorithms.

In this paper, we use a metaheuristic called Brain Storm Optimization (BSO) to construct a fuzzy classifier. BSO [8, 9] is a population-based metaheuristic algorithm, inspired by the human brainstorming process. In BSO, each person in the population is represented as an idea or a potential solution to a problem, and all ideas are grouped into several clusters at each iteration. Ideas are then updated based on one or two ideas in clusters by neighborhood search and combination. BSO is a kind of search space reduction algorithm; all solutions eventually fall into several clusters. These clusters indicate the local optima of the problem. Information about a domain containing solutions with good fitness values propagates from one cluster to another [10]. BSO has the best optimization capabilities in solving various types of complex optimization problems that are difficult to solve with classical optimization algorithms [5].

The BSO algorithm is described in Algorithm 1.

Algorithm 1: The Brain Storm Optimization algorithm

Input: Population size *PN*, number of cluster *M*, current of iteration *t*; maximum of iteration *T*;
Output: The best individual;
1: Randomly generate *PN* potential individuals and evaluate the *PN* individuals;
2: **while** $t<T$ **do**
3: Divide the *PN* individuals into *M* clusters;
4: Evaluate the population;
4: Select the best individual in each cluster as the center;
5: Generate new individuals using one of the strategies randomly;
6: Record the best individual;
7: $t = t + 1$;
8: **end while**

BSO simulates the brainstorming process by a divergent operator and a convergent operator. The first divides the population into several clusters using the clustering method, and the second generates new solutions with a certain probability. These operators greatly affect the performance of the BSO, so numerous improvements to the BSO have been proposed by changing the clustering algorithm or/and modifying the mutation strategy [6, 11].

Clustering in BSO is performed by various methods such as simple grouping method, random grouping method, k-means, affinity propagation clustering method, elitist selection mechanism. New individuals are generated based on four types of individuals: center of one cluster, the random individual of one cluster, the combination of two centers which belong to two clusters, and the combination of two random individuals which belong to two clusters [6].

BSO has been successfully applied in many real academic and engineering applications [12]. It is used to solve nonlinear optimization problems in seismic exploration [13] and dynamic economic dispatch problems [14]. Binary versions of the BSO have been developed for feature selection [15, 16] and fault diagnosis [17]. Chandrasekar and Khare [18, 19] applied BSO to solve classification problems.

The main contribution of the paper is as follows:

1. A new algorithm based on the BSO for generating a fuzzy rules base is proposed.
2. A new algorithm based on the BSO to optimize membership functions parameters is proposed.

2 Algorithms for Constructing a Fuzzy Classifier

2.1 Creating a Fuzzy Classifier Structure by BSO-Based Clustering

The basis for building a fuzzy classifier with the possibility of selecting informative features is IF-THEN fuzzy rule:

$$R_{ij}: \text{IF } s_1 \wedge x_1 = A_{1i} \text{ AND } s_2 \wedge x_2 = A_{2i} \text{ AND} \ldots \text{ AND} s_n \wedge x_n = A_{ni} \text{ THEN class} = c_j \quad (1)$$

where x_i is the ith feature of the classified object, $i \in [1, n]$, n is the number of features; s_i indicates the presence ($s_i = 1$) or absence ($s_i = 0$) of a feature in the classifier, $\mathbf{S} = (s_1, s_2, \ldots, s_n)$; A_{ij} is the fuzzy term describing the membership function of the ith attribute to the jth rule, $j \in [1, R]$, R is the number of rules, c_k is the kth class label, $k \in [1, M]$, M is the number of classes.

To determine the output variable for some object $\mathbf{x}$, the degree of confidence in the belonging of this object to each of the classes is calculated:

$$\beta_k(\mathbf{x}_p) = \sum_{r_{jk}} \prod_{i=1}^{n} \mu_{A_{ij}}(x_i) \quad (2)$$

where r_{jk} are rules with the k class output label, $\mu_{A_{ij}}(x_i)$ is the value of the membership function of the term A_{ij} for the ith attribute of object $\mathbf{x}$. The output of the classifier is the class that has received the highest degree of confidence:

$$\text{class} = c_{k*}, \quad k* = \arg \max_{1 \le k \le M} \beta_k(\mathbf{x}) \quad (3)$$

The objective function or measure of classification accuracy can be expressed as follows:

$$E(\boldsymbol{\theta}, \mathbf{S}) = \frac{\sum_{p=1}^{N} \begin{cases} 1, & \text{if } c_p = c(\mathbf{x}_p, \boldsymbol{\theta}) \\ 0, & \text{otherwise} \end{cases}, p = 1, 2\ldots, N}{N} \quad (4)$$

where N is a sample size, $\boldsymbol{\theta}$ is a vector of fuzzy classifier parameters.

The problem of constructing a fuzzy classifier essentially comes down to finding the maximum of the specified function in $\mathbf{S}$ and $\boldsymbol{\theta}$. We propose to use continuous BSO for fuzzy system structure generation and parameter tuning, and binary BSO for feature selection.

To generate the structure of a fuzzy classifier, the BSO-based clustering method is used. The solution is presented as a real vector that has the coordinates of the cluster centers $C = \{\mathbf{C}_1, \mathbf{C}_2, \ldots, \mathbf{C}_h\}$. At each iteration, the distances between the cluster centers and the experimental data are calculated; then the data are distributed to clusters with the closest centers. Further, the coordinates of the cluster centers are changed:

$$\mathbf{C}_v = \begin{cases} \mathbf{X}_1^v + \xi \cdot N(0,1), \text{ if } U(0,1) < p\_one \\ U(0,1) \cdot \mathbf{X}_1^v + (1 - U(0,1)) \cdot \mathbf{X}_2^v + \xi \cdot N(0,1), \text{ otherwise} \end{cases} \tag{5}$$

where $\mathbf{C}_v$ is the center of the vth cluster, $v \in [1, h]$, $\mathbf{X}_1^v$ and $\mathbf{X}_2^v$ are randomly selected objects from the vth cluster. $N(0, 1)$ denotes the standard normal distribution, $U(0,1)$ denotes the uniform distribution from 0 to 1. ξ is the function used for adjusting the step size, as shown in Eq. (6).

$$\xi(t) = \log sig\left(\frac{0.5 \cdot T - t}{c}\right) \cdot U(0, 1), \tag{6}$$

where t is the current iteration, and T is the maximum number of iterations.

Next, the Davis-Baldwin index is computed:

$$DB = \frac{1}{h} \sum_{i=1}^{h} \max_{i \neq j} \left(\frac{\delta_i + \delta_j}{d(\mathbf{C}_i, \mathbf{C}_j)} \right) \tag{7}$$

where $d(\mathbf{C}_i, \mathbf{C}_j)$ is the distance between the ith and j^{th} cluster centers, δ_q is the average distance between all objects of the qth cluster and the center of this cluster. If the Davis-Baldwin index decreases, the new centers overwrite the old ones.

When the iteration counter reaches its maximum value, clustering accuracy is evaluated. For its calculation, each data object is assigned a class, which has the largest representation in the cluster to which this object belongs. If the assigned class coincides with the real one, accuracy is increased by 1.0. After checking all objects, the accuracy is normalized.

Next, one additional cluster is randomly generated, the iteration counter is reset, and the entire process is repeated until the number of clusters reaches a certain specified number. In this paper, the algorithm stopped when the doubled number of classes was reached. The output of the algorithm is set C, which achieved maximum clustering accuracy. After the completion of the algorithm, one fuzzy rule with Gaussian-type terms is formed on the basis of each cluster: the center of the peak for the term coincides with the coordinate of the cluster center, and the width of the term is determined along the boundaries of the cluster. The class dominating in the corresponding cluster is written to the rule's consequent.

2.2 Feature Selection with Binary BSO

The choice of informative features allows for reducing the complexity of the resulting classifier, decreasing the likelihood of retraining, and in some cases even improving accuracy. One of the methods for feature selection is the application of a binary metaheuristic algorithm in the wrapper mode. For the initially continuous metaheuristic algorithm to be able to search in binary search space, it is enough to use the transformation function [20]. The input idea, in this case, is represented as a vector $\mathbf{S} = (s_1, s_2, \ldots, s_n)$, where $s_i = 0$ indicates that the ith attribute is not used by the classifier, and $s_i = 1$ means the inclusion of the ith feature in the classifier. In general, the binary BSO is practically the same as the continuous one, except for using the S-shaped transformation function to transfer a vector from the continuous space to the binary space:

$$s_i(t+1) = \begin{cases} 1, \ \mathit{if} \ U(0, 1) < \frac{1}{1+e^{-s_i(t+1)}} \\ \quad 0, \ \text{otherwise} \end{cases} \tag{8}$$

where t is the current iteration of the binary algorithm.

2.3 Tuning the Parameters of a Fuzzy Classifier Using BSO with the Elements of the Differential Evolution Algorithm

The parameter tuning stage is designed to search for such parameters of terms that would give maximum accuracy. In this work, we use BSO with the operator from Differential Evolution as the basis for the algorithm for tuning terms [21]. The algorithm has vector $\boldsymbol{\theta}_0$ as an input, which is a list of terms parameters obtained after fuzzy classifier structure generation. Since BSO is a population metaheuristic, by superimposing a random deviation on $\boldsymbol{\theta}_0$, the rest of the ideas (or individuals, as they were referred to earlier) that together make up a population of ideas $\Theta = \{\boldsymbol{\theta}_0, \boldsymbol{\theta}_1, \ldots, \boldsymbol{\theta}_{Q\text{-}1}\}$ are generated (Q is the population size). The following parameters are also part of an input: M is the number of groups of ideas, T is the number of iterations of the term tuning algorithm, *p_one* is the coefficient for selecting the number of groups, *p_operation* is the coefficient for selecting an idea update operator, *MSF* is mutation scaling factor. Each solution's accuracy of the classification (fitness function value) is evaluated and recorded. The best solution is stored in the variable **BestIdea**.

First, the solutions are sorted by the value of the fitness function (from best to worst) and sequentially allocated to clusters.

The second step is to select the number of clusters to work with. If *p_one* > $U(0,1)$, then the work is done with one cluster, otherwise with two.

When working with a single cluster, the cluster to update G is defined. Then a choice is made as to which of the two operators will be used in this group. If *p_operation* > $U(0,1)$, then the original BSO operator is executed, otherwise it is borrowed from the Differential Evolution algorithm.

Next, the solutions are updated. When making a choice in favor of the original operator, an intermediate vector $\boldsymbol{\theta}_{\text{select}}$ is formed as follows:

$$\boldsymbol{\theta}_{\text{select}} = \boldsymbol{\theta}_p + \xi \cdot N(0, 1) \tag{9}$$

where $\boldsymbol{\theta}_p$ is the current solution, $p \in [0, M_G\text{ -}1]$, M_G is the number of solutions in G.

In the case of choosing an operator from Differential Evolution, an intermediate vector is generated using $\boldsymbol{\theta}_{\text{center}}$, the best solution in group G:

$$\boldsymbol{\theta}_{\text{select}} = \boldsymbol{\theta}_{\text{center}} + MSF \cdot (\boldsymbol{\theta}_{p1} - \boldsymbol{\theta}_{p2}) \tag{10}$$

where $\boldsymbol{\theta}_{p1}$ and $\boldsymbol{\theta}_{p2}$ are randomly selected vectors from group G, not matching with $\boldsymbol{\theta}_p$. If the solution of $\boldsymbol{\theta}_{\text{select}}$ turns out to be worse than that of $\boldsymbol{\theta}_p$, then $\boldsymbol{\theta}_p$ is written in $\boldsymbol{\theta}_{\text{select}}$.

Then, regardless of the chosen operator, a new solution $\boldsymbol{\theta}_{\text{new}}$ is created based on the obtained intermediate vector:

$$\boldsymbol{\theta}_{\text{new}} = \boldsymbol{\theta}_{\text{select}} + \xi \cdot N(0, 1). \tag{11}$$

If $\boldsymbol{\theta}_{\text{new}}$ is better than $\boldsymbol{\theta}_p$, then $\boldsymbol{\theta}_{\text{new}}$ replaces $\boldsymbol{\theta}_p$. The actions listed apply to all idea vectors from group G.

If the choice is made in favor of working with two groups of ideas, then it is necessary to randomly select two clusters $G1$ and $G2$. Next, using *p_operation*, the update operator is defined in the same way as earlier. When the original operator is selected, an intermediate vector is generated using the following expression:

$$\boldsymbol{\theta}_{\text{select}} = U(0, 1) \cdot \boldsymbol{\theta}_p^{G1} + (1 - U(0; 1)) \cdot \boldsymbol{\theta}_p^{G2} + \xi \cdot N(0, 1). \tag{12}$$

If the Differential Evolution operator is used, the generation is performed as follows:

$$\boldsymbol{\theta}_{\text{select}} = \textbf{BestIdea} + MSF \cdot \left(\boldsymbol{\theta}_{p1}^{G1} - \boldsymbol{\theta}_{p2}^{G2}\right) \tag{13}$$

where $\boldsymbol{\theta}_{p1}^{G1}$ and $\boldsymbol{\theta}_{p2}^{G2}$ are solutions selected from clusters $G1$ and $G2$ that do not coincide with $\boldsymbol{\theta}_p^{G1}$ and $\boldsymbol{\theta}_p^{G2}$.

The creation of a new solution is performed according to the formula:

$$\boldsymbol{\theta}_{\text{new}} = \boldsymbol{\theta}_{\text{select}} + \xi \cdot N(0, 1). \tag{14}$$

The vector $\boldsymbol{\theta}_{\text{new}}$ replaces the original solutions $\boldsymbol{\theta}_{p1}^{G1}$ and $\boldsymbol{\theta}_{p2}^{G2}$, if $\boldsymbol{\theta}_{\text{new}}$ turns out to be the best compared to the original solutions. The listed actions apply to all pairs of idea vectors from clusters $G1$ and $G2$.

Next, the population is combined and sorted from the best value to the worst. If solution $\boldsymbol{\theta}_0$ exceeds **BestIdea**, then **BestIdea** is written $\boldsymbol{\theta}_0$.

In the end, the maximum value of the iteration counter is checked. If the stopping criterion is not reached, then the algorithm returns to the beginning. Otherwise, **BestIdea** is output.

3 Experimental Study

The effectiveness of BSO adaptation to the construction of fuzzy classifiers was evaluated in two experiments. The purpose of the first experiment was to compare the quality of models obtained using BSO with the results of similar advanced algorithms for constructing fuzzy classifiers. Data sets from the KEEL repository were used for testing [22]. The data represent real classification tasks from a variety of subject areas. The description of the sets is given in Table 1. Variables *#F*, *#C,* and *#Ex* denote the number of features, classes, and data instances, respectively.

The first experiment was conducted according to a 10-fold cross-validation scheme and was organized as follows for each sample: In the first stage, a clustering algorithm

Table 1. Description of the data sets used in the first experiment

Data set	Reduction	#F	#C	#Ex	Description
appendicitis	app	7	2	106	Diagnosis of appendicitis
balance	bln	4	3	625	Modeling the results of psychological experiments
cleveland	clv	13	5	297	Diagnosis of diseases of the cardiovascular system
haberman	hbr	3	2	306	Prediction of survival rate after breast surgery
heart	hrt	13	2	270	Diagnosis of diseases of the cardiovascular system
hepatitis	hpt	19	2	80	Predicting survival in patients with hepatitis
magic	mgc	10	2	19020	Image recognition
newthyroid	nthr	5	3	215	Diagnosis of hyperthyroidism or hypothyroidism
phoneme	phn	5	2	5404	Sound recognition
satimage	stm	36	7	6435	Classification of pixels
vehicle	vhc	18	4	846	Classification of vehicles
wine	wn	13	3	178	Classification of wine

based on BSO was used to create the classifier structure. The second stage used a binary BSO with the S-shaped transformation function for feature selection. In the third stage, a continuous Brain Storm algorithm adjusted the parameters of the terms using the selected set of features.

The results of constructing classifiers were averaged over all samples. The following parameters were used for the algorithms: 100 iterations, 5 groups of 5 ideas, $p\_one = 0.5$, $p\_operation = 0.5$, $MSF = 0.8$. The coefficients of group selection and operator selection (*p_one* and *p_operation*) are set to 0.5 to ensure equal priority between local and global searches. The *MSF* value in [21] is recommended to be between 0 and 2; the value of 0.8 was chosen empirically as the most appropriate for all the datasets used.

The results of the step-by-step creation of fuzzy classifiers by the BSO algorithm are presented in Table 2. The first values of accuracy and number of rules are recorded after the construction of the structure by clustering with BSO (column "Clusters"). Then, on the same classifiers, the selection of features by the binary version of BSO was carried out and the accuracy and number of selected features were taken (column "Features"). Since different samples could result in varying amounts of rules and features, the numbers in the table are real. Finally, the final accuracy value was obtained after optimizing the parameters of the classifier terms on the reduced feature sets (column "Total").

For comparison, the results of similar methods of constructing fuzzy classifiers D-MOFARC and FARC-HD, provided by the creators of the repository KEEL [23], are given. These methods do not include a feature selection step but apply weighting and rule selection. Here *#R* is the number of fuzzy rules obtained by the algorithm for creating a fuzzy classifier structure, *#F'* is the number of features after the selection, A_{tst} is the percentage of correct classification on test samples.

A comparison of the results shows that in just 100 iterations of the BSO algorithm, results comparable in classification accuracy to their counterparts were achieved. Furthermore, fuzzy classifiers constructed with BSO showed an advantage in the number of rules and features. Consequently, the investigated algorithms are relevant when it is necessary to obtain interpretable and computationally fast models.

Table 2. Results of fuzzy classifiers construction

Data	Brain storm optimization algorithm					D-MOFARC		FARC-HD	
	Clusters		Features		Total				
	#R	A_{tst}	*#F'*	A_{tst}	A_{tst}	*#R*	A_{tst}	*#R*	A_{tst}
app	3.1	80.4	2.3	81.6	86.9	–	–	6.8	84.2
bln	4.6	47.3	2.0	61.4	81.5	20.1	85.6	18.8	91.2
clv	8.4	34.6	4.1	49.2	56.9	45.6	52.9	61.3	55.2
hbr	3.4	67.3	1.4	73.1	75.2	9.2	69.4	5.7	73.5
hrt	3.2	49.4	5.7	58.4	73.0	18.7	84.4	27.8	83.7
hpt	2.3	81.2	8.8	87.7	88.3	11.4	90.0	10.4	88.7
mgc	3.9	60.0	3.9	68.8	85.3	32.2	85.4	43.8	84.8
nthr	4.1	74.2	4.1	83.5	91.2	9.5	95.5	9.6	94.4
phn	3.4	68.1	4.0	74.1	80.6	9.3	83.5	17.2	82.4
stm	10.2	64.4	11.7	71.6	84.2	56.0	87.5	76.1	87.3
vhc	6.7	35.2	6.0	51.9	64.9	22.4	70.6	31.6	68.0
wn	4.4	61.7	6.6	72.0	83.3	8.6	95.8	8.3	95.5
Mean	4.8	60.3	5.1	69.5	79.3	22.1	81.9	26.5	82.4

Analysis of the results was conducted by pairwise comparison using the Wilcoxon criterion (Table 3). The null hypothesis is that there are no statistically significant differences between the two analyzed samples. The significance level is 0.05.

Table 3. Comparison of results by Wilcoxon criterion

Algorithm	Metric	*p-value*	Null hypothesis
D-MOFARC, FARC-HD	#F и #F'	0.002	Rejected
D-MOFARC	#R	0.003	Rejected
D-MOFARC	A_{tst}	0.091	Accepted
FARC-HD	*#R*	0.002	Rejected
FARC-HD	A_{tst}	0.060	Accepted

The second experiment was to study the effectiveness of BSO in constructing fuzzy classifiers for authenticating users by handwritten signatures. A set of SVC2004 signals was used as the initial data [24]. The set has the signatures of 40 users. The signature of each user is made in 40 samples – 20 genuine and 20 qualified forgeries. The classifier is supposed to effectively separate genuine signatures from forgeries. Data preparation consisted in extracting spatio-temporal features using the Fourier transform from the signals provided in SVC2004 [20]. For each user, an observation table of 100 features and 40 data instances was generated. The instances were labeled with a class "1" or "2", where "1" corresponds to a genuine signature and "2" to a qualified forgery. Two pairs of training and test samples were created as part of the cross-validation. A separate classifier was constructed for each user.

The effectiveness of BSO in the handwritten signature user authentication task was evaluated against similar metaheuristics. The process of constructing a fuzzy classifier using BSO was the same as described in the previous experiment. The remaining fuzzy classifiers designed to compare the results were created based on the algorithm of extreme values of classes, generating one rule for each class [3]. Therefore, 2 rules were generated for these models. In them, the stages of feature selection by binary versions of metaheuristics based on the S-shaped transformation function and the tuning of the term parameters by a continuous algorithm were carried out in a similar way. The following metaheuristics were used: Shuffled frog leaping algorithm (SFLA), Grey wolf (GW), Gravitational search algorithm (GSA), and Particle swarm optimization (PSO). All algorithms were run for 250 iterations to select features and for 1000 iterations to tune parameters (for SFLA – 10 global and 25 local iterations in binary version, 40 global and 25 local iterations in continuous). The population size was set to 25 vectors (in BSO and SFLA vectors were divided into 5 groups of 5 vectors). The values of the metaheuristic parameters are shown in Table 4; GW has no specific parameters. The resulting accuracy of the created fuzzy classifiers and the dimensionality of the data after the selection of features are shown in Table 5. The results of a pairwise comparison by the Wilcoxon criterion showed that there was no significant statistical difference in accuracy (*p-value* was 0.502 for SFLA, 0.514 for GW, 0.464 for GSA, and 0.245 for PSO). However, BSO had an advantage in the number of selected features, as the *p-value* was less than 0.001 as compared to the others.

Table 4. Parameter values for metaheuristic algorithms

Algorithms and their parameters		Algorithms and their parameters	
BSO	$p\_one = 0.5$, $p\_operation = 0.5$, $F = 0.8$	SFLA	$c = 1{,}2$
GSA	$G_0 = 100$, $\alpha = 10$, $\xi = 0.001$, $k_{best} = 15$	PSO	$w = 0.5$, $c_1 = 1$, $c_2 = 1$

Table 5. Accuracy of fuzzy classifiers for each user in the data set SVC2004

User	BSO		SFLA		GW		GSA		PSO	
	Acc	*F'*	*Acc*	*F'*	*Acc*	*F'*	*Acc*	*F'*	*Acc*	*F'*
1	91.5	13.8	92.9	45.4	91.8	75.2	93.9	75.9	93.3	76.2
2	93.0	13.4	81.8	46.7	78.5	53.7	79.6	50.5	80.9	46.5
3	96.0	15.3	95.4	48.6	98.4	75.9	98.3	75.2	98.3	74.5
4	79.5	13.2	89.7	50.8	92.3	53.5	92.5	49.1	93.2	49.4
5	75.0	17.2	88.9	49.0	80.2	55.3	80.9	49.8	84.2	48.2
6	81.0	18.2	86.7	48.3	94.3	50.5	93.6	49.5	93.5	49.8
7	83.5	17.6	94.3	44.8	98.2	75.0	98.6	75.0	98.2	75.4
8	91.5	20.2	94.8	48.0	93.3	51.1	93.5	49.3	92.4	49.6
9	83.0	14.0	70.8	47.1	63.4	52.0	63.9	48.2	62.0	46.9
10	88.0	14.4	83.1	48.1	79.3	53.7	79.1	47.8	80.6	47.2
11	71.0	13.2	64.4	44.4	84.5	57.1	83.8	49.8	84.6	48.3
12	74.4	17.2	76.8	46.4	80.5	54.3	80.3	49.4	82.2	45.8
13	63.5	16.3	83.0	48.7	77.0	52.6	75.3	49.6	76.5	47.6
14	73.5	18.8	89.3	47.6	89.3	56.3	87.9	49.4	89.0	47.1
15	95.0	13.2	80.1	47.5	75.8	55.2	75.8	49.2	75.8	48.3
16	81.5	15.2	95.5	49.1	95.0	100	95.0	100	95.0	100
17	90.5	16.2	87.7	44.8	81.8	52.9	82.6	49.2	83.3	48.0
18	80.0	16.7	82.1	43.2	80.2	51.3	80.8	48.1	81.1	46.9
19	90.0	16.2	89.6	51.3	92.5	52.8	92.6	49.7	93.5	49.3
20	82.0	17.8	90.3	45.5	89.1	58.6	89.5	61.9	90.8	55.0
21	94.0	16.0	81.5	48.1	90.1	51.2	89.6	51.1	90.3	51.0
22	89.0	16.0	86.8	47.0	85.0	53.0	84.3	48.8	86.1	48.4
23	96.0	16.4	89.5	49.7	92.3	52.1	92.1	49.0	90.5	49.6
24	84.0	15.3	87.3	49.2	91.6	50.8	91.3	50.7	91.5	49.7
25	91.5	14.4	91.9	44.9	90.6	75.3	91.3	74.3	92.0	73.2
26	91.0	16.5	81.2	46.8	81.9	55.8	84.8	48.0	86.1	48.2
27	79.5	18.0	73.5	44.0	83.9	53.6	83.8	52.6	85.0	49.1
28	93.0	17.5	90.2	49.9	89.1	75.2	89.8	73.6	90.5	72.7
29	85.0	16.5	81.1	48.0	97.1	48.9	97.8	50.0	97.4	50.2
30	94.5	17.1	82.1	46.4	92.3	50.5	92.9	49.7	92.8	49.9
31	99.0	18.0	88.2	49.8	100	100	100	100	100	100

(*continued*)

Table 5. (*continued*)

User	BSO		SFLA		GW		GSA		PSO	
	Acc	*F'*	*Acc*	*F'*	*Acc*	*F'*	*Acc*	*F'*	*Acc*	*F'*
32	82.0	14.0	90.8	46.3	91.7	50.0	93.4	48.3	93.4	47.8
33	78.0	14.5	81.8	45.0	92.7	52.0	92.8	48.3	92.6	46.8
34	93.5	14.0	80.4	46.7	86.4	52.0	87.6	49.5	88.4	47.7
35	88.0	15.9	83.8	59.2	77.2	54.1	78.2	48.5	77.3	46.9
36	74.0	17.0	81.0	43.6	91.6	52.3	91.6	50.9	91.6	49.5
37	97.5	17.5	83.4	44.5	81.8	51.0	81.4	47.4	84.2	45.6
38	90.5	15.2	88.6	46.5	86.3	53.1	85.2	48.5	85.8	48.6
39	92.0	16.1	65.0	47.2	70.5	53.7	69.9	48.8	72.3	48.1
40	70.5	19.4	76.6	46.3	74.3	53.0	74.8	48.1	75.5	48.1
Mean	**85.6**	**16.1**	**84.5**	**47.3**	**86.5**	**58.1**	**86.7**	**55.3**	**87.3**	**54.3**

4 Conclusion

This paper proposes to use Brain Storm Optimization for three consecutive constructing steps of fuzzy classifiers - creating fuzzy system structure, feature extraction, and parameter tuning. BSO is used to optimize the position of cluster centers when creating a fuzzy system structure. The binary version of BSO, based on the S-shaped transformation function, searches for an optimal subset of features following the wrapper scheme. Continuous BSO using the operator from Differential Evolution refines the position of fuzzy terms to achieve a better description of the subject domain. The results of the experiments showed that BSO contributes to obtaining compact classifiers because it effectively reduces the dimensionality of the data. BSO allowed obtaining comparable accuracy to the well-known counterparts on a smaller number of rules. The compactness of the resulting models will reduce the time to compute the output, as well as achieve better interpretability.

Acknowledgements. This research was funded by Ministry of Science and Higher Education of the Russian Federation, project number FEWM-2020–0042 (AAAA-A20–120111190016-9).

References

1. Fernandez, A., Herrera, F., Cordon, O., Jesus, M.J., Marcelloni, F.: Evolutionary fuzzy systems for explainable artificial intelligence: Why, When, What for, and Where to? IEEE Comput. Intell. Mag. **14**(1), 69–81 (2019). https://doi.org/10.1109/MCI.2018.2881645
2. Svetlakov, M.O., Hodashinsky, I.A.: Clustering-based rule generation methods for fuzzy classifier using Autonomous Data Partitioning algorithm. J. Phys. Conf. Ser. **1989**(1), 012032 (2021). https://doi.org/10.1088/1742-6596/1989/1/012032

3. Hodashinsky, I., Sarin, K., Shelupanov, A., Slezkin, A.: Feature selection based on swallow swarm optimization for fuzzy classification. Symmetry **11**(11), 1423 (2019). https://doi.org/10.3390/sym11111423
4. Lavygina, A., Hodashinsky, I.: Hybrid algorithm for fuzzy model parameter estimation based on genetic algorithm and derivative based methods. In: ECTA 2011 FCTA 2011 - Proceedings of the International Conference on Evolutionary Computation Theory and Applications and International Conference on Fuzzy Computation Theory and Applications, pp. 513–515 (2011). https://doi.org/10.13140/2.1.2994.6881
5. Xue, Y., Zhang, Q., Zhao, Y.: An improved brain storm optimization algorithm with new solution generation strategies for classification. Eng. Appl. Artif. Intell. **110**, 104677 (2022). https://doi.org/10.1016/j.engappai.2022.104677
6. Cai, Z., Gao, S., Yang, X., Yang, G., Cheng, S., Shi, Y.: Alternate search pattern-based brain storm optimization. Knowl. Based Syst. **238**, 107896 (2022). https://doi.org/10.1016/j.knosys.2021.107896
7. Wolpert, D., Macready, W.: No free lunch theorems for optimization. IEEE Trans. Evol. Comput. **1**(1), 67–82 (1997)
8. Shi, Y.: An optimization algorithm based on brainstorming process. Int. J. Swarm Intell. Res. **2**(4), 35–62 (2011). https://doi.org/10.4018/978-1-4666-6328-2.ch001
9. Cheng, S., Shi, Y. (eds.): Brain Storm Optimization Algorithms. ALO, vol. 23. Springer, Cham (2019). https://doi.org/10.1007/978-3-030-15070-9
10. Cheng, S., et al.: Comprehensive survey of brain storm optimization algorithms. In: IEEE Congress on Evolutionary Computation, San Sebastian, 17013779. IEEE (2017). https://doi.org/10.1109/CEC.2017.7969498
11. Cheng, S., Qin, Q., Chen, J., Shi, Y.: Brain storm optimization algorithm: a review. Artif. Intell. Rev. **46**(4), 445–458 (2016). https://doi.org/10.1007/s10462-016-9471-0
12. Cheng, S., Shi, Y.: Thematic issue on "Brain Storm Optimization Algorithms." Memetic Computing **10**(4), 351–352 (2018). https://doi.org/10.1007/s12293-018-0276-3
13. Yan, X., Zhu, Z., Wu, Q., Gong, W., Wang, L.: Elastic parameter inversion problem based on brain storm optimization algorithm. Memetic Comput. **11**(2), 143–153 (2018). https://doi.org/10.1007/s12293-018-0259-4
14. Xiong, G., Shi, D.: Hybrid biogeography-based optimization with brain storm optimization for non-convex dynamic economic dispatch with valve-point effects. Energy **157**, 424–435 (2018). https://doi.org/10.1016/J.ENERGY.2018.05.180
15. Zhang, W.-Q., Zhang, Y., Peng, C.: Brain storm optimization for feature selection using new individual clustering and updating mechanism. Appl. Intell. **49**(12), 4294–4302 (2019). https://doi.org/10.1007/s10489-019-01513-5
16. Papa, J.P., Rosa, G.H., Souza, A.N., Afonso, L.C.S.: Feature selection through binary brain storm optimization. Comput. Electr. Eng. **72**, 468–481 (2018). https://doi.org/10.1016/j.compeleceng.2018.10.013
17. Xiong, G., Shi, D., Zhang, J., Zhang, Y.: A binary coded brain storm optimization for fault section diagnosis of power systems. Electr. Power Syst. Res. **163**, 441–451 (2018). https://doi.org/10.1016/J.EPSR.2018.07.009
18. Chandrasekar, R., Khare, N.: BGFS: design and development of brain genetic fuzzy system for data classification. J. Intell. Syst. **27**(2), 231–247 (2018). https://doi.org/10.1515/jisys-2016-0034
19. Chandrasekar, R., Khare, N.: BSFS: design and development of exponential brain storm fuzzy system for data classification. Int. J. Uncertain. Fuzz. **25**(2), 267–284 (2017). https://doi.org/10.1142/S0218488517500106
20. Hancer, E., Bardamova, M., Hodashinsky, I., Sarin, K., Slezkin, A., Svetlakov, M.: Binary PSO variants for feature selection in handwritten signature authentication. Informatica (2022). https://doi.org/10.15388/21-INFOR472

21. Cao, Z., Hei, X., Wang, L., Shi, Y., Rong, X.: An improved brain storm optimization with differential evolution strategy for applications of ANNs. Math. Probl. Eng. **2015**, 1–18 (2015). https://doi.org/10.1155/2015/923698
22. Knowledge Extraction based on Evolutionary Learning. https://sci2s.ugr.es/keel/category.php?cat=clas/. Accessed 10 Nov 2021
23. Fazzolari, F., Alcalá, R., Herrera, F.: A multi-objective evolutionary method for learning granularities based on fuzzy discretization to improve the accuracy-complexity trade-off of fuzzy rule-based classification systems: D-MOFARC algorithm. Appl. Soft Comput. **24**, 470–481 (2014). https://doi.org/10.1016/j.asoc.2014.07.019
24. SVC 2004: First International Signature Verification Competition. http://www.cse.ust.hk/svc2004/. Accessed 10 Nov 2021

Quantum-Behaved Simple Brain Storm Optimization with Simplex Search

Xi Wang[1], Wei Chen[1(✉)], Qunfeng Liu[1], Yingying Cao[1], Shi Cheng[2], and Yanmin Yang[3]

[1] Dongguan University of Technology, University Road, Dongguan 523808, China
2016025@dgut.edu.cn

[2] Shaanxi Normal University, Chang An Avenue, Xi'an 710119, China

[3] Kunming University of Science and Technology, Jingming South Road, Kunming 650500, China

Abstract. The simple brain storm optimization (SimBSO) algorithm is an adjusted algorithm to simplify the process of clustering in brain storm optimization algorithm (BSO). However, SimBSO has not significantly improved the optimization performance of BSO except for its simple algorithm structure. In this paper, a new algorithm named quantum-behaved simple brain storm optimization with simplex search (QSimplex-SimBSO) is proposed to improve the performance of SimBSO. In QSimplex-SimBSO, the quantum behavior is added into SimBSO to strengthen global searching capability and then the Nelder-Mead Simplex (NMS) method is used to enhance local searching capability. After large number of experiments on the Hedar set, the results show that QSimplex-SimBSO gets a better balance of global exploration and local exploitation by the visualizing confidence interval method. Meanwhile, QSimplex-BSO is shown to be able to eliminate the degenerated L-curve phenomenon on unimodal functions.

Keywords: Simple brain storm optimization · Nelder-Mead Simplex method · Global exploration · Local exploitation · Quantum behavior

1 Introduction

Brain storm optimization (BSO) algorithm [1], as a new swarm intelligence optimization method, has become an important method for solving global optimization problems in science and engineering. It has attracted many practical applications [2–4]. Meanwhile, BSO has also attracted more and more theoretical analysis such as [5–7] and so on.

This work was supported by the Basic and Applied Basic Research Funding Program of Guangdong Province (Grant No. 2019A1515111097), Yunnan Provincial Research Foundation for Basic Research, China (Grant No. 202001AU070041) and Guangdong Universities' Special Projects in Key Fields of Natural Science (No.2019KZDZX1005).

Y. Tan et al. (Eds.): ICSI 2022, LNCS 13344, pp. 404–415, 2022.
https://doi.org/10.1007/978-3-031-09677-8_34

With the discovery of BSO, an important progress of BSO is to transform operations in the solution space to the objective space [8]. The new version of BSO is easier in implementation and lower in computational resources on the clustering strategy at each iteration. In this paper, we refer BSO to BSO in objective space [8], unless otherwise stated.

Besides BSO, the simple brain storm optimization (SimBSO) algorithm [9] is an adjusted algorithm to simplify the process of clustering in brain storm optimization algorithm (BSO). In SimBSO, the strategy of clustering is not important and SimBSO proposed a new method to generate individuals. SimBSO has simplified the BSO and makes it easier to implement. However, SimBSO has not significantly improved the optimization performance of BSO except for its simple algorithm structure.

Inspired by the quantum mechanism [10], we know that the quantum state has the properties of superposition, entanglement and uncertainty. So, integrating the idea of quantum state into the intelligent algorithm can enrich the diversity of population, thus can improve the precocity of intelligent algorithm and make the global searching ability of the algorithm better [11–16]. On the other hand, the Nelder-Mead Simplex (NMS) algorithm [17] is an effective method to improve local search ability such as [18–21]. Therefore, in this paper, we consider introducing quantum behavior and NMS into SimBSO to enhance the performance of SimBSO.

The remainder of this paper is organized as follows. There is an introduction about the related works in Sect. 2. Then the quantum-behaved simple brain storm optimization with simplex search (QSimplex-SimBSO) is developed and analyzed in Sect. 3. In Sect. 4, large number of experimental results are shown. Finally, conclusions are shown in Sect. 5.

2 Related Works

2.1 SimBSO: Simple BSO

In order to simplify the process of BSO, the simple brain storm optimization (SimBSO) algorithm [9] was proposed in 2017. A new choosing strategy is proposed in SimBSO, which choose three different individuals from population randomly. Then SimBSO linearly combine the three individuals to generate a new individual as Eq. (1).

$$I_{n+1} = \frac{a}{a+b+c}A_n + \frac{b}{a+b+c}B_n + \frac{c}{a+b+c}C_n, \tag{1}$$

where A_n, B_n and C_n are randomly choosen in the population; a, b, c are the random numbers in $(0, 1)$. Finally, the authors add white noise to every individual, then compare the fitness of new individuals and old individuals to produce a new generation of population. The process of SimBSO can be shown in Algorithm 1 and more details can be found in [9].

The SimBSO has proposed a new way to simplify the process of BSO and its optimization performance is better than BSO. But, in limited cost of calculation, the solved problems rate of SimBSO is not significantly higher than BSO.

Algorithm 1: Simple BSO (SimBSO).

1 **Initialization**: generate the orient population randomly and calculate their fitness values; **while** *stopping conditions do not hold* **do**
2   **Select**: select three different individuals from the orient population;
3   **New individual generation**: generate a child by combining them linearly;
4   **Disruption**: create a new individual by adding white noise to the child. Record the new individual if it is better than the current individual;
5   **Update**: update the whole population.
6 **end**

2.2 NMS: Simplex Search Method

Nelder-Mead Simplex search method (NMS) was proposed by J.A. Nelder and R. Mead (1965) [17]. The NMS is a local search method designed for unconstrained optimization problems without using gradient information. The NMS try to produce a local optimal individual by a base individual and three operations (reflection, contraction and expansion). The NMS could save much cost in local exploitation and was widely used in different swarm algorithms like CTSS [22], NM-PSO [23] and so on. The more details of NMS can be found in [18,19]. In this paper, we will introduce the NMS to SimBSO to enhance the local search ability.

2.3 Quantum Behavior

In quantum theory, quantum is the smallest and indivisible basic unit in physics. Because of the superposition, entanglement and uncertainty of quantum states, quantum information processing is more effective than classical information processing methods. Therefore, the introduction of quantum behavior into intelligent algorithm can increase the diversity of population and prevent the premature phenomenon of intelligent algorithm, so that the search ability of the algorithm can become better.

In QBSO [11] and QPSO [13], a new method is proposed to get better global exploration. The method tries to simulate the quantum behavior in quantum theory and adds the solution of *Schrödinger* equation to individuals. The quantum behavior had improved the global exploration in NQAFSA [14], QIA [15], QACO [16] and so on.

In 1920s, an Austrian physicist Schrödinger proposed an equation of wave function to describe the change law of microscopic particle state. After the transformation in [13], the solution of *Schrödinger* equation is shown as Eq. (2).

$$x = p \pm \frac{L}{2} ln(\frac{1}{u}), \tag{2}$$

where u is random number in the range of (0,1), L is the characteristic length of Delta potential well, x is the measured position, p is the relative position to position x. We will consider to introduce the quantum behavior into SimBSO by the Eq. (2) to improve the global search ability.

3 Quantum-Behaved Simple Brain Storm Optimization with Simplex Search

In this part, the process of QSimplex-SimBSO is shown. We show the generation of intermediate individuals in the first part, and then show the generation of new individuals in the second part. In the third part, the process of quantum-behaved simple brain storm optimization with simplex search is shown.

3.1 Generation of Intermediate Individuals

In SimBSO, the new individuals are generated by the linearly combine of three different individuals. To get better global exploration, the method of linearly combine is improved in QSimplex-SimBSO. The new method of linearly combine in QSimplex-SimBSO is shown as Eq. (3).

$$I^{ij} = p_k^1 \times A_k^{ij} + p_k^2 \times B_k^{ij} + (1 - p_k^1 - p_k^2) \times C_k^{ij}, \tag{3}$$

where A_k^{ij}, B_k^{ij} and C_k^{ij} are random individuals from k-th population, I^{ij} is a new intermediate individual obtained from the fusion of three random individuals, i represents the i-th individual, j represents the j-th dimension of the individual, p_k^1 and p_k^2 are random numbers and $p_k^1, p_k^2 \in (0, 1)$.

3.2 Generation of New Individuals

After quantum behavior is added to the intermediate individual, the new individual is generated. The generation of every individual can be shown as Eqs. (4) and (5).

$$x_{new}^{ij} = \begin{cases} I^{ij} + b \times |\frac{A_k^{ij} + B_k^{ij} + C_k^{ij}}{3} - A_k^{ij}| \times ln(\frac{1}{u}), (r < 0.5) \\ I^{ij} - b \times |\frac{A_k^{ij} + B_k^{ij} + C_k^{ij}}{3} - A_k^{ij}| \times ln(\frac{1}{u}), (r \geq 0.5), \end{cases} \tag{4}$$

$$b = 0.5 \times (1 - 0.5 \times \frac{current\_iteration\_number}{max\_iteration}), \tag{5}$$

where $current\_iteration\_number$ is the number of current iteration, $max\_iteration$ is the number of max iterations, u is a random number in $(0, 1)$, r is a random number in $[0, 1]$. I^{ij}, A_k^{ij}, B_k^{ij} and C_k^{ij} are the individuals in Eq. (3).

3.3 Quantum-Behaved Simple Brain Storm Optimization with Simplex Search

The principle of the QSimplex-SimBSO is shown as follows:

Firstly, we initialize and generate a random population with n individuals. Then, for every original individual, three individuals are randomly selected from the population and combine them linearly by the new method. Secondly, we add quantum behavior to every intermediate individual to generate a new population.

If the new individual in new population don't jump out of the search field and the fitness of the new individual is better than the orient individual, we replace the original individual by the new individual. After that, the best individual in the new population is chosen to generate a better individual by the NMS algorithm. Finally, we get a new generation of individuals.

The difference between Eqs. (3) and (1) is that the new intermediate individuals may jump out from the search field when $p_k^1 + p_k^2 > 1$. And in many results of different experiments, the quantum behavior may cause the new individuals jumping out of the search field. Those phenomena show that the method of QSimplex-SimBSO can get better global exploration. If a new individual has jumped out from the search field, give up it.

The process of QSimplex-SimBSO is shown in the Algorithm 2.

Algorithm 2: Quantum-Behaved Simple Brain Storm Optimization with Simplex Search (QSimplex-SimBSO).

---

1 **Initialization**: generate the initial population randomly;
2 **while** *stopping conditions do not hold* **do**
3 | **Combine**: choose three individuals and combine them linearly to generate an intermediate individual ;
4 | **Disruption**: add the quantum behavior to every dimension of the intermediate individual, if the new individual jump out of the search field, give up it;
5 | **New population generation**: for every individual, record the new child if it is better than the original individual, then update the whole population and identify the best individual x_0;
6 | **The NMS algorithm**: exploit the search area around x_0 through executing the NMS algorithm. Replace x_0 by the found best point x_0' in new population;
7 | **Update**: update the whole population.
8 **end**

---

4 Main Experimental Results

The following algorithms were all implemented with MATLAB R2014a. And the program executed on a PC with an Intel Core (TM), CPU i5-3470 and 8 GB RAM. Each of those was independently executed 50 times on each of 68 Hedar benchmark functions. The maximum calculation cost was set to 20000.

All the tests in this paper are based on the Hedar set, so there is a simple introduction of the Hedar set in the first part. In the second part, our algorithm QSimplex-SimBSO is compared numerically with the Simplex-BSO, SimBSO and BSO. After many tests, an additional parameter, 40n, is introduced in QSimplex-SimBSO to get better balance between global exploration and local exploitation. The parameters of those algorithms are shown as Table 1.

In the third part, the L-curves of four different functions are shown, which tested by QSimplex-SimBSO, SimBSO, Simplex-BSO and BSO. In the fourth

Table 1. Parameter settings of the four algorithms used in this paper.

Algorithms	Parameters
BSO	slope = 25
SimBSO	slope = 25
Simplex-BSO	slope = 25, NMS_cost=40n(n is the dimension of problem)
QSimplex-SimBSO	NMS_cost=40n (n is the dimension of problem)

part, the QSimplex-SimBSO is compared numerically with the other comparative algorithms to show the global searching ability of quantum behavior in QSimplex-SimBSO.

4.1 Hedar Set

In order to test the performance of global optimization methods, Dr. Abdel-Rahman Hedar presented the Hedar set to test the global optimization of different algorithms. The Hedar set is based on the Jones set, but the Hedar set extend the dimensions in some different functions. The details can be shown as Table 2.

4.2 Process Comparison Between QSimplex-SimBSO and Others on the Hedar Set

To compare the results of QSimplex-SimBSO and other algorithms, the visualizing confidence intervals (VCI) method in [24] is used. In [24], the VCI method is shown to be convenient for benchmarking stochastic global optimization algorithms, especially when the set of benchmark functions or the number of algorithms is large. In the VCI method, the H_{upper} means the confidence upper bound matrix and the H_{lower} means the confidence lower bound matrix, and then H_{upper} and H_{lower} are analyzed statistically with the data profile technique. In this paper, we refer the "QSimplex-SimBSO-Upper" to the H_{upper} of QSimplex-SimBSO and refer "Simplex-BSO-lower" to the H_{lower} of Simplex-BSO. "BSO-lower" and "SimBSO-lower" have the similar meanings. In the figure of data profile, the vertical axis shows the proportion of solved problems and the horizontal axis represents the relative computational cost. We can see more details about the VCI method in [19,24].

Table 2. Information about the Hedar test set.

Function	Dimension n	Characteristic	Search region	Minimal function value
Beale	2	Unimodal	$[-4.5, 4.5]^2$	0
Matyas	2	Unimodal	$[-8, 12.5]^2$	0
Sphere	2,5,10,20	Unimodal	$[-4.1, 6.4]^n$	0
Sum squares	2,5,10,20	Unimodal	$[-8, 12.5]^n$	0
Trid	6	Unimodal	$[-36, 36]^6$	-50
Trid	10	Unimodal	$[-100, 100]^{10}$	-200
Zakharov	2,5,10,20	Unimodal	$[-5, 10]^n$	0
Ackley	2,5,10,20	Multimodal	$[-15, 30]^n$	0
Bohachevsky 1	2	Multimodal	$[-80, 125]^2$	0
Bohachevsky 2	2	Multimodal	$[-80, 125]^2$	0
Bohachevsky 3	2	Multimodal	$[-80, 125]^2$	0
Booth	2	Multimodal	$[-100, 100]^2$	0
Branin	2	Multimodal	$[-5, 10] * [0, 15]$	0.397887357729739
Colville	4	Multimodal	$[-10, 10]^4$	0
Dixson Price	2,5,10,20	Multimodal	$[-10, 10]^n$	0
Easom	2	Multimodal	$[-100, 100]^2$	-1
Goldstein and Price	2	Multimodal	$[-2, 2]^2$	3
Griewank	2,5,10,20	Multimodal	$[-480, 750]^n$	0
Hartman 3	3	Multimodal	$[0, 1]^3$	-3.86278214782076
Hartman 6	6	Multimodal	$[0, 1]^6$	-3.32236801141551
Hump	2	Multimodal	$[-5, 5]^2$	0
Levy	2,5,10,20	Multimodal	$[-10, 10]^n$	0
Michalewics	2	Multimodal	$[0, \pi]^2$	-1.80130341008983
Michalewics	5	Multimodal	$[0, \pi]^5$	-4.687658179
Michalewics	10	Multimodal	$[0, \pi]^{10}$	-9.66015
Perm	4	Multimodal	$[-4, 4]^4$	0
Powell	4,12,24,48	Multimodal	$[-4, 5]^n$	0
Power sum	4	Multimodal	$[0, 4]^4$	0
Rastrigin	2,5,10,20	Multimodal	$[-4.1, 6.4]^n$	0
Rosenbrock	2,5,10,20	Multimodal	$[-5, 10]^n$	0
Schwefel	2,5,10,20	Multimodal	$[-500, 500]^n$	0
Shekel 5	4	Multimodal	$[0, 10]^4$	-10.1531996790582
Shekel 7	4	Multimodal	$[0, 10]^4$	-10.4029405668187
Schkel 10	4	Multimodal	$[0, 10]^4$	-10.5364098166920
Shubert	2	Multimodal	$[-10, 10]^2$	-186.730908831024

Firstly, we compare the average behaviors of QSimplex-SimBSO, Simplex-BSO, SimBSO and BSO to determine a winner algorithm. Secondly, the winner's H_{upper} is compared with the other's H_{lower} to confirm whether the winner still performs the best at the worst case. If so, then the conclusion is significant statistically that the winner performs better than the other algorithms. Otherwise, the conclusion is that the winner performs averagely better than the other algorithms.

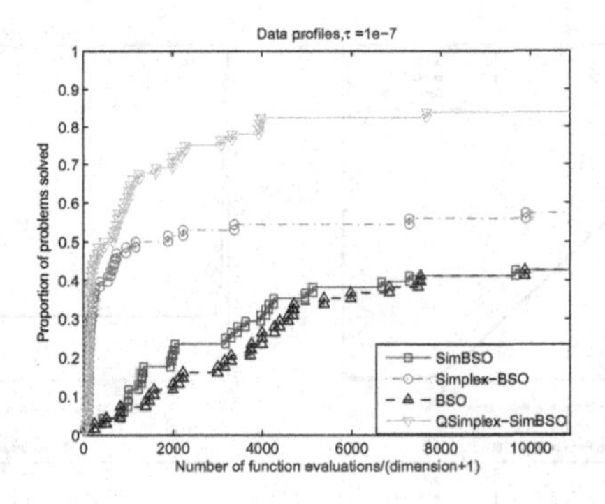

Fig. 1. Data profile average results when comparing QSimplex-SimBSO, Simplex-BSO, SimBSO and BSO on 68 functions in the Hedar set.

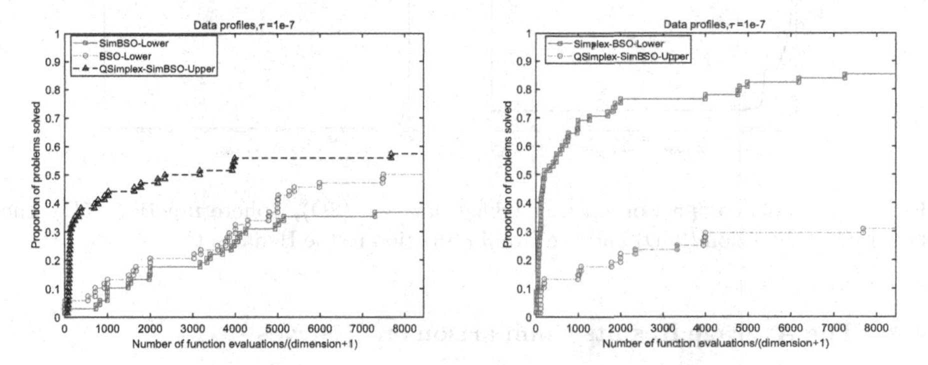

Fig. 2. Data profiles resulted from comparing SimBSO's confidence lower bounds, BSO's confidence lower bounds, Simplex-BSO's confidence lower bounds and QSimplex-SimBSO's confidence upper bounds.

Figure 1 shows QSimplex-SimBSO can solve more functions by less cost. We see that QSimplex-SimBSO solves 57 (=68 * 84%) functions, Simplex-BSO solves 39 (=68 * 59%) functions, SimBSO solves 30 (=68 * 42%) functions and BSO solves 30 (=68 * 42%) functions. Hence our algorithm QSimplex-SimBSO gets an averagely better performance than Simplex-BSO, SimBSO and BSO.

From the Fig. 2, the H_{upper} of QSimplex-SimBSO is then compared with H_{lower} of Simplex-BSO, H_{lower} of BSO and H_{lower} of SimBSO. The result confirms that the winner QSimplex-SimBSO performs better than SimBSO and BSO at the worst case. According to the VCI method, the conclusion is that QSimplex-SimBSO performs better than BSO and SimBSO. But the result also shows QSimplex-SimBSO only gets a better performance than Simplex-BSO on average.

Through adopting quantum behavior and NMS to SimBSO, our proposed algorithm QSimplex-SimBSO gets a better performance.

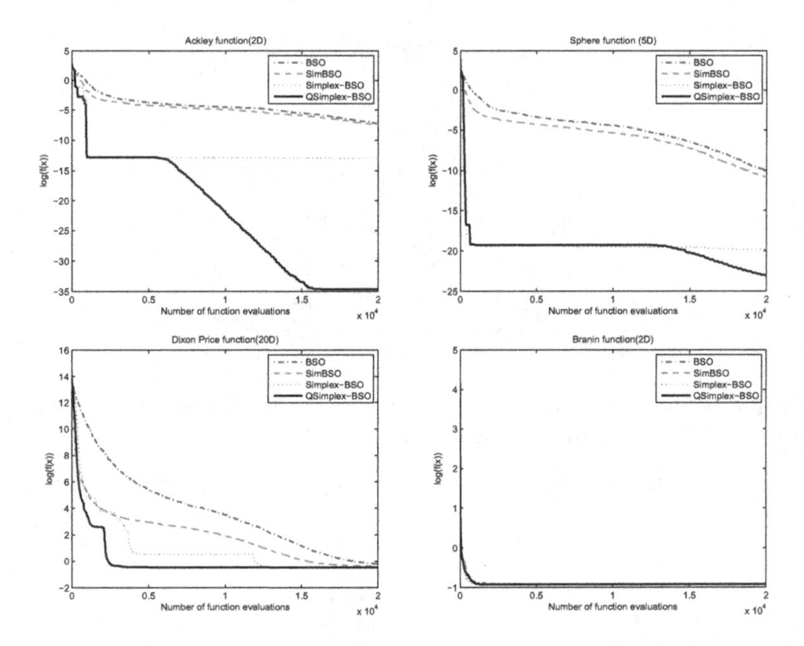

Fig. 3. L-curves comparison on the Ackley function (2D), Sphere function (5D), the Dixon Price function (20D) and the Branin function in the Hedar set.

4.3 The Average Results Comparison on L-curves

In the Fig. 3, the L-curves of four different functions are shown.

For Ackley function(2D), QSimplex-SimBSO and Simplex-BSO get a solution with accuracy of e^{-15} at a cost of less than 1000 function evolutions, BSO and SimBSO get the solution with accuracy of e^{-7} when all evolutions are consumed. At the end of function evolutions, the QSimplex-SimBSO gets the solution with accuracy of e^{-35}. We can see that QSimplex-SimBSO and Simplex-BSO have the faster convergence rate than BSO and SimBSO. Meanwhile the global optimization capability is enhanced significantly.

For Sphere function(5D), the QSimplex-SimBSO and Simplex-BSO get the solution with accuracy of e^{-20} at a cost of less than 1000 function evolutions, BSO and SimBSO get the solution with accuracy of e^{-10} when all evolutions are consumed. And at the end of function evolutions, the QSimplex-SimBSO gets the solution with accuracy of e^{-22}. The L-curves of 28 functions in Hedar set are similar as Ackley function(2D) or Sphere function(5D).

For Dixon Price function(20D), QSimplex-SimBSO gets the solution with accuracy of e^{0} at a cost of less than 3000 function evolutions, Simplex-BSO gets the solution with accuracy of e^{0} at a cost of less than 12000 function evolutions, SimBSO and BSO get the solution with accuracy of e^{0} when all evolutions are consumed. The L-curves of 19 functions in Hedar set are similar as Dixon Price function(20D).

For Branin function, the L-curves of QSimplex-SimBSO, BSO, SimBSO and Simplex-BSO are similar. QSimplex-SimBSO, Simplex-BSO, SimBSO and BSO both find a solution with accuracy of e^{-1} at a cost of less than 1000 function evolutions. The L-curves of 11 functions in Hedar set are similar as Branin function.

Among all 68 benchmark functions, there are 58 functions whose curves are similar as those in the subfigures of Fig. 3. However, there still 10 functions in Hedar set on which QSimplex-SimBSO is outperformed by the others. Therefore, we can only conclude that QSimplex-SimBSO is more efficiency to solve the most problems in Hedar set.

4.4 A Comparison Strategy to Identify the Global Exploration of Quantum Behavior in QSimplex-SimBSO

In SimBSO, the white noise is a method to get a better global exploration and is added to the intermediate individual after the linear combination of individuals. The quantum behavior in QSimplex-SimBSO is a new method to get a better global exploration and is added after the new intermediate individual generated. So, the ability of white noise is overlapped with the ability of quantum behavior. To identify the ability of global exploration between quantum behavior and white noise in QSimplex-SimBSO, three comparative algorithms based on QSimplex-SimBSO are designed.

- Comparative algorithm 1: don't add any disruption after every intermediate individual generated in QSimplex-SimBSO and then search locally by the NMS.
- Comparative algorithm 2: every new individual is generated by adding white noise in the intermediate individual of QSimplex-SimBSO and then search locally by the NMS.
- Comparative algorithm 3: every new individual is generated by adding white noise and quantum behavior in the intermediate individual of QSimplex-SimBSO and then search locally by the NMS.

The comparison between QSimplex-SimBSO and the other comparative algorithms could identify how the different methods of disruption affect the performance of QSimplex-SimBSO. The data profile result is shown in Fig. 4.

In the Fig. 4, our algorithm QSimplex-SimBSO gets a better performance than the other algorithms and solves 60(=68 * 88%) functions in limited cost. Otherwise, Comparative algorithm 2 and Comparative algorithm 3 get a better performance than Comparative algorithm 1. So, both quantum behavior and white noise could get a better global exploration. Meanwhile, the quantum behavior could get a better performance in QSimplex-SimBSO than the other methods.

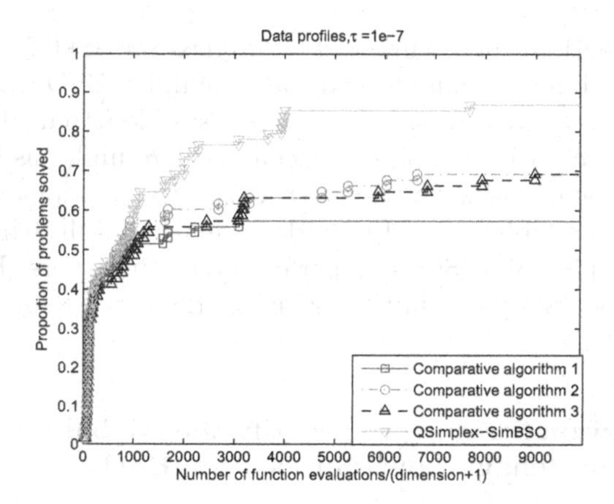

Fig. 4. Data profile average results when comparing the other comparative algorithms and QSimplex-SimBSO on 68 functions in the Hedar set.

5 Conclusions

In this paper, to get a better balance between global exploration and local exploitation of SimBSO, the quantum behavior is adopted into SimBSO and then try to search the best individual by NMS. Meanwhile, the QSimplex-SimBSO is also shown to be a simple algorithm without clustering cost, and can get a better efficiency. From the comparison of the QSimplex-SimBSO and other comparative algorithms, we find that the quantum behavior could get a better global exploration. These conclusions reveal that the global optimization ability of quantum behavior is better than that of white noise.

References

1. Shi, Y.: Brain storm optimization algorithm. In: Tan, Y., Shi, Y., Chai, Y., Wang, G. (eds.) ICSI 2011. LNCS, vol. 6728, pp. 303–309. Springer, Heidelberg (2011). https://doi.org/10.1007/978-3-642-21515-5_36
2. Jiang, Y., Chen, X., Zheng, F., Niyato, D., You, X.: Brain storm optimization-based edge caching in fog radio access networks. IEEE Trans. Veh. Technol. **70**(2), 1807–1820 (2021)
3. Duan, H., Li, S., Shi, Y.: Predator-prey brain storm optimization for DC brushless motor. IEEE Trans. Magn. **49**(10), 5336–5340 (2013)
4. Ma, X., Jin, Y., Dong, Q.: A generalized dynamic fuzzy neural network based on singular spectrum analysis optimized by brain storm optimization for short-term wind speed forecasting. Appl. Soft Comput. **54**, 296–312 (2017)
5. Zhu, H., Shi, Y.: Brain storm optimization algorithms with k-medians clustering algorithms. In: 2015 Seventh International Conference on Advanced Computational Intelligence (ICACI), pp. 107–110. IEEE, Chiang Mai (2015)
6. Zhan, Z., Zhang, J., Shi, Y., Liu, H.: A modified brain storm optimization. In: 2012 IEEE Congress on Evolutionary Computation, pp. 1–8. IEEE, Brisbane (2012)

7. Song, Z., Peng, J., Li, C., Liu, P.X.: A simple brain storm optimization algorithm with a periodic quantum learning strategy. IEEE Access **6**(19), 19968–19983 (2017)
8. Shi, Y.: Brain storm optimization algorithm in objective space. In: Congress on Evolutionary Computation (CEC). IEEE, Sendai (2015)
9. Cao, Y., et al.: A simple brain storm optimization algorithm via visualizing confidence intervals. Simul. Evol. Learn. 27–38 (2017)
10. Greiner, W.: Quantum mechanics. An introduction. 4th edn. J. Phys. Am. (2001)
11. Duan, H., Cong, L.: Quantum-behaved brain storm optimization approach to solving Loney's solenoid problem. IEEE Trans. Magn. **51**(1Pt.2), 7000,307-1–7000,307-7 (2015)
12. Narayanan, A., Moore, M.: Quantum-inspired genetic algorithms. In: Proceedings of IEEE International Conference on Evolutionary Computation, pp. 61–66. IEEE, Nagoya (1996)
13. Sun, J., Feng, B., Xu, W.: Particle swarm optimization with particles having quantum behavior. In: Proceedings of the 2004 Congress on Evolutionary Computation, pp. 325–331. IEEE, Portland (2004)
14. Zhu, K., Jiang, M., Cheng, Y.: Niche artificial fish swarm algorithm based on quantum theory. In: IEEE 10th International Conference on Signal Processing Proceedings, pp. 1425–1428. IEEE, Beijing (2010)
15. Niu, Q., Zhou, T., Shiwei, M.: A quantum-inspired immune algorithm for hybrid flow shop with makespan criterion. J. Univ. Comput. Sci. **15**, 765–785 (2009)
16. Wang, L., Niu, Q., Fei, M.: A novel quantum ant colony optimization algorithm. Bio-Inspired Computat. Intell. Appl. **4688**, 277–286 (2007)
17. Nelder, J.A., Mead, R.: A simplex method for function minimization. Comput. J. **7**, 308–313(1965)
18. Fan, S.-K.S., Zahara, E.: A hybrid simplex search and particle swarm optimization for unconstrained optimization. Eur. J. Oper. Res. **181**(2), 527–548 (2007)
19. Chen, W., Cao, Y., Cheng, S., Sun, Y., Liu, Q., Li, Y.: Simplex search-based brain storm optimization. IEEE Access **6**(75), 75997–76006 (2018)
20. Chelouah, R., Siarry, P.: Genetic and Nelder-Mead algorithms hybridized for a more accurate global optimization of continuous multiminima functions. Eur. J. Oper. Res. **148**(2), 335–348 (2003)
21. Lin, H.: Hybridizing differential evolution and Nelder-Mead simplex algorithm for global optimization. In: 2016 12th International Conference on Computational Intelligence and Security (CIS), pp. 198–202. IEEE, Wuxi (2016)
22. Chelouah, R., Siarry, P.: A hybrid method combining continuous Tabu search and Nelder-Mead simplex algorithms for the global optimization of multiminima functions. Eur. J. Oper. Res. **161**(3), 636–654 (2005)
23. Dasril, Y. B., Wen, G.K.: Modified artificial bees colony algorithm with Nelder-Mead search algorithm. In: 2016 12th International Conference on Mathematics, Statistics, and Their Applications, pp.25–30. IEEE, Hatyai, Songkhla (2017)
24. Liu, Q., et al.: Benchmarking stochastic algorithms for global optimization problems by visualizing confidence intervals. IEEE Trans. Cybern. **47**(9), 1–14 (2017)

Swarm Intelligence Approach-Based Applications

Network Community Detection via an Improved Swarm Intelligence Approach

Wei-Hsiang Sun[1] and Frederick Kin Hing Phoa[2(✉)]

[1] National Chengchi University, Taipei, Taiwan
107305008@nccu.edu.tw
[2] Institute of Statistical Science Academia Sinica, Taipei, Taiwan
fredphoa@stat.sinica.edu.tw

Abstract. This paper proposes a new nature-inspired metaheuristic optimization method for community detection. This method is improved from the Swarm Intelligence Based method proposed in Phoa (2017) in which the particle size varies during the optimization procedure. Additional changes on the particle generation is applied for the community detection problem. Simulation studies show that the proposed method outperforms all common metaheuristic methods in detecting communities, and its performance is comparable to several traditional start-of-the-art methods under some conditions. This method is applied to detect communities in six real networks.

Keywords: Community detection · Swarm intelligence · Nature-inspired metaheuristic optimisation

1 Introduction

Networks have been everywhere in our daily lives given its ability to represent relationship ties and interactions among actors. In biology, networks are used to analyze food webs within ecosystems [17], to reveal modular structures in protein-to-protein and genetic interactions [5,6]. In a society, networks are used to describe interactions between individuals, such as scientific collaborations [25] and connections on social media [14].

Communities, which are densely connected clusters that reveal important structural information of a graph, are usually of great interest. In general, members within the same community play similar roles in the network or have common properties [8]. Communities in the neuron interaction network allow us to understand functional architecture of the brain [7], whereas communities in

This work is partially supported by the Academia Sinica grant number AS-TP-109-M07 and the Ministry of Science and Technology (Taiwan) grant numbers 107-2118-M-001-011-MY3.

Y. Tan et al. (Eds.): ICSI 2022, LNCS 13344, pp. 419–431, 2022.
https://doi.org/10.1007/978-3-031-09677-8_35

social networks allow us to better understand and analyze human behavior in social science [10].

Many techniques have been developed to detect communities. Label propagation algorithm (LPA) propagates initializes every node with unique labels and let the labels propagate through the network until the termination criterion is reached [32]. [9] summarizes the extensions of LPA to directed or weighted networks and its hybrid. The Louvain algorithm iteratively moves nodes and merges community to attain maximum modularity [4], such process thus unfolds the hierarchical structure and gives access to different resolution limits. However, Louvain algorithm do not ensure graph connectivity, thus the Leiden algorithm [36] added a refinement process before aggregation of nodes to resolve the issue. Some methods adopt a more probabilistic approach, like the stochastic block model, which is a generative model with inherent community structure. Communities are discovered mainly through the assignment of community labels that yield the maximum probability [1]. Some approaches are based on information theory. A widely-adopted algorithm called the Infomap [33] detects communities by the estimation of the theoretical lower bound of the code length of a random walker given a partition, which a shorter code length implies a better partition. Other information theoretic methods can be found in [23].

2 Recent Development of Metaheuristic Methods for Community Detection

Nature-inspired metaheuristic optimization recently receives great attention due to the technology advances in parallel computing. Although the optimized solution is approximate, its efficiency still leads to great favors among practical users. Many metaheuristic approaches are ready to be implemented for community detection problems. Some classical algorithms include the Particle Swarm optimization, Genetic Algorithm, Simulated Annealing, Ant Colony Optimization, and Tabu Search.

Particle swarm optimization (PSO) was initially proposed as a continuous optimization approach [16]. [40] extended the particle swarm to discrete optimization to suit the community detection task and developed strategies to repair isolated nodes. [22] proposed a community detection algorithm that reduces the search space and increases search efficiency of multi-objective PSO. Some extensions to find overlapping communities are referred to [18].

[30] utilized the genetic algorithm and proposed the "community-score", which is a metric based on density of the community, as the fitness function to detect community. A multiobjective extension on the inter-community connections was proposed in [31]. Some other works in applying genetic algorithm to community detection are found in [3,11,20,34] and many others.

[12] applied simulated annealing (SA) to find partition with largest modularity. [21] extended SA by moving the current solution via the k-means iterative procedure. Some combine the SA with other methods, like [24,35] and [15] added

SA as a local search procedure to stabilize the genetic algorithm, memetic algorithm and ant-colony optimization respectively.

There are many other metaheuristic optimization being used in community detection, like the ant colony optimization [13], lion optimization [2], symbolic organism search [39], whale optimization [42], Parliamentary algorithm [26], social spider optimization [19], fire propagation algorithm [27], and many others.

Most existing metaheuristic methods aim at maximizing modularity, which is known to suffer resolution limits. Also, many disregard node connectivity, which communities that are internally disconnected. Here, we propose a detection algorithm based on the swarm intelligence based (SIB) optimization, which ensures the statistical significance of the detected clusters, and the connectivity of the members.

3 The Swarm Intelligence Based (SIB) Method

3.1 Preliminaries and Objectives

We first define some notations of a network. Let $G = (V, E)$ be a graph, where V is the set of all nodes in the graph, and E is the set of all links in the graph. In this paper, we focus on non-overlapping community detection in undirected and unweighted networks for simplicity. For a graph, we recall some basic properties of the Exponential Random Graph Model (ERGM). An ERGM is a statistical model for the ties in a network G. In general, It has the form

$$lClP(G) = \frac{e^{H(G)}}{Z}, \tag{1}$$

where $Z = \sum_{G=\mathcal{G}} e^{H(G)}$ is a normalized constant and $H(\cdot)$ is a graph Hamiltonian given by

$$lClH(G) = \sum_{i=1}^{r} \theta_i X_i(G), \tag{2}$$

where $\{X_i(G)\}$ are the values of the observables x_i for G and $\theta = \{\theta, \dots, \theta_r\}$ are the respective parameters. Since the ERGM follows an exponential-family distribution, the statistics $\{X_i(G)\}$ are complete and sufficient for θ. The expectation and variance of X_j are respectively, $E(X_j) = \frac{\partial \ln Z}{\partial \theta_j}$, and $V(X_j) = \frac{\partial^2 \ln Z}{\partial \theta_j^2}$

There are two major concerns in identifying the most influential community in a network: (1) how to define the most influential community, and (2) how to find the elements of the most influential community. To address the first concern, we determine the importance via the ERGM with communities. Consider a network G that is divided into k disjointed and independent parts, i.e. $G = G_0, G_1, \dots, G_k$, where G_0 is the part that does not belong to any other parts. The full likelihood of the partitioned ERGM is

$$lClL(\boldsymbol{\theta} | \{G_0, G_1, \dots, G_k\}) = \prod_{i=0}^{k} p_i^{s_i} (1 - p_i)^{N_i - s_i}, \tag{3}$$

where $p_i = \frac{\exp(\theta_i)}{1+\exp(\theta_i)}$. When the partitions are given, the maximum likelihood is obtained by replacing the p_i's with their maximum likelihood estimators (MLEs),

$$lClL(\hat{\boldsymbol{\theta}}|\{G_0, G_1, \ldots, G_k\}) = \prod_{i=0}^{k} \hat{p_i}^{s_i}(1-\hat{p_i})^{N_i - s_i}, \tag{4}$$

where $\hat{p_i} = \frac{S_i}{N_i}$. This measure is exactly the same as the measure provided in [26]. When we consider the model with only one community, the likelihood is reduced to

$$lClL(\boldsymbol{p}|\boldsymbol{s}, \boldsymbol{z}) = p_c^{s_c}(1-p_c)^{N_c - s_c} p_u^{s_u}(1-p_u)^{N_u - s_u} p_b^{s_b}(1-p_b)^{N_b - s_b}, \tag{5}$$

where $\boldsymbol{z}$ is an $n \times 1$ group membership vector with elements taking on binary values such that $z_i = 1$ when the node i is selected and $z_i = 0$ otherwise. $s|z = [s_c, s_u, s_b]$ is the observed numbers of edges among the selected and unselected nodes (these edges were commonly called "betweenness"), and $N = [N_c, N_u, N_b]$ are all possible numbers of edges with the similar definition as s. The target function is then recast as

$$lClf = \hat{p_c}^{s_c}(1-\hat{p_c})^{N_c - s_c} \hat{p_u}^{s_u}(1-\hat{p_u})^{N_u - s_u} \hat{p_b}^{s_b}(1-\hat{p_b})^{N_b - s_b} \tag{6}$$

We have formally defined the objective function for the detection task. In the following subsections, we first introduce the standard form of the proposed SIB algorithm, then elaborate on its improvements and details how it finds the optimal solution to the detection problem.

3.2 The Standard Framework of the SIB Method

The Swarm Intelligence Based (SIB) method was first proposed in [28]. Contrary to many nature-inspired metaheuristics algorithms like PSO, it intends to solve discrete optimization problems that are common in mathematics and statistics [29].

Prior to the initialization step, users are required to enter several parameters and information including (1) at least one set of stopping criteria, (2) N: the swarm size, (3) q_{LB}: the number of exchanges with the local best (LB) particle and (4) q_{GB}: the number of exchanges with the global best (GB) particle. Then N initial particles are first generated from a random pool of particle units and an objective function value of each particle is calculated. The LB and GB particles are defined accordingly.

The MIX operation is a unit exchange procedure. For each particle during an iteration, some of its units are exchanged with those of its LB particle and the GB particle respectively, resulting in two mixture candidates. More explicitly, assume that there are k units in a particle. These k units are ranked according to their contributions to the objective function value and q units with the least contributions are removed from the current particle. In return, k units with the most contributions are added from the best particle, assuring the invariant of the particle size. The order of unit addition and unit deletion is exchangeable.

The MOVE operation is a decision making procedure. After the MIX operation is complete, three candidates, including the original particle, the mixture with the LB particle and the mixture with the GB particle, are compared based on their objective function values. If one of the mixtures is the best among three, the particle is updated to the best mixture. If the original particle is still the best, q units of the original particle is substituted by q units randomly selected from the random pool of particle units. This procedure helps to avoid the particle being trapped in a local attractor.

After the MOVE operation, all particles are updated to their best candidates, so the new LB and the GB particles need to be refreshed. The iteration repeats if the set of stopping criteria is not fulfilled, or the whole method is completed otherwise.

3.3 An Improvement to Allow Particle Size Change

The standard framework of the SIB method fixes the size of the particles during the optimization process, but many problems require the particle size to be flexible in order to achieve optimum. Thus, we propose an improvement on this manner by introducing a new operation called VARY. It consists of two procedures: Unit Extension and Unit Shortening. Let X be a particle with m units, q_e and q_s be the number of units being added to and deleted from X respectively.

In the unit extension, q_e units are added to X. These added units may come from another particle or a random unit from the pool of particle units. Let Y be a better particle (either LB or GB) than X with k units, and y_i is the ith unit from Y, $i = 1, \ldots, k$. Then k extension particles $[X; y_i]$ are constructed, and the objective function values of those particles are calculated. Let y_d be the best unit among all possible choices, i.e. $[X; y_d]$ possesses the best objective function value. If this value is better than that of X, then we update X to $[X; y_d]$. If the objective function value of $[X; y_d]$ fails to improve from that of X, then random units from the random unit pool are added to X instead. The unit extension procedure continues until q_e units are added to the particle X and the size of the particle becomes $m + q_e$. An opposite procedure to reduce units are done similarly in the unit shortening, resulting in a new particle with size $m - q_s$.

We summarize the steps of the improved SIB method below.

Algorithm: An Improved SIB Method
1: Randomly generate a set of initial particles
2: Evaluate objective function value of each particle
3: Initialize the local best (LB) for each particle
4: Initialize the global best (GB) among all LBs
5: FOR EACH particle DO:
6: Perform the MIX and MOVE operations
7: IF the original particle is the best in MOVE THEN:
8: Perform the VARY and MOVE operations

9: IF the original particle is the best in MOVE THEN:
10: Perform the random jump
11: Evaluate the objective function value
12: Update the LB
13: Update the GB
14: Repeat the iteration (5–12) until converged.

Note that the VARY operation is performed under a condition that the MIX operation fails to improve the particle's objective function value. The logic of this hierarchical setup is to optimize a particle within a given size first, followed by improving the particle via changing the size. The standard SIB method has similar pseudo-code without Lines 8–9.

3.4 SIB for Community Detection

Under the context of community detection, a particle is a subgraph of a network. To generate a particle in the initialization step, we randomly select a node v_i as the center and a length of the shortest path r as the radius. This represents a subset of nodes that are within shortest-path distance r from the center v_i. Such generator had been verified as a good selection of possible community [37,38].

In the MIX and VARY operations, instead of allowing all nodes within the particle to be altered, we first identify the nodes on the boundary of the subgraph and they are the only nodes being altered. Note that these boundary nodes are distance r from v_i only in the first iteration, and the distances of these boundary nodes vary after some improvements are performed. Thus, an efficient identification on these boundary nodes is essential when a particle is given. When the algorithm is completed, the GB particle indicates all nodes being detected as a member of community from the network.

4 Simulation

We present the results of different community detection methods performed on different benchmarks and datasets. Here the methods are abbreviated as follows, Ant Colony Optimization as ACO, Particle Swarm Optimization as PSO, Genetic Algorithm as GA, Tabu Search as TABU, Simulated Annealing as SA, InfoMap as IM, semi-synchronous Label Propagation Algorithm as LPAsemi, asynchronous Label Propagation Algorithm as LPAasy, Greedy Modularity as GreedyMod, Leiden Algorithm as Leiden, Leading Eigen Vector as LeadingEV, Louvain Algorithm as Multilevel, Walktrap Algorithm as Walktrap, Standard Stochastic Block Model as SSBM, Degree Corrected Stochastic Block Model as DCSBM.

The ordered pairs in the first column of the tables are the parameters used for graph generation of that particular instance. Due to page limits, we only show the simulation results on two communities with Poisson assumption. The first column of the table indicates the sizes of two clusters and the entries are the intra-connection probability. Additional results on (1) one community with Poisson assumption; and (2) two LFR benchmark experiments are per request if needed, and will be shown in extended version or future works of this paper.

4.1 Comparison Metrics

Since multiple ground-truth communities exist, the accuracy of the partition are measured using the Cross Common Fraction (CCF) and the Jaccard index [38]. The Cross Common Fraction compares each pair of community that came from the detected partition $C^* = \{C_1^*, \ldots, C_s^*\}$, and the real partition $C = \{C_1, \ldots, C_t\}$, where s is the number of total communities detected by the community detection methods, t is the true number of communities. $CCF = 1/2 \sum_i^s \max_j |C_i^* \cap C_j| + 1/2 \sum_j^t \max_i |C_i^* \cap C_j|$. We slightly altered the definition of Jaccard index by considering all possible pairs of nodes, rather than the elements of each community only, to check if they were assigned to the same community. Let $C_{pair}^* = \{(v_i, v_j) : \delta(v_i, v_j) = 1\}$, and $C_{pair} = \{(v_i, v_j) : \xi(v_i, v_j) = 1\}$ be the sets of pairs that were assigned to the same communities with respect to the detected communities and the real ones. δ and ξ were the indicator functions that were 1 if v_i and v_j had the same label. Thus, the Jaccard index was defined as $J(C^*, C) = \frac{|C_{pair}^* \cap C_{pair}|}{|C_{pair}^* \cup C_{pair}|}$.

4.2 Results

Here we use the mean of Jaccard index and the mean of Common Cross Fraction over 10 runs to show the accuracy of the detection methods. Our proposed method performs better than other methods in both the Jaccard index and the Common Cross Fraction when the cluster size is more unbalanced, and when the community connection probability is smaller. It can be especially seen in the (25–475, 0.25) instance, where the SIB only misidentified one node while other methods lack the accuracy. Here the result of LPAasy, LPAsemi and IM tend to cluster the more unbalanced community with lower intraconnection probability into one large cluster, although the situation has improved compared to previous simulation. The results are shown in Table 1.

Table 1. Results for two communities with Poisson assumption

Jaccard index	SIB	ACO	GA	SA	PSO	TABU	IM	LPAsemi	LPAasy	GreedyMod	Leiden	LEV	Louvain	Walktrap	SSBM	DCSBM
(25-475,1.00)	0.9956	0.1735	0.7196	0.7111	0.7088	0.5627	1.0000	0.9048	0.9048	0.7925	1.0000	0.7925	0.9399	1.0000	0.7096	0.4747
(25-475,0.75)	0.7692	0.1576	0.4643	0.4115	0.4419	0.5444	0.9048	0.9048	0.9048	0.4882	1.0000	0.3300	0.1978	1.0000	0.6646	0.4746
(25-475,0.50)	0.8381	0.0581	0.3893	0.4123	0.3465	0.5171	0.9048	0.9048	0.9048	0.4209	0.1915	0.3571	0.1736	1.0000	0.5860	0.4749
(25-475,0.25)	0.9956	0.0148	0.3059	0.3032	0.3072	0.5327	0.9048	0.9048	0.9048	0.3551	0.0151	0.3611	0.1333	0.1662	0.4917	0.4768
(50-450,1.00)	0.9951	0.1716	0.7389	0.5631	0.6190	0.4715	1.0000	0.8196	0.8918	0.9854	1.0000	0.9903	0.9951	1.0000	0.7378	0.4499
(50-450,0.75)	0.9070	0.1367	0.6113	0.5471	0.5247	0.5261	1.0000	0.8196	0.8557	0.5364	1.0000	0.4910	0.2436	1.0000	0.7078	0.4510
(50-450,0.50)	0.9903	0.0619	0.4029	0.4184	0.3270	0.5711	0.8196	0.8196	0.8196	0.4046	0.1990	0.4279	0.2093	1.0000	0.7005	0.4506
(50-450,0.25)	0.9809	0.0262	0.3219	0.3369	0.3203	0.4810	0.8196	0.8196	0.8196	0.3662	0.0154	0.2428	0.1622	0.9951	0.5603	0.4502
(100-400,1.00)	0.9336	0.1547	0.6558	0.5714	0.7137	0.4478	1.0000	0.6794	0.9359	1.0000	1.0000	1.0000	1.0000	1.0000	0.6746	0.4044
(100-400,0.75)	0.9325	0.1368	0.6055	0.4778	0.5867	0.4701	1.0000	1.0000	0.9038	1.0000	1.0000	1.0000	1.0000	1.0000	0.6837	0.4048
(100-400,0.50)	0.9346	0.0767	0.4796	0.4912	0.4713	0.4717	1.0000	0.6794	0.8397	1.0000	0.1926	0.5280	0.5532	1.0000	0.6568	0.4044
(100-400,0.25)	0.9056	0.0218	0.2446	0.2837	0.3084	0.4207	0.6794	0.6794	0.6794	0.6114	0.0145	0.5006	0.2721	1.0000	0.5840	0.4048
(250-250,1.00)	0.9610	0.1106	0.6089	0.5238	0.6104	0.4216	1.0000	1.0000	1.0000	1.0000	1.0000	1.0000	1.0000	1.0000	0.3538	0.3618
(250-250,0.75)	0.8660	0.1005	0.5209	0.4799	0.5550	0.3380	1.0000	1.0000	1.0000	1.0000	1.0000	1.0000	1.0000	1.0000	0.3687	0.3964
(250-250,0.50)	0.4182	0.0594	0.4101	0.3862	0.4491	0.3317	1.0000	1.0000	1.0000	1.0000	0.1914	1.0000	1.0000	1.0000	0.4086	0.4124
(250-250,0.25)	0.5000	0.0349	0.3040	0.2983	0.3229	0.3371	1.0000	0.4990	0.7495	1.0000	0.0182	1.0000	1.0000	1.0000	0.3503	0.3550

Common cross fraction	SIB	ACO	GA	SA	PSO	TABU	IM	LPAsemi	LPAasy	GreedyMod	Leiden	LEV	Louvain	Walktrap	SSBM	DCSBM
(25-475,1.00)	499	298	448	449	447	410	500	488	488	461	500	461	486	500	447	367
(25-475,0.75)	458	293	381	371	379	409	488	488	488	386	500	338	316	500	438	367
(25-475,0.50)	467	270	366	375	346	404	488	488	488	368	318	362	312	500	421	367
(25-475,0.25)	499	252	342	345	341	411	488	488	488	347	249	363	287	309	386	372
(50-450,1.00)	499	294	447	407	426	392	500	475	485	497	500	498	499	500	442	355
(50-450,0.75)	480	288	415	404	395	402	500	475	480	416	500	386	341	500	437	357
(50-450,0.50)	498	269	364	369	340	418	475	475	475	366	323	363	330	500	436	356
(50-450,0.25)	496	253	340	344	343	395	475	475	475	360	248	306	319	499	407	356
(100-400,1.00)	488	291	424	404	439	372	500	450	490	500	500	500	500	500	426	330
(100-400,0.75)	488	285	410	381	410	385	500	500	485	500	500	500	500	500	429	332
(100-400,0.50)	488	267	371	376	370	381	500	450	475	500	320	400	422	500	421	330
(100-400,0.25)	482	243	301	309	322	363	450	450	450	442	236	398	357	500	400	332
(250-250,1.00)	495	284	430	405	428	358	500	500	500	500	500	500	500	500	271	283
(250-250,0.75)	481	280	403	390	414	320	500	500	500	500	500	500	500	500	293	311
(250-250,0.50)	344	261	353	349	377	307	500	500	500	500	321	500	500	500	314	319
(250-250,0.25)	376	240	304	299	315	305	500	375	438	500	237	500	500	500	277	283

4.3 Real Networks

We compared the triad participation ratio of the community detection methods that have yielded relatively competitive results. The triad participation ratio is used to determine the quality of a single cluster. The metric is known to be a good indicator for finding functional communities [41]. Here we formally define the triad participation ratio. Let $G(V, E)$ with an undirected graph with $n = |V|$ nodes and $m = E$ edges. Let S be a set of nodes, where $n_s = |S|$; m_S the number of edges in S, $m_S = |\{(u, v) \in E : u \in S, v \in S\}|$; c_s, the number of edges on the boundary of S, $c_s = |\{(u, v) \in E : u \in S, v \notin S\}|$; and $d(u)$ is the degree of node u. Then the triad participation ratio (TPR) is defined as

$$TPR(S) = \frac{|\{u : u \in S, \{(v, w) : v, w \in S, (u, v) \in E, (u, w) \in E, (v, w) \in E\} \neq \emptyset\}|}{n_S}$$

Here we use a total of six real-life dataset. The total number of nodes ranges from 34 to 36692. For larger networks with total number of nodes larger than 1000, our proposed method outperforms other methods in terms of triad participation ratio, and is also competitive in terms of smaller datasets, as shown in Table 2.

Table 2. Triad participation ratio for two communities with power law assumption

	SIB	IM	LPAsemi	LPAasy	GreedyMod	Leiden	LEV	Louvain	Walktrap
Karateclub	0.9701	1.0000	0.9333	0.9671	1.0000	-	0.9167	0.9524	1.0000
Dolphins	0.7143	0.7500	0.8333	0.8021	0.7391	-	0.7059	0.7389	0.8000
Polbooks	0.9783	1.0000	1.0000	1.0000	1.0000	1.0000	1.0000	1.0000	1.0000
Protein	0.2348	0.1931	-	0.1942	0.2231	-	0.2180	0.1869	0.2214
Powergrid	0.6628	0.3818	-	-	0.0985	-	0.1335	0.1018	0.4552
Enron	0.9702	0.7324	0.6656	0.6678	0.7167	-	0.6762	0.7844	0.7662

Figure 1 shows the detection result of two datasets with ground truth communities using SIB, the polbooks dataset and the karate club dataset. The results are quite promising with the F_1 score between the detected community and the officer's true community in the karate club being 0.9714 and the F_1 score between the detected community and the conservative cluster in the polbooks network being 0.8842. We observe that most misidentified community nodes have similar percentage of intra-community edges and inter-community edges, while misidentified noncommunity nodes often have a higher proportion of links connected to the community it does not belong to.

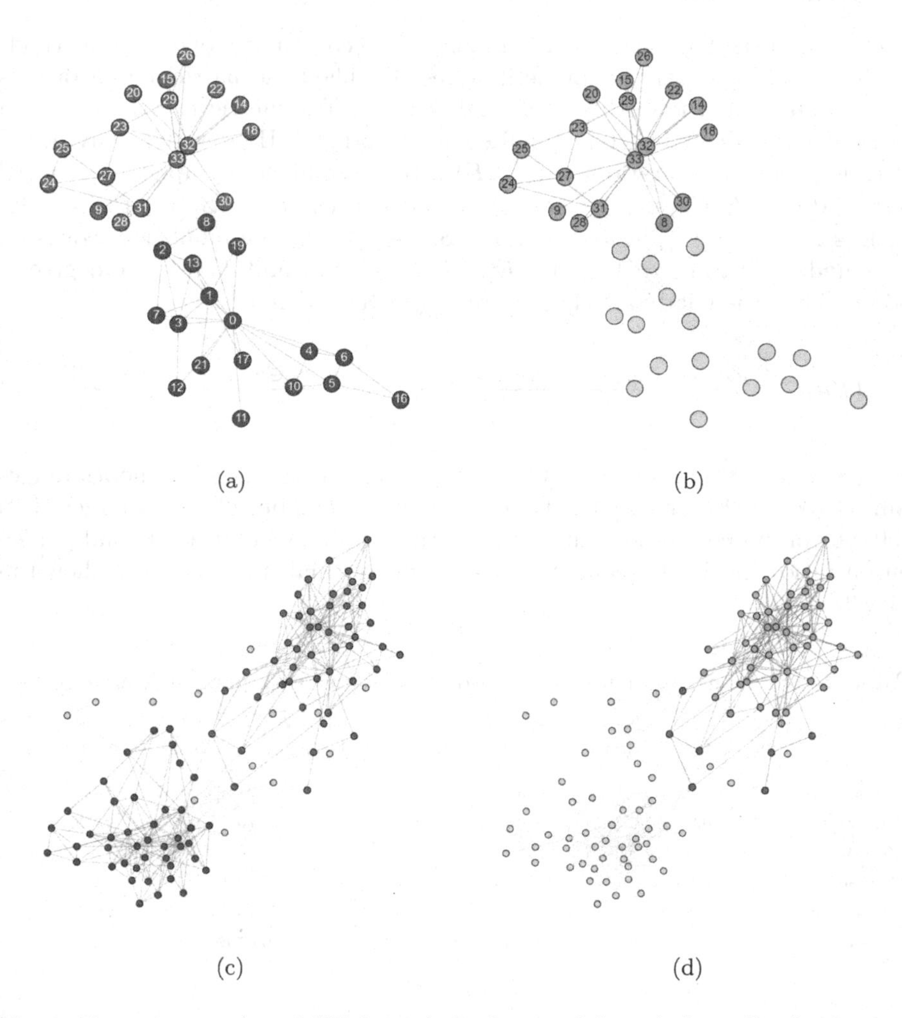

Fig. 1. Detection results of SIB in real the karate club and the polbooks datasets. (a) and (b) are respectively the ground-truth communities of the karate club network and its detection result of SIB. The two colors in (a) represent the ground-truth memberships of each node, while the colored nodes in (b) represent the detection result with highest occurrence over 10 runs using the SIB method. In particular, in (b), the orange nodes stand for correctly-detected nodes, and green nodes represent non-community nodes being incorrectly detected as community members. As shown in (b), only one node on the boundary is misidentified. Likewise, (c) and (d) are respectively the ground-truth communities of the polbooks network and its detection result of SIB. In (b), orange nodes represent nodes that are correctly detected, while red nodes are community members that SIB fails to identify as community members, and the ones in green are noncommunity nodes being incorrectly detected as community members. (Color figure online)

5 Conclusion

The main contribution of this paper is that we propose a nature-inspired metaheuristic optimization method improved from the Swarm Intelligence Based (SIB) method for community detection. With the flexibility to change particle size during the optimization, the improved SIB method outperforms all common metaheuristic methods in the simulation studies, and the performance is comparable to several traditional state-of-the-art methods. The applications to real networks ensure that our proposed method is ready for practical uses.

In specific, our method performs well for networks with more than 1000 nodes. One may argue that the largest network (Enron email network) consists of only 36692 nodes, which is still small when compared to large-scale networks of million nodes, but a clustering of large-scale networks itself is a difficult problem and it is out of the scope of this paper. At least, none of the listed methods can easily handle this task without significant modifications.

The missing LFR benchmark experiment will be shown in extended version in the near future. Nevertheless, the results show that our SIB method is the favorite in certain network conditions (cluster size is more unbalanced and community connection probability is small), and the performance is still stable and good when compared to other methods. Another potential extension from this paper is the consideration of additional complications in real networks, including the direct and weighted edges, specific network structure, and many others. As the computational burden further increases in these complications, parallel computing should be employed for computational feasibility.

References

1. Abbe, E.: Community detection and stochastic block models: recent developments. J. Mach. Learn. Res. **18**(1), 6446–6531 (2017)
2. Babers, R., Hassanien, A.E., Ghali, N.I.: A nature-inspired metaheuristic lion optimization algorithm for community detection. In: 2015 11th International Computer Engineering Conference (ICENCO), pp. 217–222. IEEE (2015)
3. Bello-Orgaz, G., Salcedo-Sanz, S., Camacho, D.: A multi-objective genetic algorithm for overlapping community detection based on edge encoding. Inf. Sci. **462**, 290–314 (2018)
4. Blondel, V.D., Guillaume, J.L., Lambiotte, R., Lefebvre, E.: Fast unfolding of communities in large networks. J. Stat. Mech. Theory Exp. **2008**(10), P10008 (2008)
5. Chen, J., Yuan, B.: Detecting functional modules in the yeast protein-protein interaction network. Bioinformatics **22**(18), 2283–2290 (2006)
6. Cline, M.S., et al.: Integration of biological networks and gene expression data using cytoscape. Nat. Protoc. **2**(10), 2366 (2007)
7. Deco, G., Corbetta, M.: The dynamical balance of the brain at rest. Neuroscientist **17**(1), 107–123 (2011)
8. Fortunato, S.: Community detection in graphs. Phys. Rep. **486**(3–5), 75–174 (2010)
9. Garza, S.E., Schaeffer, S.E.: Community detection with the label propagation algorithm: a survey. Physica A **534**, 122058 (2019)

10. Goldenberg, A., Zheng, A.X., Fienberg, S.E., Airoldi, E.M.: A survey of statistical network models (2010)
11. Guerrero, M., Montoya, F.G., Baños, R., Alcayde, A., Gil, C.: Adaptive community detection in complex networks using genetic algorithms. Neurocomputing **266**, 101–113 (2017)
12. Guimera, R., Amaral, L.A.N.: Functional cartography of complex metabolic networks. Nature **433**(7028), 895–900 (2005)
13. Honghao, C., Zuren, F., Zhigang, R.: Community detection using ant colony optimization. In: 2013 IEEE Congress on Evolutionary Computation, pp. 3072–3078. IEEE (2013)
14. Huberman, B.A., Romero, D.M., Wu, F.: Social networks that matter: Twitter under the microscope. arXiv preprint arXiv:0812.1045 (2008)
15. Ji, P., Zhang, S., Zhou, Z.P.: A decomposition-based ant colony optimization algorithm for the multi-objective community detection. J. Ambient. Intell. Humaniz. Comput. **11**(1), 173–188 (2019). https://doi.org/10.1007/s12652-019-01241-1
16. Kennedy, J., Eberhart, R.: Particle swarm optimization. In: Proceedings of ICNN 1995-International Conference on Neural Networks, vol. 4, pp. 1942–1948. IEEE (1995)
17. Krause, A.E., Frank, K.A., Mason, D.M., Ulanowicz, R.E., Taylor, W.W.: Compartments revealed in food-web structure. Nature **426**(6964), 282–285 (2003)
18. Li, Y., Wang, Y., Chen, J., Jiao, L., Shang, R.: Overlapping community detection through an improved multi-objective quantum-behaved particle swarm optimization. J. Heurist. **21**(4), 549–575 (2015). https://doi.org/10.1007/s10732-015-9289-y
19. Li, Y.H., Wang, J.Q., Wang, X.J., Zhao, Y.L., Lu, X.H., Liu, D.L.: Community detection based on differential evolution using social spider optimization. Symmetry **9**(9), 183 (2017)
20. Li, Z., Liu, J.: A multi-agent genetic algorithm for community detection in complex networks. Physica A **449**, 336–347 (2016)
21. Liu, J., Liu, T.: Detecting community structure in complex networks using simulated annealing with k-means algorithms. Physica A **389**(11), 2300–2309 (2010)
22. Liu, X., Du, Y., Jiang, M., Zeng, X.: Multiobjective particle swarm optimization based on network embedding for complex network community detection. IEEE Trans. Comput. Soc. Syst. **7**(2), 437–449 (2020)
23. Mittal, R., Bhatia, M.: Classification and comparative evaluation of community detection algorithms. Arch. Comput. Methods Eng. 1–12 (2020)
24. Mu, C.H., Xie, J., Liu, Y., Chen, F., Liu, Y., Jiao, L.C.: Memetic algorithm with simulated annealing strategy and tightness greedy optimization for community detection in networks. Appl. Soft Comput. **34**, 485–501 (2015)
25. Newman, M.E.: The structure of scientific collaboration networks. Proc. Natl. Acad. Sci. **98**(2), 404–409 (2001)
26. Ozbay, F.A., Alatas, B.: Discovery of multi-objective overlapping communities within social networks using a socially inspired metaheuristic algorithm. Int. J. Comput. Netw. Appl. **4**(6), 148–158 (2017)
27. Pattanayak, H.S., Sangal, A.L., Verma, H.K.: Community detection in social networks based on fire propagation. Swarm Evol. Comput. **44**, 31–48 (2019)
28. Phoa, F.K.H.: A swarm intelligence based (SIB) method for optimization in designs of experiments. Nat. Comput. **16**(4), 597–605 (2017)
29. Phoa, F.K.H., Chen, R.B., Wang, W., Wong, W.K.: Optimizing two-level supersaturated designs using swarm intelligence techniques. Technometrics **58**(1), 43–49 (2016)

30. Pizzuti, C.: GA-Net: a genetic algorithm for community detection in social networks. In: Rudolph, G., Jansen, T., Beume, N., Lucas, S., Poloni, C. (eds.) PPSN 2008. LNCS, vol. 5199, pp. 1081–1090. Springer, Heidelberg (2008). https://doi.org/10.1007/978-3-540-87700-4_107
31. Pizzuti, C.: A multi-objective genetic algorithm for community detection in networks. In: 2009 21st IEEE International Conference on Tools with Artificial Intelligence, pp. 379–386. IEEE (2009)
32. Raghavan, U.N., Albert, R., Kumara, S.: Near linear time algorithm to detect community structures in large-scale networks. Phys. Rev. E **76**(3), 036106 (2007)
33. Rosvall, M., Bergstrom, C.T.: Maps of random walks on complex networks reveal community structure. Proc. Natl. Acad. Sci. **105**(4), 1118–1123 (2008)
34. Said, A., Abbasi, R.A., Maqbool, O., Daud, A., Aljohani, N.R.: CC-GA: a clustering coefficient based genetic algorithm for detecting communities in social networks. Appl. Soft Comput. **63**, 59–70 (2018)
35. Shang, R., Bai, J., Jiao, L., Jin, C.: Community detection based on modularity and an improved genetic algorithm. Physica A **392**(5), 1215–1231 (2013)
36. Traag, V.A., Waltman, L., Van Eck, N.J.: From Louvain to Leiden: guaranteeing well-connected communities. Sci. Rep. **9**(1), 1–12 (2019)
37. Wang, T.C., Phoa, F.K.H.: Focus statistics for testing the degree centrality in social networks. Netw. Sci. **4**, 460–473 (2016)
38. Wang, T.C., Phoa, F.K.H.: A scanning method for detecting clustering pattern of both attribute and structure in social networks. Physica A **445**, 295–309 (2016)
39. Xiao, J., Wang, C., Xu, X.K.: Community detection based on symbiotic organisms search and neighborhood information. IEEE Trans. Comput. Soc. Syst. **6**(6), 1257–1272 (2019)
40. Xiaodong, D., Cunrui, W., Xiangdong, L., Yanping, L.: Web community detection model using particle swarm optimization. In: 2008 IEEE Congress on Evolutionary Computation (IEEE World Congress on Computational Intelligence), pp. 1074–1079. IEEE (2008)
41. Yang, J., Leskovec, J.: Defining and evaluating network communities based on ground-truth. Knowl. Inf. Syst. **42**(1), 181–213 (2013). https://doi.org/10.1007/s10115-013-0693-z
42. Zhang, Y., et al.: WOCDA: a whale optimization based community detection algorithm. Physica A **539**, 122937 (2020)

Decentralized Supply Chain Optimization via Swarm Intelligence

Karan Singh[1], Hsin-Ping Liu[2], Frederick Kin Hing Phoa[3(✉)], Shau-Ping Lin[4], and Yun-Heh Jessica Chen-Burger[1]

[1] Department of Computer Sciences, Heriot-Watt University, Edinburgh, UK
yjc32@hw.ac.uk
[2] Data Science Degree Program, National Taiwan University, Taipei, Taiwan
[3] Institute of Statistical Science, Academia Sinica, Taipei, Taiwan
fredphoa@stat.sinica.edu.tw
[4] Institute of Biotechnology, National Taiwan University, Taipei, Taiwan
shaupinglin@ntu.edu.tw

Abstract. When optimised, supply chains can bring tremendous benefits to all its participants. Supply chains therefore can be framed as a networked optimization problem to which swarm intelligence techniques can be applied. Given recent trends of globalization and e-commerce, we propose a supply chain that uses an open e-commerce business model, where all participants have equal access to the market and are free to trade with each other based on mutually agreed prices and quantities. Based on this model, we improve upon the Particle Swarm Optimization algorithm with constriction coefficient (CPSO), and we demonstrate the use of a new random jump algorithm for consistent and efficient handling of constraint violations. We also develop a new metric called the 'improvement multiplier' for comparing the performance of an algorithm when applied to a problem with different configurations.

Keywords: Swarm intelligence · Supply chain · Particle Swarm Optimization · E-commerce · Networked optimization problem

1 Introduction

In traditional supply chain models, suppliers typically sell their products to dealers or wholesalers who then supply these products to customers. Globalization and e-commerce are enabling traditional businesses to access new markets and customers via open information that was not possible previously.

Leveraging internet technologies and open access trading networks, businesses can form more robust supply chains with suppliers and traders competing in open markets. This has the potential to revolutionise and disrupt current

This work is partially supported by the Academia Sinica grant number AS-TP-109-M07 and the Ministry of Science and Technology (Taiwan) grant numbers 107-2118-M-001-011-MY3 and 109-2321-B-001-013.

Y. Tan et al. (Eds.): ICSI 2022, LNCS 13344, pp. 432–441, 2022.
https://doi.org/10.1007/978-3-031-09677-8_36

business practices. Such a model enables optimized utilisation of resources while achieving the best efficiency of supply chains.

In this paper, we propose a business model based on such an open access market to form suitable supply chains. In addition, we use swarm intelligence to enable companies in this business model to optimize the supply chain for profit maximisation.

In our previous work, we successfully optimized a traditional Taiwanese supply chain where sellers and buyers go through a middleman for trading purposes, using three different swarm intelligence algorithms [1]. In this paper, we propose a model of a decentralized supply chain; where suppliers supply products directly to customers (the direct sales model) in a many-to-many and fully connected supply chain network.

We optimize this supply chain network using the Particle Swarm Optimization algorithm with constriction coefficients (CPSO) and then improve upon its performance using our newly devised algorithms (CPSO_rj). The goal is to seek the optimized combination for the entire supply chain. In other words, we will seek the highest profit for the whole supply chain based on the different combinations of suppliers/buyers/trading products/purchase quantities and transport used. We also experiment with many different supply chain configurations to avoid bias towards any one particular configuration.

This decentralized and open supply chain should be more resilient than the traditional supply chains with fixed suppliers and buyers. The algorithms will find the most profitable combination of products that can be traded between suppliers and customers even when facing disruptions such as production shortages or unexpected decrease/increase in demand.

1.1 Supply Chain Representation

Digraph: Given the interconnected nature of a supply chain, we propose a directed graph or digraph that forms a digital representation of a supply chain. A digraph can be represented as $G(V, E)$, where G is the graph, V is the set of nodes, and E is the set of edges between nodes. Each node represents a company or entity, and the edges represent the directional flow of goods. Quantities, prices, and other edge attributes are represented as edge weights, while variables such as supplier location, customer size and customer location are represented as node attributes. Two classes of nodes are represented: $V = (S_i | i = 1, 2...n)$ represents n number of suppliers, and $V = (C_i | i = 1, 2...m)$ represents m number of customers. We denote an edge as $E_{ij} = (S_i, C_j)$, and it represents a connection between supplier i and customer j with node C_j being node S_i's successor.

1.2 Supply Chain Variables

We consider in this work the minimal and relatively arbitrary setups as described below for demonstration purpose. Our method is capable of handling problems with larger setups, and one can easily extend the setup to meet their needs.

Costs: The total cost associated with delivering products from the suppliers to the customers is the sum product of the quantity supplied and the cost associated procuring and processing the products, plus the sum product of the cost of transporting the products from suppliers to customers. This can be mathematically represented as: $Cost = \sum(transport\ cost * qty) + \sum(procurement\ cost * qty)$. We have taken 2 different location: *North* & *South* to represent the different zones, and we assume that the transport cost is greater if the supplier and the customer are in different geographical locations. We also categorize suppliers into *Small, Medium* & *Large* to differentiate their production capacities, and we assume that smaller suppliers charge more than large suppliers to simulate economies of scale in production and procurement.

Sales: Total sales is calculated by the sum product of quantity of products supplied and the price per product negotiated with the customer: $Sales = \sum(price * qty)$. We have categorised customers into *Small, Medium* & *Large* categories, with larger customers paying less than smaller customers as their demand for products is greater and to simulate the bargaining power of volume based negotiated discounts.

Profit/Loss: The profit of the supply chain is therefore the difference between the total sales and the total costs, which is used as the objective function during optimization: $Profit = Sales - Cost$. A negative profit is commonly known as the loss.

Constraints: There are two constraints that every supply chain must adhere to: (1) the supply side constraint, i.e. customers cannot buy more products than that are produced, and (2) the demand side constraints, i.e. suppliers cannot supply more products than a customer's demand. For the demonstration purpose of this paper, we consider three products *(A, B, D)* and each product has a total packed capacity 6, 10, and 12, respectively. Therefore, the supply side constraint for supplier i is $(a_i \times 6) + (b_i \times 10) + (d_i \times 12)$ where a_i, b_i, d_i are the quantities for products A, B, D supplied by supplier i. Similarly, the demand constraint is an integer value representing the total quantity of each product a customer is willing to purchase.

We assume that customers are willing to purchase the same product in different quantities from multiple suppliers. We also assume that the supply is less than the demand. The objective is to find the combination of products from suppliers to customers that maximizes the profit for the entire supply chain.

For the optimization process to be valid, it needs to meet both supply and demand constraints and should consider the cost difference paid by different categories of customers. It should also consider the difference in costs of transporting products between suppliers and customers in different geographic zones.

2 Swarm Intelligence Optimization

Given the number of different variables in a supply chain, it is infeasible to calculate and compare every possible combination. As the variables in the supply chain increases, so does the solution domain of finding the most profitable combination. Thus, a heuristic can be applied to yield an adequately good solution but not necessarily the best solution within a feasible computational time.

There are some algorithms that mimic the behavior of natural organisms and how these organisms achieve complex tasks by following simple rules [2]. The Particle Swarm Optimization (PSO) algorithm was developed to simulate social behaviors in the animal world and was later modified for optimization of continuous nonlinear functions [3]. Each particle is encoded such that its *position* represents a possible solution with the same dimensions as the problem. The particle is allowed to explore the solution domain for better solutions by adding its *velocity* vector with the same dimensions as its *position*. Later, a 'social-network' or a neighborhood of *informants* was added to delay the spread of information among the particles, in order to avoid local optima and for better exploration of the search space [4]. One such neighborhood is called the Ring neighborhood topology where every particle is connected to two other particles [4]. At every iteration, the particle attempts to improve its personal best *position*. This is done via the *velocity* vector that contains information from the best of its informants which is called the local best. Further, a certain degree of randomness is also added to the *velocity* vector to aid in exploration of the search space. As all particles are connected indirectly via the Ring topology, the information contained in the best possible *position* seen by the entire swarm of particles or the global best *position*, finds its way through the entire swarm. This results in the swarm gradually converging to the global best *position* while exploring the search space.

The PSO algorithm has been used to optimize a multi-echelon automotive supply chain with the objective of minimizing operating costs [5]. It has also been used in the biomass supply chain where the objective function was an optimal amount of biomass to be harvested [6]. A modified version of the PSO algorithm was also used to optimize a petroleum supply chain with cost reduction as the objective function [7].

2.1 PSO with Constriction Coefficient (CPSO)

CPSO is a version of the PSO algorithm with a constriction coefficient which was introduced to control the *velocity* of the particles in high dimensional search spaces [8]. The CPSO algorithm is represented using the following equation [9]:

$$v_i^{t+1} \leftarrow \chi[v_i^t + c_1 * rand(x_i^{pb,\,t} - x_i^t) + c_2 * rand(x_i^{gb,\,t} - x_i^t)] \tag{1}$$

where v_i^{t+1} is the updated velocity, v_i^t is the current velocity, *rand* is a random number between 0 & 1, $x_i^{pb,\,t}$ is particle i's personal best or best *position* visited by particle i, x_i^t is particle i's current *position*, $x_i^{pb,\,t}$ is the local best or the best *position* visited by any of particle i's informants, c_1 and $c_2 = 2.05$ and the constriction coefficient χ was set at 0.7298 as per [9]. After every iteration, the new velocity vector is added to the particle's position using the following equation:

$$x_i^{t+1} \leftarrow v_i^{t+1} + x_i^t \tag{2}$$

where x_i^{t+1} is the updated *position* for particle i. This process continues until the maximum number of iterations is reached.

2.2 Constraint Handling Strategy

When the *velocity* is added to the *position*, the returned value sometimes breaks either supply or demand constraints. To avoid these constraint violations, an extra step needs to be added to the position update process to adjust the movement of particles when they are going to break constraints.

In this section, we introduced two different constraint handling strategies: random back strategy and random back confinement strategy [10]. In addition, we introduced an explorative technique called 'random jump', which will be implemented based on the random back strategy. We will also discuss each of their advantages and disadvantages through examples.

Random Back Strategy. In this strategy, if an entry is going to break the constraints, it will be replaced with a random value that meets both demand and supply constraints. The downside of this method is that for those entries of violations, we may lose the value or information gained through the previous iterations. Moreover, under this strategy, a particle is likely to be stuck in the neighborhood of the current position. Here is an illustrative example.

Consider a three-dimensional particle whose original *position* is $[3, 9, 9]$ and *velocity* is $[5, 0, 1]$, where the demand constraint is $[10, 10, 10]$ and the supply constraint is 23, i.e., the sum of three entries should not surpass 23. Assume the optimal solution is $[10, 3, 10]$.

In the update process, the first entry $[3+5, 9, 9]$ breaks the supply constraint, so we execute a random back strategy. Since the demand for the first entry is 10 and the remaining supply is $23-9-9 = 5$, we choose the random value from 0 to 5, and we pick 4, then the updated *position* becomes $[4, 9, 9]$. The remaining two entries stay within the constraints, so they are updated by the original *velocity*. As the result, we get $[4, 9, 10]$ as the new *position* for the particle. In this case, considering the supply constraint, we have little chance to increase the first entry in the following iterations unless some other entries decrease first. Therefore, the particle gets stuck in the neighborhood of the current *position*.

Random Back Confinement Strategy [10]. Random back confinement strategy is used to make the violating *position* move along the opposite direction of its *velocity*. The *velocity* value in the violating dimension is multiplied by a negative random value between 0 and 1. The advantage of this method is that new value will be closer to original value, thus retaining most of the information from previous iterations. However, the neighborhood trap issue still exists in this method.

Consider the same example in the previous paragraph. For the first entry, since the supply constraint will be violated, we execute a modified random back procedure on it. Assume the negative random number we pick is -0.2, then the updated *velocity* becomes $5 * -0.2 = -1$, and the updated *position* is $[2, 9, 9]$. After updating the other two entries, we get $[2, 9, 10]$ as the new *position*. In this case, it is still hard for the particle to jump out from the neighborhood trap.

Random Jump. While the two methods mentioned above ensure constraints are met, they fail to address the neighborhood trap issue, which requires a value reduction in one dimension in order to allow a value increase in another dimension while meeting constraints. To address this, we propose a new method called 'random jump'. We preset a probability threshold value as a parameter of the method, and we generate random values from 0 to 1 for each dimension when we execute the random jump. If the random value for any dimension is below the threshold value, we replace the *position*'s value in that dimension with a random value that meets demand and supply constraints. For the remaining dimensions, we use the random back strategy. The downside of this strategy is that we need some expertise on the setting of the probability threshold, although the effects may not differ much among any small values. Here we suggest to set it as 0.1.

Consider the same example. We set the probability for random jump as 0.1. When the random jump is executed, we generate three random numbers between 0 to 1 from a random number generator, and we have $(0.6429, 0.0711, 0.4542)$. This result implies a random jump on the second entry. For updating the *position*, we also take the entries into consideration in order. After updating the first entry, according to the random back strategy, we will get [4, 9, 9]. For the second entry, we choose a random valid value as its new *position*. According to the demand 10, and the remaining supply $23 - 4 - 9 = 13$, we can pick an integer from 0 to 10. Assume 4 is chosen; then the *position* becomes [4, 4, 9]. As a result, the particle moves to [4, 4, 10] in this iteration. In this case, we can see that some supply quota is released by the random jump procedure, which makes it more possible for the first entry to be increased in the following iterations. Certainly, there are many other operations that are possible to execute, and our pick is relatively simple but one may choose others for specific needs.

3 Setup and Metric for Method Comparison

In this section, we compare CPSO with three different constraint strategies introduced above. In order to identify which CPSO algorithm returns the combination with the highest profit values, we set up an experiment with 50 separate supply chain configurations. All 50 configurations contain 15 customers, 15 suppliers and 3 products. Each configuration contains different demand and supply constraints and different geographical location for customers and suppliers as well as different combinations of customer and supplier categories. For each configuration, we randomly initialise 20 sets of particles. The Ring neighborhood topology is used to select *informants*. We test the three CPSO algorithms on every particle set, for 500 iterations without stopping early. As a result, we have 1000 optimized solutions, i.e., the global best *positions* at the last iteration, given by each algorithm.

In order to compare solutions from different configurations, we define an improvement multiplier to measure the progress of each algorithm from the start with random initial values to the end of the iterations. The improvement multiplier can be calculated using the following equation:

$$I = 1 + \frac{(profit_f - profit_0)}{profit_0} \tag{3}$$

where $profit_f$ and $profit_0$ are the global best profit values after the optimization process completes and at the initial stage respectively. Since the particles are chosen randomly in the initial stage, $profit_0$ stands for the profit of a random selling strategy among providers and customers. Then the improvement multiplier can be interpreted as the profit multiplying ratio that the profit of the resulting selling strategy obtained from the optimization is elevated from that of a random selling strategy. Then, we are able to fairly compare the optimization results for different configurations using this multiplier.

4 Results and Analysis

In this section, we compare the performance of CPSO using three different constraint handling strategies. The improvement multiplier is the metric for our evaluation. In all tables and figures, 'CPSO' stands for the standard CPSO with random back strategy, 'CPSO_rb' stands for CPSO with random back confinement strategy, and 'CPSO_rj' stands for CPSO with random jump strategy.

Table 1. Optimization result (improvement multiplier)

Algorithms	Mean	Median	Max	Min	StdDev.
CPSO	1.779397	1.745195	2.362166	1.467401	0.201227
CPSO_rb	1.772740	1.748541	2.346964	1.455882	**0.200838**
CPSO_rj	**1.795481**	**1.764483**	**2.383918**	**1.482471**	0.203157

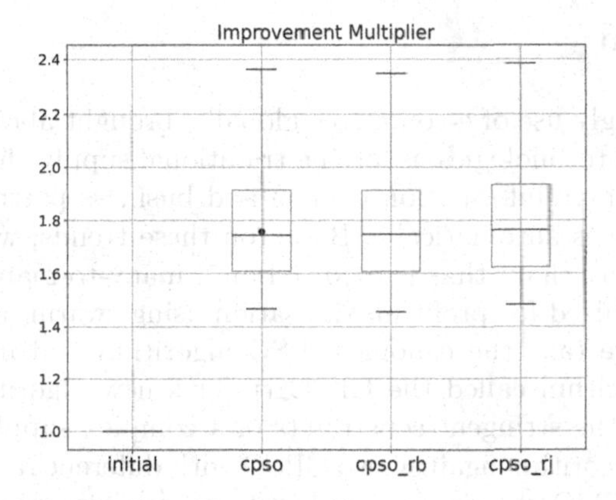

Fig. 1. Improvement multiplier. Take the red point for example, which indicates one optimization result given by CPSO algorithm. The profit given by the indicated optimal result is 413468.42, and the profit given by the initial global best particle is 235192.44. We can obtain the value 1.758 by $1 + (413468.42 - 235192.44)/235192.44$. (Color figure online)

Table 2. The number of iterations to convergence

Algorithms	Mean	Median	Max	Min	StdDev.
CPSO	**220.947**	**204**	499	17	144.759606
CPSO_rb	283.691	284.5	499	34	**125.165306**
CPSO_rj	248.221	247	499	**6**	144.835885

Results. As seen in Table 1, the mean, median, maximum and minimum of improvement multiplier of CPSO_rj are higher values than those of both the CPSO & CPSO_rb. This suggests that CPSO_rj is able to find better combinations for higher profits than CPSO & CPSO_rb even though all algorithms started from the same randomly initialised set. Figure 1 is the graphical representation of the results. The initial random value is at 1.0 and three boxplots of improvement multipliers show the capability of efficient improvement of profit values resulting from the supply chain combinations returned by the three algorithms after the optimization process.

Table 2 compares the number of iterations to convergence of three algorithms. Iterations to convergence refers to the iteration number at which the algorithm is able to find the highest profit value of the run. On average, CPSO converges faster than the other two algorithms, while on average CPSO_rb takes the highest number of iterations to converge, though the average difference is not great between the three algorithms.

In both Tables 1 and 2, CPSO_rb had the lowest standard deviations, which would suggest that its results are relatively consistent both in terms of improvement multipliers and iterations to convergence.

5 Conclusion

The ever increasingly use of e-commerce globally, brought about by the massive adoption of digital technologies, is forcing traditional supply chains to operate in an environment where information is open and business partnerships can form easily and sometimes automatically. Based on these trends, we create a decentralized open supply chain that is modeled on a many-to-many fully connected network and optimized for profit maximisation using swarm intelligence.

In doing so, we take the canonical PSO algorithm and improve it, resulting in a new algorithm called the CPSO_rj. This new algorithm guarantees a result that meets the stringent constraints of a complex supply chain. Further, we test this new algorithm against two CPSO with different random back strategies (CPSO and CPSO_rb) via a newly developed performance metric called the 'improvement multiplier'. This allows us to test these algorithms on a number of different supply chain configurations. The CPSO_rj yields the best improvement ratios out of the three algorithms and is ideally suited to optimize the decentralized supply chains of the future.

Notice that the variable setups, parameter setups, and initial datasets are all simplified for demonstration purposes only. CPSO_rj is ready to adopt similar settings and datasets of one's specific needs. For examples, one may take real-world locations and calculate the real transport costs according to the true geographic locations. One may also consider real-world customer categories and product categories associated to company's needs. Even broader, users are actually free to edit the objective by adding necessary terms to the profit.

There are a number of extensions and modifications that can be considered in the future works. First, it is of great interest to see how the increase of the problem dimension affects the performance of our proposed method. It is obvious that the numbers of customers, suppliers and products are much larger in real supply chain, so a further analysis of computational complexity will be a major work to investigate. Second, there are many state-of-the-art methods in both metaheuristics and supply chain. Among all metaheuristics algorithms, the Swarm Intelligence Based (SIB) method [11–13] is designed for optimization problems with discrete domain like the supply chain problems. A comparison between the results of SIB and CPSO_rj under different problem dimensions is another major work in the extended paper. The comparison to the results of state-of-the-art methods [14,15] in supply chain is also important. Third, PSO is known as a random algorithm, so there is a need to quantify the result variability using a formal statistical detection method in addition to the simple boxplot of improvement multiplier. Such method is certainly an important addition to CPSO_rj and any PSO-type algorithm in general. Last, the supply-chain problem has a rigorous mathematical formulation in economics, so it is of great interest to derive some underlying theoretical results prior to the optimization to help improving the optima and stabilize the variability.

References

1. Singh, K., Lin, S.-P., Phoa, F.K.H., Chen-Burger, Y.-H.J.: Swarm intelligence optimisation algorithms and their applications in a complex layer-egg supply chain. In: Jezic, G., Chen-Burger, J., Kusek, M., Sperka, R., Howlett, R.J., Jain, L.C. (eds.) Agents and Multi-Agent Systems: Technologies and Applications 2021. SIST, vol. 241, pp. 39–51. Springer, Singapore (2021). https://doi.org/10.1007/978-981-16-2994-5_4
2. Corne, D.W., Reynolds, A., Bonabeau, E.: Swarm intelligence. In: Rozenberg, G., Bäck, T., Kok, J.N. (eds.) Handbook of Natural Computing, vol. 2017, no. 6, pp. 1599–1622. Springer, Heidelberg (2012). https://doi.org/10.1007/978-3-540-92910-9_48
3. Kennedy, J., Eberhart, R.: Particle swarm optimization. In: Proceedings of ICNN 1995 - International Conference on Neural Networks, vol. 4, pp. 1942–1948 (1995)
4. Kennedy, J., Mendes, R.: Population structure and particle swarm performance. In: Proceedings of the 2002 Congress on Evolutionary Computation (CEC 2002), vol. 2, pp. 1671–1676 (2002)
5. Kadadevaramath, R.S., Chen, J.C.H., Latha Shankar, B., Rameshumar, K.: Application of particle swarm intelligence algorithms in supply chain network architecture optimization. Expert Syst. Appl. **39**(11), 10160–10176 (2012)
6. Izquierdo, J., Minciardi, R., Montalvo, I., Robba, M. and Tavera, M.: Particle swarm optimization for the biomass supply chain strategic planning. In: Proceedings of iEMSs 2008 - International Congress on Environmental Modelling and Software Integrating Sciences and Information Technology for Environmental Assessment and Decision Making, pp. 1272–1280 (2008)
7. Sinha, A.K., Aditya, H.K., Tiwari, M.K., Chan, F.T.S.: Agent oriented petroleum supply chain coordination: co-evolutionary particle swarm optimization based approach. Expert Syst. Appl. **38**(5), 6132–6145 (2011)
8. Clerc, M., Kennedy, J.: The particle swarm-explosion, stability, and convergence in a multidimensional complex space. IEEE Trans. Evol. Comput. **6**(1), 58–73 (2002)
9. Mendes, R., Kennedy, J., Neves, J.: The fully informed particle swarm: simpler, maybe better. IEEE Trans. Evol. Comput. **8**(3), 204–210 (2004)
10. Clerc, M.: Confinements and biases in particle swarm optimization. HAL-00122799 (2006)
11. Phoa, F.K.H., Chen, R.B., Wang, W.C., Wong, W.K.: Optimizing two-level supersaturated designs via swarm intelligence techniques. Technometrics **58**(1), 43–49 (2016)
12. Phoa, F.K.H.: A swarm intelligence based (SIB) method for optimization in designs of experiments. Nat. Comput. 16(4), 597–605 (2017)
13. Phoa, F.K.H., Liu, H.-P., Chen-Burger, Y.-H.J., Lin, S.-P.: Metaheuristic optimization on tensor-type solution via swarm intelligence and its application in the profit optimization in designing selling scheme. In: Tan, Y., Shi, Y. (eds.) ICSI 2021. LNCS, vol. 12689, pp. 72–82. Springer, Cham (2021). https://doi.org/10.1007/978-3-030-78743-1_7
14. Campuzano, F., Mula, J.: Supply Chain Simulation. A System Dynamics Approach for Improving Performance. Springer, London (2011). https://doi.org/10.1007/978-0-85729-719-8
15. Llaguno, A., Mula, J., Campuzano, F.: State of the art, conceptual framework and simulation analysis of the ripple effect on supply chains. Int. J. Prod. Res. **60**(6), 2044–2066 (2022)

Swarm Enhanced Attentive Mechanism for Sequential Recommendation

Shuang Geng[1], Gemin Liang[1], Yuqin He[1], Liezhen Duan[1], Haoran Xie[2(✉)], and Xi Song[1]

[1] College of Management, Shenzhen University, Shenzhen, China
gs@szu.edu.cn, {lianggemin2021,heyuqin2021}@email.szu.edu.cn
[2] Department of Computing and Decision Science, Lingnan University, Hong Kong, China
hrxie@ln.edu.hk

Abstract. Recommendation system facilitates users promptly obtaining the information they need in this age of data explosion. Research on recommendation models have recognized the importance of integrating user historical behavior sequence into the model to alleviate the matrix sparsity. Although deep learning algorithm with attentive mechanism exhibits competitive performance in sequential recommendation, the searching for optimal attentive factors still lack effectiveness. In this work, we redesign the sequential recommendation model by employing swarm intelligence for optimization in the attentive mechanism thus to improve the algorithm accuracy. We conduct extensive comparative experiments to evaluate performance of four swarm intelligence algorithms and traditional recommendation methods. Our work is the first attempt to integrate swarm intelligence into sequential recommendation algorithm. Experimental results confirmed the superior performance on AUC score of the proposed approach.

Keywords: Sequential recommendation · Swarm intelligence · Attentive mechanism

1 Introduction

Recommendation system (RS) is playing an important role in this era of in-formation explosion by assisting users to obtain most relevant content or service. Traditional recommendation techniques, such as collaborative filtering, utilizes the user-item interaction to recognize most similar items that user may have interest in [1]. However, conventional methods often suffer from data sparsity and cold start problem. Recent research attempted to leverage user sequential behavior for their profile learning, for instance, sum pooling of user sequential behaviors to represent user preference [2, 3]. Despite the improved performance of sequential recommendation, treating historical items equally may constrain the learning process and require higher computational cost. To tackle this difficulty, attention weight is proposed for each historical item to differentiate their importance in user profile, which is named attentive mechanism [4, 5]. While attentive mechanism achieves effectively to learn user representation, the attention calculation is time-consuming and sometimes only obtains sub-optimal attentive factors.

Y. Tan et al. (Eds.): ICSI 2022, LNCS 13344, pp. 442–453, 2022.
https://doi.org/10.1007/978-3-031-09677-8_37

Swarm intelligence (SI) is a collection of naturally inspired methods that show overwhelming superiority in solving optimization problems. SI-based algorithms have the advantage of increased flexibility, simplicity, easiness of implementation and robustness. Prior research has demonstrated the effectiveness of SI in enhancing multi-criteria recommendation techniques and achieves good balance between multiple recommendation objectives [6]. SI has also been used for assigning optimal weights of different features in similarity computation [8]. In ensemble and hybrid RS, SI helps to identify the optimal combination of sub-algorithms [9].

Considering the searching capability for optimal solutions of SI, this research leverages SI algorithms for learning the optimal attentive factor in sequential recommendation assuming that users have different preferences for historical items. We first leverage DeepWalk [15] and Skipgram [16] to generate item embedding, and then integrate SI in the user profile learning. Instead of validating one SI algorithm for the factor learning, we also compare the capability of different swarm algorithms, namely, Particle Swarm Optimization (PSO), Artificial Bee Colony (ABC), Bacterial Foraging Optimization (BFO), and Firefly Algorithm (FA) in this optimization task. Experiment results demonstrate the promising performance of SI in optimizing attentive factors compared with average pooling and stochastic gradient descent method. To brief, the main contributions of this work are as follows.

1. We propose an SI based method for learning optimal attentive factors for user's historical item in user preference learning, named Swarm Enhanced Attentive Mechanism (SEAM). To the best of our knowledge, this is the first attempt to optimize attentive factors with swarm intelligence algorithms in sequential recommendation.
2. We conduct extensive experiments based on a real-world dataset to demonstrate the effectiveness of proposed method. The result shows that SEAM has outperformed selected state-of-the-art recommendation algorithms and optimization methods.

2 Related Work

2.1 Sequential Recommendation

User sequential behavior involves dynamic user interest information, which is valuable for modeling user preferences. Sequential recommendation is to learn user profile through the historical interactions with candidate items. Sum pooling as a popular sequential recommendation method learns the item-item similarity by the product of two low-dimension matrices, and then pool the latent factors of historical items to obtain user representation [2]. These models overcome the limitation of sparse matrix, but fail to address the timing of sequence information.

To more effectively capture user preference, CNN [10], RNN [11] and self-attention mechanism [12] have also been proposed. These methods can transfer user sequential behavior into fixed dimensional vectors. However, they lack the ability to extract user's dynamic interests since they treat historical items equally. In that light, some recent researchers [4, 5] leverage attention mechanism to model user behavior by differentiating the importance of historical items with attentive functions. Whereas the typical constraint is that calculating the similarity by attention mechanism is computationally expensive with low efficiency. Thus, there is still need for more effective optimization methods.

2.2 Swarm Intelligence Based Recommendation

SI algorithms simulate the cooperative behavior of natural phenomena, such as the foraging of insects, the flight of birds or the crowding of fish. The main idea of SI is to use the intelligence of the population for collaborative search, so as to find the optimal solution in the solution space. SI has the advantages of flexibility, robustness and parallelism. It is highly compatible with different mathematical form of the target problem, either simple or complex, and stands out in solving complicated optimization problems. In the recent years, SI has been applied to improve the accuracy of recommendation systems.

SI techniques were used to assign optimal weights to features in order to find better neighborhoods for users, such as bacterial foraging optimization [6], gravitational search algorithm [7] and bat algorithm [8]. To be noted, clustering-based models are able to reduce the time complexity caused by larger datasets, especially in memory-based recommendation. In extant research, SI techniques are commonly used to improve the performance of clustering algorithms. For instance, ABC-KM [13] utilized artificial bee colony to regulate the optimal center points of K-Means algorithm, and then aggregated users into different clusters. HMRS [14] proposed to generate a ranked list by graph-based approach and re-ranked the list using PSO to obtain highly optimized results.

3 Swarm Enhanced Attentive Mechanism Model

3.1 Problem Definition

As shown in Fig. 1, we assign attentive factors α_{uj} for each user-item pair, which can differentiate the attached importance of user historical items and figure out the most interested historical items of users. Our objective is to obtain the optimal attentive factors for each user to minimize the loss function of recommendation.

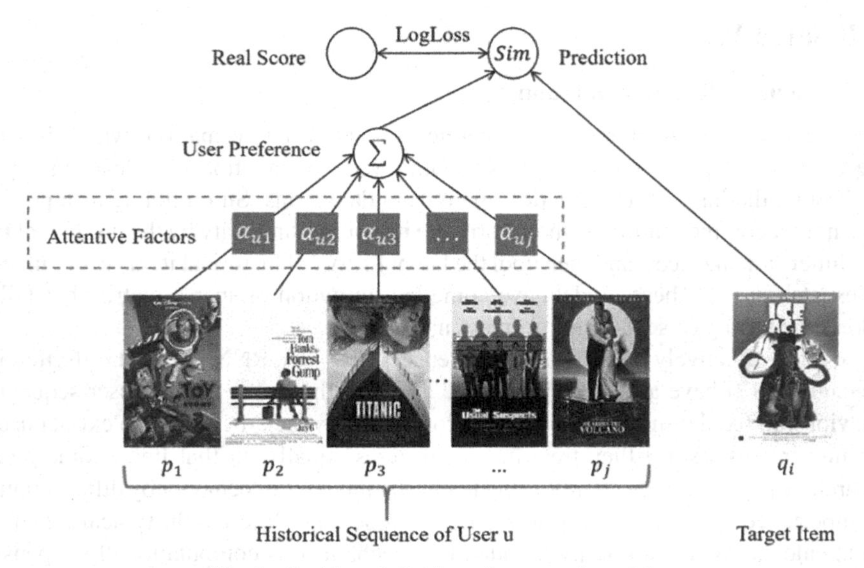

Fig. 1. Problem definition of proposed model.

3.2 General Framework

The general framework is presented in Fig. 2. At the first step, we construct item graph and utilize DeepWalk [15] to generate item embedding, which can represent the features of each item. Subsequently, we apply SI algorithms to obtain the attentive factors of user sequential behavior. Details are explained as follows.

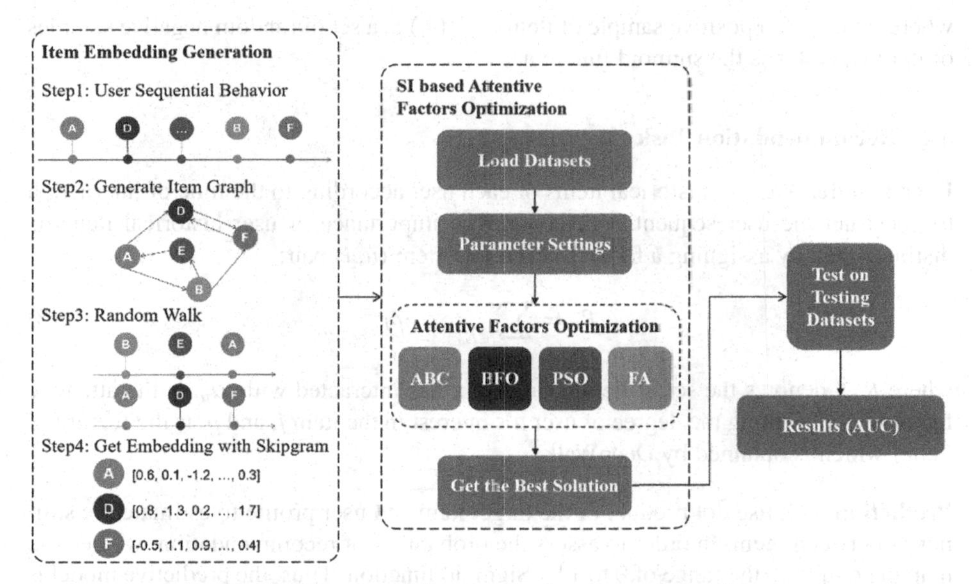

Fig. 2. The general framework of our proposed model.

3.3 Item Embedding

We use DeepWalk to generate item embedding, which enables representing the latent information and relationship between items. Given the users' historical sequence that have been sorted by time, we generate item graph, denoted as $G = (V, E)$, where V is the set of nodes and E is the set of edges. Then, random walk is performed on the item graph G, which selects the starting points randomly and walk the specific number of steps. Finally, we get new item sequences.

We employ Skipgram [16] to learn the latent representation of items. The goal of the Skipgram model is to maximize the probability of items that appear in the same item sequence. This yields the optimization problem:

$$\underset{\phi}{\text{minimize}} - logPr(\{v_{i-t}, \ldots, v_{i-1}, v_{i+1}, \ldots, v_{i+t}\}|\phi(v_i)) \tag{1}$$

where t is the window size, ϕ is a mapping function that maps v_i to weights matrix $W \in R^{|V|\times D}$, and W represents the embedding of items that can be obtained by optimizing the objective function in the training process.

In order to improve the training efficiency of Skipgram, we adopt negative sampling [17] that computes the loss function for a few random negative samples, not all negative samples. Equation (1) can be transformed:

$$\underset{\phi}{\text{minimize}} - log\sigma\left(\phi\left(v_i^{'}\right)^T \phi(v_i)\right) - \sum\nolimits_{j\in N(v_i)} \log\sigma(-\phi(v_j)^T\phi(v_i)) \quad (2)$$

where item $v_i^{'}$ is a positive sample of item v_i, $N(v_i)$ is a set of random negative samples of item v_i, and σ is the sigmoid function.

3.4 Recommendation Task

User Profile. We sort historical items of each user according to the time of interaction to construct the user sequential behavior. The importance of user historical items is distinguished by assigning a factor to each user item (u, j) pair:

$$P_u = \sum\nolimits_{j\in R_u{}^+} \alpha_{uj} p_j \quad (3)$$

where $R_u{}^+$ denotes the set of items that user u has interacted with, α_{uj} is the attentive factor in contributing the degree of user u's interest in the item j, and p_j is the feature of item j which is obtained by DeepWalk.

Prediction. We use dot product of the target item and user profile to estimate the similarity between them. In order to assess the probability of recommendation, we need to map the output to the range of 0 to 1 by Sigmoid function. Thus, the predictive model is as follows:

$$\hat{y}_{ui} = sigmoid(P_u \odot q_i) \quad (4)$$

where q_i is the feature of target item i, $\hat{y}_{ui}$ is the prediction of probabilities recommending target item i to user u.

Loss Function. We denote a user-item interaction matrix as $Y \in Y^{M\times N}$. for users' implicit feedback, where M and N denote the number of users and items respectively. Formally, the user-item interaction matrix Y is as follow:

$$y_{ui} = \begin{cases} 1, \textit{if user u has interacted with item i.} \\ 0, \textit{if user u has not interacted with item i.} \end{cases} \quad (5)$$

The objective function is the negative log-likelihood function as follows:

$$LOSS = -\frac{1}{N}(\sum\nolimits_{(u,i)\in T} y_{ui}\ln(\hat{y}_{ui}) + (1 - y_{ui})\ln(1 - \hat{y}_{ui})) \quad (6)$$

where T is training set, N is the number of instances in T.

3.5 Attentive Factors Optimization

In this section, we use four SI algorithms to optimize the attentive factors α_{uj} in user sequential behavior in Eq. (3) by minimizing the loss function in Eq. (6).

Particle Swarm Optimization (PSO). PSO [18] is a SI-based algorithm by simulating the migration behavior of birds. The movement of particle i depends on its inertia ω, its local best location $pbest_i$ and the global best position $gbest$. Its velocity update and location update are as follow:

$$v_i = \omega v_i + c_1 \times rand() \times \left(pbest_i - x_i\right) + c_2 \times rand() \times (gbest - x_i) \quad (7)$$

$$x_i = x_i + v_i \quad (8)$$

where v_i and x_i are the velocity and position of particle i respectively, c_1 and c_2 are cognitive coefficient and social coefficient respectively and $rand()$ is a random number between 0 and 1.

Artificial Bee Colony (ABC). ABC [19] is an optimization algorithm based on the behavior of three kinds of bees looking for the best food source. To be specific, employed bees memorize information about their food sources and seek the better food source in their neighbors. Onlooker bees search food sources with probability on the basis of the fitness given by employed bees. Scout bees randomly search for new food sources as long as the fitness of employed bees stop increasing.

Bacterial Foraging Optimization (BFO). BFO [20] mimics the chemotaxis, reproduction and dispersion behavior of bacterial population. After each step of chemotaxis, bacteria i updates its position as follow:

$$\theta^i(j+1,k,l) = \theta^i(j,k,l) + c(i)\phi(i) \quad (9)$$

where $\theta^i(j,k,l)$ represents the position of bacteria i after the j-th chemotaxis, the k-th reproduction and the l-th dispersion, $c(i)$ is the step size, $\phi(i)$ is the randomly selected unit directional vector.

Firefly Algorithm (FA). FA [21] is a SI algorithm based on the mutual attraction between fireflies. Attraction depends on the brightness perceived by the firefly, which decreases with distance. When firefly i find a brighter firefly j, firefly i moves towards firefly j as follow:

$$x_i = x_i + \frac{\beta_0}{1+\gamma \cdot {r_{ij}}^2}\left(x_j - x_i\right) + \alpha r \quad (10)$$

where x_i and x_j are the position of firefly i and j respectively, r_{ij} is the distance between firefly i and j, β_0 and γ are the mutual attraction and light absorption coefficient respectively.

4 Experiments

4.1 Experimental Settings

Datasets. We evaluated our proposed model based on MovieLens, a popular rating dataset. We chose the latest small version of MovieLens, containing 100836 ratings and each user has evaluated at least 20 movies. The length of user sequential behavior varies greatly, so we limit the maximum to 100. More details are shown in Table 1.

Table 1. Information of the latest small MovieLens

Dataset	Interaction#	Item#	User#	Sparsity
MovieLens	100,836	9,724	610	98.30%

Evaluation Protocols. We sorted user interactions by time. We took the early 80% interactions as training sets and use the latest 20% for testing. For each training and testing instance, we randomly sampled 4 items that have not been interacted by the user. In the training session, we used the early 80% interactions to construct user historical sequence and optimize its attention factors through SI algorithms. Then, we can learn user features from these early 80% interactions. In the testing session, we estimated the similarity between users and target items by dot product. We applied AUC to measure the accuracy of recommendation.

Comparison Algorithms. We named our proposed model Swarm Enhanced Attentive Mechanism (SEAM) and in this study, we carry out experiments for four SI algorithms, namely, AF-ABC, AF-BFO, AF-PSO and AF-FA respectively. We also designed AF-AVG and AF-SGD as comparison methods to obtain the attentive factors by average pooling and stochastic gradient descent, rather than SI-based methods. The rest con-figurations of AF-AVG and AF-SGD are as the same as those of the proposed model. Additionally, four conventional recommendation methods are selected as baseline methods:

- User-CF [22]: It calculates the user-user similarity using Pearson correlation.
- ItemKNN [1]: It is a well-known item-based recommendation. We utilized adjusted cosine similarity and considered all item neighborhoods.
- MF [23]: It is a latent factor model that factorizes user-item matrix into low dimensional user vectors and item vectors. The dimension of latent factor is set to 16.
- SLIM [24]: It directly learns item-item similarity matrix from the data.

Parameter Settings. The dimension of candidate solution depends on the length of user historical interactions. In order to reduce the complexity of optimization algorithms, we set the maximum length of user historical sequence by [100, 80, 60, 40, 20]. Before the optimization, we randomly initialized the candidate solutions between 0 and 1. The lower and upper boundaries of the candidate solutions were set to −2 and 2 respectively, so that the attentive factors of historical items are relatively balanced after Softmax. The population size of four SI-based algorithms is set to 50. We searched from extant literature and found the most employed and recommended values to set parameters of four SI algorithms. The settings of parameters are shown in Table 2.

Table 2. Parameter settings of four SI algorithms

Swarm Intelligence Algorithms	Parameters	Value
Particle Swarm Optimization	Inertia weight	0.49
	Cognitive coefficient	0.72
	Social coefficient	0.72
Artificial Bee Colony	Abandonment criteria	10
	Number of employed bees	50
	Number of onlooker bees	50
Bacterial Foraging Optimization	Step size (c)	0.1
	Swimming length (N_s)	4
	Chemotactic steps (N_c)	100
	Reproduction steps (N_{re})	5
	Elimination-dispersal events (N_{ed})	2
	Elimination-dispersal probability (p_{ed})	0.25
Firefly Algorithm	Mutual attraction	1
	light absorption coefficient	1
	Initial/final randomization parameter alpha	1/0.1
	First/second Gaussian parameter	0/0.1

4.2 Experimental Results

Comparison Between Four SI-Based Algorithms. Figure 3 displays the training loss of four SI-based methods along with iteration process. AF-PSO is featured by the fast convergence while it obtains the highest training loss, which may be caused by trapping into local optimum since all particles move towards the best position and diversity is undermined. The training loss of AF-ABC, AF-BFO and AF-FA converge to lower fitness values. As shown in Fig. 4, AF-FA, AF-ABC and AF-BFO also obtain better testing performance than AF-PSO. Furthermore, the convergence speed of AF-FA is faster than AF-ABC and AF-BFO. It needs to note that Fig. 3 only provides the fitness optimization of the first four users, and the training loss for most users shows the similar trend.

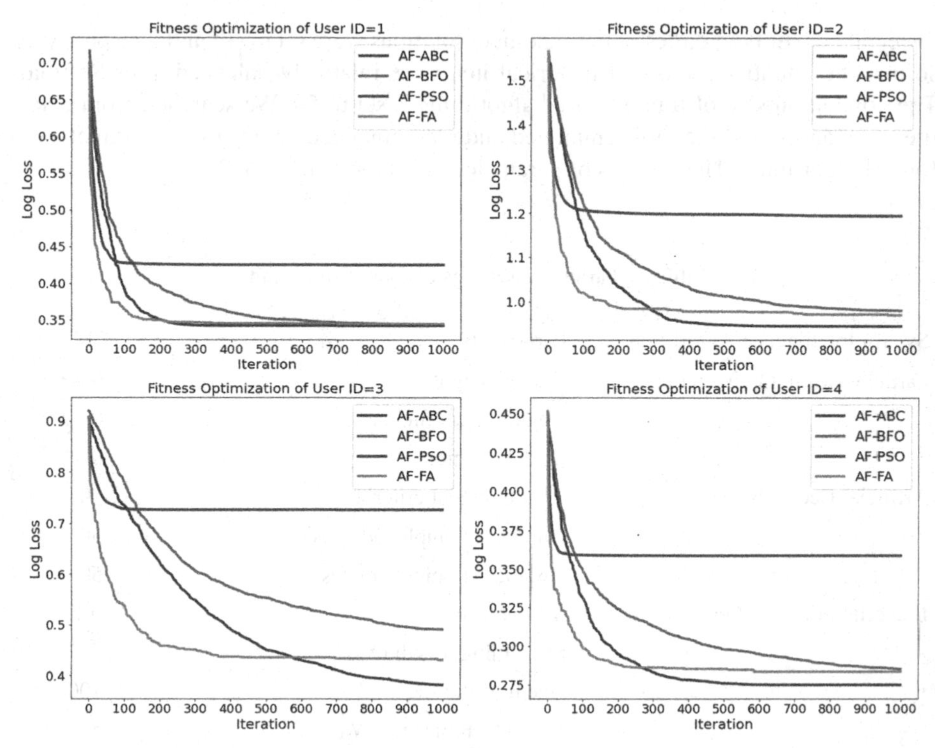

Fig. 3. Training loss of four SI-based methods at embedding size 16, maximum length 100.

Recommendation Accuracy Performance Comparison. Figure 4 shows the overall performance of ten algorithms. As we found, firstly, AF-SGD and the four proposed SI-based methods achieve higher AUC scores than AF-AVG. It confirms the effectiveness of attentive factors for learning user preferences. Moreover, three SI-based methods (AF-FA, AF-ABC and AF-BFO) perform better than the conventional optimization algorithm (AF-SGD), evident from the positive effect of swarm intelligence to optimize the complicated recommendation problem. Furthermore, model-based methods (MF, SLIM, AF-SGD and SEAM) outperform heuristic-based methods (User-CF and ItemKNN). It reveals that model-based methods are more capable in extracting larger volume of information from given datasets, so as to alleviate the problem of sparse matrix.

Hyper-parameter Study. When the embedding size is too small, item embedding may not represent latent information effectively due to the information loss. Further, it would lead to over fitting with high embedding size. As shown in Table 3, the AUC scores of all methods rise significantly as the embedding size increases to 16, and then their test performance has a slight change when embedding size is higher than 16.

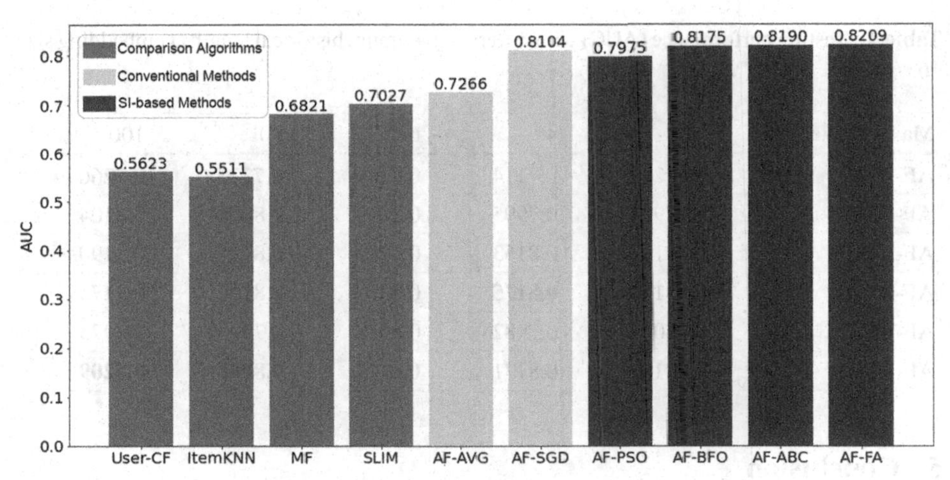

Fig. 4. Testing performance (AUC) of comparison algorithms at embedding size 16, maximum length 100.

Moreover, longer historical sequence obtains more user interest information, which complicates the optimization problem at the same time. As shown in Table 4, four SI-based methods achieve the similar AUC score at any maximum length. It reveals that four SI-based methods are less affected by the length of user historical sequence. Among four SI based methods, AF-FA achieves the best performance in most cases, except for the maximum length 40.

Table 3. Testing performance (AUC) with different embedding size at maximum length 100.

Embedding size	4	8	16	32	64
AF-AVG	0.6895	0.7219	0.7266	0.7231	0.6951
AF-SGD	0.7436	0.7800	0.8104	0.8078	0.8115
AF-ABC	**0.7551**	0.7856	0.8190	0.8193	**0.8257**
AF-BFO	0.7520	0.7878	0.8175	0.8160	0.8169
AF-PSO	0.7354	0.7731	0.7975	0.7924	0.7781
AF-FA	0.7547	**0.7886**	**0.8209**	**0.8207**	0.8236

Table 4. Testing performance (AUC) with different maximum historical length at embedding size 16.

Maximum length	20	40	60	80	100
AF-AVG	0.7223	0.7114	0.7179	0.7220	0.7266
AF-SGD	0.7997	0.7995	0.8059	0.8096	0.8104
AF-ABC	0.8109	0.8153	0.8207	0.8194	0.8190
AF-BFO	0.8120	**0.8175**	0.8213	0.8190	0.8175
AF-PSO	0.8025	0.7982	0.8018	0.7990	0.7975
AF-FA	**0.8127**	0.8171	**0.8215**	**0.8206**	**0.8209**

5 Conclusion

In this work, we proposed to integrate swarm intelligence into attentive mechanism in learning more effective user profile for sequential recommendation. The experimental results show that four swarm intelligence algorithms (PSO, BFO, AF, ABC) outperform average pooling and SGD in searching for optimal attentive factors. Moreover, optimization-based methods outperform traditional heuristic recommendation methods (User-CF, ItemKNN, MF, SLIM). Among the four swarm intelligence algorithms, firefly algorithm has relatively faster convergence speed and better global search ability to optimize the recommendation problem. This study provides theoretical insights in swarm intelligence based sequential recommendation.

Acknowledgement. This study is supported by National Natural Science Foundation of China (71901150, 71901143), Natural Science Foundation of Guangdong (2022A1515012077), Guangdong Province Innovation Team "Intelligent Management and Interdisciplinary Innovation" (2021WCXTD002), Shenzhen Higher Education Support Plan (20200826144104001).

References

1. Sarwar, B., Karypis, G., Konstan, J.: Item-based collaborative filtering recommendation algorithms. In: Proceedings of the 10th International Conference on World Wide Web, pp. 285–295 (2001)
2. Koren, Y.: Factorization meets the neighborhood: a multifaceted collaborative filtering model. In: Proceedings of the 14th ACM SIGKDD International Conference on Knowledge Discovery and Data Mining, pp. 426–434 (2008)
3. Kabbur, S., Ning, X., Karypis, G.: FISM: factored item similarity models for top-N recommender systems. In: Proceedings of the 19th ACM SIGKDD International Conference on Knowledge Discovery and Data Mining, pp. 659–667 (2013)
4. Zhou, G., Zhu, X., Song, C.: Deep interest network for click-through rate prediction. In: Proceedings of the 24th ACM SIGKDD International Conference on Knowledge Discovery and Data Mining, pp. 1059–1068 (2018)
5. He, X., He, Z., Song, J.: NAIS: neural attentive item similarity model for recommendation. IEEE Trans. Knowl. Data Eng. **30**(12), 2354–2366 (2018)

6. Geng, S., He, X., Wang, Y.: Multicriteria recommendation based on bacterial foraging optimization. Int. J. Intell. Syst. **37**(2), 1618–1645 (2022)
7. Choudhary, V., Mullick, D., Nagpal, S.: Gravitational search algorithm in recommendation systems. In: Tan, Y., Takagi, H., Shi, Y., Niu, B. (eds.) ICSI 2017. LNCS, vol. 10386, pp. 597–607. Springer, Cham (2017). https://doi.org/10.1007/978-3-319-61833-3_63
8. Yadav, S., Nagpal, S.: An improved collaborative filtering based recommender system using bat algorithm. Procedia Comput. Sci. **132**, 1795–1803 (2018)
9. Xia, X., Wang, X., Li, J.: Multi-objective mobile app recommendation: a system-level collaboration approach. Comput. Electr. Eng. **40**(1), 203–215 (2014)
10. Yuan, F., Karatzoglou, A., Arapakis, I.: A simple convolutional generative network for next item recommendation. In: Proceedings of the Twelfth ACM International Conference on Web Search and Data Mining, pp. 582–590 (2019)
11. Hidasi, B., Karatzoglou, A.: Recurrent neural networks with top-k gains for session-based recommendations. In: Proceedings of the 27th ACM International Conference on Information and Knowledge Management, pp. 843–852 (2018)
12. Kang, W.C., McAuley, J.: Self-attentive sequential recommendation. In: IEEE International Conference on Data Mining (ICDM), pp. 197--206. IEEE (2018)
13. Katarya, R.: Movie recommender system with metaheuristic artificial bee. Neural Comput. Appl. **30**(6), 1983–1990 (2018)
14. Katarya, R., Verma, O.P.: Efficient music recommender system using context graph and particle swarm. Multimed. Tools Appl. **77**(2), 2673–2687 (2018)
15. Perozzi, B., Al-Rfou, R., Skiena, S.: Deepwalk: online learning of social representations. In: Proceedings of the 20th ACM SIGKDD International Conference on Knowledge Discovery and Data Mining, pp. 701–710 (2014)
16. Mikolov, T., Sutskever, I., Chen, K.: Distributed representations of words and phrases and their compositionality. In: Advances in Neural Information Processing Systems, pp. 3111–3119 (2013)
17. Goldberg, Y., Levy, O.: Word2vec explained: deriving Mikolov et al.'s negative-sampling word-embedding method. arXiv preprint. arXiv:1402.3722 (2014)
18. Kennedy, J., Eberhart, R.: Particle swarm optimization. In: Proceedings of ICNN 1995-International Conference on Neural Networks, vol. 4, pp. 1942–1948. IEEE (1995)
19. Karaboga, D., Basturk, B.: On the performance of artificial bee colony (ABC) algorithm. Appl. Soft. Comput. **8**(1), 687–697 (2008)
20. Passino, K.M.: Biomimicry of bacterial foraging for distributed optimization and control. IEEE Control Syst. Mag. **22**(3), 52–67 (2002)
21. Yang, X.S.: Nature-Inspired Metaheuristic Algorithms. Luniver Press (2010)
22. Schafer, J.B., Frankowski, D., Herlocker, J., Sen, S.: Collaborative filtering recommender systems. In: Brusilovsky, P., Kobsa, A., Nejdl, W. (eds.) The Adaptive Web. LNCS, vol. 4321, pp. 291–324. Springer, Heidelberg (2007). https://doi.org/10.1007/978-3-540-72079-9_9
23. Koren, Y., Bell, R., Volinsky, C.: Matrix factorization techniques for recommender systems. Comput. **42**(8), 30–37 (2009)
24. Ning, X., Karypis, G.: SLIM: sparse linear methods for top-n recommender systems. In: 2011 IEEE 11th International Conference on Data Mining, pp. 497–506. IEEE (2011)

Multi-objective Optimization

Non-dominated Sorting Based Fireworks Algorithm for Multi-objective Optimization

Mingze Li and Ying Tan(✉)

Key Laboratory of Machine Perception (MOE), School of Artificial Intelligence, Institute for Artificial Intelligence, Peking University, Beijing, China
{mingzeli,ytan}@pku.edu.cn

Abstract. Multi-objective optimization is one of the most important problem in the mathematical optimization. Some researchers have already proposed several multi-objective fireworks algorithms, of which *S*-metric based multi-objective fireworks algorithm (*S*-MOFWA) is the most representative work. *S*-MOFWA takes the hypervolume as the evaluation criterion of external archive updating, which is easy to implement but ignores the landscape information of the population. In this paper, a novel multi-objective fireworks algorithm named non-dominated sorting based fireworks algorithm (NSFWA) is proposed. The proposed algorithm updates the external archive with the selection operator based on the fast non-dominated sorting approach, which is specially designed for the spark generation characteristic of FWA to improve the diversity. A multi-objective guided mutation operator is also designed to enhance the efficiency of population information utilization and improve the search capability of the algorithm. Experimental results on the benchmarks demonstrate that NSFWA outperforms other multi-objective swarm intelligence algorithms of *S*-MOFWA, NSGA-II and SPEA2.

Keywords: Fireworks algorithm · Multi-objective optimization · Swarm intelligence · Non-dominated sorting based fireworks algorithm

1 Introduction

Fireworks algorithm (FWA) proposed by Tan et al. in 2010 is a novel swarm intelligence algorithm [14]. FWA has a double-layer structure, in which the higher layer is the global coordination between the firework populations and the lower one is the local search of a certain population. This hierarchical structure ensures that FWA can solve kinds of optimization problems with different landscape and shows a significant performance on the single-objective optimization problem. In recent years, guided fireworks algorithm (GFWA) [9], loser-out tournament fireworks algorithm (LoTFWA) [7] and other new variants [2,8,10,11,17,18] further

Y. Tan et al. (Eds.): ICSI 2022, LNCS 13344, pp. 457–471, 2022.
https://doi.org/10.1007/978-3-031-09677-8_38

enhance the performance of FWA from the aspects of global coordination and local exploitation.

Multi-objective optimization problem (MOP) is a kind of mathematical optimization problem with multiple conflicting objective functions to be optimized at the same time. Multi-objective optimization algorithm is aimed to find an optimal solution set composed of Pareto optimal solutions, which covering the whole Pareto front as completely as possible. Naturally, convergence and diversity are two main measures for MOP. Convergence mainly indicates the distance between the Pareto front and solutions obtained by the algorithm. And diversity can be roughly regarded as a ratio of the section covered by the solution set to the entire Pareto front.

According to the method of solution set updating, multi-objective optimization algorithms could be classified as two mainstream categories. Pareto dominance based methods such as non-dominated sorting genetic algorithm-II (NSGA-II) [4] and the improved strength Pareto evolutionary algorithm (SPEA2) [20] calculate the Pareto dominance between individuals in each iteration and update the solution set accordingly. Hypervolume indicator based methods like SMS-EMOA [1] use the volume covered by individuals instead of the Pareto dominance as the criterion to update the solution set.

Some researchers also proposed multi-objective FWA, and one of the most representative work is S-MOFWA proposed by Liu and Tan [12]. S-MOFWA adopted the hypervolume based framework and designed a novel external archive updating methods. The framework reduces the difficulty of multi-objective optimization significantly and makes it possible to inherit mechanisms of single-objective FWA. However, due to the limitation of the framework, S-MOFWA also ignores the information of dominated solutions and has a relatively low information utilization efficiency.

In this paper, a novel multi-objective FWA named non-dominated sorting based fireworks algorithm is proposed. NSFWA adopts a non-dominated sorting based external archive updating methods as the selection operator, and extends the idea of GFWA to MOP. In order to accelerate the convergence of MOFWA without affecting diversity, the multi-objective guided mutation operator is designed to generate guiding sparks with two different methods according to certain characteristic of fireworks. The adaptive amplitude mechanism and mapping rule are also revised.

The remaining parts is organized as follows. Some related works are introduced in Sect. 2. Our proposed algorithm is described in detail and the improved mutation operator is analyzed in Sect. 3. Then Sect. 4 presents the experimental results to present the good performance of NSFWA. Section 5 gives the conclusion.

2 Related Works

NSGA-II is one of the most influential multi-objective swarm intelligence algorithm. Swarm intelligence algorithms usually have a large number of populations and individuals, and thus it is necessary to calculate the dominance between individuals efficiently in MOP. Deb et al. proposed a fast non-dominated sorting algorithm in NSGA-II. The algorithm divides the population into several disjoint fronts $\{F_1, F_2, ..., F_m\}$ with the acceptable time complexity, and these fronts satisfies the dominance relation $F_1 \succ F_2 \succ ... \succ F_m$. Then the external archive or solution set could be updated accordingly. In order to keep the diversity of solutions, NSGA-II also introduced a density indicator named crowding distance as the other updating criterion. The fast non-dominated sorting algorithm provides efficient evaluation and updating framework for many algorithms. However, directly applying them on FWA would obtain a solution set with lower diversity.

Liu et al. adopted another mainstream framework in S-MOFWA. S-MOFWA update the external archive according to the S-metric which is a kind of hypervolumes indicator. Intuitively, S-metric could be regarded as the space that only dominated by a certain solution in the entire solution set, and the solution with better S-metric usually locates in the area with lower density and closer to the Pareto front. Thus, S-metric could unify convergence measure and diversity measure into one indicator, and simplify the framework of multi-objective swarm intelligence algorithm. Whereas, the calculation method of S-metric in S-MOFWA is only applicable for the non-dominated solutions, and the S-metrics of dominated solutions are assigned as 0. This characteristic leads to the lost of population information and reduce the information utilization efficiency.

Based on the previous works, this paper redesigns the operators in FWA, and proposes a Non-dominated Sorting Based Fireworks Algorithm with higher information utilization ratio.

3 Non-dominated Sorting Based Fireworks Algorithm

3.1 Framework

NSFWA is mainly composed of explosion operator, non-dominated sorting based selection operator, multi-objective guided mutation operator, mapping rule and adaptive explosion amplitude mechanism, and its principle to improve the convergence and diversity of the algorithm with the population information.

Initialization. The initialization of NSFWA is same as the single-objective FWA. NSFWA generates N fireworks randomly in the decision space D:

$$\mathbf{x}_i = (x_{i1}, x_{i2}, ..., x_{in}),\ i = 1, 2, ..., N, \tag{1}$$

where n is the dimension of decision space.

Explosion Operator. The explosion operator of NSFWA also follows the single-objective FWA and randomly generates a certain number of explosion sparks in the hyperspace with firework $\mathbf{x}_i$ as the center and explosion amplitude

A_i as the radius. If the generated explosion spark is out of the bound, it would be remapped into the feasible region according to a certain rule. The mapping rule used in NSFWA is the midpoint mapping, and it would be introduced in the following section.

Mapping Rule. Traditional mapping rule is random mapping, that is, if some dimensions of the spark is out of the bound, the values of the corresponding dimensions would be randomly generated again until the spark is completely within the feasible region. The explosion amplitude is usually decreased during the search process, thus, FWA is tend to have a poor performance on the problem that the global optimum locates near the bound. And the random mapping would exacerbate the problem sometimes. Shown as the Algorithm 1, midpoint mapping rule would reset the dimension that out of the bound as the midpoint of the bound and firework.

Algorithm 1. Midpoint Mapping Rule

Input: Firework $\mathbf{x}_{ij}$, explosion spark $\mathbf{s}_{ij}$, upper bound ub, lower bound lb
Output: Explosion spark $\mathbf{s}_{ij}$
1: **for** $k = 1$ to n **do**
2: **if** $s_{ij}^{(k)} > ub$ **then**
3: $s_{ij}^{(k)} \leftarrow \frac{1}{2}(x_{ij}^{(k)} + ub)$
4: **end if**
5: **if** $s_{ij}^{(k)} < lb$ **then**
6: $s_{ij}^{(k)} \leftarrow \frac{1}{2}(x_{ij}^{(k)} + lb)$
7: **end if**
8: **end for**
9: **return** explosion spark $\mathbf{s}_{ij}$

Compared with other mapping rules, midpoint mapping could help the population find the optimum near the bound of feasible region, and ensure that population also has the ability to escape from the bound.

Selection Operator. Non-dominated sorting based selection operator is used to update the external archive and select new fireworks. In NSFWA, fireworks, explosion sparks and the individuals in the external archive compose the candidate pool, and the selection operator would select N_R individuals from the candidate pool in to the external archive of the next generation, where the top N individuals would be identified as the new fireworks. Concretely, the operator divides the candidate pool C into several disjoint fronts $\{F_1, F_2, ..., F_m\}$ according to the fast non-dominated sorting algorithm proposed by Deb, and these fronts satisfied the definition and dominance listed as the following:

$$F_k = \{\mathbf{x} | n_{\mathbf{x}} = k, \mathbf{x} \in C\}, \quad (2)$$

$$F_1 \succ F_2 \succ ... \succ F_m, \tag{3}$$

where $n_{\mathbf{x}}$ is the number of individuals that dominate $\mathbf{x}$. Then, the candidate solutions would be put into the external archive from front F_1 successively, until a certain front F_k cannot be entirely put in. In order to determine which candidate solutions in F_k are supposed to be put into archive, NSGA-II introduces the crowding distance as the indicator. For solution $\mathbf{x}_m$ in front F_k, its crowding distance can be defined as the following:

$$D(\mathbf{x}_m) = \sum_{i=1}^{r} |f_i(\mathbf{x}_{m+1}) - f_i(\mathbf{x}_{m-1})|, \tag{4}$$

where $\mathbf{x}_{m-1}$ and $\mathbf{x}_{m+1}$ are the neighbors of $\mathbf{x}_m$, and r is the number of objective function. Fig. 1 shows the calculation of the crowding distance.

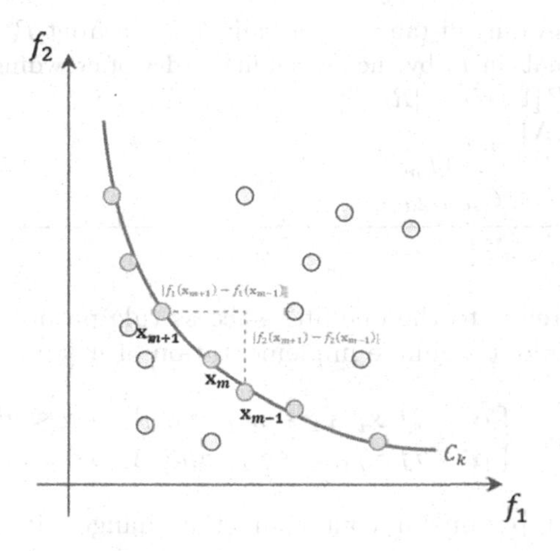

Fig. 1. The calculation of crowding distance.

The solutions with larger crowding distance tend to locate in a low density area, and would be selected into the archive. As mentioned above, explosion sparks are generated in a specific hyperspace centered on fireworks, which means that the diversity of the entire solution group is highly related to the density of fireworks' location. Therefore, NSFWA not only sorts the front P_k but also the first front P_1 to ensure diversity of fireworks. The entire process of selection operator is shown as the Algorithm 2.

Adaptive Amplitude Mechanism. Explosion amplitude is one the decisive factors of the global exploration and local exploitation. In NSFWA, amplitude is adjusted adaptively according to the dominance relation between the current fireworks and the previous fireworks in each generation. The parameter setting

Algorithm 2. Selection Operator

Input: Fireworks $X_t = \{\mathbf{x}_1, \mathbf{x}_2, ...\mathbf{x}_N\}$, sparks $S_t = \{S_{1,t}, S_{2,t}, ..., S_{N,t}\}$, external archive R_t, size of archive N_R

Output: external archive R_{t+1}, fireworks X_{t+1}, sorted candidate pool $C_{sorted,t}$

1: Calculate the fitness values of X_t and S_t
2: Generate $C_t = \{X_t \cup S_t \cup R_t\}$
3: Sort the candidate pool $P = \text{FastNonDominatedSorting}(C_t), P = \{P_1, P_2, ..., P_m\}$
4: Declare the external archive $R_{t+1} = \emptyset$ and the counter $i = 1$
5: **while** $|R_{t+1}| + |P_i| \leq N_R$ **do**
6:     **if** i=1 **then**
7:         Calculate the crowding distance D_i of individuals in front P_i
8:         Sort the individuals in P_i by the descending order of crowding distance
9:     **end if**
10:     $R_{t+1} = R_{t+1} \cup P_i$
11:     $i = i + 1$
12: **end while**
13: Calculate the crowding distance D_i of individuals in front P_i
14: Sort the individuals in P_i by the descending order of crowding distance
15: $R_{t+1} = R_{t+1} \cup P_i[1 : N_R - |R_{t+1}|]$
16: $X_{t+1} = R_{t+1}[1 : N]$
17: $C_{sorted,t} = P_1 \cup P_2 \cup ... \cup P_m$
18: **return** R_{t+1}, X_{t+1}, $C_{sorted,t}$

of the mechanism refers to the one-fifth success rule proposed by Schumer and Steiglitz [13], and adopt a simple implementation of it [6]:

$$A_{i,t+1} = A_{i,t} \cdot \begin{cases} \alpha & if\ \mathbf{x}_{i,t+1} \succ \mathbf{x}_{i,t}\ and\ A_{i,t} \cdot \alpha \leq ub - lb \\ \alpha^{-\frac{1}{4}} & if\ \mathbf{x}_{i,t+1} \preceq \mathbf{x}_{i,t}\ and\ A_{i,t} \cdot \alpha \geq \beta \cdot (ub - lb), \end{cases} \tag{5}$$

where α is a hyper parameter controlling the change rate of amplitude, and β is a hyper parameter used to set the minimum of amplitude. In the early phase of search, there is a relatively high probability for population to find a better solution, and the amplitude tends to increase. On the contrary, it is more difficult for the population to make progress, and the amplitude tends to decrease. Therefore, the amplitude usually changes from large to small, which means that the population is encouraged to explore globally in the early phase and exploit a certain local area in the later.

Mutation Operator. In order to further enhance the local search capability of the algorithm by using the population information, NSFWA designs a novel operator named multi-objective guided mutation operator. Different from the single-objective FWA, the guided mutation operator in MOP must be executed after the selection operator obtaining the fitness information of populations. The main idea of the mutation operator is to calculate the difference between solutions and the solutions dominated by them, and add the difference on the location of fireworks to generate the mutation sparks. These mutation sparks usually has a better fitness compared with the fireworks. The entire process is shown as the Algorithm 3.

Algorithm 3. Multi-objective Guided Mutation Operator

Input: Firework $\mathbf{x}_{i,t+1}$, sorted candidate pool $C_{sorted,t} = \{\mathbf{c}_1, \mathbf{c}_2, ..., \mathbf{c}_{|C|}\}$, group ratio σ, group size μ
Output: Guided mutation spark $\mathbf{g}_{i,t+1}$
1: **if** Firework $\mathbf{x}_{i,t+1}$ is not from the external archive **then**
2: Extract the population of $\mathbf{x}_i$ from $C_{sorted,t}$ and keep the relative order
3: Calculate the guiding vector $\Delta_i = \frac{1}{\sigma\lambda_i}(\sum_{j=1}^{\sigma\lambda_i} \mathbf{s}_{ij} - \sum_{j=\lambda_i-\sigma\lambda_i+1}^{\lambda_i} \mathbf{s}_{ij})$
4: **end if**
5: **if** Firework $\mathbf{x}_i$ is from the external archive **then**
6: $\Delta_i = \frac{1}{\mu}(\sum_{j=1}^{\mu} \mathbf{c}_{rand(0,\sigma|C|)} - \sum_{j=1}^{\mu} \mathbf{c}_{rand(\sigma|C|-\mu+1,\sigma|C|)})$
7: **end if**
8: Generate the explosion spark $\mathbf{g}_i = \mathbf{x}_i + \Delta_i$
9: **return** $\mathbf{g}_i$

To accelerate the convergence, the mutation operator directly use new firework selected by the selection operator to calculate the guiding spark (GS). However, the firework might come from the external archive and have already lost its population, thus the mutation calculate the guiding vector (GV) with two different methods:

1. If the firework $\mathbf{x}_{i,t+1}$ is not from the external archive: the firework have its own population in the candidate pool, and the population is already sorted after the selection operator. The guiding vector would be calculated as the difference between the centroid of the top $\sigma\lambda_i$ sparks and the bottom $\sigma\lambda_i$ sparks in the population.
2. If the firework $\mathbf{x}_{i,t+1}$ is from external archive: the operator calculate the difference between two groups that formed by μ solutions in the top $\sigma|C|$ solutions and μ solutions in the bottom $\sigma|C|$ solutions respectively to generate the guiding vector. Here, the μ solutions in two groups are selected randomly, so as to avoid that fireworks share a same guiding vector and affect the diversity of the final solution set.

Guiding sparks would be put into the external archive and participate the selection in the next generation, but not replace fireworks directly.

The framework of the Non-dominated Sorting Based Fireworks Algorithm is shown as the Algorithm 4.

3.2 Principle and Analysis

Analysis of Multi-objective Guided Mutation Operator. The main purpose of multi-objective guided muation opeartor is to improve the search capability of NSFWA without affecting the diversity of the solution set.

Algorithm 4. Non-dominated Sorting Based Fireworks Algorithm

Input: upper bound ub, lower bound lb, number of firework$N$, number of spark λ, external archive size N_R, group ratio σ, group size μ, explosion amplitude A, change rate α, minimum parameter β
Output: Optimal solution set
1: Initialize N fireworks in the decision space D bounded by lb and ub
2: Declare the external archive $R = \{X_1\}$ and generation counter $t = 1$
3: **while** termination condition not met **do**
4: Generate explosion sparks $S_t = \text{ExplosionOperator}(X_t, A_t, \lambda_t)$
5: Update external archive, candidate pool and fireworks $R_{t+1}, C_{sorted,t}, X_{t+1} = \text{SelectionOperator}(X_t, S_t, R_t)$
6: Adjust explosion amplitude $A_{t+1} = \text{AdaptiveAmplitude}(X_t, X_{t+1}, A_t, \alpha, \beta)$
7: Generate mutation sparks $G_t = \text{MultiObjectiveMutationOperator}(X_t, C_{sorted,t}, \sigma, \mu)$
8: Update external archive $R_{t+1} = R_{t+1} \cup G_t$
9: $t = t + 1$
10: **end while**
11: **return** $\{R_t \backslash G_{t-1}\}$

Different with the single-objective optimization, the multi-objective optimization algorithm searches for a entire Pareto front composed of several Pareto optimums rather than a certain global optimum. Therefore, all directions that make the new individual generated on the direction better than the original fireworks are acceptable related directions. The following visualization will explain why the guiding spark could guide the population to search along the relevant direction.

Suppose the optimization problem is ZDT1 test function, which includes two convex objective functions. Set the dimension of decision vector as 30, and the Pareto optimal solutions of ZDT1 is 0 on all dimensions except x_1. The objective functions could be visualized on the first two dimensions as the Fig. 2a, where the pink segment is the projection of Pareto front on the objective function. And the guiding vector could be approximately decomposed as the weighted sum of negative gradients of two objective functions (See Fig. 2b–Fig. 2d):

$$\Delta = w_1 \nabla_1 + w_2 \nabla_2, \; w_1 \leq 0 \; and \; w_2 \leq 0, \tag{6}$$

where ∇_1 and ∇_2 are gradients.

Thus, guiding spark is likely to obtain a better fitness value on at least one objective function compared with the firework, which means these non-dominated elite individuals might be selected as the firework in the next generation. Guided by the elite individuals, populations could approach the Pareto front stably and the search ability of NSFWA are also enhanced.

Selection of Parameters. Now, suppose there are two fireworks that selected from the external archive and locate in a same region. If they share a same guiding vector, then their population is tend to move towards a same region

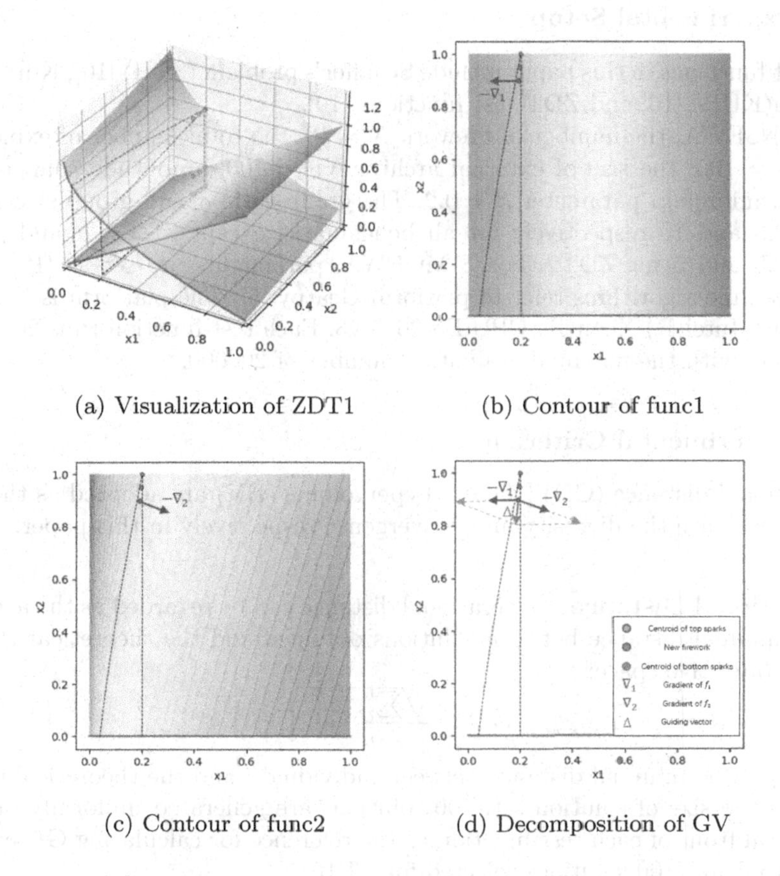

(a) Visualization of ZDT1 (b) Contour of func1

(c) Contour of func2 (d) Decomposition of GV

Fig. 2. Principle of multi-objective guided mutation function.

of the Pareto front, and this would be harmful for the diversity. The random mechanism is introduced to alleviate the problem. Generally speaking, the larger group ratio σ and smaller group size μ means stronger randomness, and the diversity of the solution set could be better. On the contrary, the convergence could be accelerated but the diversity might be weakened. The analysis above could be a basis for selection of parameters.

4 Experiments

To illustrate the performance of NSFWA, experiments on several test functions were conducted, and S-MOFWA, NSGA-II, SPEA2 and RVEA [3] are selected as baselines. Besides, the ablation experiments were also conducted to verify the effectiveness of operators and mechanisms in NSFWA.

4.1 Experimental Setup

The test functions in this paper include Schaffer's problem (SCH) [16], Kursawe's problem(KUR) [16] and ZDT test functions [19].

For NSFWA, the number of firework $N = 10$, the total number of explosion spark $\lambda = 100$, the size of external archive $N_R = 100$, amplitude change rate $\alpha = 1.2$, minimum parameter $\beta = 0.2$. The group ratio σ and group size μ are set as 0.3 and 10 respectively for all benchmarks except ZDT2. σ and μ are set as 0.5 and 5 for ZDT2. For S-MOFWA, parameters are set as [12]. And other baseline algorithms refer to platform Geatpy [5]. The platform is Ubuntu 18.04 with Intel(R) Xeon(R) CPU E5-2675 v3. Each test function runs 20 times repeatedly with the maximal evaluation number of 200000.

4.2 Experimental Criterion

Generational distance (GD) [15] and hypervolume (HV) are adopted as the criteria to evaluate the diversity and convergence respectively in this paper.

Generational Distance. Generational distance can be regarded as the average of the minimal distance between solutions obtained and the theoretical Pareto front in objective space:

$$GD = \frac{\sqrt{\sum_{i=1}^{n} d_i^2}}{n}, \tag{7}$$

where d_i is the minimal distance between individual i and the theoretical front, and n is the size of solution set. 500 solutions are generated uniformly on the theoretical front of each test function as the reference for calculating GD except KUR problem. (100 solutions selected for KUR.)

Hypervolume. Hypervolume is one of the most applied criterion of MOP. HV is the volume of the objective space that covered by optimal solution set obtained:

$$S(M) = \Lambda(\cup_{i=1}^{n}\{\mathbf{x}|\mathbf{x}_i \succ \mathbf{x} \succ \mathbf{x}_{ref}\}), \tag{8}$$

where Λ represents Lebesgue measure, and $\mathbf{x}_{ref}$ is a reference point that dominated by all solutions. Actually, HV can not only evaluate the diversity but also the convergence. The reference point selected for SCH, KUR and ZDT1-6 are $(4, 4)$, $(-14, 1)$, and $(1, 1)$ respectively.

4.3 Experimental Results

Ablation Experiments. To verify the effectiveness of the selection operator, mutation operator and mapping rule proposed in this paper, ablation experiments take the following algorithm as comparison: (i) NSFWA - CD: NSFWA without crowding distance sorting of firework in selection operator, (ii) NSFWA - GS: NSFWA without multi-objective guided mutation operator and (iii) NSFWA

Table 1. Generational distance of ablation experiments.

	NSFWA		NSFWA - CD		NSFWA - GS		NSFWA + RM	
Func.	Mean	Std	Mean	Std	Mean	Std	Mean	Std
SCH	3.27E−03	3.45E−04	**3.17E−03**	1.74E−04	3.16E−03	**2.88E−04**	3.29E−03	**1.10E−04**
KUR	5.10E−02	2.91E−03	**5.04E−02**	**1.39E−03**	5.46E−02	2.45E−03	5.02E−02	1.84E−03
ZDT1	**1.10E−03**	**3.69E−05**	1.21E−03	1.15E−04	8.22E−02	3.49E−03	8.10E−01	2.88E−02
ZDT2	8.03E−04	**3.42E−05**	**7.84E−04**	5.88E−05	1.41E−01	4.27E−03	1.44E−01	2.24E−03
ZDT3	**1.07E−03**	8.82E−05	1.19E−03	**5.21E−05**	4.72E−02	6.99E−04	8.25E−01	3.15E−02
ZDT6	**5.92E−04**	**2.32E−05**	7.43E−03	7.41E−04	1.24E+00	3.54E−01	1.48E+00	1.61E−01

Table 2. Hypervolume of ablation experiments.

	NSFWA		NSFWA - CD		NSFWA - GS		NSFWA + RM	
Func	Mean	Std	Mean	Std	Mean	Std	Mean	Std
SCH	1.33E+01	**3.79E−05**	**1.45E+01**	3.30E−03	1.32E+01	2.98E−03	1.32E+01	1.44E−03
KUR	**3.68E+01**	**4.03E−02**	3.66E+01	5.47E−02	3.65E+01	4.31E−02	3.68E+01	7.69E−02
ZDT1	**6.61E−01**	**3.24E−05**	6.59E−01	1.83E−04	5.48E−01	4.51E−03	1.23E−01	2.78E−02
ZDT2	**3.28E−01**	1.18E−04	3.26E−01	**9.61E−05**	1.68E−01	2.81E−03	1.67E−01	2.67E−03
ZDT3	**1.04E+00**	**9.71E−06**	1.04E+00	6.22E−05	8.91E−01	8.67E−03	3.51E−01	3.80E−02
ZDT6	**3.21E−01**	**7.93E−04**	3.18E−01	3.30E−03	2.20E−01	9.79E−02	1.81E−01	9.97E−02

+ RM: NSFWA using random mapping rule. The experimental results are shown as Table 1 and Table 2.

Complete NSFWA wins a better HV than NSFWA without crowding distance sorting, and it could be seen that sorting fireworks in selection operator could improve the diversity. The better GD indicates that guided mutation operator improves the search capability of NSFWA significantly. And the HV curve (See Fig. 3) also proves that the mutation operator could accelerate the convergence of NSFWA. NSFWA using midpoint mapping outperforms NSFWA using random mapping obviously on ZDT test functions. It is worthy noting that most of optimal solutions of ZDT locate near the lower bound, which means that midpoint mapping improve the performance of FWA on the kind of problem.

Comparison with Other Algorithms. Table 3 and Table 4 gives the results of NSFWA and other algorithms. The average rank of GD of NSFWA is 1.83 and the average rank of HV is 1.50, which is the best compared with other algorithms. The difference between the average rank of GD and HV indicates that NSFWA performs better on the diversity and could obtain a solution set covering larger target space.

Among the benchmarks, SCH and ZDT1 have convex Pareto fronts. The means and standard deviations on these two problems show that NSFWA has a stable and good performance on problem with a convex front. KUR and ZDT2 has non-convex Pareto fronts. The performance of NSFWA is slightly worse than *S*-MOFWA and NSGA-II on KUR, but better on ZDT2. As mentioned above, the setting of group ratio σ and group size μ for ZDT2 is different with other problems. It is inferred that populations is easily to be trapped in a certain part

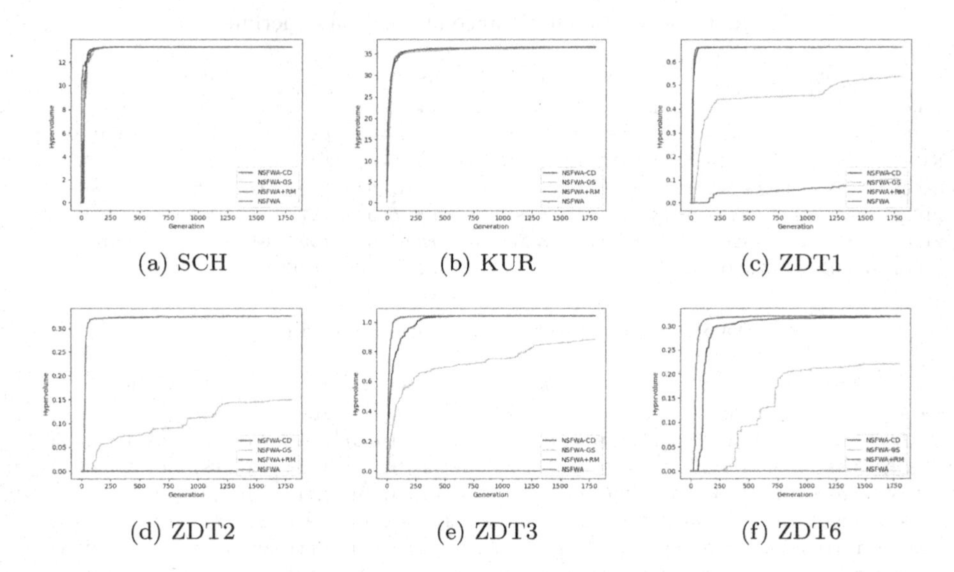

(a) SCH (b) KUR (c) ZDT1

(d) ZDT2 (e) ZDT3 (f) ZDT6

Fig. 3. Hypervolume curve of ablation experiments.

of global optimum in the test of ZDT2, and thus there is a need of stronger randomness to help populations get rid of the area. The Pareto front of ZDT3 is composed of several non-contiguous convex parts, which also requires better diversity of population, and NSFWA outperforms other algorithms on both GD and HV. The decision space and Pareto front of ZDT6 is non-uniform. The density of individuals is gradually lower when they locate closer to the Pareto front. NSFWA ranked second on both GD and HV with a small gap from the best. Generally speaking, the results indicate that NSFWA performs well on kinds of functions with different characteristics of Pareto front. The solution set obtained by NSFWA is visualized as Fig. 4.

Table 3. Generational distance of NSFWA and other algorithms.

	NSFWA		*S*-MOFWA		NSGA-II		SPEA2		RVEA	
Func.	Mean	Std	Mean	Std	Mean	Std	Mean	Std	Mean	Std
SCH	3.27E−03	3.45E−04	3.32E−03	9.62E−05	3.33E−03	**7.82E−05**	4.16E−03	4.27E−04	**3.03E−03**	1.83E−04
KUR	5.10E−02	2.91E−03	**3.57E−02**	1.83E−03	3.77E−02	3.06E−03	6.84E−01	**1.12E−03**	4.05E−02	8.25E−03
ZDT1	**1.10E−03**	**3.69E−05**	1.47E−03	4.37E−05	1.41E−03	8.55E−05	1.69E−03	1.90E−04	1.76E−03	3.48E−04
ZDT2	**8.03E−04**	3.42E−05	1.17E−03	6.22E−05	1.06E−03	1.54E−04	1.04E−03	**1.14E−05**	1.21E−03	2.06E−04
ZDT3	**1.07E−03**	8.82E−05	4.01E−03	1.88E−03	1.09E−03	**6.96E−05**	1.96E−01	1.95E−01	1.64E−03	1.15E−04
ZDT6	5.92E−04	2.32E−05	**5.66E−04**	**1.83E−05**	6.30E−04	9.65E−05	1.09E−01	6.87E−02	6.44E−04	1.86E−05
AR	1.83	2.50	2.67	2.33	2.67	2.83	4.33	3.50	3.50	3.83

Table 4. Hypervolume of NSFWA and other algorithms.

	NSFWA		*S*-MOFWA		NSGA-II		SPEA2		RVEA	
Func.	Mean	Std	Mean	Std	Mean	Std	Mean	Std	Mean	Std
SCH	**1.33E+01**	**3.79E−05**	1.32E+01	7.62E−03	1.33E+01	1.21E−03	1.30E+01	7.52E−02	1.32E+01	3.09E−03
KUR	3.68E+01	4.03E−02	**3.71E+01**	9.02E−02	3.70E+01	**1.20E−02**	2.84E+01	2.92E−02	3.66E+01	1.31E−01
ZDT1	**6.61E−01**	**3.24E−05**	6.54E−01	1.03E−04	6.60E−01	2.77E−04	6.52E−01	2.23E−03	6.60E−01	4.84E−04
ZDT2	**3.28E−01**	**1.18E−04**	3.27E−01	4.00E−04	3.27E−01	2.24E−04	3.20E−01	2.62E−03	3.27E−01	4.09E−04
ZDT3	**1.04E+00**	**9.71E−06**	1.04E+00	3.20E−04	1.04E+00	1.44E−04	6.74E−01	3.58E−01	1.04E+00	3.76E−04
ZDT6	3.21E−01	7.93E−04	3.20E−01	5.32E−04	3.21E−01	3.26E−04	3.13E−01	3.43E−03	**3.22E−01**	**5.26E−05**
AR	1.50	1.83	2.50	3.16	2.67	2.00	5.00	4.50	3.33	3.50

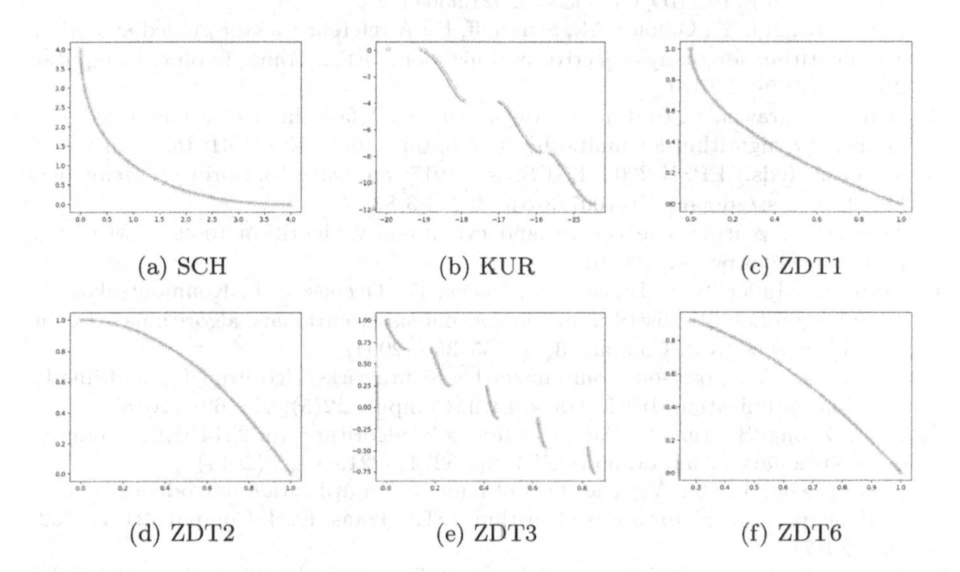

(a) SCH (b) KUR (c) ZDT1

(d) ZDT2 (e) ZDT3 (f) ZDT6

Fig. 4. Solution set obtained by NSFWA.

5 Conclusion

In this paper, a novel multi-objective FWA named non-dominated sorting based fireworks algorithm is proposed. Non-dominated sorting based selection operator updates the external archive and selects fireworks according to the dominance relation and density to improve the diversity of solution set. Then, a multi-objective guided mutation operator is used to generate elite individuals for each populations to accelerate the convergence and enhance the stability. In order to further boost the performance of NSFWA on the problem that optimums locate near the bound, a novel mapping rule named midpoint mapping was proposed. Experiments on several test functions with different properties indicate that NSFWA has good performance on kinds of multi-optimization problems.

Acknowledgments. This work is supported by the National Natural Science Foundation of China (Grant No. 62076010), and partially supported by Science and Technology

Innovation 2030 - "New Generation Artificial Intelligence" Major Project (Grant Nos.: 2018AAA0102301 and 2018AAA0100302).

References

1. Beume, N., Naujoks, B., Emmerich, M.: SMS-EMOA: multiobjective selection based on dominated hypervolume. Eur. J. Oper. Res. **181**(3), 1653–1669 (2007)
2. Chen, M., Tan, Y.: Exponentially decaying explosion in fireworks algorithm. In: 2021 IEEE Congress on Evolutionary Computation (CEC), pp. 1406–1413 (2021). https://doi.org/10.1109/CEC45853.2021.9504974
3. Cheng, R., Jin, Y., Olhofer, M., Sendhoff, B.: A reference vector guided evolutionary algorithm for many-objective optimization. IEEE Trans. Evolut. Computat. **20**(5), 773–791 (2016)
4. Deb, K., Agrawal, S., Pratap, A., Meyarivan, T.: A fast elitist non-dominated sorting genetic algorithm for multi-objective optimization: NSGA-II. In: Schoenauer, M., et al. (eds.) PPSN 2000. LNCS, vol. 1917, pp. 849–858. Springer, Heidelberg (2000). https://doi.org/10.1007/3-540-45356-3_83
5. Jazzbin, e.: geatpy: The genetic and evolutionary algorithm toolbox with high performance in python (2020)
6. Kern, S., Müller, S.D., Hansen, N., Büche, D., Ocenasek, J., Koumoutsakos, P.: Learning probability distributions in continuous evolutionary algorithms - a comparative review. Nat. Comput. **3**(3), 355–356 (2004)
7. Li, J., Tan, Y.: Loser-out tournament-based fireworks algorithm for multimodal function optimization. IEEE Trans. Evol. Comput. **22**(5), 679–691 (2018)
8. Li, J., Zheng, S., Tan, Y.: Adaptive fireworks algorithm. In: 2014 IEEE Congress on evolutionary computation (CEC), pp. 3214–3221. IEEE (2014)
9. Li, J., Zheng, S., Tan, Y.: The effect of information utilization: introducing a novel guiding spark in the fireworks algorithm. IEEE Trans. Evol. Comput. **21**(1), 153–166 (2017)
10. Li, Y., Tan, Y.: Multi-scale collaborative fireworks algorithm. In: 2020 IEEE Congress on Evolutionary Computation (CEC), pp. 1–8. IEEE (2020)
11. Li, Y., Tan, Y.: Enhancing fireworks algorithm in local adaptation and global collaboration. In: Tan, Y., Shi, Y. (eds.) ICSI 2021. LNCS, vol. 12689, pp. 451–465. Springer, Cham (2021). https://doi.org/10.1007/978-3-030-78743-1_41
12. Liu, L., Zheng, S., Tan, Y.: S-metric based multi-objective fireworks algorithm. In: 2015 IEEE Congress on Evolutionary Computation (CEC), pp. 1257–1264 (2015)
13. Schumer, M., Steiglitz, K.: Adaptive step size random search. IEEE Trans. Autom. Control **13**(3), 270–276 (1968). https://doi.org/10.1109/TAC.1968.1098903
14. Tan, Y., Zhu, Y.: Fireworks algorithm for optimization. In: Tan, Y., Shi, Y., Tan, K.C. (eds.) ICSI 2010. LNCS, vol. 6145, pp. 355–364. Springer, Heidelberg (2010). https://doi.org/10.1007/978-3-642-13495-1_44
15. Van Veldhuizen, D.A., Lamont, G.B., et al.: Evolutionary computation and convergence to a pareto front. In: Late Breaking Papers at the Genetic Programming 1998 Conference, pp. 221–228. Citeseer (1998)
16. Van Veldhuizen, D.A.: Multiobjective evolutionary algorithms: classifications, analyses, and new innovations. Air Force Institute of Technology (1999)
17. Zheng, S., Janecek, A., Li, J., Tan, Y.: Dynamic search in fireworks algorithm. In: Proceedings of the IEEE Congress on Evolutionary Computation, CEC 2014, Beijing, China, 6–11 July 2014, pp. 3222–3229. IEEE (2014). https://doi.org/10.1109/CEC.2014.6900485, https://doi.org/10.1109/CEC.2014.6900485

18. Zheng, S., Janecek, A., Tan, Y.: Enhanced fireworks algorithm. In: 2013 IEEE Congress on Evolutionary Computation, pp. 2069–2077. IEEE (2013)
19. Zitzler, E., Deb, K., Thiele, L.: Comparison of multiobjective evolutionary algorithms: empirical results. Evol. Comput. **8**(2), 173–195 (2000)
20. Zitzler, E., Laumanns, M., Thiele, L.: SPEA 2: improving the strength pareto evolutionary algorithm. TIK-report 103 (2001)

Multi-objective Evolutionary Algorithm with Adaptive Fitting Dominant Hyperplane

Zhiqi Zhang, Limin Wang, Xin Yang, Xuming Han(✉), and Lin Yue

Jinan University, Guangzhou, China
hanxuming@jnu.edu.cn

Abstract. Most of the existing multi-objective optimization algorithms try to evenly distribute all solutions in the objective space. But for the irregular Pareto front(PF), it is difficult to find the real PF. Aiming at the multi-objective optimization problem with complex PF, a multi-objective evolutionary algorithm for adaptive fitting dominant hyperplane ($MOEA\_DH$) is developed. Before each iteration, non-dominated sorting is applied on all candidate solutions. Solutions in the first front are used to fit a hyperplane in the objective space, which is called the current dominant hyperplane(DH). DH reflects the evolution trend of the current generation of non-dominanted solutions and guides the rapid convergence of dominanted solutions. A new partial ordering relation determined by front number and crowding distance on DH is set. When solving CF benchmark problems from multi-objective optimization in IEEE Congress on Evolutionary Computation 2019, the experiments validate our advantages to get the PF with better convergence and diversity.

Keywords: Multi-objective optimization · Evolutionary algorithm · Dominant hyperplane · Crowding distance

1 Introduction

The problems that the number of optimization objectives is greater than or equal to 2 and the objectives conflict with each other is called multi-objective optimization problems ($MOPs$) [10]. Mops widely appear in practical problems of industry and engineering, such as wireless sensor network [8,9],data mining [1,2] and resource allocation [6,11]. The balance between diversity and convergence [7] of $MOPs$ is a problem that many scholars are committed to solving. More and more multi-objective evolutionary algorithms ($MOEA$) have been proposed to solve $MOPs$. These algorithms can be roughly divided into the Pareto dominance based $MOEA$, the decomposition based $MOEA$ and the indicator based $MOEA$.

The Pareto dominance based $MOEA$ has the advantages of simple principle and few parameters. Pareto dominance based mechanisms are used to

Y. Tan et al. (Eds.): ICSI 2022, LNCS 13344, pp. 472–481, 2022.
https://doi.org/10.1007/978-3-031-09677-8_39

select candidate solutions. A fast non-dominated sorting algorithm is proposed in $NSGA-II$ [5] to ensure the diversity of population with crowding distance. The non dominanted sorting genetic algorithm III ($NSGA-III$) [4] introduces the reference points to retain the population individuals which are non-dominated and close to the reference points. However, it generates uniform reference points in the objective space. It is hard to get a complex PF.

The decomposition based $MOEA$ divides the traditional mop into several single objective optimization subproblems, which are solved by using the information between subproblems. For example: Multiobjective EA based on decomposition ($MOEA/D$) [14], but this kind of method can not guarantee to completely generate a set of uniformly distributed solution sets. $RVEA$ [3] is an algorithm with a reference vector to guide the population evolutionary. The reference vector can not only be used to decompose the original multi-objective optimization problem into multiple single objective subproblems, but also clarify user preferences, aiming at the preferred subset of the whole PF. A scalarization method called angle penalty distance is used to balance the convergence and diversity of solutions. And the distribution of reference vector is dynamically adjusted according to the scale of objective function.

The indicator based $MOEA$ is mainly used to test the performance of solutions based on indicators and select high-quality solutions,such as indicator based EA ($IBEA$) [15]. Also a grid based multi-objective optimization algorithm($GrEA$) [12] is proposed, which needs to complex parameters. A novel preference based dominion relation for evolutionary multiobjective ($ARMOEA$) [13] proposes a method based on an enhanced inverted general distance indicator, in which an adaptation method is proposed to adjust a group of reference points according to the index contribution of candidate solutions in external files. However, in some situations, such methods need to spend a lot of time to calculate the value of performance indicators.

Although the above algorithms show good results in $MOPs$, different kinds of algorithms have certain defects. Through the comparative analysis of the three types of algorithms and taking their advantages, we propose a multi-objective evolutionary algorithm for adaptive fitting the dominant hyperplane. The contributions are as follows:

1) Before each iteration, the solution of the first layer of non dominated ranking is fitted into the dominant hyperplane in the objective space, which can well reflect the shape of the real PF.
2) The crowding distance between the points projected to the dominant hyperplane is proposed as a new index to screen the last front. Since the solution before the last front is also projected onto the dominant hyperplane, all projected points are covered when calculating the congestion distance, and the truly evenly distributed last front can be selected into the next generation.
3) This paper constructs a new partial order relation. Firstly the front numbers of individuals are judged, the one with the smallest front number is the best solution. When the front numbers are the same, the crowding distance of the projected points on the dominant hyperplane is considered. The one with

largest crowding distance is the best solution. The order relation speeds up the convergence speed of the algorithm under the condition of ensuring the diversity of the population.

2 Related Work

2.1 Multi Objective Optimization Problems (*MOPs*)

Usually two or more conflicting objectives are optimized at the same time. The mathematical definition of *MOPs* is as Eq. 1:

$$\min_{x} F(x) = (f_1(x), ..., f_m(x)),\ s.t.\ x \in X, \tag{1}$$

where $X = (x_1, x_2, ..., x_n)$ is the decision vector. X is the decision space. M is the number of objective problems, and F is the objective vector. Generally speaking, no optimal solution can optimize all objectives at the same time. In order to balance multiple objectives, a set of optimal solutions can be obtained, which is called Pareto optimal solution. In most cases, Pareto optimal solution is also called Pareto set (PS) in decision space, and its corresponding objective vector is called Pareto front (PF).

2.2 The Coordinate of an Arbitrary Point Projected to a Hyperplane

An n-dimensional hyperplane H can be represented by Eq. 2:

$$\lambda_0 + \lambda_1 \cdot x_1 + \lambda_2 \cdot x_2 + ... + \lambda_n \cdot x_n = 0. \tag{2}$$

A point $x_a = (x_a1, ..., x_an)$ in n-dimensional space is projected onto H, and its projected coordinate is $x_p = (x_p1, ...x_pn)$, because the line $x_a x_p$ is perpendicular to H,the following relationship can be obtained according to the vertical constraints:

$$\frac{(x_{p1} - x_{a1})}{\lambda_1} = \frac{(x_{p2} - x_{a2})}{\lambda_2} = ... = \frac{(x_{pn} - x_{an})}{\lambda_n}. \tag{3}$$

By solving the Eq. 2 and Eq. 3, we can get Eq. 4

$$\begin{aligned} x_{pi} =& ((\lambda_0 + \lambda_1^2 + ... + \lambda_{i-1}^2 + \lambda_{i+1}^2 + ... + \lambda_n^2) \cdot x_{ai} - \lambda_i(\lambda_0 + \lambda_1 x_{a1} + ... \\ &+ \lambda_{i-1} x_{a-1} + \lambda_i x_a + 1 + ... + \lambda_n x_{an}))/\left(\lambda_1^2 + ... + \lambda_n^2\right). \end{aligned} \tag{4}$$

3 Multi-objective Evolutionary Algorithm with Adaptive Fitting Dominant Hyperplane

3.1 Algorithm Flow

The overall process is shown in Algorithm 1. Mating selection generates mating pool. Genetic algorithm(GA) generates the offspring. Environment selection is to get the population into the next generation.

Algorithm 1. MOEA_DH

Require: N: the population size; M: the number of objectives; T_{max}: the maximum number of iterations
Ensure: the final population;
1: Initialize the population
2: **while** $t < T_{max}$ **do**
3: MatingPool = MatingSelection(FrontNo,crowding distance)
4: Offspring = GA(Population(MatingPool))
5: [FrontNo,dist,Population]=EnvironmentSelection([Population,Offspring])
6: $t = t + 1$
7: **end while**
8: **return** Population

MatingSelection is shown in Algorithm 2. We randomly selects K solutions after non-dominated sorting, setting front number as the first index. If there are more than two solutions with the smallest front number, CD is token as the second index. And the solution with the largest CD enters the mating pool. Similar to the tournament selection algorithm of $NSGAII$, the difference is that our algorithm takes the distance between the points projected onto the DH as CD. Since all solutions are projected onto the same DH, individuals with different front numbers will squeeze each other, which further ensures the diversity of the population.

Algorithm 2. MatingSelection

Require: k:number of randomly selected points;$FrontNo$:the numbers of non-dominated sorting $FrontNo$,CD:crowding distance
Ensure: MatingPool
1: MatingPool = *Empty*
2: *Pop_k* = Select k individuals randomly
3: **while** size(MatingPool)¡N **do**
4: Best_index = min($FrontNo$ of *Pop_k*)
5: **if** there are more than 2 Best_indexes **then**
6: Best_index = max(crowding dist of *Pop_k*)
7: **end if**
8: MatingPool = $MatingPool \cup Best\_index$
9: **end while**
10: **return** MatingPool

Environment Selection is shown in Algorithm 3. First, all individuals including parents and offspring are non-dominated sorted to find the last front $F(C)$. Then individuals of the first front are picked, and fit it as DH. All solutions are projected to DH. We calculate the distance between the projected points. Finally, the points with the smallest CD are removed until the number of remaining individuals equals to N.

For the calculation of CD in Algorithm 4, all individuals are projected to the DH. Except for the edge points, the sum of the distances of each point to the nearest two points is the crowding distance. The crowding distance of the edge points is ∞, so as to ensure that the edge points can enter the next generation.

Algorithm 3. EnvironmentSelection

Require: Population;Offspring;M
Ensure: $FrontNo$;CD;Population
1: Pop_temp = $Population \cup Offspring$
2: Normalize the Pop_temp
3: FrontNo=NDsort(Pop_temp)
4: i = 1
5: Pop_F(i)=Pop_temp(FrontNo=i)
6: **while** size of Pop_F(i) ¡ M **do**
7: i = i+1
8: **end while**
9: Dominant Hyperplane(DH) = Linear fit a hyperplane with Pop_F(i)
10: Find the critical layer F(C),m satisfies length(F(1)+F(2)+...+F(C-1))¡ N and length(F(1)+F(2)+...+F(C)) ¡ N
11: Pop_temp = Pop_temp(F(1),F(1),F(2),...,F(C))
12: CD=Calculate the crowding distance on DH
13: **while** length($F(1) + F(2) + ... + F(C)$) $> N$ **do**
14: Remove the individual with min(CD) from Pop_temp($F(C)$)
15: **end while**
16: return Pop_temp
17: MatingPool =
18: Pop_k = Select k individuals randomly
19: **while** size(MatingPool)¡N **do**
20: Best_index = min($FrontNo$ of Pop_k)
21: **if** there are two Best_indexes **then**
22: Best_index = max(crowding distitance of Pop_k)
23: **end if**
24: $MatingPool = MatingPool \cup Best_index$
25: **end while**
26: **return** MatingPool

Algorithm 4. Calculate the crowding distance on DH

Require: Pop_temp
Ensure: CD
1: $Pop\_temp = Population \cup offspring$
2: $Pop\_Projection$ = Project Pop_temp onto DH
3: Calculate the distance between each other
4: **for** $i = 1 \rightarrow N$ **do**
5: CD(j) = min(distance(j) + secondmin(distance(j))
6: **end for**
7: **for** $j = 1 \rightarrow M$ **do**
8: $sorted\_num$=sort($Pop\_Projection(i)$)
9: $CD(Sorted\_num = 1) = CD(Sorted\_num = N) = \infty$
10: **end for**
11: **return** CD

3.2 Computational Complexity

To select N parent solutions for offspring generation, the MatingSelection of Algorithm requires a time complexity of $O(Nlogk)$. k is the number of randomly selected solutions each time and N is the population size. The non-dominated sorting requires a time complexity $O(NlogN)$. The time complexity of fitting hyperplane is related to the number of samples S, and S is usually small. Therefore, the least square method is used to fit hyperplane. The time complexity is O (SM^2). M is the number of objectives. And the time complexity of projecting to DH is 0(N). The calculation of CD in Algorithm 2 requires a time complexity of $O(N^2 + MNlogN)$. GA operation for the population requires a time complexity of $O(MN)$. Because $SM^2 << N^2$, the time complexity of $MOEA\_DH$ is $O(TN^2)$, where T is the maximum number of generations.

4 Experiment

4.1 Experimental Settings

The test problems were $CF1-7$ benchmark problems of IEEE CEC 2019. The experimental platform adopts MATLAB 2016b PlatEMO. The comparison algorithms are $ARMOEA, GrEA, MOEAD, RVEA, NSGAIII$. All algorithms adopt the population size $N = 100$, the maximum number of generations T_{max} $= 10000$. α and fr of $RVEA$ were set as 2 and 0.1. The computation was conducted on a personal computer with an Intel Core i7-3770, 3.40 GHz CPU, 8 GB RAM.

4.2 Comparison with 5 Algorithms on IGD

Inverted Generational Distance (IGD) is a comprehensive performance evaluation index. The performance of the algorithm is evaluated by calculating the minimum distance between the individual and the real set. The smaller the value, the better the comprehensive performance of the algorithm, including convergence and distribution performance.

$$IGD\,(P,Q) = \frac{\Sigma_{v\epsilon P} d\,(v,Q)}{|P|}, \tag{5}$$

where P is the point set evenly distributed on the real PF, and —P— is the number of individuals in the point set distributed on the real PF. Q is the Pareto optimal solution set obtained by the algorithm. $D(v,Q)$ is the minimum Euclidean distance from individual v to population Q in P. Therefore, IGD evaluates the comprehensive performance of the algorithm by calculating the average value of the minimum distance from the point set on the real PF to the obtained population. From the above formula, it can be seen that when the convergence performance of the algorithm is relatively good, $D(v,Q)$ is relatively small, so the convergence performance of the algorithm can be evaluated. However, when the distribution performance of the algorithm is very poor and most individuals in the population are concentrated in a narrow area, it can be seen from the formula that the$D(v,Q)$ of many individuals will be very large, so we can evaluate the distribution performance of the algorithm. Table 1 shows the minimum value of IGD obtained by each algorithm running 20 times independently. $CF2-7$ on $MOEA\_DH$ are the best, which shows that $MOEA\_DH$ has good convergency and diversity.

Table 1. .

Problem	MOEA_DH	ARMOEA	GrEA	MOEAD	RVEA	NSGAIII
CF1	8.84E−02	**7.56E−02**	4.43E−01	2.61E−01	1.61E−01	8.64E−02
CF2	**4.73E−02**	5.46E−02	5.57E−02	4.67E−01	8.01E−02	7.15E−02
CF3	**3.60E−01**	3.75E−01	3.71E−01	5.53E−01	4.85E−01	4.51E−01
CF4	**1.01E−01**	1.76E−01	1.06E−01	6.38E−01	1.59E−01	1.45E−01
CF5	**2.40E−01**	4.63E−01	4.24E−01	6.20E−01	4.52E−01	2.45E−01
CF6	**1.16E−01**	2.79E−01	1.33E−01	3.38E−01	1.17E−01	1.43E−01
CF7	**2.49E−01**	3.00E−01	3.65E−01	4.06E−01	2.84E−01	2.62E−01

4.3 PF Analysis

Figure 1 and Fig. 2 show PF pf $MOEA\_DH$ and other 5 algorithms on $CF1$ and $CF2$. The last picture is the real PF. As can be seen from Fig. 1, the real PF of $CF1$ is regular and evenly distributed in the objective space. Other algorithms

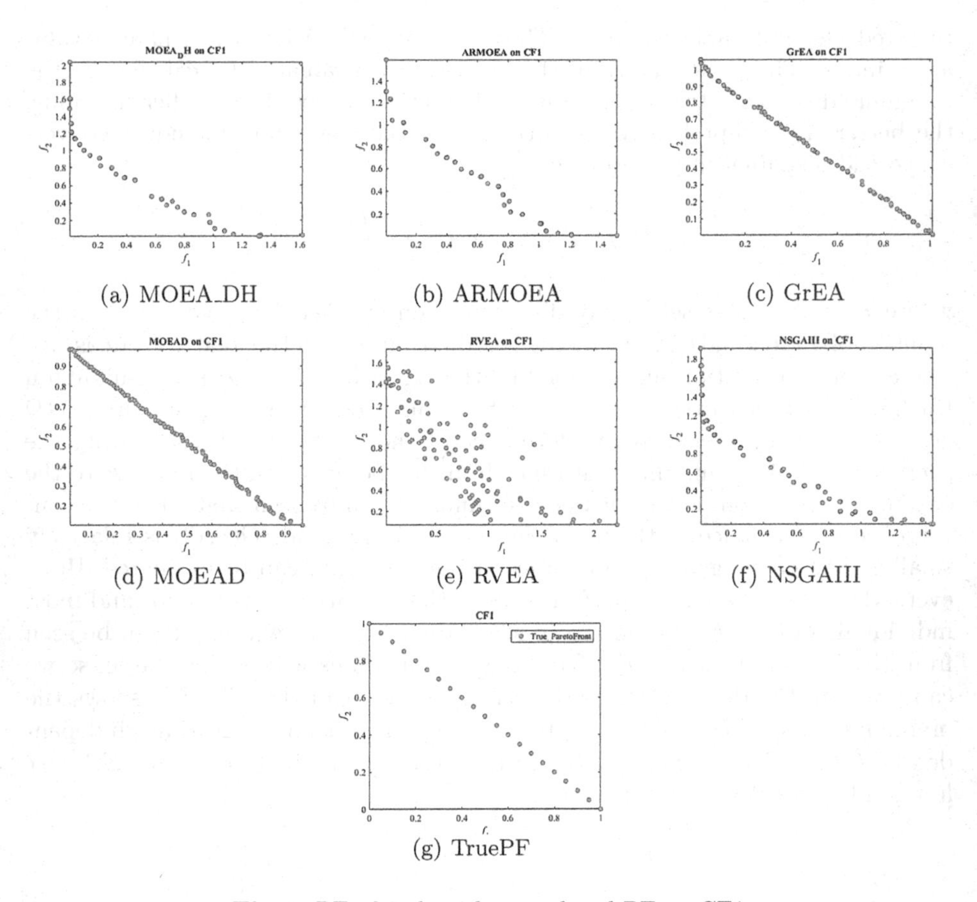

(a) MOEA_DH (b) ARMOEA (c) GrEA (d) MOEAD (e) RVEA (f) NSGAIII (g) TruePF

Fig. 1. PF of 6 algorithms and real PF on CF1

can also find a better PF. While the shape of the real PF in $CF2$ is irregular and unevenly distributed in the objective space, so it is difficult for the other 5 algorithms to find the real PF. Only $MOEA\_DH$ is close to real PF.

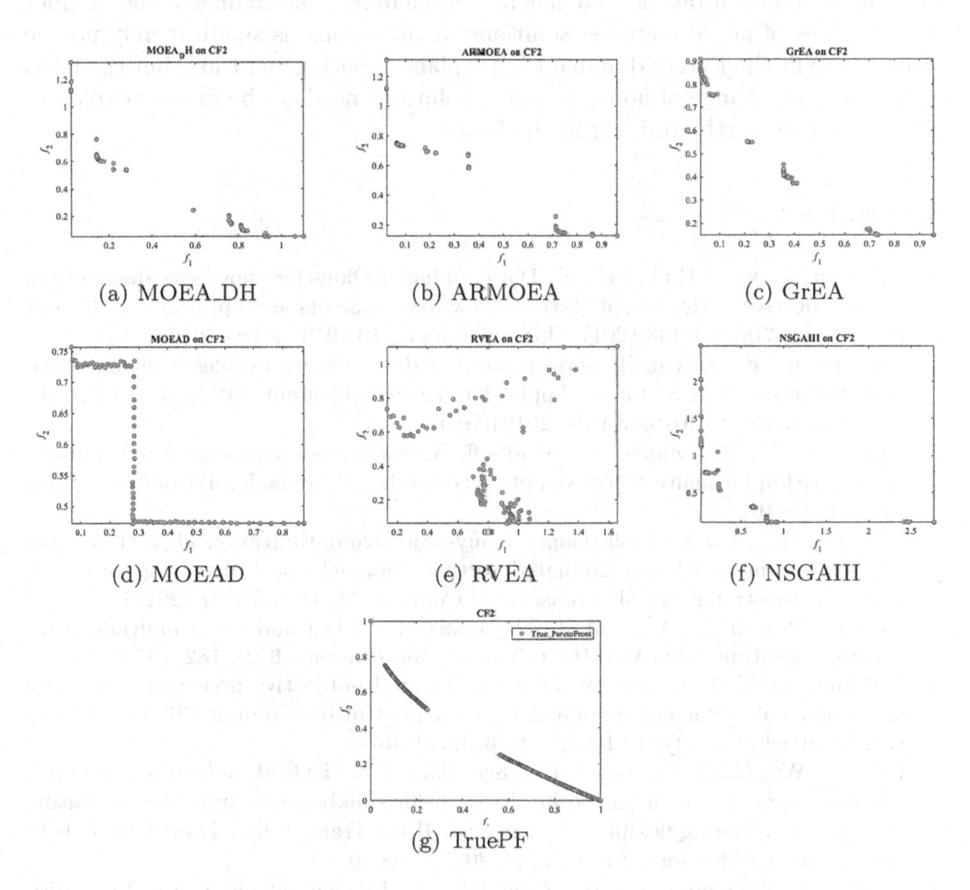

(a) MOEA_DH (b) ARMOEA (c) GrEA

(d) MOEAD (e) RVEA (f) NSGAIII

(g) TruePF

Fig. 2. PF of 6 algorithms and real PF on CF2

5 Conclusion

For MOPs with complex PF, MOEA_DH is proposed. It suggests that the solution with the smallest front number in each generation be fitted as a dominant hyperplane. Then all candidate solutions are projected onto the hyperplane, and the distance between projection points is calculated. The distance from each point to its nearest point is the crowding distance of the dominant hyperplane of the point. The crowding distance is used to screen the points in the last front, and the dominant solution participates in the screening of non dominant solutions. Then all candidate solutions are projected onto the hyperplane, and the distance between projection points is calculated. The results show that $MOEA\_DH$ has better IGD on complex PF. There is also more work to do. The time of fitting Pareto dominant hyperplane is large. We want to judge the change degree of non-dominated solutions of each generation, including the change of the number and positions of non-dominated solutions. If the change is small, it may not be necessary to fit the Pareto dominant hyperplane in each generation, but the basis for judging the change of non-dominated solutions needs to be further explored. It is a problem worth studying in the future.

References

1. Bandaru, S., Ng, A.H.C., Deb, K.: Data mining methods for knowledge discovery in multi-objective optimization: part B - new developments and applications. Expert Syst. Appl. **70**, 119–138 (2017). https://doi.org/10.1016/j.eswa.2016.10.016
2. Bourennani, F.: Solving heterogeneous big data mining problems using multi-objective optimization. Int. J. Appl. Metaheuristic Comput. **10**(4), 18–37 (2019). https://doi.org/10.4018/IJAMC.2019100102
3. Cheng, R., Jin, Y., Olhofer, M., Sendhoff, B.: A reference vector guided evolutionary algorithm for many-objective optimization. IEEE Trans. Evol. Comput. **20**(5), 773–791 (2016)
4. Deb, K., Jain, H.: An evolutionary many-objective optimization algorithm using reference-point-based nondominated sorting approach, part I: solving problems with box constraints. IEEE Trans. Evol. Comput. **18**(4), 577–601 (2013)
5. Deb, K., Pratap, A., Agarwal, S., Meyarivan, T.: A fast and elitist multiobjective genetic algorithm: NSGA-II. IEEE Trans. Evol. Comput. **6**(2), 182–197 (2002)
6. Fathalla, A., Li, K., Salah, A.: Best-KFF: a multi-objective preemptive resource allocation policy for cloud computing systems. Cluster Comput. **25**(1), 321–336 (2021). https://doi.org/10.1007/s10586-021-03407-z
7. Luo, Q., Wu, G., Ji, B., Wang, L., Suganthan, P.N.: Hybrid multi-objective optimization approach with pareto local search for collaborative truck-drone routing problems considering flexible time windows. IEEE Trans. Intell. Transp. Syst. 1–15 (2021). https://doi.org/10.1109/TITS.2021.3119080
8. Mazloomi, N., Gholipour, M., Zaretalab, A.: Efficient configuration for multi-objective QoS optimization in wireless sensor network. Ad Hoc Netw. **125**, 102730 (2022). https://doi.org/10.1016/j.adhoc.2021.102730
9. Tam, N.T., Hung, T.H., Binh, H.T.T., Vinh, L.T.: A decomposition-based multi-objective optimization approach for balancing the energy consumption of wireless sensor networks. Appl. Soft Comput. **107**, 107365 (2021). https://doi.org/10.1016/j.asoc.2021.107365
10. Tian, Y., Cheng, R., Zhang, X., Cheng, F., Jin, Y.: An indicator-based multiobjective evolutionary algorithm with reference point adaptation for better versatility. IEEE Trans. Evol. Comput. **22**(4), 609–622 (2017)
11. Wang, L., Pan, X., Shen, X., Zhao, P., Qiu, Q.: Balancing convergence and diversity in resource allocation strategy for decomposition-based multi-objective evolutionary algorithm. Appl. Soft Comput. **100**, 106968 (2021). https://doi.org/10.1016/j.asoc.2020.106968
12. Yang, S., Li, M., Liu, X., Zheng, J.: A grid-based evolutionary algorithm for many-objective optimization. IEEE Trans. Evol. Comput. **17**(5), 721–736 (2013)

13. Yi, J., Bai, J., He, H., Peng, J., Tang, D.: ar-MOEA: a novel preference-based dominance relation for evolutionary multiobjective optimization. IEEE Trans. Evol. Comput. **23**(5), 788–802 (2019). https://doi.org/10.1109/TEVC.2018.2884133
14. Zhang, Q., Li, H.: MOEA/D: a multiobjective evolutionary algorithm based on decomposition. IEEE Trans. Evol. Comput. **11**(6), 712–731 (2007)
15. Zitzler, E., Künzli, S.: Indicator-based selection in multiobjective search. In: Yao, X., et al. (eds.) PPSN 2004. LNCS, vol. 3242, pp. 832–842. Springer, Heidelberg (2004). https://doi.org/10.1007/978-3-540-30217-9_84

Multi-objective Evolutionary Algorithm for Reactive Power Optimization of Distribution Network Connecting with Renewable Energies and EVs

Biao Xu[1], Guiyuan Zhang[1], Xiaoshun Zhang[2], Ke Li[1](✉), Yao Yao[1], Zhun Fan[1], and Yonggang Zhang[3]

[1] Shantou University, Shantou 515063, China
ericlee@stu.edu.cn
[2] Foshan Graduate School of Northeastern University, Foshan 528311, China
[3] Key Laboratory of Symbolic Computation and Knowledge Engineering of Ministry of Education, Jilin University, Changchun 130012, People's Republic of China

Abstract. With the high penetration of new energy and electric vehicle (EV) charging and battery change stations connected to the distribution network, the negative impact on the power quality such as voltage deviation and voltage fluctuation of distribution network is also enhanced. Therefore, it is necessary to tap the potential of reactive power regulation of new energy and EVs, reduce the pressure of reactive power optimization of distribution network and improve its voltage quality. Firstly, the reactive power regulation model of new energy and charging and battery change stations will be established, so that the dynamic evaluation method of reactive power adjustable capacity can be proposed. Then, the three objectives included distribution network voltage deviation, line loss and static voltage stability margin will be focused. Finally, a variety of multi-objective algorithms are used to optimize the model above, and a series of examples are extended in this paper.

Keywords: Renewable energies · Electric vehicles (EVs) · Reactive power optimization · Pareto front (PF) · Multi-objective optimization

1 Introduction

With the rapid development of science and technology, people have correctly realized that new energies have the characteristics of recyclability, cleanliness and infinity. High penetration of new energies connected to the distribution network may have a great impact on the voltage quality of network nodes and transmission loss [1]. Factors closely related to voltage in distribution network can be identified as reactive power. Then, how to make rational use of the reactive power regulation capacity of new energy connected to the distribution network is the key to alleviate such problems [2]. However, a new problem that there will be a great uncertainty with new energies due to the natural environment

Y. Tan et al. (Eds.): ICSI 2022, LNCS 13344, pp. 482–490, 2022.
https://doi.org/10.1007/978-3-031-09677-8_40

impact, such as the hurricane, cloud cover, dust diffusion and so on. Coincidentally, electric vehicles (EVs) that now are popular in large scale of cities may promote the reactive power regulation ability of new energies to the distribution network. The phenomenon of bi-directional transmission of electric energies between EVs and distribution network is produced after the huge power system and transportation system cross. Specifically, this is due to the recent introduction of an emerging technology called Vehicle to Grid (V2G) for EVs. V2G allows EVs to participate in regulating the output curve of the power grid and trading clean energy power, to realize the benign interaction between EVs and the power grid [3]. Therefore, to accelerate the transformation of energy structure and improve the overall reactive power regulation ability of distribution network, the innovation that applying EVs is proposed to optimize the reactive power of distribution network.

In sum, the remaining of this paper is given as follows. In Sect. 2, EV is presented to the distribution network model. Section 3 introduce different algorithms which are applied on the model and the flow chart of the reactive power optimization. In Sect. 4, the model mentioned above would be employed to the various nodes systems. Finally, the work will be concluded in Sect. 5.

2 Reactive Power Optimization Model of PV and EVs Connected to Distribution Network

2.1 Reactive Power Regulation Model of Wind Power Generator

The presented model of wind turbine in this paper is modified from [4]. The input mechanical power, P_m, and the injection of active power into the distribution network, P_g, are related to the wind speed via the Betz equation. The calculation of wind turbine converting wind speed into mechanical power, P_m, is as follows:

$$P_m = \frac{1}{2} C_p \rho \pi R^2 V^3 \tag{1}$$

where ρ means the air density; R is the radius of the wind turbine; V is the current wind speed through the wind generators; C_P is related to blade tip speed ratio, λ, and pitch angle, β.

The power injected into the distribution network has some connections to the current wind speed. Thus, P_g will be presented as the following formula (2):

$$P_g = \begin{cases} 0,\ v_w < v_w^{in}\ or\ v_w > v_w^{out} \\ P_w^{base} \frac{v_w - v_w^{in}}{v_w^{base} - v_w^{in}},\ v_w^{in} \le v_w < v_w^{base} \\ P_w^{base},\ v_w^{base} \le v_w \le v_w^{out} \end{cases} \tag{2}$$

where P_w^{base} represents the rated wind generators output; v_w^{base} denotes the rated wind speed; v_w^{out} and v_w^{in} mean the maximum and minimum wind speed to start grid connected power generation, respectively.

The reactive power output regulation range of wind turbine is directly related to the reactive power regulation capability of stator side and grid side converters, as shown in the following formula (3):

$$\begin{cases} Q_{g,\max} = Q_{s,\max} + Q_{c,\max} \\ Q_{g,\min} = Q_{s,\min} + Q_{c,\min} \end{cases} \tag{3}$$

where $Q_{g,\min}$ and $Q_{g,\max}$ mean the lower and upper limits of reactive power regulation range, respectively;$Q_{s,\min}$ and $Q_{s,\max}$ are the lower and upper limits of the reactive power regulation range of the stator side, respectively; $Q_{c,\min}$ and $Q_{c,\max}$ are the lower and upper limits of reactive power regulation range of grid side converter, respectively.

2.2 Reactive Power Regulation Model of PV System

The power generation principle of PV connected to the distribution network is relevant the current solar irradiation and temperature. Therefore, the active power output P_{pv} will be shown as follows [5]:

$$P_{pv} = P_{pv}^{base}\left[1 + \alpha_{pv} \cdot (T - T_{ref})\right] \cdot \frac{s_{pv}}{1000} \tag{4}$$

where P_{pv}^{base} means the total rated power; α_{pv} is a temperature conversion coefficient; T denotes the current temperature; T_{ref} is reference temperature; s_{pv} is the current solar irradiation.

PV usually generates direct current (DC). Thus, DC will be converted to sinusoidal alternating current (AC) with the same frequency as the grid. When analysing the reactive power regulation range of PV, the inverter capacity and active power output need to be considered as the following formula (5):

$$\begin{cases} Q_{pv,\max} = \sqrt{(S_{pv})^2 - (P_{pv})^2} \\ Q_{pv,\min} = -\sqrt{(S_{pv})^2 - (P_{pv})^2} \end{cases} \tag{5}$$

where $Q_{pv,\min}$ and $Q_{pv,\max}$ represent the lower and upper limits of reactive power regulation range, respectively; S_{pv} means the inverter capacity.

2.3 Reactive Power Regulation Model of EVs

EVs with new technology may promote the reactive power optimization of the distribution network. They can act as generators if the remaining energies exceeds a threshold. Otherwise, EVs will absorb energies from distribution network [6]. The active power of EVs can be presented as follows (6):

$$P_{car} = \frac{V_s V_c \sin(\delta)}{\omega L_c} \tag{6}$$

where P_{car} means the EVs input or output active power; V_s is the grid voltage; V_c refers charging piles voltage; δ is the phase difference between V_c and V_s; ω denotes to the angular frequency; L_c refers the simplified inductance.

Similar to PV, the adjustable range of EVs reactive power output mainly depends on the current active power input or output and the capacity of inverter, as shown in the following formula (7):

$$\begin{cases} Q_{\text{car,max}} = \sqrt{(S_{\text{car}})^2 - (P_{\text{car}})^2} \\ Q_{\text{car,min}} = -\sqrt{(S_{\text{car}})^2 - (P_{\text{car}})^2} \end{cases} \tag{7}$$

where $Q_{\text{car,min}}$ and $Q_{\text{car,max}}$ are the lower and upper limits of reactive power regulation range, respectively; S_{car} means the charging piles inverter capacity.

2.4 Objective Function

In this work, line loss, voltage deviation and static voltage stability margin are three objectives [7]. Especially, the maximized static voltage stability margin needs to be replaced by solving the reciprocal of the minimum singular value of the minimum convergent power flow Jacobian matrix. Therefore, the objective functions are shown as formula (8):

$$\begin{cases} min\, f_1 = \sum_{i,j\in N_L} g_{ij}(V_i^2 + V_j^2 - 2V_iV_j\cos\theta_{ij}) \\ min\, f_2 = \sum_{j\in N_i} (V_j - V_j^*)^2 \\ min\, f_3 = 1/\delta_{\min} \end{cases} \tag{8}$$

where f_1, f_2 and f_3 are the line loss, voltage deviation and singular value reciprocal of Jacobian matrix of system, respectively; V_i and V_j mean the ith and jth nodes voltage amplitude, respectively; θ_{ij} refers the phase angle difference between the ith and jth nodes; g_{ij} represents the admittance between the ith and jth nodes; N_i is the total node set; N_L is the all branch set; V_j^* denotes the jth node rated voltage; $\delta_{\min}$ refers the system Jacobian matrix minimum singular value.

2.5 Constraint Condition

Power Flow Equality Constraints

$$\begin{cases} P_{Gi} - P_{Di} - V_i \sum_{j\in N_i} V_j(g_{ij} \cos\theta_{ij} + b_{ij} \sin\theta_{ij}) = 0, i \in N_0 \\ Q_{Gi} - Q_{Di} - V_i \sum_{j\in N_i} V_j(g_{ij} \sin\theta_{ij} - b_{ij} \cos\theta_{ij}) = 0, i \in N_{PQ} \end{cases} \tag{9}$$

where P_{Gi} refers the active power of the ith node; Q_{Gi} is the reactive power of the ith node; P_{Di} means the active power demand of the ith node; Q_{Di} is the reactive power demand of the ith node; b_{ij} is the susceptance between the ith and jth nodes; N_0 means the node set except the balance node; N_{PQ} is the PQ node set.

Generator Constraints

$$\begin{cases} Q_{Gi}^{\min} \le Q_{Gi} \le Q_{Gi}^{\max}, i \in N_G \\ V_{Gi}^{\min} \le V_{Gi} \le V_{Gi}^{\max}, i \in N_G \end{cases} \tag{10}$$

where Q_{Gi}^{max} and Q_{Gi}^{min} are the upper and lower limits of reactive power range of the ith generator, respectively; Q_{Gi} is the reactive power input into the grid by the ith generator; V_{Gi}^{min} and V_{Gi}^{max} represent the lower and upper limits of output voltage of the ith generator, respectively; V_{Gi} is the output voltage of the ith generator; N_G is the generator set.

Reactive Power Compensation Device and Transformer Tap Constraints

$$\begin{cases} Q_{Ci}^{min} \leq Q_{Ci} \leq Q_{Ci}^{max}, i \in N_c \\ T_h^{min} \leq T_h \leq T_h^{max}, h \in N_T \end{cases} \quad (11)$$

where Q_{Ci}^{max} and Q_{Ci}^{min} refer the upper and lower capacity of the ith reactive power compensation device, respectively; T_h^{min} and T_h^{max} mean the lower and upper limit of the hth transformer tap, respectively; N_c denotes the set of reactive power compensation devices; N_T is the set of transformer taps.

3 Process of Reactive Power Regulation Model

3.1 Overview of c-DPEA, SPEA2 and NSGAII-ARSBX

C-DPEA [8] which was proposed in recent years can be employed to solve the constraints multi-objective problems for reactive power optimization. This algorithm presents two populations to deal with the infeasible solutions. Population 1 proposes an idea that the solutions with better objective function value but low constraint violation can be preserved. However, feasible solution in population 2 can get more advantages than infeasible solutions. Thus, the two different populations provide complementary capabilities each other. Population 1 reaches better points more quickly in infeasible regions, while population 2 approaches the PF faster. Finally, a win-win result will be shown, then the offspring of two populations can cooperate by exchanging information.

SPEA2 [9] stores the nondominated solutions in another continuously updated population. It computes the fitness according to the number of nondominated solutions that an individual independently dominates. A cluster analysis process is added for the sake of reducing the nondominated solution set without destroying its characteristics. Besides, Pareto dominance is used to preserve population diversity. Path planning simulation can be addressed by the traditional hybrid target method and the improved SPEA2 based on local search.

NSGAII-ARSBX [10] was presented as an improved version based on NSGAII. Elite strategy is introduced and the sampling space is expanded in NSGA-II. Such approaches are conducive to maintaining the better individuals in the parent generation. Thus, the accuracy of optimization results will be highly improved. The best individuals will not be lost and the population level can be rapidly improved by storing all individuals in layers. Finally, researchers proposed the rotation-based simulated binary crossover (RSBX) to improve the performance of MOEAs on rotated problems.

3.2 Application of Different Algorithms in Reactive Power Optimization of Distribution Network

Both continuous variables and discrete variables will exist in the process of reactive power optimization. Continuous variables will be normally iterated, while discrete variables would be rounded by continuous spatial values. The fitness function needs to meet the above constraints by adding a penalty mechanism, as follow (12):

$$f_{\text{fit},t}(x^i) = f_t(x^i) + \eta q, t \in T \tag{12}$$

where T and t represent the set of objective functions and the tth value in T, respectively; $f_t(x^i)$ means the objective function value; $f_{\text{fit},t}(x^i)$ refers the fitness function value; η represents penalty coefficient which is usually regarded as a large constant; q means the number of objectives without catering the constrain.

4 Example Analysis

4.1 Data Setting

In this work, 33-bus system is mainly applied as an example. The total installed capacity of wind turbines and PV is set as 20 KW and 300 KW, respectively. The optimization variables include 7 new energies output, 3 eV charging stations, 2 reactive power compensation devices and 5 different tap gears of transformer. There are 3 charging areas in this work. Then the number of EVs in each charging area is 10 and the remaining power of a single EV is 0–10 KW. The population size and iteration steps will be set to 50 and 50 by considering the line loss and voltage deviation. However, they population size and iteration steps increase to 200 and 50 respectively in three-objective optimization.

4.2 Analysis of Optimization Results

Figure 1 shows the PF obtained by different algorithms in 33-bus system. It can be found that the three algorithms can find excellent PF in two-objective. SPEA2 has the smaller PF than NSGAII-ARSBX and c-DPEA. Most of results are distributed in small voltage deviation with a trend of priority of single objective optimization. Besides, the optimization degree is not high enough for NSGAII-ARSBX because of some dominated solutions. By contrast, the prominent distributed PF can be seen in c-DPEA. In three-objective optimization, the optimized surface of SPEA2 seems concentrated in the middle small area while NSGAII-ARSBX has the advantage of smooth solutions. However, the characteristic of c-DPEA presents rapid speed, uniform surface and deep degree in optimization from Table 1. In this regard, the equilibrium and stability will be well expressed incisively and vividly in c-DPEA.

The results of whether reactive power compensation devices exist are shown in Fig. 2(a) and (b). Population and iteration are set as 5000 and 50 in two different situations. The former is the solutions without reactive power compensation devices. The points form an even and smooth surface. The latter presents the solutions by adding reactive power compensation device. Evidently, the result has the feature of narrow and long. Thus, c-DPEA can be used in distribution network in different situations.

Table 1. Statistical results of PF obtained in the 33-bus system

Objective	Standard	c-DPEA	SPEA2	NSGAII-ARSBX
Line loss/MW	Minimum	0.0599	0.0614	0.0596
	Maximum	0.0658	0.0672	0.0658
	Average	0.0624	0.0639	**0.0619**
Voltage deviation/pu	Minimum	0.0064	0.0066	0.0062
	Maximum	0.0101	0.0102	0.0100
	Average	**0.0079**	0.0082	0.0080
Static voltage stability margin/pu	Minimum	0.2725	0.2728	0.2726
	Maximum	0.2734	0.2733	0.2731
	Average	**0.2727**	0.2730	0.2728

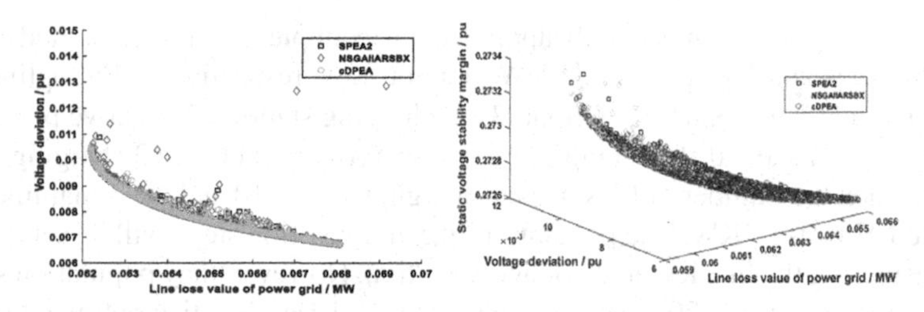

Fig. 1. PF comparison of different algorithms for bi-objective and three-objective

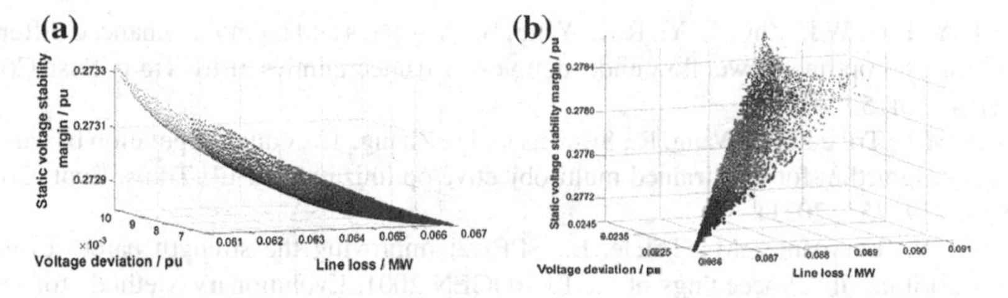

Fig. 2. (a) Reactive power compensation device not connected to distribution network, (b) Reactive power compensation device connected to distribution network

5 Conclusion

(1) The model of distribution network with new energies and EVs is proposed under the circumstance of explosive spread of V2G. The reactive power regulation potential of new energies and EVs can be fully seen in this work.

(2) The proposed model is simulated in 33-bus systems. Then c-DPEA can obtain better optimization solutions and smooth PF compared with other algorithms in different number of objectives situations.

Acknowledgement. This work was jointly supported by the Key Lab of Digital Signal and Image Processing of Guangdong Province; the Science and Technology Planning Project of Guangdong Province of China (180917144960530); the State Key Lab of Digital Manufacturing Equipment & Technology (DMETKF2019020); the Scientific Research Staring Foundation of Shantou University (NTF19028, NTF20009); the Natural Science Foundation of Guangdong Province of China (2021A1515011709); the Fundamental Research Funds for the Central Universities, JLU, (93K172021K13).

References

1. Zheng, W.Y., Wu, W.C.: Distributed multi-area load flow for multi-microgrid systems. IET Gener. Transm. Dis. **13**(3), 327–336 (2019)
2. Molina-Garcia, A., Mastromauro, R.A., Garcia-Sanchez, T., Pugliese, S., Liserre, M., Stasi, S.: Reactive power flow control for PV inverters voltage support in LV distribution networks. IEEE Trans. Smart Grid **8**(1), 447–456 (2017)
3. Lassila, J., Haakana, J., Tikka, V., Partanen, J.: Methodology to analyze the economic effects of electric cars as energy storages. IEEE T. Smart Grid **3**(1), 506–516 (2012)
4. Jiang, T.X., Putrus, G., Gao, Z.W., Donald, S.M., Wu, H.: Analysis of the combined impact of small-scale wind generators and electric vehicles on future power networks. In: 2012 47th International Universities Power Engineering Conference, pp. 1–5 (2012)
5. Brini, S., Abdallah, H.H., Ouali, A.: Economic dispatch for power system included wind and solar thermal energy. Leonardo J. Sci. **8**(14), 204–220 (2009)
6. Ma, Y.J., Liu, C., Zhou, X.S., Gao, Z.Q.: Reactive power compensation method for distribution network from electric vehicles. In: 2018 37th Chinese Control Conference, pp. 8826–8830 (2018)

7. Bu, C.Y., Luo, W.J., Zhu, T., Yi, R.K., Yang, B.: A species and memory enhanced differential evolution for optimal power flow under double-sided uncertainties. IEEE Trans. Sust. Comput. **5**(3), 403–415 (2020)
8. Ming, M.J., Trivedi, A., Wang, R., Srinivasan, D., Zhang, T.: A dual-population based evolutionary algorithm for constrained multi-objective optimization. IEEE Trans. Evol. Comput. **25**(4), 739–753 (2021)
9. Zitzler, E., Laumanns, M., Thiele, L.: SPEA2: improving the strength pareto evolutionary algorithm. In: Proceedings of the EUROGEN 2001. Evolutionary Methods for Design, Optimization and Control with Applications to Industrial Problems (2001)
10. Pan, L.Q., Xu, W.T., Li, L.H., He, C., Cheng, R.: Adaptive simulated binary crossover for rotated multi-objective optimization. Swarm Evol. Comput. **60**, 100759 (2021)

Multi-objective Evolutionary Ensemble Learning for Disease Classification

Nan Li, Lianbo Ma(✉), Tian Zhang, and Meirui He

College of Software, Northeastern University, Shenyang, China
malb@swc.neu.edu.cn

Abstract. Ensemble learning (EL) is a paradigm, involving several base learners working together to solve complex problems. The performance of the EL highly relies on the number and accuracy of weak learners, which are often hand-crafted by domain knowledge. Unfortunately, such knowledge is not always available to interested end-user. This paper proposes a novel approach to automatically select optimal type and number of base learners for disease classification, called Multi-Objective Evolutionary Ensemble Learning (MOE-EL). In the proposed MOE-EL algorithm, a variable-length gene encoding strategy of the multi-objective algorithm is first designed to search for the weak learner optimal configurations. Moreover, a dynamic population strategy is proposed to speed up the evolutionary search and balance the diversity and convergence of populations. The proposed algorithm is examined and compared with 5 existing algorithms on disease classification tasks, including the state-of-the-art methods. The experimental results show the significant superiority of the proposed approach over the state-of-the-art designs in terms of classification accuracy rate and base learner diversity.

Keywords: Ensemble learning · Evolutionary multi-objective optimization · Variable-length encoding · Disease classification

1 Introduction

Ensemble learning (EL) has demonstrated its exceptional superiority in prediction tasks [1]. It is a paradigm of building a strong classifier from several weak classifiers (e.g., Support Vector Machine (SVM) [2], Naïve Bayes (NB) [3], and Multilayer Perceptron (MLP) [4]), where predicted results of the weak classifier are combined via voting [5] or averaging [6] method. The key of EL is that by combining various models, the loss of the single weak learner will likely be compensated by other weak learners and as a result [1, 5, 6]. Thus, the overall prediction accuracy of the EL's model is better than that of a single weak learner.

Although EL has demonstrated promising performance in many applications, constructing EL's model is not an easy task. It heavily depends on the accuracy and the number of weak learners and demands considerable domain knowledge [7]. It would be a complex multi-objective optimization problem, where the two objectives are to maximize prediction performance and minimize the number of learners. In addition, the

Y. Tan et al. (Eds.): ICSI 2022, LNCS 13344, pp. 491–500, 2022.
https://doi.org/10.1007/978-3-031-09677-8_41

weak classifiers may differ in the hyperparameters, which are required to ensure the learner's performance through continuous trial and error. it will lead to a multi-level multi-objective optimization problem.

Multi-Objective Evolutionary Algorithms (MOEA) [8–10], which can obtain a set of Pareto optimal solutions, do not require abundant domain expertise. A few work use MOEA to solve the EL construction problem [11–13]. [14] tried to search for an optimal trade-off between diversity and accuracy; [12] proposed an ensemble deep learning algorithm for remaining useful life prediction through combining accuracy and diversity. However, the above work that 1) does not consider the optimization of weak classifier hyperparameters, 2) solutions are limited in terms of diversity due to fixed-length gene encoding strategy. For this drawback, we propose a new approach to solve the multi-level multi-objective optimization problem and enhance the diversity of Pareto optimal solutions.

The main contributions of the proposed method are as follow:

1) Develop a multi-level multi-objective optimization framework for EL. The dynamic population strategy can effectively reduce the computational overhead
2) Design a new variable-length gene encoding strategy that can represent optimal components of the EL with arbitrary weak classifier types.
3) Show the superiority of the proposed method compared with the peer competitors via undertaking the sufficient experiments.

The rest of this paper is organized as follows. Section 2 presents some related work. Section 3 describes the framework of MOE-EL. In Sect. 4, experimental results are shown to validate the superiority of the MOE-EL. Finally, the conclusion is written in Sect. 5.

2 Related Work

2.1 Multi-objective Evolutionary Algorithms

MOEA, which aims to simultaneously optimize more than two often conflicting objective functions [15], is widely applied to various real applications, such as industrial scheduling, robotics, and aircraft formation. It can be mathematically defined as follows:

$$\begin{array}{ll} \min & F(\mathrm{x}) = (f_1(x), f_2(x), \ldots, f(x)_m) \\ s.t. & x \in X \end{array}, \tag{1}$$

where m is the number of objective functions, x is the decision vector, and X is decision space. MOEA can be roughly divided into three main classes: 1) Pareto-dominance-based approaches (e.g., NAGA-II [16]); 2) Decomposition-based approaches (e.g., MOEA/D [17]); 3) Indicator-based approaches (e.g., IBEA [18]).

Toward a better construction of the machine learning (ML) model, many works [19, 20] have proven that combining MOEA and ML is an effective approach. For example, [21] presented multi-objective-oriented algorithm called MoreMNAS (Multi-Objective Reinforced Evolution in Mobile Neural Architecture Search) via leveraging good virtues from EA and RL. In this way, model accuracy and complexity (e.g., the number of model parameters and multiply-adds operators) are well traded off [22], which proposed a nonlinear ensemble method, obtained a set of high accuracy and strong diversity weak learners.

2.2 Ensemble Learning

Inspired by the thought that two minds are better than one, multiple weak learners are strategically combined for solving prediction problems (e.g., classification) [1, 23]. In general, EL can be categorized into homogeneous ensembles and heterogeneous ensembles according to the type of weak learner. For homogeneous ensembles, weak learners are the same type. Heterogeneous ensembles are built using learners of different types [2].

The most famous EL includes the following methods: 1) Stacking includes classifiers of multiple layers, where the primary learners are trained using k-fold cross-validation on the same dataset, and the secondary learners utilize the outputs of the primary learner to make predictions [1]; 2) Bagging (e.g., random forest) obtains the prediction result using uniform averaging or voting. In this method, weak learners are trained by special training dataset that is built by put-back sampling from an existing sample set [2]; 3) Boosting is different from Stacking and Bagging. It is a sequential ensemble method, which exploit the dependence between the weak learners via weighing previously mislabeled instances with higher weight [1, 2].

It is well-known that EL performance depends on both type and number of weak learner. However, in practice, it poses huge challenges to search for the most suitable EL's configuration in terms of weak learner to best solve a given problem because it corresponds to address a highly complex optimization problem of non-convex and black-box nature. Thus, many researchers [25, 26] have turned their attention to how to design efficient EL models without human intervention. For example, Zhang et al. [26] developed an automatic ensemble learning strategy to ensure the robustness of algorithms.

3 The Proposed Method

3.1 Problem Formulation

In EL construction, the first step is to address the number and type of weak learners. Then, the hyperparameters of the weak learners can be selected automatically from hyperparametric collections. Finally, each built strong learner should be recorded accuracy and the number of weak learners as two conflicting objectives. The above process can be modeled as a multi-level multi-objective optimization problem as follows:

$$EL_{construction} = \begin{cases} H_p = (Learners_1(h_1), Learners_2(h_2), \ldots, Learners_m(h_m)) \\ s.t.\;\; h \in \text{Hyperparametric}Collections \quad m \in [1, |Learners|] \\ \begin{cases} Learners = (learner_1(S_1, n_1), learner_2(S_2, n_2), \ldots, learner_t(S_t, n_t)) \\ s.t.\;\; S \in [0, 1] \quad n \in [0, N_{\max}] \quad t \in [1, T] \end{cases} \end{cases} \tag{2}$$

where *Learners* indicate the set of selected weak learners; S_t is the selection marker, if equal to "1" means the learner of that type is selected and vice versa, it is not selected; n_t indicates the number of learners; T is the number of learner types; H_P indicates the configuration of the learner hyperparameters; h_m represents the hyperparameter configuration of the *m-th* learner.

3.2 Framework of MOE-EL

The framework of MOE-EL is presented in Algorithm 1, which involves the following procedures:

1) **Population Initialization**: This process plays an important role in maintaining the population diversity and improving the search convergence for MOEA algorithms. An initial population P_0 with variable-length is generated using Latin hypercube sampling. The encoding method is described in Sect. 3.3.
2) **Offspring Generation**: The crossover and mutation are common genetic operators, where crossover performs the local search and mutation exercises the global search. We design the genetic operators to match the encoding method, as shown in Sect. 3.4.
3) **Dynamic Strategy**: The diversity and convergence of solutions are the main indicators of the performance of multi-objective algorithms. In order to improve the efficiency of the algorithm while reducing the damage of reducing population diversity, we propose a dynamic population strategy, as shown in Sect. 3.5.

The MOE-EL's fitness evaluation, non-dominated sorting and others follow the original NSGA-II algorithm.

Algorithm 1: Framework of MOE-EL

Input: *N*: population size; *T*: number of iterations;
Output: P_t: a set of promising EL's configurations;
1: $P_0 \leftarrow$ Initialize a population of size *N* according to variable-length encoding strategy;
2: $t = 0$;
3: **while** $t < T$ **do**
4: $F \leftarrow$ Evaluate P_t on validation set;
5: $S \leftarrow$ Select parent solutions with non-dominated sorting;
6: $Q_t \leftarrow$ Generate offspring with the genetic operators from *S*;
7: $P_{t+1} \leftarrow$ Environmental selection from $P_t \cup Q_t$;
8: $P_{t+1} \leftarrow$ Modify the population size according to the dynamic population strategy;
9: $t = t + 1$;
10: **end**
11: **Return** P_t;

3.3 Encoding Strategy

The critical step of using MOEA/D is to represent the potential solutions of the optimization problem to be addressed by chromosomes via the appropriate encoding strategy. In the proposed method, each chromosome represents a potential EL's configuration.

In order to more demonstrate how to indicate the EL's configuration, we divided the entire EL into various weak learner units, as shown in Fig. 1 (Top), where each unit include type, marker, number, hyperparameter set (i.e., H_1, H_2, H_3). For population initialization, we first generate the number of weak learner units, then randomly initialize the each weak learner according to the configuration table (see Sect. 4.1). In order to give a better explanation of the designed variable-length gene encoding strategy. Figure 1 (Bottom) describe three chromosomes with different lengths.

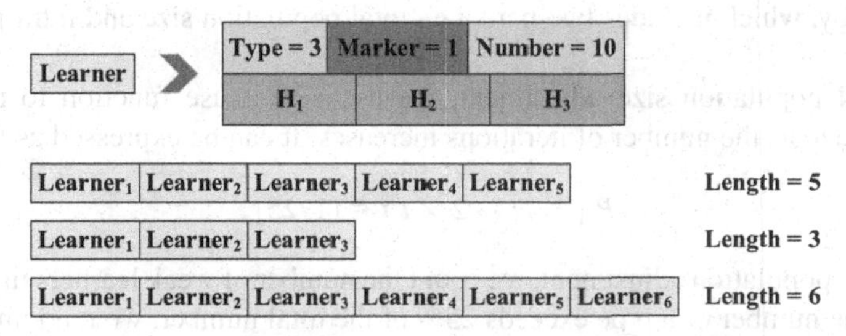

Fig. 1. An example of three chromosomes with different lengths (i.e., 3, 5, 6) in the proposed MOE-EL algorithm.

3.4 Genetic Operators

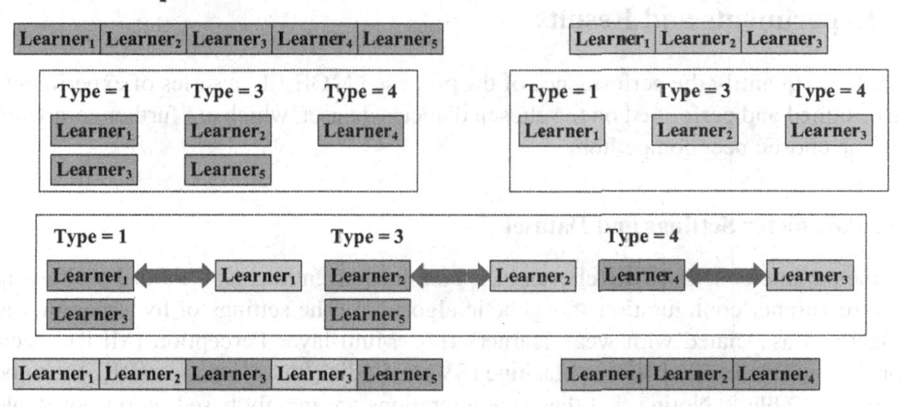

Fig. 2. An example to show the entire crossover operation.

Genetic operators (i.e., crossover and mutation) collectively give rise to the prominent performance of genetic algorithms. In the MOE-EL, we use the following genetic operator in order to match the proposed coding.

Crossover: we use Unit Alignment (UA) [27] for reconstructing two chromosomes with different lengths. In the crossover process, different types of weak learners are firstly collected into different lists based on their orders in the corresponding chromosome, where these different lists are aligned at the top, and learner units on the same positions are conducted the crossover. Figure 2 shows crossover operation.

Mutation: It can be performed at different positions, where a learner unit can be added or deleted, and the internal configuration of a learner can be modified, each of which is with the same probability.

3.5 Dynamic Population Strategy

In order to ensure diversity and convergence of population, we design a dynamic population strategy, which includes two parts i.e., total population size and intra-population adjustment.

For total population size adjustment, we use a staircase function to reduce the population size as the number of iterations increases. It can be expressed as follows:

$$|P_t| = N - 2 \times t \; t \in [1, 25], \tag{3}$$

For intra-population adjustment, we count the number of weak learners in each type, and when the number of a type exceeds 25% of the total number, we randomly remove the z individuals containing that learner and then randomly initialize the z individuals that do not contain that learner. This strategy can improve convergence (i.e., adjustment of the total number of populations) and diversity (i.e., adjustment of the number of learners).

4 Experiments and Results

In order to quantify the performance of the proposed MOE-EL, a series of experiments are designed and performed on the chosen diabetes dataset, which are further compared to some chosen peer competitors.

4.1 Parameter Settings and Dataset

All the parameter settings are chosen based on the conventions (see Table 1) following most researcher configurations for genetic algorithm. The settings of hyperparametric collections associated with weak learners (i.e., Multi-layer Perception (MLP), Decision Tree (DT), Support Vector Machine (SVM), Radial Basis Function (RBF)) can be viewed in Table 2. Noting that these configurations are mainly based on our available computational resources (including the selection of data sets).

Table 1. The parameter settings

Parameter	Setting
Population size (N)	100
Crossover probability	0.9
Mutation probability	0.1

(continued)

Table 1. (*continued*)

Parameter	Setting
Number of iterations	100
Maximum number of single learners	20

Here, diabetes dataset, which has 768 instances and 8 classes, is used to test all EL's models. The attributes of this dataset include the following information of a pregnant man, e.g., Age, Body Mass Index (BMI), Blood Pressure, and Insulin. The training set include 70% of the total dataset; the validation and test set contain 15% each.

Table 2. The hyperparametric collections

Type	Name	Value
MLP	Learning rate	[0.01, 0.05, 0.07]
	Activation function	[Sigmoid, Tanh, ReLu]
	Optimizer	[BGD, SGD, Adam]
DT	Criterion	[Entropy, Gini]
	Splitter	[Best, Random]
	Max_depth	[5, 9, 10, None]
SVM	Kernel	[Sigmoid, Linear]
	Gamma	[Poly, RBF, Sigmoid]
	Degree	[2, 4, 6]
RBF	Kernel	[Polynomial, Laplacian, Gaussian]
	Number of hidden layer neurons	[3, 5, 7]

4.2 Result and Discussion

Table 3 reports the comparison between MOE-EL and the peer competitors (i.e., Random Forest, AdaBoost, XGBoost, GBDT, LightGBM) on the diabetes dataset. In this table, training, validation, test accuracy obtained by all methods are listed. The models obtained by the algorithm achieved an advantage in all three types of accuracy. Training accuracy is 2% higher than the best competitor (i.e., LightGBM); Verification accuracy and test accuracy higher than LightGBM 1.7%, 1.6% respectively.

Table 3. Comparison with other EL's baselines on diabetes dataset

Method	Accuracy		
	Training set (%)	Validation set (%)	Test set (%)
Random Forest	94.3	94.2	93.9
AdaBoost	95.6	95.4	95.2
XGBoost	92.1	91.3	92.3
GBDT	95.9	95.4	95.2
LightGBM	96.8	96.7	97.0
MOE-EL-A	97.3	97.1	97.0
MOE-EL-B	98.1	98.2	98.1
MOE-EL-C	98.8	98.4	98.6

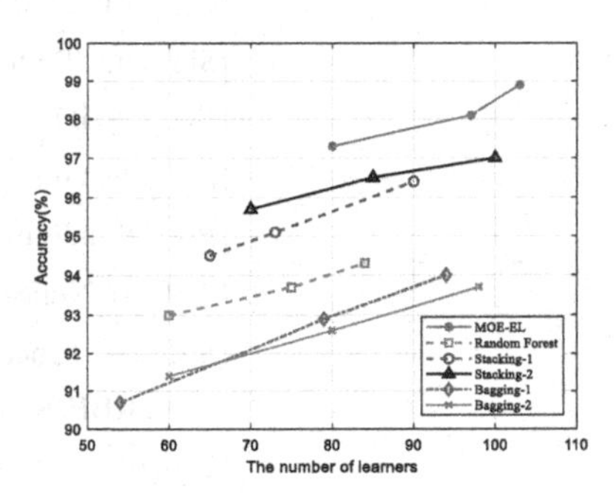

Fig. 3. The number of learners vs. Accuracy trade-off curves comparing MOE-EL existing approaches on diabetes dataset.

Figure 3 shows the number of learners and accuracy trade-off for diabetes dataset. MOE-EL-{A, B, C} outperform other models (e.g., Random Forest, Stacking, Bagging) in both objectives. In particular, MOE-EL-C achieves a state-of-the-art accuracy of 98.8% under 103 learners. MOE-EL-A improves Bagging-1 accuracy by 4.5% with similar number of learners.

5 Conclusion

In this work, a novelty MOE-EL is proposed. A new variable-length encoding method is used to denote the weak classifier and the corresponding hyperparameters. In MOE-EL, the dynamic population strategy is developed by current population information, i.e., population size adaptive. The operation can enhance the diversity and convergence of the population.

Experiments have been run on the disease dataset where MOE-EL is compared with several EL algorithms. Experimental results show that the MOE-EL generally performs better than other EL algorithms regarding the accuracy and number of weak classifiers. Meanwhile, the hyperparameters of each weak classifier are optimized.

In the future, the MOE-EL will use state-of-the-art algorithms [29] to improve further performance [29] on complex high-dimensional datasets. In addition, applying the MOE-EL algorithm to solve practical issues is also the next research direction [30].

References

1. Sagi, O., Rokach, L.: Ensemble learning: a survey. Wiley Interdiscip. Rev. Data Min. Knowl. Discov. **8**(4), e1249 (2018)
2. Cervantes, J., Garcia-Lamont, F., Rodríguez-Mazahua, L., et al.: A comprehensive survey on support vector machine classification: applications, challenges and trends. Neurocomputing **408**, 189–215 (2020)
3. Chen, S., Webb, G.I., Liu, L., et al.: A novel selective naïve Bayes algorithm. Knowl.-Based Syst. **192**, 105361 (2020)
4. Heidari, A.A., Faris, H., Mirjalili, S., et al.: Ant lion optimizer: theory, literature review, and application in multi-layer perceptron neural networks. Nat.-Inspired Optim. 23–46 (2020)
5. Zhou, K., Yang, Y., Qiao, Y., et al.: Domain adaptive ensemble learning. IEEE Trans. Image Process. **30**, 8008–8018 (2021)
6. Zhou, Z.-H.: Ensemble learning. In: Machine Learning, pp. 181–210. Springer, Singapore (2021). https://doi.org/10.1007/978-981-15-1967-3_8
7. Bi, Y., Xue, B., Zhang, M.: An automated ensemble learning framework using genetic programming for image classification. In: Proceedings of the Genetic and Evolutionary Computation Conference, pp. 365–373 (2019)
8. Ma, L., Huang, M., Yang, S., et al.: An adaptive localized decision variable analysis approach to large-scale multiobjective and many-objective optimization. IEEE Trans. Cybern. (2021). https://doi.org/10.1109/TCYB.2020.3041212
9. Ma, L., Li, N., Guo, Y., et al.: Learning to optimize: reference vector reinforcement learning adaption to constrained many-objective optimization of industrial copper burdening system. IEEE Trans. Cybern. (2021). https://doi.org/10.1109/TCYB.2021.3086501
10. Ma, L., Wang, X., Huang, M., et al.: Two-level master–slave RFID networks planning via hybrid multiobjective artificial bee colony optimizer. IEEE Trans. Syst. Man Cybern. Syst. **49**(5), 861–880 (2019)
11. Zhang, S., Chen, Y., Zhang, W., et al.: A novel ensemble deep learning model with dynamic error correction and multi-objective ensemble pruning for time series forecasting. Inf. Sci. **544**, 427–445 (2021)
12. Ma, M., Sun, C., Mao, Z., et al.: Ensemble deep learning with multi-objective optimization for prognosis of rotating machinery. ISA Trans. **113**, 166–174 (2021)
13. Zhang, C., Sun, J.H., Tan, K.C.: Deep belief networks ensemble with multi-objective optimization for failure diagnosis. In: 2015 IEEE International Conference on Systems, Man, and Cybernetics, pp. 32–37. IEEE (2015)
14. Chandra, A., Yao, X.: Ensemble learning using multi-objective evolutionary algorithms. J. Math. Model. Algorithms **5**(4), 417–445 (2006)
15. Deb, K.: Multi-objective evolutionary algorithms. In: Kacprzyk, J., Pedrycz, W. (eds.) Springer handbook of computational intelligence, pp. 995–1015. Springer, Heidelberg (2015). https://doi.org/10.1007/978-3-662-43505-2_49
16. Deb, K., Pratap, A., Agarwal, S., et al.: A fast and elitist multiobjective genetic algorithm: NSGA-II. IEEE Trans. Evol. Comput. **6**(2), 182–197 (2002)

17. Zhang, Q., Li, H.: MOEA/D: a multiobjective evolutionary algorithm based on decomposition. IEEE Trans. Evol. Comput. **11**(6), 712–731 (2007)
18. Xue, Y., Li, Y.F.: Multi-objective integer programming approaches for solving optimal feature selection problem: a new perspective on multi-objective optimization problems in SBSE. In: Proceedings of the 40th International Conference on Software Engineering, pp. 1231–1242 (2018)
19. Ma, L., Li, N., Yu, G., et al.: How to simplify search: classification-wise pareto evolution for one-shot neural architecture search. arXiv preprint arXiv:2109.07582 (2021)
20. Karagoz, G.N., Yazici, A., Dokeroglu, T., et al.: A new framework of multi-objective evolutionary algorithms for feature selection and multi-label classification of video data. Int. J. Mach. Learn. Cybern. **12**(1), 53–71 (2021)
21. Chu, X., Zhang, B., Xu, R.: Multi-objective reinforced evolution in mobile neural architecture search. In: Bartoli, A., Fusiello, A. (eds.) ECCV 2020. LNCS, vol. 12538, pp. 99–113. Springer, Cham (2020). https://doi.org/10.1007/978-3-030-66823-5_6
22. Wang, X., Hu, T., Tang, L.: A multiobjective evolutionary nonlinear ensemble learning with evolutionary feature selection for silicon prediction in blast furnace. IEEE Trans. Neural Netw. Learn. Syst. **33**, 2080–2093 (2021)
23. Oza, N.C., Russell, S.: Online Ensemble Learning. University of California, Berkeley (2001)
24. Krawczyk, B., Minku, L.L., Gama, J., et al.: Ensemble learning for data stream analysis: a survey. Inf. Fusion **37**, 132–156 (2017)
25. Zhang, Y., Jin, X.: An automatic construction and organization strategy for ensemble learning on data streams. ACM SIGMOD Rec. **35**(3), 28–33 (2006)
26. Zhang, X.: Automatic ensemble learning for online influence maximization. arXiv preprint arXiv:1911.10728 (2019)
27. Sun, Y., Xue, B., Zhang, M., et al.: Evolving deep convolutional neural networks for image classification. IEEE Trans. Evol. Comput. **24**(2), 394–407 (2019)
28. Abdollahi, J., Nouri-Moghaddam, B.: Hybrid stacked ensemble combined with genetic algorithms for prediction of diabetes. arXiv preprint arXiv:2103.08186 (2021)
29. Ma, L., Cheng, S., Shi, Y.: Enhancing learning efficiency of brain storm optimization via orthogonal learning design. IEEE Trans. Syst. Man Cybern. Syst. **51**(11), 6723–6742 (2020)
30. Ma, L., Wang, X., Wang, X., et al.: TCDA: truthful combinatorial double auctions for mobile edge computing in industrial Internet of Things. IEEE Trans. Mob. Comput. (2021). https://doi.org/10.1109/TMC.2021.3064314

An Improved Multi-objective Optimization Algorithm Based on Reinforcement Learning

Jun Liu[1], Yi Zhou[1](✉), Yimin Qiu[1], and Zhongfeng Li[2]

[1] School of Information Science and Engineering, Wuhan University of Science and Technology, Wuhan 430081, China
zhouyi83@wust.edu.cn

[2] School of Electrical Engineering, Yingkou Institute of Technology, Yingkou 115014, China

Abstract. Multi-objective optimization (MOP) has been widely applied in various applications such as engineering and economics. MOP is an important practical optimization problem, and finding approaches to better solve it is of both practical and theoretical significance. The core of solving an MOP is to find the global optimal solution set efficiently and accurately. The current MOP algorithm has premature convergence or poor population diversity, and the solution set obtained falls easily into the local optimal or clustering phenomenon. In this study, an MOP algorithm based on the non-dominated sorting genetic algorithm based on reinforcement learning (RL-NSGA-II) is proposed; the algorithm is adopted based on the prediction, forecasting, Monte Carlo method, which is based on the action of population genetic information, and environment interaction information and Markov decision process in mathematical modeling. This is because using a non-dominated solution contains valuable information, which can be used to guide the evolution direction of the population and search the optimal solution set more accurately. The proposed RL-NSGA-II algorithm was evaluated on the ZDT and DTLZ test sets, and the experimental results verified the effectiveness of the proposed algorithm in solving MOPs.

Keywords: Reinforcement learning · NSGA-II · Monte Carlo method · Multi-objective optimization problem

1 Introduction

In real life, the vast majority of real-world problems, such as recommendation systems, scheduling problems, and practical problems of combinatorial optimization, can be transformed into multi-objective optimization (MOP) problems involving multiple objective functions and mutual constraints. As there are often multiple contradictory objective functions in the abovementioned problems, solving MOPs presents a significant challenge. Owing to the mutual restriction of multiple objects, the goal of the MOP solution is to find a compromised solution set for each target value in the global space. The searchability of improved algorithms in the global space is one of the key problems in the field of MOP.

Y. Tan et al. (Eds.): ICSI 2022, LNCS 13344, pp. 501–513, 2022.
https://doi.org/10.1007/978-3-031-09677-8_42

In recent years, many researchers have realized that various multi-objective evolutionary algorithms (MOEA) can effectively solve MOPs. At present, there are a series of evolutionary algorithms, including the genetic algorithm (GA), particle swarm optimization (PSO), and ant colony optimization (ACO). Specific algorithms include the fast non-dominated sorting algorithm with elite strategy (NSGA-II) [1], decomposition based NSGA-II [2], chaos-based "micro-variation" adaptive genetic algorithm [3], local search strategy based on Gaussian variation improved NSGA-II [4], and a dynamic MOP reinforcement learning (RL) method [5]. NSGA-II is a multi-objective genetic algorithm based on non-dominated sorting proposed by Deb et al. The algorithm is based on a fast non-dominated sorting method and a selection operator, that is, the parent population and offspring population merge to select the best population as the next generation. Elarbi et al. proposed a factory-based NSGA-II [2], which addresses MOP problems and new diversity factors of penalization-based boundary intersection methods based on a new factor-based dominance relation. GA can solve complex combinatorial optimization problems, but it has two shortcomings; first, its search efficiency is lower than those of other optimization algorithms. Second, it easily converges prematurely and falls into a local optimum. Therefore, Xu et al. [3] proposed an adaptive GA based on chaos "variation", and applied it to the radio fuze interference cluster analysis. The algorithm used the characteristics of the randomness of the chaotic optimization algorithm to solve the premature convergence. Furthermore, chaos perturbation was added to the algorithm selection operator, crossover operator, and mutation operator adaptive adjustment, which improved the performance of the GA. Zhang et al. [4] proposed an improved non-dominated sorting GA NSGA-II-GLS, which introduced a Gaussian mutation operator in the genetic operation to make individuals focus on using their own nearby space. Furthermore, the improved algorithm employed the jitter local search strategy to jump out of the non-global Pareto front. Hussein et al. [6] used the RL-based meme particle swarm optimization (RLMPSO) in the entire search process. Researchers believe that RL can promote the evolution of populations by utilizing prior information and Markov decision process (MDP).

Machine learning technology combined with evolutionary algorithms is a recent interest of research, which shows good performance in multi-objective evolution. Their purpose is to establish a prediction model based on the history and existing population information in the process of population evolution and guide the population to approach the Pareto optimal front (POF) or Pareto optimal sets (POS) continuously.

In this study, we propose a combination of RL and an MOP algorithm, resulting in a fast and elitist multi-objective GA based on RL (RL-NSGA-II). The algorithm is adopted based on prediction, forecasting, and the Monte Carlo method, which is based on the action of population genetic information and the environment interaction information, and the MDP in mathematical modeling, which better uses information about non-dominant solutions. The main contributions of this study are as follows:

1. In this study, a fast non-dominated sorting algorithm based on RL (RL-NSGA-II) is designed, which will guide the evolution process of the population through the optimal strategy model of RL and improve the convergence efficiency of the population;

2. The genetic information of the population, interaction information between the population and the environment, and Markov decision process were modeled by the environment prediction, action prediction, and Monte Carlo method;
3. Given the optimization task, the state transition probability of the MDP $P_r[s', r|s, a]$ and reward function $r(s, a)$ are unknown. Thus, by the Monte Carlo method in environmental samples, according to the sample solution of the MDP, the optimal strategy.

2 A Fast and Elitist Multi-objective Genetic Algorithm Based on Reinforcement Learning (RL-NSGA-II)

In this study, the idea of RL is introduced into the crossover operator of the algorithm, and the information between the dominant solutions is reasonably used to provide the evolution direction of the crossover operator.

2.1 Markov Decision Model

When the population genetics change, the agent and the environment will change, the agent changes the current overall population, and the environment changes the individual genes in a population. In the process of iteration generation population, the individual index in the population is denoted as $\{S_i | i = 1, 2, \ldots, m\}$, where m represents the number of populations, and the following events will occur successively:

1. The environment where the agent analyzes is $S_t \in \dot{S}$; the observed $O_t \in \dot{O}$ is the collection of observed values. S is the population gene space, the collection of all individual genes.
2. According to the analysis in Step 1, the agent renders $A_t \in \dot{A}$; $\dot{A}$ represents the set of all individuals performing crossover actions at different crossover sites.
3. The environment renders rewards $R_{t+1} \in \dot{R}$ according to the action A in Step 2. Simultaneously, the environment state changes $S_t \rightarrow S_{t+1} \in S$, where R is the reward space, representing the set of reward values.
4. After the completion of Step 3, the environment changes, and the agent needs to re-analyze the information of the environment, repeat step 1 until $S_{T-1} \rightarrow S_T$.

Finally, the trajectory form of the task can be described as:

$$S_1, O_1, A_1, R_2, S_2, O_2, A_2, R_3, \ldots, S_m = s_t \tag{1}$$

If the environmental state is completely observable and the observed environmental state of the agent is the same as the actual environmental state, then $O_i = S_i, (i = 1, 2, \ldots)$. All individuals need to not only perform cross actions, but also make individual gene (S), action (A) and reward (R) corner marks uniform, the interaction trajectory should be changed to:

$$S_1, A_1, R_1, S_2, A_2, R_2, \ldots, S_m, A_m, R_m = R_t \tag{2}$$

The above interaction trajectory is represented as an individual S_1 who makes the cross A_1 and gets the reward R_1, traverses all the individuals until the individual S_m completes the cross A_m and gets the reward R_m, thus obtaining an MDP trajectory.

2.2 Agent and Policy

On the basis of the above, strategies are introduced. In fact, in the process of optimizing GA, the strategy π need not only provide evolution direction for the population (agent), but also to improve the population (agent) search efficiency. To enhance the search ability and search scope, this study will introduce the ε-greedy policy. The corresponding ε-greedy policy can be expressed as Eq. (3).

$$\pi(a|s)\begin{cases} 1-\varepsilon+\frac{\varepsilon}{|A(s)|}, a=a^* \\ \frac{\varepsilon}{|A(s)|}, a\neq a^* \end{cases} s\in S, a\in A(s) \tag{3}$$

The ε-greedy strategy distributes the ε probability equally among all actions and assigns the remaining $(1-\varepsilon)$ probability to the best action a^*.

2.3 Reward System

In the optimization task of the GA, after adopting the ε-greedy policy, the parent $S_i(i = 1, 2, 3, \ldots, m)$ randomly select parent $S_r(r = randperm(m))$ for crossover and select a different crossover site (action) $A_j(j = 1, 2, 3, \ldots, n)$, which will produce different offspring $X_j(j = 1, 2, 3, \ldots, n)$, where N is the number of individual gene segments, as shown in Fig. 1.

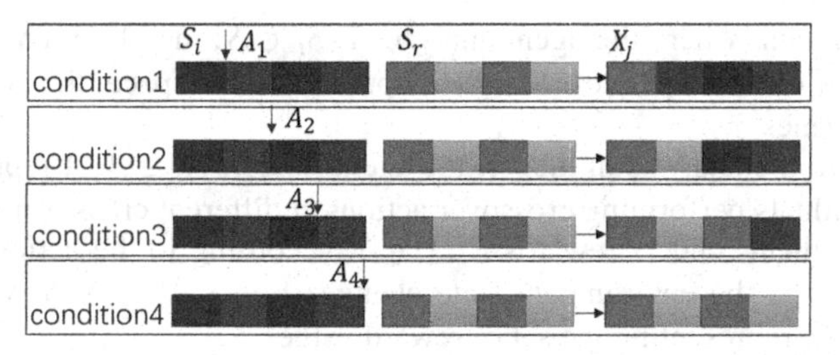

Fig. 1. Population crossing.

Calculate the $rank(X_j)$ of child X_j in population X_{merge}, ($X_{merge} = X_j \cup X_{rank=1}$, where $X_{rank=1}$ represents the non-dominant solution in the population) to determine the reward R_j for the cross A_j. The optimization process of GA is a process in which feasible solutions approach the theoretical Pareto front (PF_{true}) gradually, so it only needs to make the feasible solution set evolve continuously towards the theoretical Pareto frontier. The crossover action gets the offspring with higher ranking, and the reward R_j should be larger, as shown in Fig. 2.

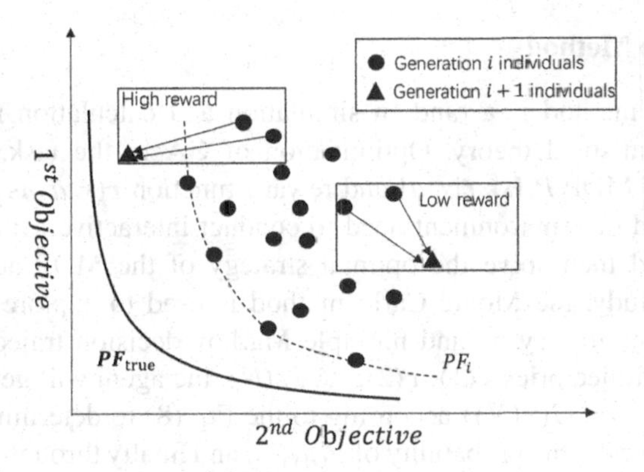

Fig. 2. Reward mechanism.

Therefore, $1/rank(X_j)$ is used as the reward of cross A_j, which can be expressed as Eq. (4).

$$\begin{cases} rank(X_j) = g(X_j, X_{merge}) \\ R_j = \frac{1}{rank(X_j)} \end{cases} \tag{4}$$

where $g(\cdot)$ represents the function that executes the non-dominant algorithm.

2.4 Return

In the optimization task, when the agent performs t actions with the environment within a period of time T, it will get t rewards. Subsequently, in the Markov decision-making process mentioned above, the agent's return G_t can be defined as the sum of rewards obtained, which can be calculated from Eq. (5):

$$G_t = R_1 + R_2 + \ldots + R_t \tag{5}$$

2.5 Action Value Function

The action value function $q_\pi(s, a)$ represents the expected return of adopting strategy π after action a is taken in state s, which can be calculated by Eq. (6) as follows

$$q_\pi(s, a) = E_\pi[G_t|S_t = s, A_t = a] \tag{6}$$

In the GA, the action value $q_\pi(s, a)$ is expressed as the probability of action a being selected after individual s takes cross action a, which is determined according to the rank of the offspring. It can be calculated by Eq. (7).

$$\pi(a|s) = \begin{cases} 1 - \varepsilon + \frac{\varepsilon}{|A(s)|}, a = arg \max\limits_{a} rank(X_j, a) \\ \frac{\varepsilon}{|A(s)|}, a \neq arg \max\limits_{a} rank(X_j, a) \end{cases} \tag{7}$$

2.6 Monte Carlo Method

The Monte Carlo method is a random simulation and calculation method based on probability and statistical theory. Optimization of GA in the task, the state transition probability of MDP $P_r[s', r|s, a]$ and reward function $r(s, a)$ is unknown. In this case, the agent and the environment need to conduct interactive sampling and collect some samples, and then solve the optimal strategy of the MDP according to these samples. In this study, the Monte Carlo method is used to explore the environment randomly based on strategy π, and multiple Markov decision trajectories are generated. For different trajectories $\tau(1), \tau(2), \ldots, \tau(N)$, the agent will get different returns $G(\tau(1)), G(\tau(2)), \cdots, G(\tau(N))$ according to the Eq. (8) to determine the population selection trajectory $\tau(i)$, the probability of $P_{G(\tau(i))}$, and finally through the selected track $\tau(i)$, guidelines to cross species reproducing offspring.

$$\begin{cases} P_{G(\tau(i))} = \frac{G(\tau(i)) - G_{min}}{G_{max} - G_{min}} \\ G_{max} = \max\{G(\tau(1)),\ G(\tau(2)),\ \cdots,\ G(\tau(N))\} \\ G_{min} = \min\{G(\tau(1)),\ G(\tau(2)),\ \cdots,\ G(\tau(N))\} \end{cases} \quad (8)$$

2.7 Rl-NSGA-II

For every parent population randomly assigned female parent to cross, traverse all cross action, and produce offspring group, crossover operator, and the current population of solutions for efficient non dominated sorting based on their offspring sorting level to generate the corresponding offspring cross action reward and the action of the selected probability (according to Eqs. (4) and (7) can be calculated separately as rewards and probability). After all the parents have completed the above operations, a matrix space b (containing the parent index and the parent index) and an action value space q (containing the probability of different cross actions being selected) are obtained. First, sample N trajectories $\tau(1), \tau(2), \ldots, \tau(N)$, and then according to Eq. (5) calculate the return $G(\tau(1)), G(\tau(2)), \cdots, G(\tau(N))$, and then according to Eq. (8) determine the population selection trajectory $\tau(i)$ the probability of $P_{G(\tau(i))}$, and finally through the selected track $\tau(i)$, guidelines to cross species reproducing offspring. The specific steps of the RL-NSGA-II are as follows:

- Step 1: Initial population. Random generation of initial population R_t; the population size of R_t is $2N$, where $t = 0$ represents iterations.
- Step 2: The RL method outputs all trajectories and the probability that the trajectories are selected.
- Step 3: Start the iteration.
- Step 4: The choice of locus $\tau(i)$ guides the selection of the crossover action and the generation of offspring Q_t.
- Step 5: A new population Q_t' is generated by using the mutation operator (real mutation) for progeny Q_t.
- Step 6: Fast non-dominated sort. According to the target vector of each individual in Q_t', the population is quick sorted according to the dominant situation among individuals.

After the sorting is completed, the individuals are assigned to a different non-dominant plane F_i (i is non-dominant order).

- Step 7: Define crowded degree calculation. An empty set P_{t+1} is calculated according to the order of the values i of childhood F_i within individual crowding distance, into a collection P_{t+1}, until $|P_{t+1}| + |F_i| > N$ ($|P|$ element in the set P number).
- Step 8: Select suitable individuals to form a new population. Sort the crowded F_i individuals according to the distance from big to small before finally taking $N - |P_{t+1}|$ individuals, incorporated into a collection of P_{t+1}.
- Step 9: Determine whether the number of iterations $t + 1$ is greater than the maximum number of iterations; otherwise, return Step 4; otherwise, end the iteration and output the non-dominated solution set of the current population.

The RL method only learns the strategy of the initial population, and when the population iterates, the strategy is not updated, but still guides the subsequent population evolution by relying on previous experience. Therefore, the algorithm needs to be improved, and the strategy can be further studied in the direction of updating the evolution of the population.

2.8 Theoretical Analysis of RL-NSGA-II

There are many theoretical paradigms in RL. RL-NSGA-II adopts the ε greedy strategy [11], which distributes the ε probability evenly among different crossovers, and allocates the remaining $(1 - \varepsilon)$ probability to the best crossovers a^*. Therefore, the RL-NSGA-II strategy can effectively guide the population to approach the target solution space all the time. The proof is as follows:

For a certain ε greedy strategy, use

$$\pi(a|s) = \begin{cases} 1 - \varepsilon + \frac{\varepsilon}{|A(s)|}, a = arg \max_{a'} q_\pi(s, a') \\ \frac{\varepsilon}{|A(s)|}, a \neq arg \max_{a'} q_\pi(s, a') \end{cases} \tag{9}$$

If the improved strategy is π', then

$$q_\pi(s, a') = \sum_a \pi'(a|s) q_\pi(s, a) = \frac{\varepsilon}{|A(s)|} \sum_a q_\pi(s, a) + (1 - \varepsilon) \max_a q_\pi(s, a) \tag{10}$$

Note that $(1 - \varepsilon) > 0$, and

$$1 - \varepsilon = \sum_a \left(\pi(a|s) - \frac{\varepsilon}{|A(s)|}\right) \tag{11}$$

So,

$$\begin{aligned} (1 - \varepsilon) \max_a q_\pi(s, a) &= \sum_a \left(\pi(a|s) - \frac{\varepsilon}{|A(s)|}\right) \max_a q_\pi(s, a) \\ &\geq \sum_a \left(\pi(a|s) - \frac{\varepsilon}{|A(s)|}\right) q_\pi(s, a) \\ &= \sum_a \pi(a|s) q_\pi(s, a) - \frac{\varepsilon}{|A(s)|} \sum_a q_\pi(s, a) \end{aligned} \tag{12}$$

and then

$$
\begin{aligned}
q_{\pi'}(s,a) &= \frac{\varepsilon}{|A(s)|}\sum_a q_\pi(s,a) + (1-\varepsilon)\max_a q_\pi(s,a) \\
&\geq \frac{\varepsilon}{|A(s)|}\sum_a q_\pi(s,a) + \sum_a \pi(a|s)q_\pi(s,a) - \frac{\varepsilon}{|A(s)|}\sum_a q_\pi(s,a) \\
&= \sum_a \pi(a|s)q_\pi(s,a)
\end{aligned}
\tag{13}
$$

The above procedure verifies the conditions of the policy improvement theorem. To ensure that the updated policy is still the ε soft policy, the algorithm needs to be initialized as ε soft policy. In the subsequent iterations, the strategy π is the ε soft strategy, so it can cover all reachable target solution space or cross-action pairs theoretically. In this way, the global optimal strategy can be obtained, which provides a theoretical basis for the algorithm to converge to the target solution space.

3 Experimental Results

To verify the performance of RL-NSGA-II, the test functions selected in this study are the ZDT [7] test functions: ZDT1–ZDT4, ZDT6, and DTLZ1–5 [8]. The three classical algorithms RL-NSGA-II and NSGA-II [1], MOEA/D [9] and SPEA2 [10] were compared in simulation experiments. The simulation experiment was carried in an AMD-RYZEN7-5800HCPU-@3.2 GHz environment.

3.1 Parameter Settings

To verify the effectiveness of RL method, this study did not tune the algorithm for the selected test functions.

Furthermore, to ensure the fairness of the comparative experiment, the common parameter settings (see Table 1) and test function settings (see Table 2) of all algorithms are as follows:

Population size: Set all test functions to 500. Crossover operator: the variation probability $P_m = 1/n$, where n is the number of decision variables of the test function; crossover probability $P_c = 1$, and the distribution indices of the crossover operation and variation operation are $\eta_c = 20$ and $\eta_m = 20$, respectively. Termination condition: The maximum number of iterations of all test functions is set to 500 generations. The settings of unique parameters of each algorithm are consistent with literature [1, 9, 10].

3.2 Performance Evaluation

This study adopts two indicators to measure algorithm performance, as follows:

Inverted Generational Distance (IGD): measures the performance of the algorithm approaching the theoretical Pareto front according to the average distance between each point in theoretical Pareto front and the nearest solution in actual Pareto front. To avoid contingency, the mean value of IGD of 10 repeated experiments is calculated and recorded as IGD_{mean} as one of the performance evaluation indexes.

Table 1. Parameter settings of the algorithm.

Name of parameter	Parameter meaning	Parameter values
Popsize	Population size	500
Varnumber	Variable dimension	n
Iteration	Iterations	500
P_c	Crossover probability	1
P_m	Mutation probability	$1/n$
η_c	Cross distribution index	20
η_m	Variation distribution index	20

Table 2. The function setting of the algorithm.

Function name	Dimensions of decision variables	Number of objective functions
ZDT1–ZDT3	$n = 30$	$m = 2$
ZDT4, ZDT6	$n = 10$	$m = 2$
DTLZ1	$n = m + 4$	$m = 3$
DTLZ2–DTLZ5	$n = m + 9$	$m = 3$

Hypervolume (HV): measures the volume enclosed between the optimization result and the reference point and measures the approximation degree between the optimization solution and the theoretical solution. To avoid contingency, the mean value of HV of 10 repeated experiments is calculated and denoted as HV_{mean} as one of the performance evaluation indexes.

3.3 Experimental Results of Competing Algorithms on the ZDT Functions

The four algorithms are compared on ZDT1–4 and ZDT 6 reference functions, and the performances of the algorithm are evaluated by the two index values of IGD_{mean} and HV_{mean}. Best results are displayed in bold font.

It can be seen from Table 3 that the IGD_{mean} values of the algorithms on ZDT1–4 function are both less than 10–1, but the IGD_{mean} values of them on ZDT6 function are too large. This is because the parameter settings of all functions are the same. For ZDT6, the maximum number of iterations currently set is too small and does not converge to the optimal solution. By comparing the IGD_{mean} values of RL-NSGA-II and NSGA-II, the IGD_{mean} values of RL-NSGA-II in ZDT1–4 and 6 functions are all lower than those of NSGA-II, indicating that the optimization performance of RL-NSGA-II in ZDT1–4 and 6 functions is better than that of NSGA-II. According to the data in Table 4, the HV_{mean} of RL-NSGA-II and NSGA-II on ZDT6 function is negative for two reasons: 1)

In the maximum number of iterations set, the algorithm cannot converge to the optimal solution. 2) The reference point is not set properly.

Table 3. IGD_{mean} comparison of algorithms on ZDT functions.

Function	RL-NSGA-II	NSGA-II	MODA-D	SPEA2
ZDT1	**1.398E−2**	2.397E−2	1.736E−1	5.502E−2
ZDT2	**2.291E−2**	4.631E−2	1.483E−0	8.121E−2
ZDT3	**9.840E−3**	1.769E−2	3.010E−1	8.484E−2
ZDT4	**4.568E−3**	2.485E−2	1.904E−0	1.365E−0
ZDT6	1.909E−0	2.306E−0	4.455E−0	**3.323E−3**

Table 4. HV_{mean} comparison of algorithms on ZDT functions.

Function	RL-NSGA-II	NSGA-II	MODA-D	SPEA2
ZDT1	**6.452E−1**	6.299E−1	4.191E−1	5.842E−1
ZDT2	**2.976E−1**	2.631E−1	−1.312E−0	2.134E−1
ZDT3	**1.024E−0**	1.006E−0	3.982E−1	8.705E−1
ZDT4	**6.592E−1**	6.288E−1	−1.661E−0	−1.177E−0
ZDT6	−9.061E−1	−1.171E−0	−1.571E−0	**3.216E−1**

Since RL-NSGA-II is an improvement on NSGA-II, RL-NSGA-II and NSGA-II are compared separately to verify the effectiveness of RL. In Fig. 3, The black point is RL-NSGA-II, and the red point is NSGA-II. As can be seen from Fig. 3, RL-NSGA-II is closer to the theoretical Pareto frontier on functions ZDT1–4 and 6, although RL-NSGA-II is closer to theoretical Pareto frontier on ZDT6 function with poor performance.

3.4 Experimental Results of Competing Algorithms on DTLZ Functions

To verify the performance of RL-NSGA-II in high-dimensional problems, DTLZ1–5 functions are used as the problem to be optimized, and IGD_{mean} and HV_{mean} are also used as indicators to test the performance of the algorithm.

Table 5 and Table 6, respectively, record IGD_{mean} and HV_{mean} values of the algorithms optimized for different DTLZ functions. It is worth noting that the algorithms have some difficulties in the DTLZ1 and DTLZ3 functions.

It can be seen from the data in Table 5 and Table 6 that RL-NSGA-II does not perform well in high-dimensional problems, and only performs better than NSGA-II in DTLZ1 and 3 functions. The reason lies in the deviation in the judgment of the return mechanism of the RL method.

In Fig. 4, The bule point is NSGA-II, and the red point is RL-NSGA-II. for the DTLZ1 function, the solution set of NSGA-II is not evenly distributed, and there is

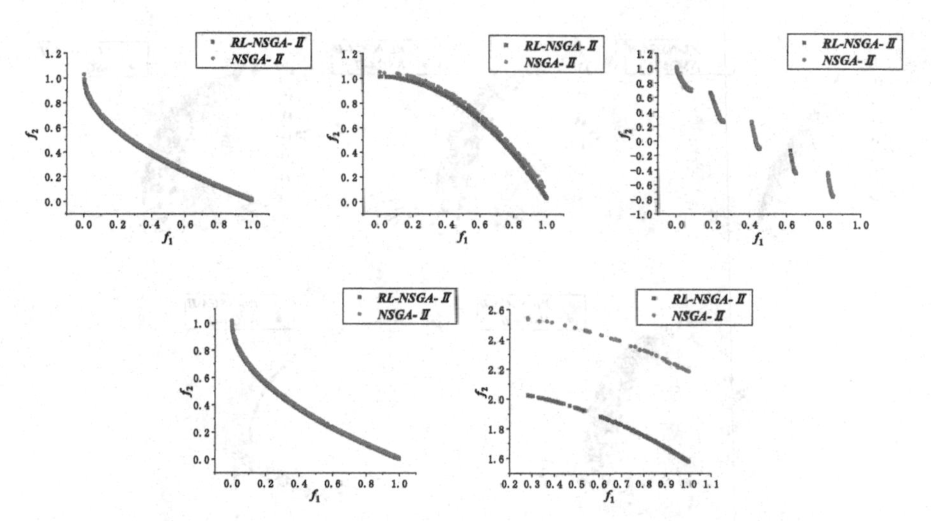

Fig. 3. Pareto frontier comparison graph of algorithms on ZDT function. (Color figure online)

Table 5. IGD_{mean} comparison of algorithms on DTLZ functions.

Function	RL-NSGA-II	NSGA-II	MODA-D	SPEA2
DTLZ1	**2.137E−2**	1.354E−1	26.52E−0	4.195E−0
DTLZ2	1.344E−3	**1.313E−3**	1.231E−1	8.366E−2
DTLZ3	**3.304E−1**	1.124E−0	1.562E+2	1.697E+2
DTLZ4	8.126E−3	**8.094E−3**	1.135E−1	7.618E−2
DTLZ5	8.126E−3	**1.390E−3**	3.137E−2	6.715E−3

Table 6. HV_{mean} comparison of algorithms on DTLZ functions.

Function	RL-NSGA-II	NSGA-II	MODA-D	SPEA2
DTLZ1	**9.238E−1**	7.952E−1	−5.145E+2	−8.686E+1
DTLZ2	6.049E−1	**6.058E−1**	5.535E−1	4.268E−1
DTLZ3	**3.069E−2**	−2.946E−0	−1.142E+4	−2.17E+4
DTLZ4	6.049E−1	**6.0526E−1**	5.845E−1	3.325E−1
DTLZ5	4.429E−1	4.432E−1	5.447E−1	**5.477E−1**

a certain distance from the theoretical Pareto front. In the current maximum iteration number, the solution set of RL-NSGA-II is more evenly distributed and closer to the theoretical Pareto front in the same iteration number.

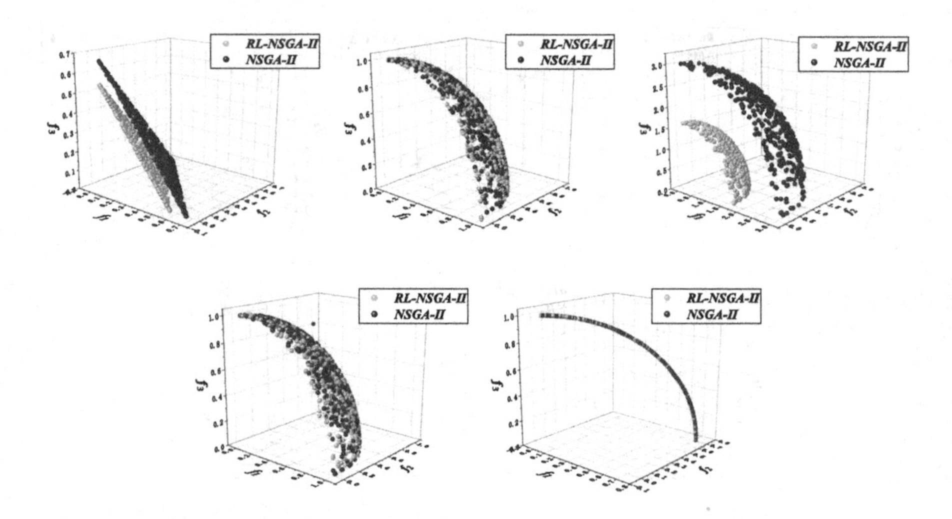

Fig. 4. Pareto frontier comparison graph of algorithms on DTLZ function.(Color figure online)

4 Conclusions

In this study, an RL-NSGA-II algorithm was designed to solve the multi-objective optimization problem. Different from other algorithms, this algorithm adopts RL technology to improve the search efficiency of the algorithm, accelerate the convergence speed, and maintain the diversity of the population. This design is based on environment prediction, action-based prediction, and the Monte Carlo method for mathematical modeling of population genetic information, interaction information with the environment, and MDP, and better employs non-dominant information. Owing to the fact that the global non-dominant solution contains valuable information, it can be used to guide the evolution direction of the population, and accurately reflect the information of the optimal solution set. The algorithm can sense the state of the current environment and select the appropriate strategy to guide the evolution according to the information and experience obtained by the strategy. Finally, the algorithm NSGA-II is used for comparison experiments, and the experimental results verify the effectiveness of RL-NSGA-II on ZDT and DTLZ benchmark problems. Experimental results show that RL-NSGA-II can effectively find the POF on most test problems.

Since RL-NSGA-II is still in the early stage of research, the subsequent optimization of RL-NSGA-II can be started from the perspective of RL method and combined with more advanced theoretical paradigm of reinforcement learning, which will be one of the optimization directions for further research.

Acknowledgement. This work is supported by the National Science Foundation of China (Grant No. 62173259) and 2021 scientific research fund project (key project) of Liaoning Provincial Department of Education (Grant No. LJKZ1199).

References

1. Deb, K., Pratap, A., Agarwal, S., Meyarivan, T.: A fast and elitist multiobjective genetic algorithm: NSGA-II. IEEE Trans. Evol. Comput. **6**, 182–197 (2002). https://doi.org/10.1109/4235.996017
2. Elarbi, M., Bechikh, S., Gupta, A., et al.: A new decomposition-based NSGA-II for many-objective optimization. IEEE Trans. Syst. Man Cybern.: Syst. **48**(7), 1191–1210 (2017). https://doi.org/10.1109/TSMC.2017.2654301
3. Xu, G., Tong, X., Pan, W., Ren, T.: Clustering analysis based on chaos micro variation adaptive genetic algorithm for radio fuze jamming. In: 29th Chinese Control and Decision Conference (CCDC), pp. 616–620 (2017). https://doi.org/10.1109/CCDC.2017.7978287
4. Zhang, Z., Lu, B.: Improving NSGA-II by a local search strategy with Gaussian mutation. In: 40th Chinese Control Conference (CCC), pp. 1628–1633 (2021). https://doi.org/10.23919/CCC52363.2021.9550337
5. Zou, F., Gary, Y., Tang, L., Wang, C.: A reinforcement learning approach for dynamic multi-objective optimization. Inf. Sci. **546**, 815–834 (2021). https://doi.org/10.1016/j.ins.2020.08.101
6. Samma, H., Lim, J., Saleh, M.: A new reinforcement learning-based memetic particle swarm optimizer. Appl. Soft Comput. **43**, 276–297 (2016). https://doi.org/10.1016/j.asoc.2016.01.006
7. Wang, Z., Zhang, Q., Zhou, A., Gong, M., Jiao, L.: Adaptive replacement strategies for MOEA/D. IEEE Trans. Cybern. **46**(2), 474–486 (2016). https://doi.org/10.1109/TCYB.2015.2403849
8. Sun, Y., Yen, G., Yi, Z.: Improved regularity model-based EDA for many-objective optimization. IEEE Trans. Evol. Comput. **22**(5), 662–678 (2018). https://doi.org/10.1109/TEVC.2018.2794319
9. Zhang, Q., Li, H.: MOEA/D: a multiobjective evolutionary algorithm based on decomposition. IEEE Trans. Evol. Comput. **11**(6), 712–731 (2007). https://doi.org/10.1109/TEVC.2007.892759
10. Eckart, Z., Marco, L., Lothar, T.: Improving the strength Pareto evolutionary algorithm for multiobjective optimization. EUROGEN, Evol. Method Des. Optim. Control Ind. Problem, 1–21 (2001). https://doi.org/10.3929/ethz-a-004284029
11. Sutton, R., Barto, A.: Reinforcement Learning: An Introduction. MIT Press, Cambridge (2008)

Fuzzy Multi-objective Particle Swarm Optimization Based on Linear Differential Decline

Nana Li[1], Lian Yuan[2], Xiaoli Shu[1], Huayao Han[3], Jie Yang[2], and Yanmin Liu[2](✉)

[1] School of Data Science and Information Engineering, Guizhou Minzu University, Guizhou 550025, China
[2] Zunyi Normal College, Guizhou 563002, China
Yanmin7813@163.com
[3] Guizhou Xingqian Information Technology Co. Ltd., Guizhou 563002, China

Abstract. To improve convergence performance of the algorithm and prevent the algorithm from falling into local optimal location, we proposes a novel fuzzy multi-objective particle swarm optimization based on linear differential decline (LDDF-MOPSO). In LDDFMOPSO, the fuzzy control strategy is applied to the inertia weight, so that the search ability of the global and local can be flexibly adjusted, thereby improving convergence performance of the algorithm. At the same time, in order to prevent the algorithm from falling into local optimal location, the strategy of linear differential decline is used to adjust the position change of particles. The experimental results illustrate that LDDFMOPSO has good performance compared to four state-of-the-art multi-objective particle swarm optimizations.

Keywords: Particle swarm optimization · Multi-objective optimization problems · Fuzzy control · Linear differential decline

1 Introduction

Many of the problems faced in engineering practice and scientific research are multi-objective optimization problems (MOPs) [14]. Compared with the existing multi-objective optimization, particle swarm optimization (PSO) [4] converges quickly and easy to implement, so it has been widely studied and applied in MOPs [6]. In 2002, Coello et al. proposed the multi-objective particle swarm optimization (MOPSO) [2]. At the same time, the search process of the PSO is non-linear and very complicated, so it is easy to fall into local optimal location. In recent years, many scholars have proposed some ways to solve these problems. For example, in [11], a novel dynamic weighting method based on chaotic sequence was proposed to select global optimal particle, so as to improve the diversity of solutions. In [9], a scheduled competition learning based MOPSO was proposed, which combined MOPSO and competition learning mechanism

Y. Tan et al. (Eds.): ICSI 2022, LNCS 13344, pp. 514–523, 2022.
https://doi.org/10.1007/978-3-031-09677-8_43

[12]. The competition learning mechanism was used in every certain iterations to maintain the diversity of the population. In MOPSO/GMR [5], Li et al. proposed Global Marginal Ranking strategy (GMR) to rank particles in the swarm, which combined the population distribution information and the individuals' information, and GMR can make the solution distribution obtained by the algorithm better when compared to traditional Pareto dominance. Most of the main tasks in the above algorithms are enhance diversity of particles. In order to improve convergence performance of the algorithm and prevent the algorithm from falling into local optimal location, we proposes a novel fuzzy MOPSO based on linear differential decline (LDDFMOPSO), which uses a linear differential decline strategy to adjust the positional change of particles and use fuzzy control for inertia weights.

2 Related Work

2.1 Multi-objective Optimization Problems

A minimization MOPs can be formulated as follows:

$$\min y = F(x) = (f_1(x), f_2(x), \cdots, f_m(x)) \tag{1}$$

$$s.t. \begin{cases} g_j(x) \geq 0 \; i = 1, 2, \cdots, p \\ h_k(x) = 0 \; k = 1, 2, \cdots, q. \end{cases}$$

where x is a vector with n decision variables; m is the number of objective functions; g_j is constraint of p inequality constraints; and h_k is constraint of q equality constraints.

2.2 Multi-objective Particle Swarm Optimization

Assuming there are N particles in the D-dimensional objective space that make up a swarm. The position and velocity of particle $i - th$ are denoted as x_i and v_i respectively. The velocity and position of each particle in swarm updates according to the following equations:

$$v_i(t+1) = w \cdot v_i(t) + c_1 r_1 (pbest_i(t) - x_i(t)) + c_2 r_2 (gbest_i(t) - x_i(t)) \tag{2}$$

$$x_i(t+1) = x_i(t) + v_i(t+1). \tag{3}$$

where t is the iteration number; w is the inertial weight; c_1 and c_2 are the learning factors; r_1 and r_2 are random numbers generated uniformly in (0, 1),which $c_1 = c_2 = 2$ is usually taken; *pbest* represents optimal position of the individual; and *gbest* represents global optimal position.

3 The Details of LDDFMOPSO

3.1 Strategy of Fuzzy Control

In [13], the generalized bell-shaped membership function μ was used to fuzzy control the inertia weight, so that the search space of the population changes from small to large. After the search reaches a certain number of iterations, the search space is from large to small, thus flexible adjustment of global and local search capabilities. Therefore, in order to flexibly adjust the global and local search ability, the fuzzy control strategy of inertia weight is used in MOPSO, and the velocity update formula of the particles is adjusted to the formula (5) by using formula (4).

$$w' = \mu \cdot w, \mu = \left(1 + \left|\frac{t-b}{a}\right|^4\right)^{-1} \tag{4}$$

$$v_i(t+1) = w' \cdot v_i(t) + c_1 r_1 (pbest_i(t) - x_i(t)) + c_2 r_2 (gbest_i(t) - x_i(t)). \tag{5}$$

where t is the current iteration; a and b are parameters, for details, please refer to [13].

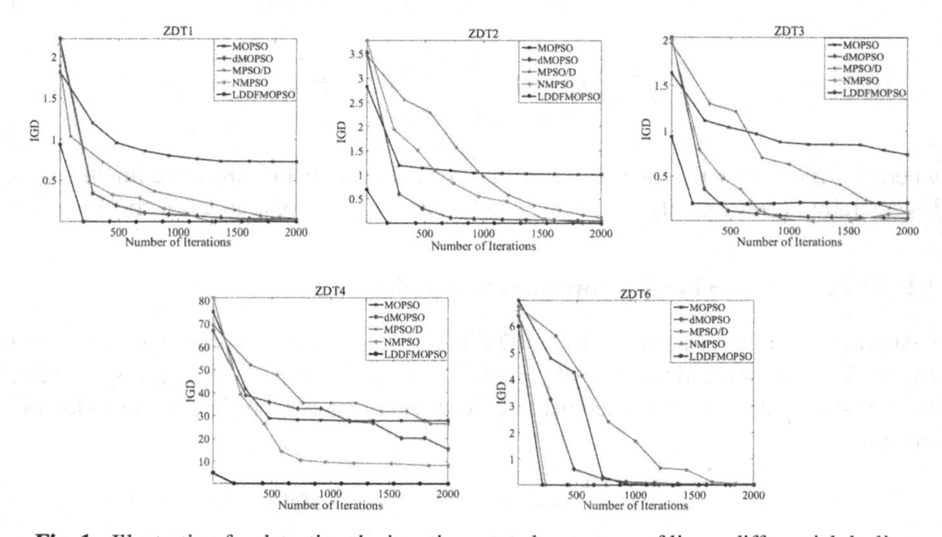

Fig. 1. Illustration for detecting the iterations state by strategy of linear differential decline

Figure 1 shows LDDFMOPSO used strategy of linear differential decline to detect the iterations state. We use it to compared with four late-of-the-art MOPSOs on ZDT1–ZDT4, and ZDT6. The experimental results indicate that it has a the promising convergence performance of the proposed LDDFMOPSO in comparison with four late-of-the-art MOPSOs on ZDT1–ZDT4, and ZDT6.

Table 1. The influence of α value on IGD.

Functions	IGD	0.2	0.3	0.4	0.5	0.6	0.7	0.8
ZDT1	Mean	4.12e−01	1.86e−01	6.02e−03	**4.36e−03**	4.49e−03	4.42e−03	4.62e−03
ZDT2	Mean	3.80e−01	1.47e−01	6.89e−03	4.52e−03	**4.10e−03**	4.39e−03	4.89e−03
ZDT3	Mean	4.36e−01	3.03e−01	2.25e−01	**1.97e−01**	1.99e−01	2.00e−01	2.01e−01
ZDT4	Mean	3.94e+00	5.81e−01	3.24e−02	4.59e−03	4.50e−03	**4.38e−03**	4.56e−03
ZDT6	Mean	2.21e+00	2.16e−02	4.57e−03	**2.04e−03**	2.08e−03	2.09e−03	2.10e−03

3.2 Strategy of Linear Differential Decline

In the early stage of algorithm evolution, the particle has a large range of position changes, and has a strong global search ability, which is beneficial to the particle to seek global optimal position. As the number of iterations increases, the closer the particle is to the optimal solution, the easier it is to fall into the local optimum. To prevent the algorithm from falling into local optimal location, a position update strategy of linear differential decline is proposed in LDDFMOPSO. t is the current iteration, when the number of iterations is low, the range of particle position change is large. α is a constants, and the value of α is determined by the sensitivity analysis experiment in Table 1 and Table 2. With the increase of iterations, the range of particle position change is reduced rapidly to prevent particles from approaching the global optimal position too fast. Such a position update strategy can effectively avoid the particle falling into local optimal location. In LDDFMOPSO, the update formula for the particles' position is as follows:

$$\frac{d\sigma}{dt} = \frac{2(1-\alpha)}{t_{\max}^2} \cdot t \tag{6}$$

$$\sigma(t) = 1 - \frac{1-\alpha}{t_{\max}^2} \cdot t^2 \tag{7}$$

$$x_i(t+1) = x_i(t) + \sigma(t) \cdot v_i(t+1). \tag{8}$$

In this paper, we set the values of α to 0.2, 0.3, 0.4, 0.5, 0.6, 0.7, and 0.8, and run independently 30 times on the five test functions, then analyzing the sensitivity of the different α values to the algorithm. From the experimental results, when $\alpha = 0.5$, the performance of LDDFMOPSO is best. It can be seen that when $\alpha = 0.4$, the mean values of HV obtained by algorithm is best on ZDT3; when $\alpha = 0.5$, the mean values of IGD obtained by algorithm is best on ZDT1, ZDT3 and ZDT6, the mean values of HV obtained by algorithm is best on ZDT1, ZDT2, ZDT4, and ZDT6; when $\alpha = 0.6$,

Table 2. The influence of α value on HV.

Functions	HV	0.2	0.3	0.4	0.5	0.6	0.7	0.8
ZDT1	Mean	3.94e−01	5.98e−01	7.16e−01	**7.19e−01**	7.18e−01	7.18e−01	7.18e−01
ZDT2	Mean	1.37e−01	3.03e−01	4.41e−01	**4.44e−01**	4.44e−01	4.44e−01	4.43e−01
ZDT3	Mean	4.76e−01	7.25e−01	**7.37e−01**	6.58e−01	6.57e−01	6.57e−01	6.56e−01
ZDT4	Mean	–	3.29e−01	6.94e−01	**7.19e−01**	7.19e−01	7.19e−01	7.19e−01
ZDT6	Mean	2.53e−01	3.07e−01	3.82e−01	**3.89e−01**	3.89e−01	3.87e−01	3.86e−01

Table 3. IGD of different algorithms on five test functions.

Functions		MOPSO [2]	dMOPSO [8]	MPSO/D [3]	NMPSO [7]	LDDFMOPSO
ZDT1	Mean	7.1799e−1	5.7847e−2	9.3183e−2	3.5090e−2	**4.3554e−3**
	Std	1.94e−1	1.68e−2	4.15e−2	2.49e−2	**4.86e−4**
	Wilcoxon	–	–	–	–	
ZDT2	Mean	1.3758e+0	4.2019e−2	1.1169e−1	3.2642e−2	**4.5162e−3**
	Std	2.83e−1	1.58e−2	7.92e−2	4.92e−2	**8.07e−4**
	Wilcoxon	–	–	–	–	
ZDT3	Mean	7.9358e−1	**3.6270e−2**	2.0157e−1	9.3162e−2	1.9730e−1
	Std	1.85e−1	**8.60e−3**	3.69e−2	2.65e−2	4.17e−3
	Wilcoxon	–	+	=	+	
ZDT4	Mean	1.4572e+1	5.9152e+0	3.6456e+1	1.5571e+1	**4.5921e−3**
	Std	5.18e+0	6.25e+0	7.26e+0	6.76e+0	**5.64e−4**
	Wilcoxon	–	–	–	–	
ZDT6	Mean	4.7995e−2	4.1046e−3	1.7750e−2	2.2710e−3	**2.0431e−3**
	Std	1.25e−1	3.44e−3	8.84e−3	1.83e−4	**3.72e−4**
	Wilcoxon	–	–	–	–	
±/=		0/5/0	1/5/0	0/5/1	1/5/0	
Best/all		0/5	1/5	0/5	0/5	4/5

The mean values of IGD obtained by algorithm is best on ZDT2; when $\alpha = 0.7$, the mean values of IGD obtained by algorithm is best on ZDT4; when $\alpha = 0.2, \alpha = 0.3$, and $\alpha = 0.8$, the performance of the algorithm is the worst. From the experimental results, when $\alpha = 0.5$, the performance of LDDFMOPSO is best. Therefore, the value of α is set as 0.5.

Table 4. HV of different algorithms on ZDT1–ZDT4, and ZDT6.

Functions		MOPSO [2]	dMOPSO [8]	MPSO/D [3]	NMPSO [7]	LDDFMOPSO
ZDT1	Mean	8.8943e−2	6.5184e−1	5.8657e−1	6.9153e−1	**7.1922e−1**
	Std	9.72e−2	1.89e−2	5.63e−2	1.39e−2	**8.94e−4**
	Wilcoxon	–	–	–	–	
ZDT2	Mean	0.0000e+0	3.8229e−1	3.0932e−1	4.2110e−1	**4.4486e−1**
	Std	0.00e+0	2.43e−2	8.08e−2	4.81e−2	**1.42e−3**
	Wilcoxon	–	–	–	–	
ZDT3	Mean	9.1199e−2	6.0600e−1	4.6355e−1	5.7286e−1	**6.5828e−1**
	Std	8.31e−2	1.52e−2	4.50e−2	9.87e−3	**1.39e−3**
	Wilcoxon	–	–	–	–	
ZDT4	Mean	0.0000e+0	4.4632e−2	0.0000e+0	0.0000e+0	**7.1874e−1**
	Std	0.00e+0	7.82e−2	0.00e+0	0.00e+0	**1.04e−3**
	Wilcoxon	–	–	–	–	
ZDT6	Mean	3.5835e−1	3.8789e−1	3.7455e−1	**3.8982e−1**	3.8876e−1
	Std	7.58e−2	3.45e−3	7.78e−3	**1.25e−4**	3.14e−4
	Wilcoxon	–	=	–	=	
±/=		0/5/0	0/5/1	0/5/0	0/5/1	
Best/all		0/5	0/5	0/5	1/5	4/5

3.3 Procedure of LDDFMOPSO

Step 1 Initialize the population, and set acceleration constants c_1 and c_2 other parameters;
Step 2 Determine whether the terminal conditions are met. If met, output the results and terminate the algorithm. Otherwise, continue to the next step;
Step 3 An external archive is established;
Step 4 The global optimal position (*gbest*) and individual optimal position (*pbest*) are selected;
Step 5 The velocity and position of particles are updated according to formula (5) and (8);
Step 6 The *gbest* and *pbest* are update;
Step 7 They enter the next iteration, then move to step 3.

4 Experimental Results

4.1 Test Functions and Experiment Parameters

To demonstrate the performance of LDDFMOPSO, we have chosen ZDT series test functions [16]. The population size N in this paper is set to 200; The number of objective is 2; The dimension of the decision space of ZDT1–ZDT3 is 30, and the dimension of the decision space of ZDT4 and ZDT6 is 10; The maximum number of iterations is 2000. For fair comparison, all relevant parameters of other algorithms are set according to the suggestions in the original literature.

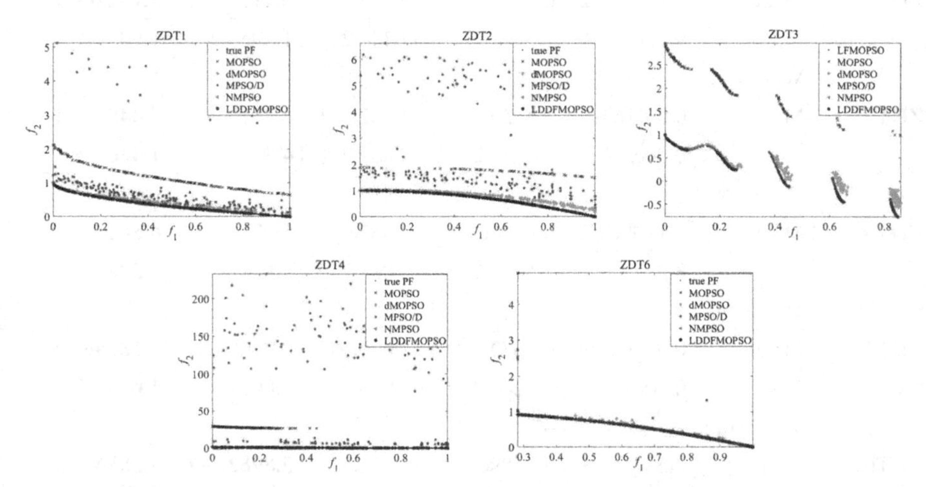

Fig. 2. The simulation diagram of LDDFMOPSO and comparative algorithms on ZDT1–ZDT4, and ZDT6.

The goal of MOPs is to find a uniformly distributed set that is as close to the true Pareto fronts as possible. To evaluate the performance between different algorithms, we adopt Inverse Generation Distance (IGD) [1] and Hyper-Volume Metric (HV) [15] in this paper. They are believed that this performance index can not only explain the convergence effects of the algorithm, but also explain the distribution of the final solution. The true Pareto front for were downloaded from http://jmetal.sourceforge.net/problems.html. In order to draw statistical conclusions, the number of independent runs of each test experiment is set to 30.

In addition, in order to determine the statistical significance, a Wilcoxon rank sum test was further carried out to test the statistical significance of the difference between the results obtained by LDDFMOPSO and the results obtained by other algorithms at $\alpha = 0.05$. All experimental results are obtained on PC with 2.3 GHz CPU and 8 GB memory. All source codes of these competing algorithms are provided in the PlatEMO [10].

4.2 Experiments Results and Analysis

In order to visually compare the performance of MOPSO [2], dMOPSO [8], MPSO/D [3], NMPSO [7], and LDDFMOPSO, the convergence characteristics of non-dominant solutions obtained by the five algorithms are shown in Fig. 2. It can be seen that MOPSO cannot obtain a good convergence to the true Pareto front on ZDT1–ZDT4, and the search results of MOPSO only can convergence to the Pareto front region on ZDT3. The dMOPSO cannot obtain a good convergence to the true Pareto front on ZDT1–ZDT3. The MPSO/D cannot obtain a good convergence to the true Pareto front on ZDT1, ZDT2, and ZDT4. In contrast, LDDFMOPSO and NMPSO can convergence the true Pareto front of the five test functions quite well and are evenly distributed.

Moreover, the Wilcoxon rank-sum test is adopted at a significance level of 0.05, where the symbols "+", "−", and "=" in the last row of the tables indicate that the result is significantly better than it, significantly worse than it, and statistically similar to that obtained by LDDFMOPSO, respectively. The best average for each test instance is shown in bold.

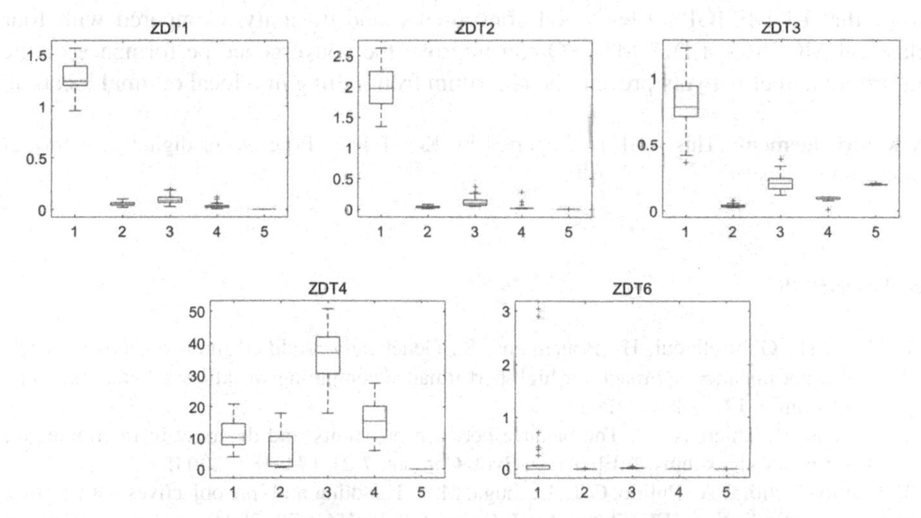

Fig. 3. Statistical boxplot of IGD indicator of different algorithms on ZDT1–ZDT4, and ZDT6, respectively.

It can be observed directly that the performance of LDDFMOPSO is superior to the comparative algorithm than MOPSO, dMOPSO, MPSO/D, and NMPSO. For example, the number of optimal IGD for MOPSO, MPSO/D, and NMPSO is zero, the number of optimal IGD for dMOPSO is one, and LDDFMOPSO has four optimal IGD values. At the same time, the number of optimal HV for MOPSO, dMOPSO, and MPSO/D is zero, the number of optimal HV for NMPSO is one, and LDDFMOPSO has four optimal HV values.

At the same time, when different algorithms are run independently for 30 times, the partial statistical block diagram of the evaluation index IGD of LDDFMOPSO and comparative algorithm is shown in Fig. 3 (1, 2, 3, 4, and 5 represent MOPSO, dMOPSO, MPSO/D, NMPSO, and LDDFMOPSO respectively). As shown in Fig. 3, LDDFMOPSO recorded the minimum values on ZDT1–ZDT4 and ZDT6. It can be clearly seen from Fig. 3 that LDDFMOPSO can obtain better non-dominated solutions compared with other algorithms. The results are consistent with the qualitative analysis in Table 3 (Table 4).

5 Conclusion

We propose novel fuzzy MOPSO based on linear differential decline. The main purpose of this algorithm is to prevent the algorithm from falling into local optimal location and improve convergence performance of the algorithm. In the proposed algorithm, the fuzzy control of the inertia weight is used to flexibly adjust the size of the search space, so that the algorithm can balance the global and local search capabilities, thus improving the convergence performance of the algorithm. At the same time, the algorithm uses the linear differential decline strategy to adjust the position change of the particles and prevent the algorithm from falling into local optimal location. The experimental results show that LDDFMOPSO has good convergence and diversity. Compared with four classical MOPSOs, LDDFMOPSO can improve the convergence performance of the algorithm and effectively prevent the algorithm from falling into local optimal location.

Acknowledgement. This work is supported by Key Talents Program in digital economy of Guizhou Province.

References

1. Hafsi, H., Gharsellaoui, H., Bouamama, S.: Genetically-modified multi-objective particle swarm optimization approach for high-performance computing workflow scheduling. Appl. Soft Comput. **122**, 108791 (2022)
2. Bosman, P., Thierens, D.: The balance between proximity and diversity in multiobjective evolutionary algorithms. IEEE Trans. Evol. Comput. **7**(2), 174–188 (2003)
3. Coello Coello, C.A., Pulido, G.T., Lechuga, M.S.: Handling multiple objectives with particle swarm optimization. IEEE Trans. Evol. Comput. **8**(3), 256–279 (2004)
4. Dai, C., Wang, Y., Ye, M.: A new multi-objective particle swarm optimization algorithm based on decomposition. Inf. Sci. **325**(1), 541–557 (2015)
5. Kennedy, J.: Particle swarm optimization. Encycl. Mach. Learn. **1**, 760–766 (2010)
6. Li, L., Wang, W., Xu, X.: Multi-objective particle swarm optimization based on global margin ranking. Inf. Sci. **375**, 30–47 (2017)
7. Lin, Q., et al.: Particle swarm optimization with a balanceable fitness estimation for many-objective optimization problems. IEEE Trans. Evol. Comput. **22**(1), 32–46 (2018)
8. Martnez, S., Coello, C.: A multi-objective particle swarm optimizer based on decomposition. In: Genetic and Evolutionary Computation Conference, GECCO 2011, vol. 1, pp. 69–76 (2011)
9. Ming, L., Minggang, D., Chao, J.: Scheduled competition learning based multiobjective particle swarm optimization algorithm. J. Comput. Appl. **30**(2), 26–31 (2019)
10. Tian, Y., Cheng, R., Zhang, X., Jin, Y.: Platemo: A matlab platform for evolutionary multi-objective optimization [educational forum]. IEEE Comput. Intell. Mag. **12**(4), 73–87 (2017)
11. Yang, J., Ma, M., Che, H., Xu, D., Guo, Q.: Multi-objective adaptive chaotic particle swarm optimization algorithm. Control Decis. **30**(12), 2168–2174 (2015)
12. Zhang, X., Zheng, X., Cheng, R., Qiu, J., Jin, Y.: A competitive mechanism based multi-objective particle swarm optimizer with fast convergence. Inf. Sci. **427**, 63–76 (2018)
13. Zhou, H., Ouyang, C., Liu, X., Zhu, P.: An adaptive fuzzy particle swarm optimization. Comput. Eng. Appl. **46**(33), 46–48 (2010)

14. Zhu, S., Wu, Q., Jiang, Y., Xing, W.: A novel multi-objective group teaching optimization algorithm and its application to engineering design. Comput. Ind. Eng. **155**(1), 107198 (2021)
15. Zitzler, E., Thiele, L.: Multiobjective evolutionary algorithms: a comparative casestudy and the strength Pareto approach. IEEE Trans. Evol. Comput. **3**(4), 257–271 (1999)
16. Zitzler, E., Deb, K., Thiele, L.: Comparison of multiobjective evolutionary algorithms: empirical results. Evol. Comput. **8**(2), 173–195 (2000)

Multi-objective Bacterial Colony Optimization Based on Multi-subsystem for Environmental Economic Dispatching

Hong Wang[1,2], Yixin Wang[1,2], Mengjie Wan[2], Sili Wen[2], and Shan Wei[3](✉)

[1] College of Management, Shenzhen University, Shenzhen 518060, China
[2] Greater Bay Area International Institute for Innovation, Shenzhen University, Shenzhen 518060, China
[3] School of Mechanical Engineering, Anhui Polytechnic University, Wuhu 241000, China
ws@ahpu.edu.cn

Abstract. When addressing the multi-objective optimization, bacterial colony optimization algorithms are easy to fall into local optimum, which leads to the insufficient diversity and convergence. To overcome this drawback, in this study, a new multi-objective bacterial colony optimization based on multi-subsystems, abbreviated as MOBCOMSS, is proposed. The MOBCOMSS uses a hierarchical clustering approach to adapt the colony into multiple sub-colony systems based on evolutionary state. Each subsystem in the colony searches and stores information independently. Then, the diversity and convergent information from subsystems are returned to the elite archive for the whole colony. Besides, information suitable for the development of diverse subsystems is extracted from the elite archive for adaptive updating to eventually balance global and local search and achieve problem adaptation. Finally, the proposed MOBCOMSS is compared with 4 popular algorithms in the environmental economic dispatch of power systems (EED) on the standard IEEE 30-bus test system. The results demonstrate that MOBCOMSS can find optimal solutions with better convergence and diversity than other comparison algorithms in solving the EED problem with lower computational consumption, showing good feasibility and effectiveness.

Keywords: Multi-objective optimization · Environmental economic dispatching · Bacterial colony optimization · Multi-subsystem

1 Introduction

Environmental/Economic Dispatch (EED) has become an important optimization problem in power system operation with the increasing concern for environmental pollution. According to EED, economic maintenance and pollutant emissions are both kept as low as possible while satisfying all equality and inequality constraints [1]. Nonetheless, minimizing total emissions and economic maintenance costs are inherently contradictory, and they cannot be addressed just using traditional single-objective optimization techniques simply due to their multiple nonlinear constraints. Therefore, it is necessary

Y. Tan et al. (Eds.): ICSI 2022, LNCS 13344, pp. 524–533, 2022.
https://doi.org/10.1007/978-3-031-09677-8_44

to transform EED problem into a multi-objective optimization problem (MOP) while handling multiple equality and inequality constraints.

MOP means two or more contradictory goals are optimized concurrently. Moreover, these objective functions always contradict each other. Numerous evolutionary algorithms have been used to solve the multi-objective EED problem successfully, attracting the interest of many scholars [1, 2]. Many optimization algorithms based on bacteria were proposed in recent years, where the prominent examples are bacterial foraging algorithm (BFO) [3], bacterial colony optimization (BCO) [4], slime mould algorithm (SAM) [5]. On the one hand, most bacterial algorithms could be highly efficient in solving single-objective optimization problems for its global search ability [5, 6].On the other hand, bacterial optimization algorithms showed adaptive behavior of intelligent emergence facing high computational consumption and inefficient utilization of prior knowledge in multi-objective optimization problems [3, 7]. For EED problem, Panigrahi et al. [8] applied a fuzzy method for BFO to solve the EED problem. Tan et al. [9] proposed a discrete BFO that used the health classification method to control the reproduction and elimination opportunities on EED problem.

Simulation results show the effectiveness of above algorithms. However, these multi-objective BFO algorithms are based on a complex three-layer nested computing structure, effective calculations are at the cost of sacrificing a large amount of computing power. In addition, the capability to balance global search and local search is still needed to enhance for multi-objective BFO algorithms. The disequilibrium leads to local Pareto or even stops convergence prematurely. The BCO further proposed a life cycle model instead of the three-layer nested structure that enhances computing effectiveness. However, BCO is updated and iterated with the guidance of individual bacteria which leads to trapping into local optimal easily.

Given the above considerations, a new multi-objective bacterial colony optimization based on multi-subsystems, abbreviated as MOBCOMSS, is developed in this paper. The MOBCOMSS newly proposed to consider not only the behavior in the evolutionary structure but also multi-subsystems search strategy for enhancing the diversity of population and avoiding trapping in local Pareto front.

2 Background

2.1 Environmental/economic Power Dispatch (EED)

EED is to find a dispatching scheme that solves for the optimal value of both objective functions (fuel cost and pollution emissions) while satisfying the power supply-demand balance and unit capacity constraints. The EED is a non-linear and high-dimensional optimization problem that must also satisfy both equation and inequality constraints, making it difficult to find a globally optimum solution using traditional gradient-based optimization methods.

In this paper, the IEEE 6 machine 30-bus standard system is chosen for verifying the performance of MOBCOMSS, More detailed parameters can check [9].

2.2 Bacterial Colony Optimization (BCO)

Bacterial Colony Optimization (BCO)is a new evolutionary algorithm proposed by Niu et al. [4] that simulates bacterial life-cycle behaviors in the swarm intelligence way. The main improvement in BCO is the way to forage that bacterium usually towards nutrients by exchanging information between individuals instead of random walks. More information about BCO can refer to [4, 6].

3 Multi-objective Bacterial Colony Optimization Based on Multi-subsystems

From previous multi-objective optimization algorithms based on bacteria, it seems that there are generally problems such as insufficient population diversity and poor convergence, which in turn lead to failure to obtain a good Pareto front [3, 9]. In order to enable populations to preserve and extract information with diversity and convergence, this paper proposes a multi-subsystem search strategy with adaptive colony behaviors. For a specific algorithmic framework see Fig. 1 and Algorithm 1.

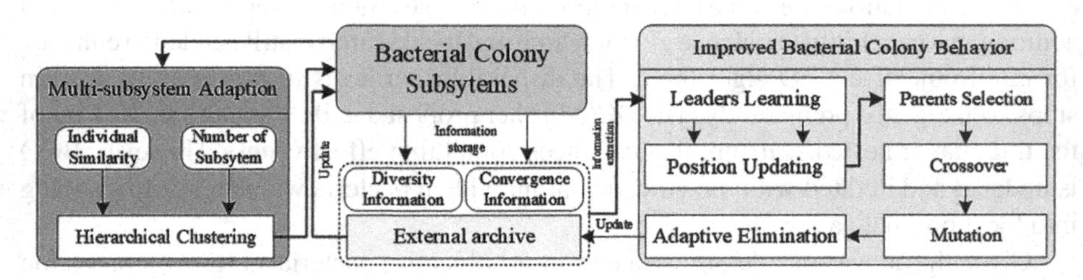

Fig. 1. The overall framework of MOBCOMSS.

3.1 Multi-subsystems Search Strategy

The main idea of bacterial colony optimization is to first initialize the colony $X_i = [x_{i_1}, x_{i_2}, \ldots, x_{i_n}]^T$, $i = 1, 2, 3, \ldots, m$, and perform random foraging behavior. The whole population is updated through continuous iteration with each bacterium updating its position through group communication [4]. Traditional bacterial colony optimization algorithms typically set a global optimum individual and drives the entire population towards found global optimum [6, 9]. The global optimum oriented search allows the algorithm to converge more quickly than a random search. Nevertheless, a single global optimum is not necessarily effective in multi-objective optimization problems. Multi-objective optimization is often not optimal for all objectives due to conflicts between objectives, which drives us to explore how to obtain the information that drives the evolution of the entire population.

Inspired by the biological swarm phenomenon of system-subsystem-individual system, we explored the influence of multiple subsystems in a bacterial colony system and devised a multi-subsystem search strategy. As shown in the Algorithm 1 on lines 3–10, the similarity of the population is calculated firstly with the metric that can be used as positional similarity, convergent similarity relative to the origin, and diversity similarity. The whole population is sliced by means of hierarchical clustering to obtained multiple sub-colony systems, each of which includes multiple bacteria.

Multiple bacterial colony subsystems operate independently and an external archive of a central information hub is designed to store the optimality search information. For multi-objective optimization problems, diversity and convergence information is stored in the external archive. During independent optimization searches, subsystems extract information from the central information hub that is appropriate for the development of that subsystem and proceeds to the next step of the adaptive optimization process until a specified number of iterations.

Algorithm 1. Overview of MOBCOMSS

01: **Input**: *npop; MaxFEs; learning rate α; Genetic Parameters*
02: **Initialization**: ***Pop*** (Population)
03: **while** *Fes* ≤ *MaxFEs* **do**
04: | Calculate the individual similarity with crowding distance and position;
05: | Hierarchical Clustering;
06: | Store non-dominated solutions to ***EA*** (*External archive*);
07: | **for** *each subsystem* ∈ ***Pop*** **do**
08: | | **for** *each bacterium* ∈ *subsystem* **do**
09: | | | Position updating using Eq.(1)
10: | | **end**
11: | **end**
12: | Parents selection;
13: | Crossover;
14: | Mutation;
15: | **if** *meet elimination condition* **then**
16: | | Adaptive Elimination;
17: | **else**
18: | | Continue;
19: | **end**
20: | Update the elite archive
21: **end**
22: **Output**: ***EA*** (External archive)

3.2 Improved Bacterial Colony Behaviors

The previous bacterial colony optimization had a high reliance on individual optimum and single global optimum, which did not satisfy the requirements for diversity and convergence in multi-objective optimization. To enhance the ability to improve the diversity of the population and accelerate the convergence of the algorithm, a new updating method is proposed as shown in Eq. (1).

$$x_i^t = w \cdot x_i^{t-1} + C_i \cdot \left\{ r_1 \cdot \left(x_c - x_i^{t-1} \right) + r_1 \cdot \left(x_d - x_i^{t-1} \right) \right\} \tag{1}$$

where w is the initial weight, C_i represents the chemotaxis steps and the x_c and x_d are convergent leaders and diversity leaders suit for each of subsystem. As shown in lines 11–13, in order to further enhance population diversity, the proposed algorithm introduces operations such as selection, crossover and mutation in genetic algorithms instead of the traditional replication operations of colony optimization. Furthermore, to avoid the population falling into a local optimum, an adaptive elimination strategy is proposed, see lines 14–18. Adaptive elimination refers to the fact that if the current convergent optimum stored in the central information hub does not change for a long time which means that the whole algorithm is not further improved. If the convergence information remains unchanged for a long time, as shown in Eq. (2), the probability of elimination of the population is increased as the number of iterations increases.

$$Ped^t = Ped^{t-1} + 0.1, \text{ if } x_d \text{ not changed} \tag{2}$$

A timer is put up in the adaptive elimination adjustment to keep track of the time it takes for the convergence to stagnate. Whenever the counter hits a predetermined value, the likelihood of elimination rises in lockstep with the growth in the counter. Similarly, the eliminated bacteria are replaced to some new position.

4 Simulation Analysis

4.1 Experimental Setup

In this paper, MOBCOMSS is applied to the EED optimization problem and the energy consumption parameters, emission parameters and loss factors of the generating units are referred to the relevant literature [9]. In this paper, MOPSOCD [10], MMOPSO [11], NSGAII [12], PESAII [13] are selected as comparative algorithms. The simulation analysis is carried out for the two cases of considering network losses and not considering network losses respectively. All experiments are carried out on a PC with Intel Core i-5 10210U @1.60 GHz and 16 GB memory, windows 11 system and Matlab 2020b. Among all comparison algorithms, the population size is set to 100 and the maximal number of fitness evaluations (FEs) is set at 10000. All the experimental results are obtained after 30 independent runs. In the experiments, the Hypervolume (HV) [14] and Spread [15] metric are used to evaluate the optimization performance of the algorithm, and the reference point for HV is set as $[1.1,1.1,\ldots,1.1]^d$.

4.2 Results and Analysis

Table 1 gives the best solutions for economic cost in case 1 and case 2 obtained by diverse algorithms. The proposed MOBCOMSS and the MMOPSO get the minimum value of economic cost is 605.9984 ($/h), significantly better than other algorithms. As shown in Table 1, The proposed MOBCOMSS gets the minimum value of economic cost is 605.9984 ($/h) while other algorithms getting results above it, which means the MOBCOMSS is much better than other algorithms. Table 2 gives the best solutions for environmental emission in case 1 and case 2 by selected algorithms. From Table 2, the minimum value of case 1 emissions obtained by the proposed MOBCOMSS is 0.194180 (t/h), while the minimum values of that obtained by MOPSOCD, MMOPSO, NSGAII, PESAII are higher than that of MOBCOMSS. Table 2 shows that MOBCOMSS, MMOPSO and NSGAII reach 0.194179 simultaneously in case 2. However, the proposed algorithm outperforms other algorithms in terms of emission at higher precision.

Table 1. Best solutions for cost ($/h) in case 1/2. (30 trials).

Methods	Case	P_1	P_2	P_3	P_4	P_5	P_6	Cost	Emission
MOBCOMSS	C1	**0.121165**	**0.286481**	**0.583648**	**0.992943**	**0.523379**	**0.351946**	**605.9984**	0.220724
	C2	**0.120808**	**0.286384**	**0.583565**	**0.992423**	**0.524187**	**0.352195**	**605.9984**	0.220702
MOPSOCD	C1	0.114755	0.288016	0.590255	0.988176	0.525345	0.352914	606.0074	0.220654
	C2	0.118497	0.288106	0.582691	0.988334	0.526146	0.35583	606.0028	0.22043
MMOPSO	C1	0.121026	0.286232	0.584042	0.992663	0.523847	0.351736	**605.9984**	0.220722
	C2	0.121004	0.286407	0.583672	0.9929	0.523653	0.351926	**605.9984**	0.220729
NSGAII	C1	0.121191	0.283844	0.58349	0.994651	0.526673	0.349685	606.0002	0.220925
	C2	0.121343	0.284665	0.583159	0.992522	0.525649	0.352203	605.9989	0.220723
PESAII	C1	0.125551	0.288536	0.583595	0.988879	0.523704	0.349215	606.0034	0.22031
	C2	0.122422	0.286405	0.584904	0.991566	0.51855	0.355684	606.0016	0.220542

Table 2. Best solutions for emission (ton/h) in case 1/2. (30 trials).

Methods	Case	P_1	P_2	P_3	P_4	P_5	P_6	Cost	Emission
MOBCOMSS	C1	**0.411291**	**0.465579**	**0.543524**	**0.390158**	**0.54634**	**0.512457**	646.2336	**0.19418**
	C2	**0.410987**	**0.461506**	**0.543599**	**0.391264**	**0.546415**	**0.51553**	646.0564	**0.194179**
MOPSOCD	C1	0.404013	0.466756	0.546965	0.392052	0.540709	0.518472	645.8045	0.194184
	C2	0.410904	0.466762	0.537149	0.395849	0.54337	0.515441	645.9386	0.194183
MMOPSO	C1	0.413928	0.464214	0.546861	0.391304	0.53896	0.514232	646.3356	0.194181
	C2	0.410271	0.464083	0.545808	0.388842	0.546933	0.513304	646.2185	**0.194179**
NSGAII	C1	0.412433	0.462986	0.543986	0.392253	0.546065	0.511635	646.0202	0.19418
	C2	0.411563	0.461548	0.546942	0.389859	0.545276	0.514091	646.1526	**0.194179**
PESAII	C1	0.409714	0.45389	0.555662	0.388831	0.548949	0.511831	645.6954	0.194193
	C2	0.413379	0.459561	0.553073	0.389989	0.539954	0.513335	646.16	0.194185

To demonstrate further the distribution of solutions on the obtained Pareto front, Fig. 2 displays the graphical results produced by the MOBCOMSS algorithm and other four algorithms for case 1 and case 2. At the same time, the hypervolume HV and Spread are applied to measure the performance of algorithm. As shown in Fig. 2, the Pareto front obtained by MMOPSO on case 1/2, MOPSOCD on case 1, NSGAII on case 1 and PESAII on case 1/2 can be seen to be unevenly distributed, with vacant Pareto fronts. In addition, the Pareto front of MOPSOCD on case 2, NSGAII on case 2 and PESAII on case 1/2 have overlapping solutions. In contrast, the pareto fronts obtained by the proposed MOBCOMSS on case 1 and case 2 are smoother and more uniform, with a wider distribution and no overlapping solutions.

From Table 3, the proposed MOBCOMSS is able to achieve the highest HV value compared to the other algorithms in Case 1 and Case 2, respectively, and Table 4 shows that the lowest Spread value can be achieved for the diversity metric, which proves the effectiveness of the proposed algorithm in improving diversity as well as convergence.

Table 3. Statistical results of the metrics HV for Case 1/2 (30 trials).

HV	Case	Best	Worst	Median	Average	STD
MOBCOMSS	C1	0.128396	0.128356	0.128387	**0.128385**	8.03E-06
	C2	0.128394	0.128355	0.12839	**0.128387**	8.38E-06
MOPSOCD	C1	0.128371	0.128327	0.128354	0.128352	1.16E-05
	C2	0.12837	0.128296	0.128351	0.128347	2.01E-05
MMOPSO	C1	0.128391	0.128366	0.128387	0.128384	7.65E-06
	C2	0.128394	0.128372	0.128388	0.128387	5.25E-06
NSGAII	C1	0.128385	0.128366	0.128379	0.128378	5.38E-06
	C2	0.128386	0.12837	0.128379	0.128378	4.34E-06
PESAII	C1	0.128348	0.128157	0.128311	0.128306	4.14E-05
	C2	0.128352	0.128169	0.128304	0.128292	5.07E-05

Table 4. Statistical results of the metrics Spread for Case 1/2 (30 trials).

HV	Case	Best	Worst	Median	Average	STD
MOBCOMSS	C1	0.555919	0.665939	0.627925	**0.620376**	3.03E-02
	C2	0.548758	0.670397	0.619972	**0.621405**	3.07E-02
MOPSOCD	C1	0.607426	0.757483	0.694027	0.69153	3.63E-02
	C2	0.597029	0.776177	0.684779	0.682769	0.045266
MMOPSO	C1	0.548858	0.714839	0.652339	0.651771	4.77E-02
	C2	0.515596	0.750139	0.652181	0.650466	0.049113
NSGAII	C1	0.568894	0.792423	0.689091	0.694539	5.33E-02
	C2	0.632373	0.840626	0.705666	0.712683	0.044622
PESAII	C1	0.887064	1.133629	0.997456	0.987289	5.25E-02
	C2	0.744185	1.169833	0.927624	0.940232	0.079911

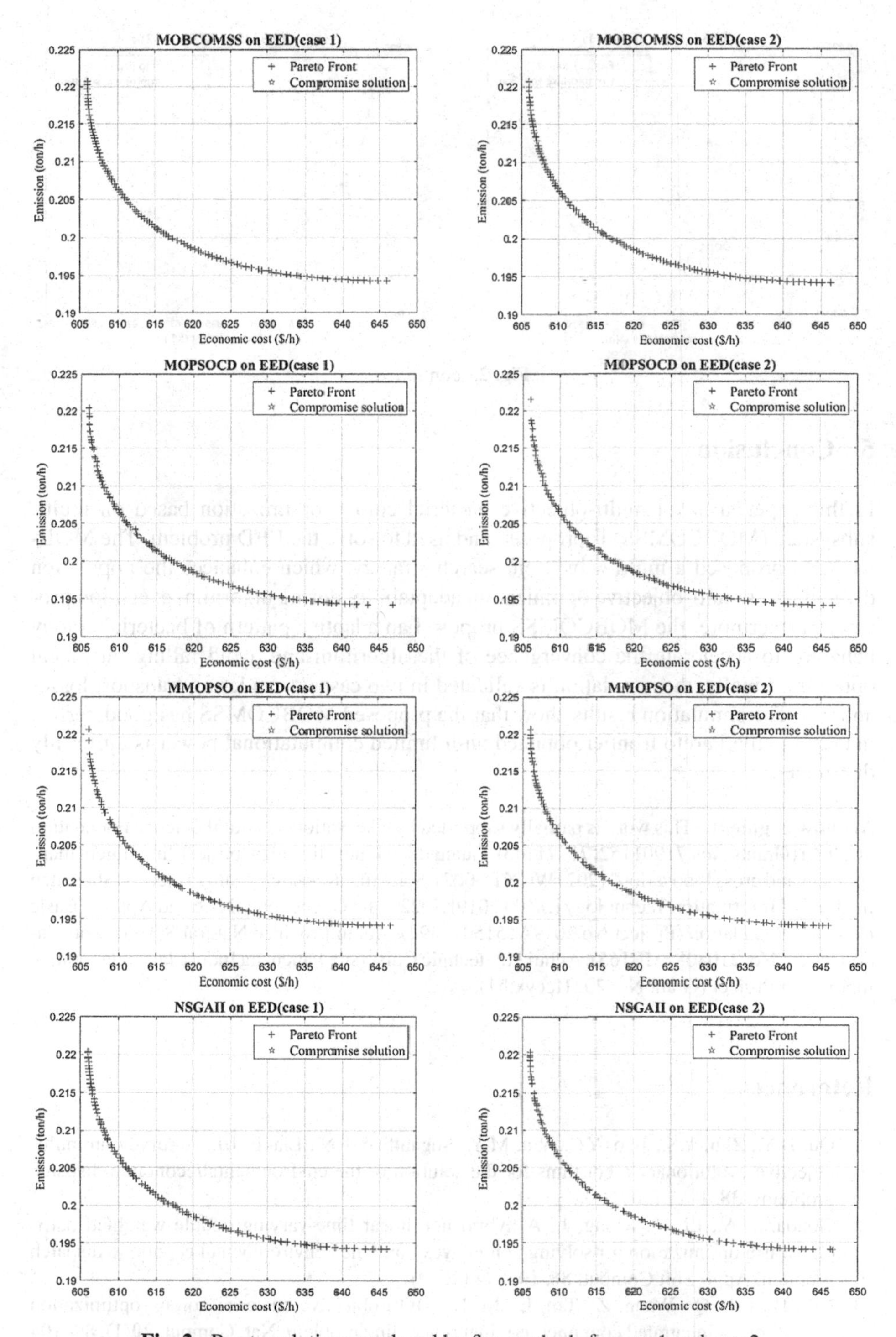

Fig. 2. Pareto solutions produced by five methods for case1 and case2

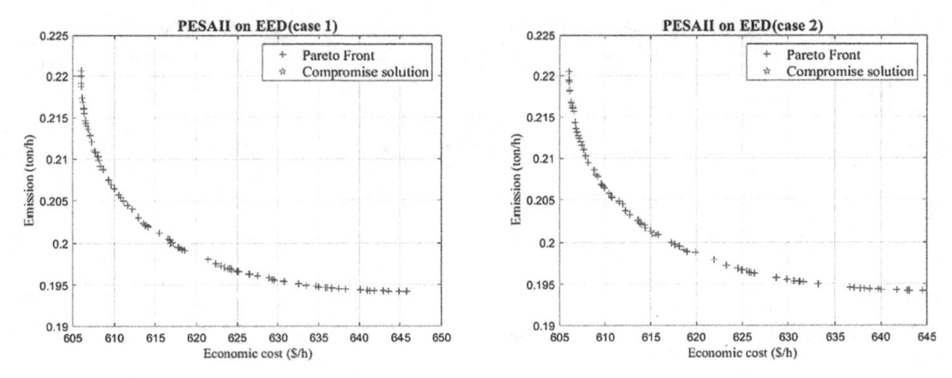

Fig. 2. continued

5 Conclusion

In this paper, a novel multi-objective bacterial colony optimization based on multi-subsystem (MOBCOMSS) is proposed and used to solve the EED problem. The MOBCOMSS proposed a multi-subsystem search strategy, which enhances the population diversity and multi-objective optimization adaptability during algorithm execution process. Furthermore, the MOBCOMSS proposed an adaptive pattern of bacterial colony behavior to accelerate the convergence of the algorithm and avoid falling into local optimum. Finally, the simulation is validated in two cases considering transport losses and not. The simulation results show that the proposed MOBCOMSS has good performance and the Pareto frontier obtained with limited computational power is uniformly distributed.

Acknowledgment. This work is partially supported by The National Natural Science Foundation of China (Grants Nos.71901152, 71971143), Guangdong innovation team project "intelligent management and cross innovation" (2021WCXTD002), Scientific Research Team Project of Shenzhen Institute of Information Technology (SZIIT2019KJ022), and Guangdong Basic and Applied Basic Research Foundation (Project No.2019A1515011392), Anhui Province Natural Science Foundation (Grant No. 2108085ME165), Anhui Polytechnic University Fanchang Industrial Collaborative Innovation Project (Grant No. 2021fccyxtb7).

References

1. Qu, B.Y., Zhu, Y.S., Jiao, Y.C., Wu, M.Y., Suganthan, P.N., Liang, J.J.: A survey on multi-objective evolutionary algorithms for the solution of the environmental/economic dispatch problems. **38**, 1–11 (2017)
2. Goudarzi, A., Li, Y., Xiang, J.: A hybrid non-linear time-varying double-weighted particle swarm optimization for solving non-convex combined environmental economic dispatch problem. Appl. Soft Comput. **86**, 105894 (2019)
3. Niu, B., Liu, Q., Wang, Z., Tan, L., Li, L.: Multi-objective bacterial colony optimization algorithm for integrated container terminal scheduling problem. Nat. Comput. **20**(1), 89–104 (2021)

4. Niu, B., Hong, W.: Bacterial colony optimization. Disc. Dyn. Nat. Soc. 2012 (2012)
5. Li, S., Chen, H., Wang, M., Heidari, A.A., Mirjalili, S.: Slime mould algorithm: a new method for stochastic optimization. Futur. Gener. Comput. Syst. **111**, 300–323 (2020)
6. Wang, H., Niu, B., Tan, L.: Bacterial colony algorithm with adaptive attribute learning strategy for feature selection in classification of customers for personalized recommendation. Neurocomputing (2020)
7. Yi, D., Huang, S., Fu, H., He, J.: Multi-objective bacterial foraging optimization algorithm based on parallel cell entropy for aluminum electrolysis production process. IEEE Trans. Ind. Electron. **63**, 2488–2500 (2015)
8. Panigrahi, B.K., Pandi, V.R., Das, S., Das, S.: Multiobjective fuzzy dominance based bacterial foraging algorithm to solve economic emission dispatch problem. Energy **35**, 4761–4770 (2010)
9. Tan, L., Wang, H., Yang, C., Niu, B.: A multi-objective optimization method based on discrete bacterial algorithm for environmental/economic power dispatch. Nat. Comput. **16**, 549–565 (2017)
10. Dai, D., Fang, Y., Rui, L., Liu, Z., Wei, L.: Performance analysis and multi-objective optimization of a Stirling engine based on MOPSOCD. Int. J. Therm. Sci. **124**, 399–406 (2018)
11. Liu, R., Liu, J., Li, Y., Liu, J.: A random dynamic grouping based weight optimization framework for large-scale multi-objective optimization problems. Swarm Evol. Comput. **55**, 100684 (2020)
12. Deb, K., Pratap, A., Agarwal, S., Meyarivan, T.: A fast and elitist multiobjective genetic algorithm: NSGA-II. IEEE Trans. Evol. Comput. **6**, 182–197 (2002)
13. Jain, A., Lalwani, S., Lalwani, M.: A comparative analysis of MOPSO, NSGA-II, SPEA2 and PESA2 for multi-objective optimal power flow. In: International Conference on Power, Energy and Environment: Towards Smart Technology (Year)
14. Leng, R., Ouyang, A., Liu, Y., Yuan, L., Wu, Z.: A multi-objective particle swarm optimization based on grid distance. Int. J. Patt. Recogn. Artif. Intell. **34** (2019)
15. Soheyl, K.: Bahman, naderi, saman, khalilpourazary: multi-objective stochastic fractal search: a powerful algorithm for solving complex multi-objective optimization problems. Soft. Comput. **24**, 3037–3066 (2020)

Layout Optimization of Indoor Obstacle Using a Multimodal Multi-objective Evolutionary Algorithm

Tianrui Wu, Qingqing Liu, Weili Wang, and Qinqin Fan(✉)

Logistics Research Center, Shanghai Maritime University, Shanghai 201306, China
forever123fan@163.com

Abstract. To improve evacuation efficiency and safety, indoor obstacles are usually used in different scenes. However, layout of indoor obstacles may have many actual constraints; thus more equivalent or alternative schemes should be provided to decision-makers. To carry out the above objective, a multimodal multi-objective layout optimization problem of indoor obstacles is proposed in the present study. Moreover, a state-of-the-art multimodal multi-objective evolutionary algorithm is used to solve this problem. Simulation results show that the current study can provide more schemes to decision-makers for different scenario constraints when the evacuation efficiency and safety are the same.

Keywords: Multimodal multi-objective optimization · Evolutionary computation · Crowd evacuation · Obstacle layout · Evacuation efficiency

1 Introduction

With the increase in number of large-scale buildings, indoor crowd safety has been widely concerned. Moreover, the evacuation efficiency and safety are great impacted by environmental factors, such as exits and obstacles [1]. Actually, a short evacuation time and a low evacuation risk are two main objectives in the crowd evacuation. Therefore, using obstacle is a feasible method to implement the above objectives. Note that exit obstacle may hinder crowd evacuation in initial studies [2]. Subsequently, many studies show that exit obstacle can improve the crowd evacuation efficiency, i.e., reduce the crowd time [3]. However, Koo et al. [4] pointed out that the crowed time and the crowed safety may be conflicted due to the exit obstacle. In other words, the efficiency and the safety of crowd evacuation cannot be satisfied at the same time. Additionally, many environmental constraints need to be considered in layout of indoor obstacles, such as the cost, building structure, and so on. Therefore, more alternative schemes should provide when the efficiency and safety of crowd evacuation are the same. Clearly, the layout optimization of indoor obstacle is a multimodal multi-objective optimization problem (MMOP), in which at least one objective vector has multiple equivalent solutions in the decision space [5, 6].

To solve the above issue, a multimodal multi-objective layout optimization model of indoor obstacle is introduced in the present study. In this model, the evacuation time

Y. Tan et al. (Eds.): ICSI 2022, LNCS 13344, pp. 534–544, 2022.
https://doi.org/10.1007/978-3-031-09677-8_45

and the evacuation risk are two objectives, and three optimized variables are the obstacle length, the distance from obstacle to exit, and the deviate distance from obstacle to the center of exit. Moreover, a self-organized speciation based multi-objective particle swarm optimizer (SS-MOPSO) [7] is used to solve this problem. Compared with previous studies, the experimental results indicate that the SS-MOPSO can provide more alternative layout schemes, which permit users to choose the most suitable one based on actual scenarios or their preferences.

The remaining study is described as follows. Section 2 introduces the related work of crowd evacuation. Section 3 introduces the social force model (SFM) and multimodal multi-objective optimization (MMO). The proposed model and the multimodal multi-objective optimization evolutionary algorithm (MMOEA) used in this study are presented in Sect. 4. Section 5 shows the experimental results and analysis. Finally, the study is concluded in Sect. 6.

2 Related Work

In early emergency management studies, indoor obstacles are often regarded as hindering the pedestrian flow, which has a negative influence on evacuation efficiency and safety [2]. However, many recent studies have shown that indoor obstacles can reduce the total evacuation time in some cases. Helbing et al. [8] pointed out that obstacles can relieve the pressure on pedestrians and alleviate congestion, which could improve the evacuation efficiency. Chen et al. [9] found that the impact of obstacles on the evacuation efficiency is not all positive in the evacuation environment with obstacles. Zuriguel et al. [10] speculated that obstacles can relieve the pressure of the exit through the actual soldier evacuation experiment. Wang et al. [11] investigated the influence of placing obstacles in front of the corner exit, wherein the distance from the obstacle to the exit played an obvious role in evacuation. Sitcco et al. [12] suggested that the three-entry vestibule structure can considerably improve real-life emergency evacuations. Liu et al. [13] proposed to make full use of rigid obstacles' guiding function close to hazard sources. Ding et al. [14] indicated that, if the flexible obstacle is too close to the exit and its height is very low, then it can reduce evacuation time compared to the rigid obstacle. To discover the effect of the obstacles near the door, a game-theoretical model of pedestrian evacuation was built by Chen et al. [15]. Therefore, layout optimization of indoor obstacles is important.

To tackle the above issues, a large number of meta-heuristics algorithms have been proposed in previous studies. For example, Zhao et al. [16] proposed a crowed evacuation simulation method for planning the route in terms of the dynamic changes of evacuation, which is solved by an improved artificial bee colony algorithm. Experiments show that the improved model can efficiently evacuate a dense crowd in multiple scenes. Considering attractive force of target position, repulsive forces of crowd and obstacles, Zong et al. [17] employed a visual guidance-based artificial bee colony algorithm to optimize the evacuation process with a large number of obstacles and evacuees. A modified particle swarm optimizer [18] with a dual-strategy adaptive control method was introduced to alleviate the message divergence between leaders and agents. Evacuation experiments are set in a rectangle venue with multiple exits, the result shows that the movement of leaders is different from other agents and the setting of doors can significantly affect the evacuation time. For the crowd dynamics problems, Cui et al. [19] managed to analyze optimal initial individual evacuation condition. In experiments, a genetic algorithm is incorporated into the floor field cellular automata model, and the simulation results indicate that the initial condition including a mixture of patient and impatient pedestrians has a great influence on the evacuation efficiency.

3 Preliminary Knowledge

3.1 The Social Force Model

Based on Newton's second law, the resultant force received by pedestrians is converted into acceleration and speed, which changes the position of pedestrians. The basic form of SFM [20] is:

$$m_\alpha \frac{\mathrm{d}\vec{v}_\alpha(t)}{\mathrm{d}t} = \vec{f}_\alpha^0 + \sum_{\beta(\neq\alpha)} \vec{f}_{\alpha\beta} + \sum_W \vec{f}_{\alpha w}, \tag{1}$$

where m_α is the quality of pedestrian α; $\vec{f}_\alpha^0$ represents the driven force of pedestrian α, and names the desired force; $\vec{f}_{\alpha\beta}$ represents the repulsive force between pedestrian α and pedestrian β; $\vec{f}_{\alpha w}$ represents the repulsive force of the wall to pedestrians.

$$\vec{f}_\alpha^0 = m_\alpha \frac{v_\alpha^0(t)\vec{e}_\alpha^0(t) - \vec{v}_\alpha(t)}{\tau_\alpha}, \tag{2}$$

where $v_\alpha^0(t)$ is the desired speed value of pedestrian α at t time; $\vec{e}_\alpha^0$ is the target direction of pedestrian α, and it directs the exit; $\vec{v}_\alpha(t)$ is the actual speed of pedestrian α; τ_α is the reaction time of pedestrian α.

$$\vec{f}_{\alpha\beta} = A_\alpha e^{\frac{r_{\alpha\beta}-d_{\alpha\beta}}{B_\alpha}} \vec{n}_{\alpha\beta} + kg(r_{\alpha\beta} - d_{\alpha\beta})\vec{n}_{\alpha\beta} + \kappa g(r_{\alpha\beta} - d_{\alpha\beta})\Delta v_{\beta\alpha}^t \vec{t}_{\alpha\beta}, \tag{3}$$

where $kg(r_{\alpha\beta} - d_{\alpha\beta})\vec{n}_{\alpha\beta}$ and $\kappa g(r_{\alpha\beta} - d_{\alpha\beta})\Delta v_{\beta\alpha}^t \vec{t}_{\alpha\beta}$ collectively referred to as the granular force $\vec{f}_{\alpha\beta}^P$; $A_\alpha e^{\frac{r_{\alpha\beta}-d_{\alpha\beta}}{B_\alpha}} \vec{n}_{\alpha\beta}$ is the psychological repulsion force $\vec{f}_{\alpha\beta}^s$; A_α and B_α are two constants; $r_{\alpha\beta}$ represents the sum of psychological exclusion distance between pedestrian α and pedestrian β; $d_{\alpha\beta}$ indicates the actual distance between pedestrian α and pedestrian β; k and κ are fixed coefficients; $\vec{n}_{\alpha\beta}$ denotes the normalized vector, which includes the direction and distance from pedestrian α to pedestrian β; $\Delta v_{\beta\alpha}^t t_{\alpha\beta} = (v_\beta - v_\alpha)\vec{t}_{\alpha\beta}$ is the speed difference in tangential direction between pedestrian α and pedestrian β; $\vec{t}_{\alpha\beta} = \left(-n_{\alpha\beta}^2, -n_{\alpha\beta}^1\right)$ refers to the normalized vector which is perpendicular to $\vec{n}_0$.

$$\vec{f}_{\alpha w} = A_\alpha e^{\frac{r_\alpha-d_{\alpha w}}{B_\alpha}} \vec{n}_{\alpha w} + kg(r_\alpha - d_{\alpha w})\vec{n}_{\alpha w} + \kappa g(r_\alpha - d_{\alpha w})\Delta v_{w\alpha}^t \vec{t}_{\alpha w}, \tag{4}$$

$$g(x) = \begin{cases} x, x > 0 \\ 0, x \leq 0 \end{cases},$$

In Eq. (4), $\vec{f}_{\alpha w}$ is similar to $\vec{f}_{\alpha\beta}$.

3.2 Multimodal Multi-objective Optimization

The MMOP can be defined as follows [5, 6]:

$$\min_{s.t. x \in \Omega} F(x) = (f_1(x), f_2(x), ..., f_m(x))^T, \tag{5}$$

where $x = (x_1, x_2, ..., x_N)^T$ represents *n*-dimensional decision vector; $\Omega \in R_N$ denotes n-dimensional decision space; *m*-dimensional objective space is composed of all possible values of $F(x)$.

Some basic concepts in MMO are presented as follows [21]:

Definition 1 (*Dominance relation*): For a minimization optimization problem, there are two vectors $\boldsymbol{u}$ and $\boldsymbol{v}$, if $\forall n \in \{1, 2, \cdots, m\}$, $u_n \leq v_n$ and $\boldsymbol{u} \neq \boldsymbol{v}$, then $\boldsymbol{v}$ is considered to dominate $\boldsymbol{v}$, summarized as $\boldsymbol{u} \succ \boldsymbol{v}$.

Definition 2 (*Pareto optimal set*): A solution $\boldsymbol{x}^* \in R^D$ is called a Pareto optimal solution of a MMOP if and only if there is no other solution $\boldsymbol{x}$ such that $\boldsymbol{F}(\boldsymbol{x}) \succ \boldsymbol{F}(\boldsymbol{x}^*)$. The Pareto set (*PS*) is regarded as the set of all the Pareto optimal solutions, noted as $\boldsymbol{X}^*$.

Definition 3 (*Pareto front*): Pareto Front of a MMOP can be defined as $PF = \{F(\boldsymbol{x}^*) | \boldsymbol{x}^* \in \boldsymbol{X}^*\}$.

In the MMO, there may exist two or more distinct *PS*s corresponding to the same *PF* [5, 6]. To effectively solve MMOPs, a large number of MMOEAs have been proposed in previous studies [6, 7, 21–23].

4 Proposed Methodology

4.1 The Proposed Model

The present study aims to optimize the layout of obstacles, thus the SFM based on the collision prediction is used [24].

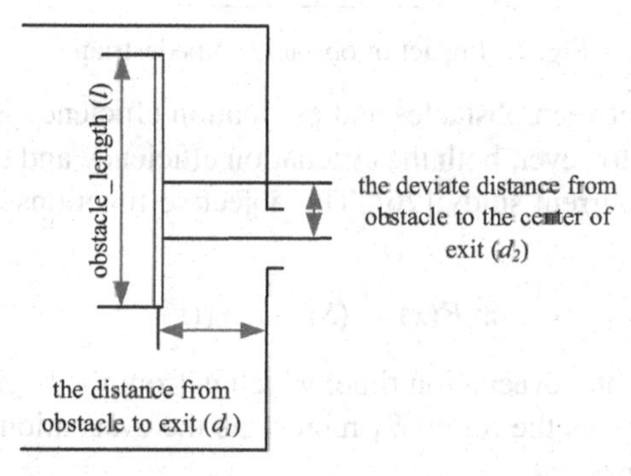

Fig. 1. Three optimized variables of obstacle

Three optimized variables are shown in Fig. 1, which are the obstacle length, the distance from obstacle to exit, and the deviate distance from obstacle to the center of exit.

Based on Ref. [24], the SFM with obstacles can be defined as follows:

$$m_\alpha \frac{d\vec{v}_\alpha(t)}{dt} = \vec{f}_\alpha^0 + \sum_{\beta(\neq\alpha)} \vec{f}_{\alpha\beta} + \sum_W \vec{f}_{\alpha w} + \sum_O \vec{F}_{\alpha o}, \tag{6}$$

where $\vec{F}_{\alpha o}$ represents the force of the obstacle to pedestrian α.

$$\vec{F}_{\alpha o} = \frac{2m_\alpha\left(\vec{B}_{\alpha o} - \vec{D}_{\alpha o} - \vec{v}_{\alpha o}\Delta t\right)}{(\Delta t)^2}, \tag{7}$$

where $\vec{B}_{\alpha o}$ denotes the vector, which includes the buffer distance and the direction between pedestrian α and the obstacle; $\vec{D}_{\alpha o}$ represents the vector from the obstacle to pedestrian α; $\vec{v}_{\alpha o}$ is the speed of the pedestrian perpendicular to the obstacle; Δt is the time interval of pedestrian movement.

Other specific rules are shown below (Fig. 2).

(1) The pedestrian target is the exit unless the obstacle is sensed.
(2) The pedestrian target is changed to channel A or B due to the resultant force $\vec{F}_{\alpha o}$.
(3) The pedestrian target will be the exit after passing through the obstacle.

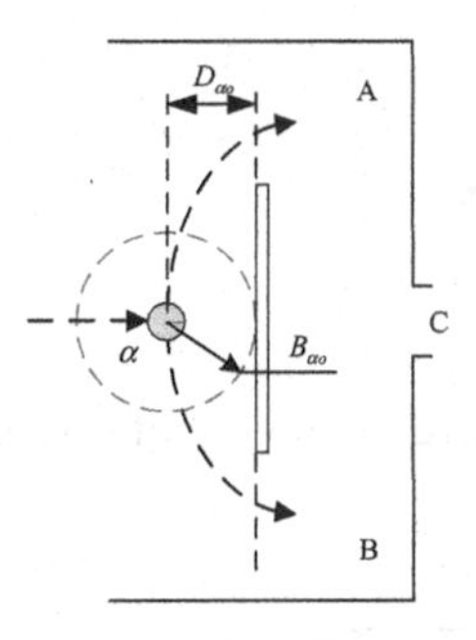

Fig. 2. Impact of obstacle on pedestrian

The relationship between obstacles and evacuation efficiency is only considered in existing studies [25]. However, both the evacuation efficiency and the evacuation safety are considered in the current study [26]. The objective functions of the model can be defined as follows:

$$\min F(x) = (S_T(x), S_A(x)), \tag{8}$$

where S_T denotes the total evacuation time, which is from the beginning of evacuation until the last person leaves the room; S_A represents the evacuation risk coefficient and can be defined as follows.

$$S_A(x) = \max\left(\frac{\max(F_c(x)) - F_{cr\alpha t}(x)}{\max(F_c(x))}\right), \tag{9}$$

$$F_c(x) = 20.254 * \rho_0 + 846.97 * v_\alpha^0(t) + 0.846 * C - 120.84.,$$

$$F_{crat}(x) = 1050 * v_\alpha(t) - 53.33.,$$

where F_c is the crowing pressure value that expresses the risk of emergency; ρ_0 represents the crowd density; C denotes the congestion level based on the number of pedestrians in a certain space; $v_\alpha(t)$ is the speed of pedestrian; F_{crat} means the critical squeeze force in the crowd to maintain the stability of the body under the action of external force.

4.2 A Self-organized Speciation Based Multi-objective Particle Swarm Optimizer

In this study, a self-organized speciation based multi-objective particle swarm optimizer (SS-MOPSO) is used to solve the multimodal multi-objective layout optimization of indoor obstacle. The pseudocode of SS-MOPSO is described in **Algorithm 1** [7]. In lines 1 to 2, the entire population ***POP***(0) and ***POA*** are initialized. In lines 3 to 7, the good particles are selected and input into ***POA***. In line 8, the self-organization speciation is applied to form multiple species. In lines 9 to 17, the particles are evaluated so that the best particle can be chosen to store by using the Non-dominated-SCD-sort method. Eventually, these steps will not stop until reaching the stopping condition.

Algorithm 1: SS-MOPSO	
1	Initialize the population ***POP***(0)
2	Initialize ***POA***
3	**for** i = 1: ParticleNumber
4	***POA***{*i*} = ***POP***$_i$(0)
5	**end for**
6	**while** *Fes* < *Max Fes* **do**
7	Sort all the particles in the ***POP*** by the Non-dominated-SCD-sort method
8	Formulate species using the self-organized speciation
9	**for** i = 1 : ParticleNumber
10	//Select *pbest* and *nbest*
11	*pbesti* = the first particle in sorted ***POA***{*i*}
12	*nbesti* = the seed of its own species
13	Update ***POP***$_i$(*t*) to ***POP***$_i$(*t*+1)
14	Evaluate (***POP***$_i$(*t*+1)
15	*Fes*=*Fes*+1
16	//Update ***POA***
17	Put ***POP***$_i$(*t*+1) into ***POA***{*i*} and sort the solutions in ***POA***{*i*}
18	**end for**
19	**end while**
20	Output the non-dominated particles in ***POA***

5 Experimental Results and Analysis

5.1 Environment Settings

The room is set to be 18 m × 12 m, and the width of the exit is set to be 1.5 m [24]. The pedestrian parameters are set as below: the diameter is set to be 0.6 m, the mass is set to be 80 kg, the expected speed is set to be 1.5 m/s, and the reaction time τ is set to be 0.5 s. The number of all pedestrians is 100. All pedestrians are evenly distributed in the 12 m × 12 m area, which is illustrated in Fig. 3. Additionally, the parameters of the SFM are given in Table 1.

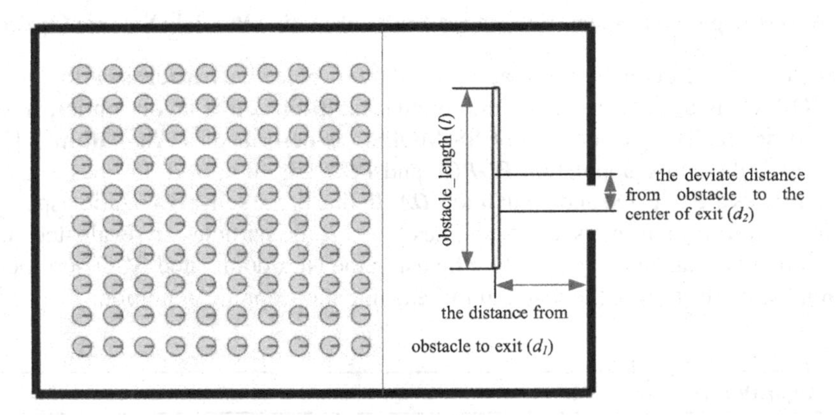

Fig. 3. Evacuation environment setting

Table 1. Parameter setting in SFM

Parameter	Value	Unit
A_α	2000	N
B_α	0.08	m
κ	1.0×10^5	$kg \cdot s^{-2}$
k	3.0×10^4	$kg \cdot s^{-2}$
r	3	m
$B_{\alpha o}$	1.5	m
obstacle_width	0.2	m

5.2 Experimental Results

In this experiment, the total evacuation time and the emergency risk are considered as the two objectives. For the SS-MOPSO algorithm, the population size and the maximum fitness evaluation are set to 800 and 120000, respectively. Additionally, the algorithm is executed for 5 independent times on the problem.

The *PF* approximation obtained by SS-MOPSO is illustrated in Fig. 4. It can be observed from Fig. 4 that the evacuation risk increases with the shortening of evacuation time. It means that the evacuation safety and the evacuation efficiency are conflicted with each other. Consequently, the decision makers cannot achieve the optimal scheme when the evacuation efficiency and safety are high at the same time, but they can balance these two objectives based on the obtained solution set. For example, if the evacuation efficiency is a priority, such as subway stations and airports, decision makers can select some schemes with a short evacuation time. Moreover, these schemes should satisfy with the minimum evacuation safety. Additionally, if the evacuation safety is a priority, decision makers can choose some schemes with a low evacuation risk. Moreover, the evacuation efficiency can be considered via the achieved solution set.

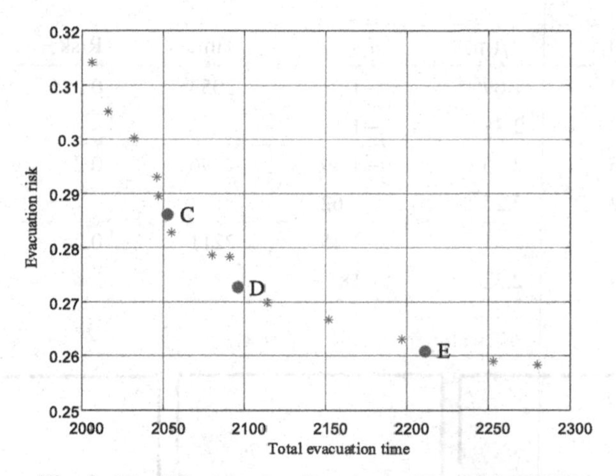

Fig. 4. The *PF* approximation obtained by SS-MOPSO

To further display the obtained results, the solutions of three points (i.e., *C*, *D*, and *E*) shown in Fig. 4 are presented in Table 2. Moreover, six obstacle layouts of these cases given in Table 3 are illustrated in Fig. 5.

It can be seen from Table 2 that the obstacle length of C_1 is shorter than that of C_2, but the obstacle of C_2 is relatively closer to the room center than that of C_1. Therefore, decision makers can select a suitable scheme to meet the requirement of scenarios.

According to the results shown in Table 2 , D2 is longer than D_1. However, their positions in the room are not much different. When cost becomes the main consideration, D_1 is more likely to be selected.

As can be shown in Table 2, the most significant distinction between E_1 and E_2 is the deviation from the exit center. In this case, planners can choose E_1 or E_2 according to the actual utilization needs of space.

Considering the layout of these 6 solutions, it is the length of obstacle that plays a major role in evacuation efficiency and safety. Note that the shorter the length of obstacle, it is more inclined to improve the evacuation efficiency but increase the risk of evacuation at the same time. That is because obstacle will separate the pedestrian flow so as to change the path and alleviate the aggregation degree of pedestrians, which improve the safety in evacuation proceeding. While the obstacle increases the walking time of pedestrians and reduces the evacuation efficiency. Furthermore, The longer the obstacle, the more obvious the effect.

Based on the all results, it can be observed that there is a reverse relationship between evacuation efficiency and evacuation safety. For decision makers, they have to weigh up two objectives to make the most realistic judgment. Due to the MMOP analysis in layout of indoor obstacles, the solutions can not only assist decision makers to balance the contradiction between evacuation efficiency and risk, but also choose different obstacle layout schemes according to the actual situation.

Table 2. Obstacle layout information

PS	l(m)	d_1(m)	d_2(m)	Time	Risk	*PF**
C_1	1.62	1.69	−1.93	2053	0.286	*C*
C_2	2.33	2.38	−1.18			
D_1	1.55	2.15	−1.49	2096	0.272	*D*
D_2	2.82	2.25	−1.62			
E_1	3.92	2.31	−1.45	2211	0.261	*E*
E_2	4.19	2.32	1.18			

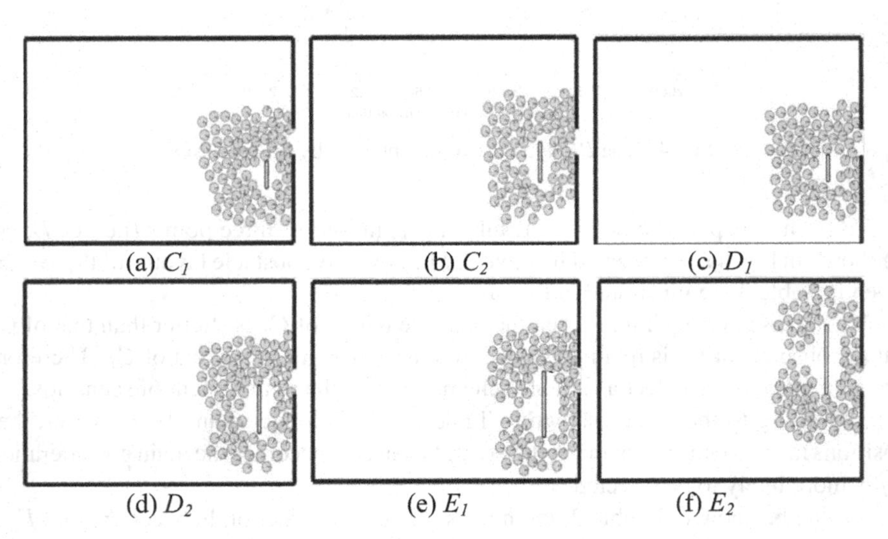

Fig. 5. The layout of indoor obstacles in six cases

6 Conclusion

The evacuation efficiency and the evacuation safety are two main objectives need to be optimized in the field of crowd evacuation. To improve the effect of evacuation process with obstacles, a multimodal multi-objective layout optimization problem of indoor obstacles is proposed in the present study. Moreover, the state-of-the-art SS-MOPSO is utilized to achieve more layout schemes of indoor obstacle. The experimental results show that the decision makers can easily choose a satisfactory scheme from the obtained solution set based on actual indoor layout. Additionally, the proposed algorithm can provide users with more equivalent obstacle layout schemes when the evacuation efficiency and the evacuation safety remain unchanged.

References

1. Zhang, D., Huang, G., Ji, C., Liu, H., Tang, Y.: Pedestrian evacuation modeling and simulation in multi-exit scenarios. Physica a-Stat. Mech. Appl. **582**, 126272 (2021)
2. Gipps, P.G., Marksjö, B.: A micro-simulation model for pedestrian flows. Math. Comput. Simul. **27**, 95–105 (1985)
3. Xu, H., et al.: The effect of moving obstacle on regulation of pedestrian flow in a single exit room. J. Stat. Mech: Theory Exp. **2022**, 023407 (2022)
4. Koo, J., Kim, Y.S., Kim, B.-I., Christensen, K.M.: A comparative study of evacuation strategies for people with disabilities in high-rise building evacuation. Expert Syst. Appl. **40**, 408–417 (2013)
5. Liang, J.J., Yue, C.T., Qu, B.Y.: Multimodal multi-objective optimization: a preliminary study. In: 2016 IEEE Congress on Evolutionary Computation (CEC), pp. 2454–2461 (2016)
6. Fan, Q., Ersoy, O.K.: Zoning search with adaptive resource allocating method for balanced and imbalanced multimodal multi-objective optimization. IEEE/CAA J. Automatica Sinica **8**, 1163–1176 (2021)
7. Qu, B.Y., Li, C., Liang, J., Yan, L., Yu, K.J., Zhu, Y.S.: A self-organized speciation based multi-objective particle swarm optimizer for multimodal multi-objective problems. Appl. Soft Comput. **86**, 105886 (2020)
8. Helbing, D., Farkas, I., Vicsek, T.: Simulating dynamical features of escape panic. Nature **407**, 487–490 (2000)
9. Chen, L., Zheng, Q., Li, K., Li, Q.R., Zhang, J.L.: Emergency evacuation from multi-exits rooms in the presence of obstacles. Physica Scripta **96**, 115208 (2021)
10. Zuriguel, I., Echeverria, I., Maza, D., Hidalgo, R.C., Martin-Gomez, C., Garcimartin, A.: Contact forces and dynamics of pedestrians evacuating a room: the column effect. Saf. Sci. **121**, 394–402 (2020)
11. Wang, J.H., et al.: Performance optimization of the obstacle to corner bottleneck under emergency evacuation. J. Build. Eng. **45**, 103658 (2022)
12. Sticco, I.M., Frank, G.A., Dorso, C.O.: Improving competitive evacuations with a vestibule structure designed from panel-like obstacles in the framework of the Social Force Model. Safety Sci. **146**, 105544 (2022)
13. Liu, Q.J., Lu, L.J., Zhang, Y.J., Hu, M.Q.: Modeling the dynamics of pedestrian evacuation in a complex environment. Physica a-Stat. Mech. Appl. **585**, 126426 (2022)
14. Ding, Z., Shen, Z., Guo, N., Zhu, K., Long, J.: Evacuation through area with obstacle that can be stepped over: experimental study. J. Stat. Mech: Theory Exp. **2020**, 023404 (2020)
15. Chen, Z.H., Wu, Z.X., Guan, J.Y.: Twofold effect of self-interest in pedestrian room evacuation. Phys. Rev. E **103**, 062305 (2021)

16. Zhao, Y., Liu, H., Gao, K.: An evacuation simulation method based on an improved artificial bee colony algorithm and a social force model. Appl. Intell. **51**(1), 100–123 (2020). https://doi.org/10.1007/s10489-020-01711-6
17. Zong, X., Liu, A., Wang, C., Ye, Z., Du, J.: Indoor evacuation model based on visual-guidance artificial bee colony algorithm. Build. Simul. **15**(4), 645–658 (2021). https://doi.org/10.1007/s12273-021-0838-z
18. Li, F., Zhang, Y.F., Ma, Y.Y., Zhang, H.L.: Modelling multi-exit large-venue pedestrian evacuation with dual-strategy adaptive particle swarm optimization. IEEE Access **8**, 114554–114569 (2020)
19. Cui, G., Yanagisawa, D., Nishinari, K.: Incorporating genetic algorithm to optimise initial condition of pedestrian evacuation based on agent aggressiveness. Physica a-Stat. Mech. Appl. **583**, 126277 (2021)
20. Helbing, D., Molnár, P.: Social force model for pedestrian dynamics. Phys. Rev. E **51**, 4282–4286 (1995)
21. Yue, C., Qu, B., Liang, J.: A multiobjective particle swarm optimizer using ring topology for solving multimodal multiobjective problems. IEEE Trans. Evol. Comput. **22**, 805–817 (2018)
22. Fan, Q., Yan, X.: Solving multimodal multiobjective problems through zoning search. IEEE Trans. Syst. Man Cybern. Syst. **51**, 4836–4847 (2021)
23. Zhang, K., Chen, M., Xu, X., Yen, G.G.: Multi-objective evolution strategy for multimodal multi-objective optimization. Appl. Soft Comput. **101**, 107004 (2021)
24. Han, Y., Wang, W., Fan, Q., Hu, Z.: Optimization of building evacuation exits based on genetic algorithm and pedestrian evacuation model. Comput. Eng. Appl. **56**, 254–261 (2020)
25. Shukla, P.K.: Genetically optimized architectural designs for control of pedestrian crowds. In: Korb, K., Randall, M., Hendtlass, T. (eds.) ACAL 2009. LNCS (LNAI), vol. 5865, pp. 22–31. Springer, Heidelberg (2009). https://doi.org/10.1007/978-3-642-10427-5_3
26. Wu, W., Chen, M., Li, J., Liu, B., Wang, X., Zheng, X.: Visual information based social force model for crowd evacuation. Tsinghua Sci. Technol. **27**, 619–629 (2022)

Author Index

Printed in the United States
by Baker & Taylor Publisher Services